VOLUME FORTY SEVEN

ADVANCES IN INSECT PHYSIOLOGY

VOLUME FORTY SEVEN

ADVANCES IN INSECT PHYSIOLOGY

Insect Midgut and Insecticidal Proteins

Edited by

TARLOCHAN S. DHADIALLA

Dow AgroSciences, LLC., Indianapolis, Indiana, USA

SARJEET S. GILL

Department of Cell Biology and Neuroscience,
University of California, Riverside, California, USA

ELSEVIER

AMSTERDAM • BOSTON • HEIDELBERG • LONDON
NEW YORK • OXFORD • PARIS • SAN DIEGO
SAN FRANCISCO • SINGAPORE • SYDNEY • TOKYO
Academic Press is an imprint of Elsevier

Academic Press is an imprint of Elsevier
32 Jamestown Road, London NW1 7BY, UK
The Boulevard, Langford Lane, Kidlington, Oxford OX5 1GB, UK
225 Wyman Street, Waltham, MA 02451, USA
525 B Street, Suite 1800, San Diego, CA 92101-4495, USA

First edition 2014

ISBN: 978-0-12-800197-4
ISSN: 0065-2806

For information on all Academic Press publications
visit our website at store.elsevier.com

CONTENTS

CONTRIBUTORS

Michael J. Adang
Department of Entomology, and Department of Biochemistry and Molecular Biology, University of Georgia, Athens, Georgia, USA

Md. Shohidul Alam
Division of Chemistry and Structural Biology, Institute for Molecular Bioscience, The University of Queensland, St. Lucia, Queensland, Australia

James A. Baum
Monsanto Company, Chesterfield, Missouri, USA

Niraj S. Bende
Division of Chemistry and Structural Biology, Institute for Molecular Bioscience, The University of Queensland, St. Lucia, Queensland, Australia

Colin Berry
Cardiff School of Biosciences, Cardiff University, Cardiff, United Kingdom

Neil Crickmore
School of Life Sciences, University of Sussex, Falmer, Brighton, United Kingdom

Rhoel R. Dinglasan
Department of Molecular Microbiology and Immunology, The Johns Hopkins Bloomberg School of Public Health, Baltimore, Maryland, USA

Andrea J. Dowling
Biosciences, University of Exeter, Cornwall, United Kingdom

Richard H. ffrench-Constant
Biosciences, University of Exeter, Cornwall, United Kingdom

Volker Herzig
Division of Chemistry and Structural Biology, Institute for Molecular Bioscience, The University of Queensland, St. Lucia, Queensland, Australia

Juan Luis Jurat-Fuentes
Department of Entomology and Plant Pathology, University of Tennessee, Knoxville, Tennessee, USA

Robert M. Kennedy
Vestaron Corporation, Kalamazoo, Michigan, USA

Glenn F. King
Division of Chemistry and Structural Biology, Institute for Molecular Bioscience, The University of Queensland, St. Lucia, Queensland, Australia

Paul J. Linser
The University of Florida Whitney Laboratory, St. Augustine, Florida, USA

Thomas Meade
Dow AgroSciences, LLC., Indianapolis, Indiana, USA

Kenneth E. Narva
Dow AgroSciences, LLC., Indianapolis, Indiana, USA

Lêda Regis
Centro de Pesquisas Aggeu Magalhães-Fiocruz, Recife-Pernambuco, Brazil

James K. Roberts
Monsanto Company, Chesterfield, Missouri, USA

Maria Helena Neves Lobo Silva Filha
Centro de Pesquisas Aggeu Magalhães-Fiocruz, Recife-Pernambuco, Brazil

Nicholas P. Storer
Dow AgroSciences, LLC., Indianapolis, Indiana, USA

H. William Tedford
Vestaron Corporation, Kalamazoo, Michigan, USA

Yidong Wu
College of Plant Protection, Nanjing Agricultural University, Nanjing, China

PREFACE

The idea for this volume on "Insect midgut and insecticidal proteins" was conceived from the realization that not a single source of reviews covers the insect midgut and insecticidal proteins isolated from bacteria or arthropods. This volume benefits anyone researching to find solutions for insect pest control in agriculture and in public health.

The first chapter reviews "Insect gut structure, function, development and target of biological toxins". The insect midgut is the first barrier or a target for ingested toxophores (small-molecule insecticides or insecticidal proteins). For insecticidal proteins from the bacteria, *Bacillus*, *Lysinibacillus* and *Photorhabdus*, the midgut provides several target sites by which these proteins manifest their toxic action. However, for many target sites in other tissues, the midgut can be a barrier for efficient delivery, like peptides from spider venom (Chapter 8). Although a lot of the information reviewed here is from mosquito and *Drosophila* midguts, these approaches and understanding can help draw parallels and differences between phytophagous insects (agriculturally important) versus hematophagus insects (medical importance). Additional proteomic studies on the midgut to identify and characterize putative target sites would be beneficial for developing or discovering alternate mechanisms of action. Linser and Dinglasan have provided an excellent review of the insect midgut with a discussion of possible target sites.

Chapters 2–5 review various aspects of insecticidal proteins from *Bacillus* and *Lysinibacillus*. In Chapter 2, Adang et al. review the diversity of insecticidal proteins (three domain crystal (Cry), Cytolytic (Cyt), Binary Cry and other parasporal toxins) from *Bacillus*. They review the mode of action of these proteins, providing similarities and differences in the receptors used for manifesting toxicity. The identification and characterization of toxin receptors is important not only to create opportunities for discovering newer toxins but also to modify known toxins to target insect pests that are less or non-susceptible. Moreover, such investigations allow the development of strategies to overcome or delay the development of resistance to insecticidal proteins.

In Chapter 3, Filha et al. review the Binary (Bin) proteins from *Lysinibacillus sphaericus* (*Ls*) that are mosquitocidal. The authors discuss the structure, function and mechanisms by which these proteins cause toxicity in mosquito larvae. Unlike genes encoding insecticidal proteins from *Bacillus* species, which are now used as transgenes in crops for creating insect pest resistance, the *Ls* bacteria have been used as biolarvicides.

The potential of using *Bacillus thuringiensis* (Bt) as bioinsecticides was recognized in the early twentieth century, and subsequently many Bt products from Bt strains were developed for commercial use. However, these products suffered from a lack of stability in the sprayed environment and reduced efficacies. It was not until the first genes encoding Cry insecticidal proteins were cloned that research to their use as transgenes was initiated. Between 1995 and 1996, the first transgenic crop (potato, corn and cotton) carrying a Cry gene for controlling an insect species was developed. Since then, there has been a rapid adoption of transgenic crops worldwide, increasing from 1.7 million hectares in 1996 to just over 175 million hectares in 2013. This trend for reliance on transgenic crops will continue to grow until newer, better and more effective approaches to prevent damage from insect pests are discovered and developed.

In Chapter 4, Narva et al. review the discovery and use of genes encoding insecticidal Bt Cry proteins for developing transgenic crops that provide control of insect pests. In this chapter, the use of multiple Cry genes (gene stacking or pyramiding) is also reviewed to describe approaches to not only broaden the spectrum of insect pests controlled within a crop but also providing an approach to delay the development of insect resistance to a single Bt gene product. The authors not only review the various Bt genes that have been used for developing transgenic crops but also provide an overview of approaches used for transferring genes into crops, selection of transgenic events and what needs to be done to register and the deployment of such transgenic crops in different geographic regions. Every time a new mechanism for insect pest control is developed, it comes with the possibility of the target insect developing resistance, making the product less efficacious. The authors provide a brief overview of insect resistance management strategies, which is reviewed more extensively in Chapter 6 by Wu.

In Chapter 5, Baum and Roberts review yet another approach that relies on knocking down or down regulating genes encoding proteins essential for target insect pest survival. The use of double-stranded RNAi (dsRNAi) has been very effectively used in non-arthropods and plants to knock down genes to understand gene function in specific pathways. This approach has now been used for inactivating specific genes critical to the survival of insect pests. This approach is an alternative to the use of chemical insecticides for interfering with the function of target site proteins. However, the use of dsRNAi provides a much higher level of selective toxicity to insect pests and serves as an attractive approach. Although there are no commercial products harnessing this approach as yet, it will not be long before such products are commercially available.

In the last chapter (Chapter 6) related to Bt insecticidal proteins, Wu reviews resistance development and resistance management strategies for transgenic crops carrying Bt genes. The development of resistance is inevitable, and the challenge faced is how strategies can be deployed to delay the targeted insect pests from developing resistance to the insecticidal proteins in host transgenic crops. In this respect, it is also important to understand the mechanisms and target site receptors/proteins these insect toxins use for manifesting insecticidal activity and the mechanisms that lead to resistance development. This aspect of resistance ties very well with the review in Chapter 1 on mode of action of Bt proteins.

Chapters 7 and 8 review alternate sources of insecticidal proteins or peptides. The discovery of insecticidal proteins from the bacteria *Photorhabdus* and *Xenorhabdus* created a lot of interest among academic labs and industry to understand the structure–function and mode of action of these very large (molecular size) and complex proteins. This is reviewed in Chapter 7. Although genes encoding these proteins or their peptides have not been used as transgenes in crops to control specific insect pests, the information generated can be leveraged with new approaches and capabilities to possibly make use of such genes (modified or unmodified). ffrench-Constant and Dowling have provided an extensive review of the many proteins from the two bacteria, high-resolution structures and possible mechanisms of action of the insecticidal proteins.

In Chapter 8 on "Methods for deployment of spider venom peptides as bioinsecticides", the authors describe a novel source of peptides from spider venom that show very interesting and selective toxic activities in insects. Most of these act on neuropeptide targets and provide a challenging opportunity as how to make use of these peptides as biopesticides or use knowledge of their structures to invent new small-molecule toxophores that can interact at the same target sites used by the peptides from the spider venom.

Chapters in this volume were chosen to provide a single comprehensive review of structure and function of the insect midgut and the insecticidal proteins and genes that have been used as alternatives to chemical insecticides for controlling insect pests of agricultural and medical importance.

Discovery of newer insect control approaches and their use will be an important component of increasing crop yields in an ever-shrinking arable land and continued insect transmission of many human diseases in an increasing world population that is projected to increase to 9 billion by 2050.

Tarlochan S. Dhadialla
Sarjeet S. Gill

CHAPTER ONE

Insect Gut Structure, Function, Development and Target of Biological Toxins

Paul J. Linser*, Rhoel R. Dinglasan†
*The University of Florida Whitney Laboratory, St. Augustine, Florida, USA
†Department of Molecular Microbiology and Immunology, The Johns Hopkins Bloomberg School of Public Health, Baltimore, Maryland, USA

Contents

Abstract

Insects as vectors of disease to humans and domesticated animals and as direct agricultural pests are a source of tremendous economic and health-related challenge. The eating habits of insects can provide the bases for disease transmission or the outright destruction of crops. The alimentary canal of insects is a common target and often barrier for pest control strategies. Recent advances in technology have made it possible to develop ever better understanding of the structure/function of the insect gut and hence provide new and better targets for developing novel methods for limiting the burdens that insects can present to humanity. In this review, we focus attention on recent developments in our understanding of insect gut structure/function with particular emphasis on a few of the most challenging groups of insects: mosquitoes (dipterans), caterpillars (lepidopterans), beetles (coleopterans) and aphids (hemipterans).

1. INTRODUCTION

The alimentary canal of any higher organism is part of that organism's first order environmental contact. Consequently insects have evolved highly

Advances in Insect Physiology, Volume 47
ISSN 0065-2806
http://dx.doi.org/10.1016/B978-0-12-800197-4.00001-4

specialised capacities to live in many varied ecological niches ranging from aquatic to terrestrial to airborne. In all cases, gut function is crucial for survival and hence is specifically adapted to the life style of the insect. Details associated with the ingestion of biological substrate (food), digestion of that material into useable small molecules and finally the absorption of the liberated nutrients into the cells, tissues and hemolymph of the animal are complex and varied from specie to specie. The purpose of this review is to address structural details of the insect alimentary canal with commentary on the structural interface for targeting of the gut with biological toxins. The importance of developmental changes and lifestyle differences between life stages will also be addressed. It is beyond the scope of any single review to address specific details for the wide variety of insects and their specialisations of the gut. Therefore, we have selected a few representative model systems for discussion.

The importance of insects to life on earth including human existence is indisputable. For us as co-inhabitants of the planet, insects have particular relevance in their capacity to interfere with aspects of our health and wellbeing. Many insects have evolved complex relationships with organisms and viruses that can cause human disease. Hematophagy has evolved in arthropods over 20 × (Black and Kondratieff, 2005). The propensity to take blood meals from vertebrates in general has been accompanied by the development of the capacity to harbour and transmit disease microbes and viruses. This reality creates numerous challenges for human beings ranging from negative impacts on domesticated animal stocks as well as the vectoring of human pathogens directly. Therefore, one of the most important groups of insects for the purposes of this review is mosquitoes, which transmit some of the deadliest known human pathogens. The morbidity and mortality brought about by hematophagy of mosquitoes results in incalculable losses of life and human potential. Our efforts to control mosquito populations with various pesticides and integrated strategies are continuously thwarted by the capacity of mosquitoes to adapt and evolve rapidly under selective pressure. Hence, a deep understanding of mosquito biology is essential for the development of new disease control strategies. The gut of the mosquito (Dipterans) in both larval and adult stages is a productive target for control strategies and hence a point of emphasis in this review.

Human development and the use of agriculture has provided a basis for the expansion of our species from hunter-gatherers dependent on the whim of Mother Nature to the truly dominating natural force on our planet. Agricultural development has been continuously challenged by opportunistic and natural competitors for the crops in the field. Relevant to this review of course are a range of insect "pests" which consume or damage crops in a variety of ways including the transmission of plant diseases. The impact of pest insects on the world economy and the security of the human food supply is gigantic. Hence, we will also review the structural biology of certain groups of agricultural threats: Lepidopteran and coleopteran larvae and hemipteran life stages that impact crops. Of course, there are many and diverse insects that will not be covered in this review but we hope to present structural considerations that can be instructive and of fairly generalised relevance.

2. MOSQUITO LARVAL ALIMENTARY CANAL

For the purpose of this review, we will not go into detailed discussion of structure/function analyses that have been reviewed in great detail previously. There are superb resources for examining the depth of analyses performed with the foundational techniques of traditional microscopy and biochemistry (e.g. Billingsley, 1990; Billingsley and Lehane, 1996; Lehane and Billingsley, 1996). Herein, we will focus on a broad structural view associated with fairly recent applications of newer techniques for structure/function analysis.

Development of the insect alimentary canal has been investigated exhaustively and excellent reviews and reference texts are available (e.g. Klowden, 2007). A generalised summary of the embryological origins of the cells of the gut is shown in Fig. 1.1. Posterior and anterior invaginations of the embryonic ectoderm give rise to the anus and mouth respectively. Masses of endodermal cells emerge from the invaginating epithelium and give rise to the endodermal tube that will eventually connect forming the midgut. The invaginating ectodermal cells will become the hindgut and foregut. Fusion of the epithelial primordia eventually produces the continuity of the alimentary canal and all of its subdivisions (Klowden, 2007).

Dipterans such as mosquitoes are holometabolous. This term means that they exhibit very distinct larval developmental stages, pupation and the emergence of an adult imago that does not resemble the larval stages (Klowden, 2007). Similarly lepidopterans that includes butterflies and moths such as *Manduca sext* and coleopterans (beetles) have very distinct larval and adult stages such that casual observation might lead one to believe the different life stages are actually different organisms. Differences in organismal structures are quite severe such that environmental niches of dissimilar stages of a given organism can be vastly different representing very distinct selective

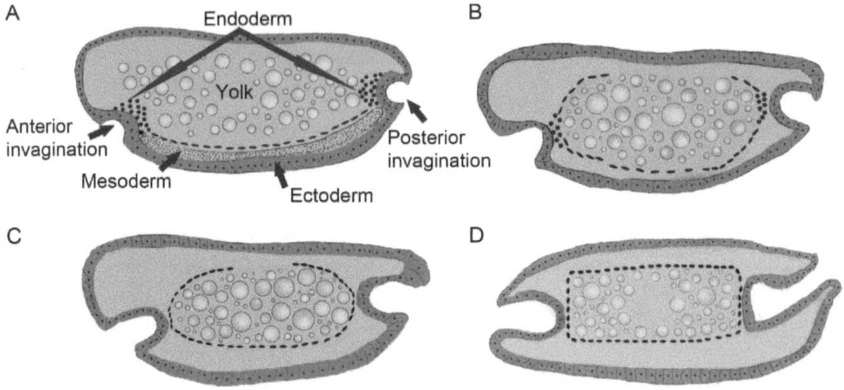

Figure 1.1 Embryonic development of the main components of the alimentary canal in insects. Invaginations of the ectoderm at the anterior and posterior poles give rise to the foregut and hindgut, respectively. The midgut forms from endodermal cells adjacent to the invaginations, proliferating and migrating to enclose the central yolk. The ectodermal and endodermal tubes eventually fuse to form a contiguous alimentary canal. The sequence of developmental steps progress from "A" through "D". *Redrawn by Gabriela Marie Ferguson after several sources including Johansen and Butt (1941).* (See the colour plate.)

pressures. Mosquitoes are an excellent example of this phenomenon in that larvae are completely aquatic whereas adults are winged and live aloft or resting on terrestrial surfaces. The alimentary canal of larval mosquitoes (and others) is nearly completely autolysed and replaced during pupation so that the adult digestive apparatus is largely built anew.

These differences in environmental niche and the associated structural adaptations naturally produce significant distinctions in supplying control agents to a targeted pest or disease vector species. The genomic era that has captured us all has shown that the huge physical differences that distinguish embryonic, larval, pupal and adult stages of holometabolous insects are the product of surprisingly subtle modifications in gene expression rather than the exposure of batteries of stage-specific genes (Goltsev et al., 2009; Marinotti et al., 2006).

In contrast, hemimetabolous insects (hemipterans, e.g. aphids) show much less dramatic structural remodelling during development from nymph stages to adults. The alimentary canal though varied in structure between species is retained and expanded as the insect matures.

In general considerations, the insect alimentary canal is a contiguous epithelial tube with the typical anterior oral opening (mouth) and posterior anus of higher metazoans. Figure 1.2A shows a scanning electron micrograph

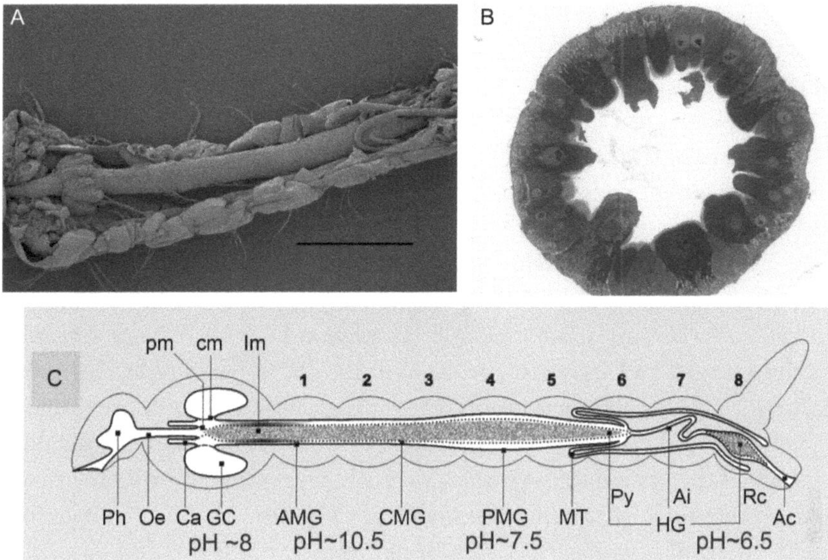

Figure 1.2 Architecture of the larval mosquito alimentary canal. Panel A: a scanning electronmicrograph of a fourth instar larva of *Aedes aegypti* that has been dissected to reveal the full length of the midgut amidst the exoskeleton and integument. Panel B: a cross section of the gut epithelium in the region of the posterior midgut (PMG) showing that it is a single cell thick epithelial tube but with varying morphological characteristics of the individual cells. Panel C: a diagrammatic rendering of the larval mosquito alimentary canal with labels applied for orientation. Numbers 1–8 indicate abdominal segments. GC, gastric caecum; AMG, anterior midgut; CMG, central midgut; PMG, posterior midgut; MT, malpighian tubules; HG, hindgut; Py, pyloris; Ai, anterior intestine; Rc, rectum; Ac, anal canal; Ph, pharynx; Oe, oesophagus; Ca, cardia; pm, peritrophic membrane; cm, caecal membrane; lm, gut lumen. The approximate pH of the lumen of the gut is shown at the bottom of the cartoon. *Taken from Linser et al. (2007) with permission.* (See the colour plate.)

(SEM) of a fourth instar *Aedes aegypti* larvae (Linser et al., 2007). In this image, the cuticular exoskeleton has been peeled back revealing the gross architecture of nearly the entire alimentary canal. Figure 1.2B shows a cross-sectional histological view that highlights one of the important characteristics of the insect gut: it is a tubular organ system made up of a single cell thick epithelium of highly polarised cells in terms of structure (i.e. apical, lateral and basal structural distinctions) which presumably reflect distinctions in function as well. Figure 1.2C provides a cartoon rendering of the major organisational features or functional zones of the gut. A major subdivision of the alimentary canal not shown in this cartoon is the pair of salivary glands (SGs) that extend laterally from the oesophagus but these will enter the discussion later. Also, the structural components of the mouth and oral cavity leading to the pharynx are not covered herein but the reader can find many details in the literature (e.g. Clements, 1992). Figure 1.2C also depicts one of the key structure/function relationships in many larval insect alimentary canals: the lumen of the gut exhibits a range of pH values that can reach extreme levels of basicity (Boudko et al., 2001; Corena et al., 2004; Dadd, 1975; Terra et al., 1996; Zhuang et al., 1999). In mosquito larvae, the anterior midgut (AMG) lumen pH can be as high as 11 (Boudko et al., 2001; Corena et al., 2004; Dadd, 1975; Terra et al., 1996; Zhuang et al., 1999). In certain other insect larvae such as caterpillars (e.g. *Manduca sexta*) the luminal pH may exceed 12 (Cioffi, 1979; Harvey et al., 1983; Wieczorek, 1992). The evolution of a digestive strategy that employs extremely high pH is a subject of considerable interest both from the detailed physiology of the system to the impact such pH extremes have on the implementation of control strategies that target gut function such as the bacterial toxins from *Bacillus thuringiensis* and *B. thuringensis israeliensis* (BT and BTI respectively; Gill et al., 1992; Chapter 2). This will be a recurring theme within this review and volume.

The gross architecture of the larval mosquito gut is depicted in greater detail in Fig. 1.3. We will address certain details for most of the specialised functional zones. The first zone in this figure is the pair of bi-lobed (anterior and posterior lobe) SGs. Although relatively little research has been done on larval SGs, much is known about adult SGs as they are part of the infection pathway for the transmission of viral and protozoan pathogens (Black and Kondratieff, 2005). The SGs of larvae do in fact produce some of the earliest

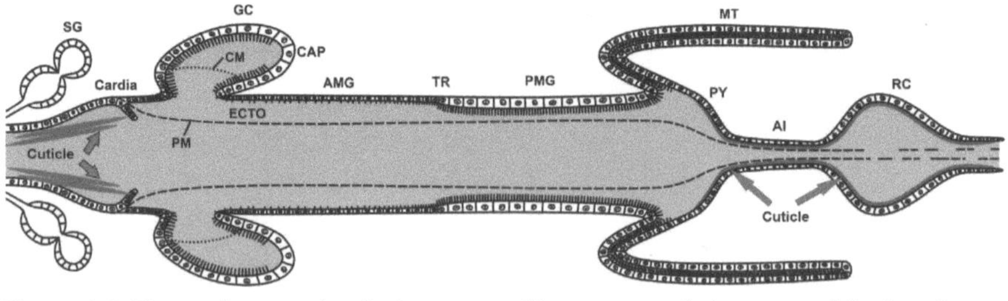

Figure 1.3 Figure shows a detailed cartoon of key structural elements of the larval mosquito alimentary canal from the foregut including the salivary glands (SGs) to the rectum (RC). Abbreviations are as in Fig. 1.2 except that the CMG is called the TR (transition region) in this figure. Note the extent and location of the ectoperitrophic space (ECTO), the cuticular lining of the foregut and hindgut and the variable distribution and size of brush border membranes (microvilli) on the apical aspects of the gut cells at the various regions of functional specialisation.

effectors of digestion as defined by micro array-based transcriptomic analyses (Neira Oveido et al., 2009) as is true for SGs of adult mosquitoes and other organisms (e.g. Chagas et al., 2013). The structure implies a cascade of functionalities from the inline organisation of the two lobes of each gland. The dynamics and contents of larval saliva are areas with little basic information but may affect the effectiveness of control agents that encounter the saliva once consumed by the larva.

Posterior to the point in the oesophagus at which the SGs connect is a complex structure (much simplified in Fig. 1.3) called the cardia that is the junction between the foregut and the midgut. Epithelial layers of the foregut and midgut overlap for a short distance creating a crypt in which the peritrophic matrix (PM) is secreted and assembled. The PM of larval mosquitoes is referred to as a Type II PM and is constitutively and continuously synthesised by the complex arrangement of cells of the cardia (Clements, 1992). This acellular material, sometimes referred to as the peritrophic membrane, is similar in function to dialysis tubing and even looks very similar to dialysis tubing at the macroscopic level of examination. The PM provides a physical barrier between the ingested bolus of food and the actual epithelial cells of the midgut. It also provides a barrier to macromolecular complexes that may be secreted by gut cells or sloughed by cells. The PM is composed of a complex mixture of proteins, and chitin microfibrils in a proteo glycan matrix (Hegedus et al., 2009; Lehane, 1997). The tubular PM lines the midgut and is eventually passed from the rectum in various stages of disintegration and possibly reabsorption during faecal elimination. The PM is a microporous barrier and the porosity limits diffusion through the membrane to specific size limits (Edwards and Jacobs-Lorena, 2000; Hegedus et al., 2009; Lehane, 1997). This barrier function may affect the access of certain toxic materials and biological materials to the gut cells. The cuticular lining of the oral cavity and foregut ends as the PM manifests at the anterior extreme of the midgut. From this point posteriorly, the midgut epithelial cells are separated from the food bolus by the PM. This separation defines a specific compartment, which is termed the ectoperitrophic space (Clements, 1992; Smith et al., 2007). This fluid-filled compartment is very dynamic and provides the medium from which ions, solutes and nutrients are trafficked into and out of the midgut cells (Terra, 1990; Terra et al., 1996). Any macromolecules, solutes or even biological control toxins that will contact the epithelial cells directly will do so from this active space. The water movement within the ectoperitrophic space is also dynamic and tracer studies have shown that there is a net movement of water from the posterior midgut through this compartment to absorption in more anterior positions (Terra, 1990). We will address more details of the PM and the ectoperitrophic space in a later discussion.

Posterior to the cardia, the gut tube of mosquito larvae flares laterally into eight diverticuli termed the gastric caeca (GC). Such lateral pouches off the main structure of the gut tube are common in many insects but not all. The pouches define an interior space surrounded by epithelial cells. The interior space is also set apart from the ectoperitrophic space by the existence of another acellular membrane similar to the PM, which is termed the caecal membrane (CM). Like the PM, the CM has a barrier function and presumably allows for traffic into and out of the caecal cavity on a size-exclusion basis and perhaps other forms of selectivity (Edwards and Jacobs-Lorena, 2000). Various types of transport physiology studies, as well as micro array analyses of gene expression, indicate that caecal cells are involved in both the secretion of various gene products including certain digestive enzymes as well as the uptake of specific solutes and small-molecule nutrients (e.g. amino acids, Harvey et al., 2009; Volkman and Peters, 1989a,b). It is apparent that control agents that would target the cells within the caecal cavity would have to pass through both the PM as well as the CM. The use of fluorescently labelled plant lectins which discriminate the glycoconjugate patterns on a variety of structural macromolecules show that the CM and PM are not biochemically identical (Linser et al., 2008) which is also inferred by ultrastructural details (Hegedus et al., 2012; Lehane, 1997). The caecal cavity is fluid filled but that fluid is viscous and also exhibits strong labelling with certain fluorescent lectins. The caecal fluid is likely composed of a rich mixture of proteins, glycoproteins and proteoglycans (Linser et al., 2008). Any molecules such as small molecule nutrients or ingested toxins must traverse the CM and the caecal fluid to make contact with the caecal epithelial cells.

The endodermal epithelial cells that comprise the caeca exhibit striking structural characteristics. As a transporting epithelium, the caecal cells possess extensive expansion of the plasma membrane both on the apical and basal aspects of the cells. Two major cell types have been described, each possessing extensive microvilli patterns at the apical surface. What have been termed "ion-transporting cells" exhibit very long and densely packed microvilli, usually containing long tubular mitochondria within each microvillus (Seron et al., 2004). The second principal cell type, which has been called a "resorbing/secreting cell", also possesses extensive microvilli at the apical surface that usually lack internal (microvillar) mitochondria

(Seron et al., 2004). The basal aspect of both aforementioned types of caecal cell exhibit varied but extensive plasma membrane infoldings (labyrinth), again indicative of extensive expansion of the cell surface area. Figure 1.4 shows a representative transmission electron microscopic (TEM) image of caecal cells from *Aedes aegypti* larva with these specific characteristics evident. In addition to the two main types of caecal cell described above and numerous times in the classical literature, a third type of epithlelial cell has been noted. These cells typically occur at the posterior extreme of each caecal diverticulus and thus have been referred to as "CAP" cells (Seron et al., 2004). CAP cells show very small or no microvilli on the apical surface and generally appear to be less broad in the apical to basal dimension. In subsequent

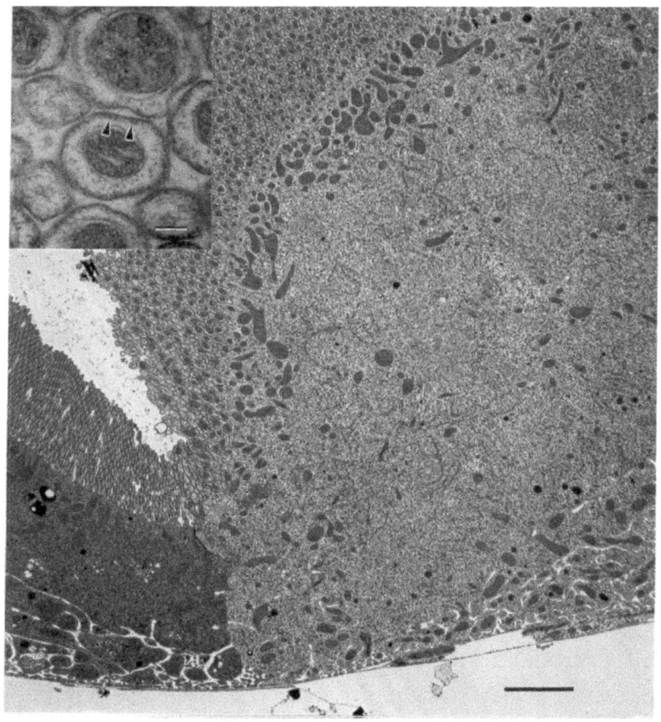

Figure 1.4 Gastric caeca cells as seen with transmission electron microscopy. Two different cells are shown: a lightly staining "ion transporting cell" (on the right) has microvilli (shown in cross section) containing mitochondria. The darker staining of the "resorptive cell" (left) is due to the presence of extensive rough endoplasmic reticulum. The scale bar represents 5 μm. The inset is a high-magnification electron micrograph of a transverse section from an ion-transporting cell with microvilli that contain mitochondria. Portasomes (arrowheads) are prominent on the cytoplasmic face of the membrane. Scale bar represents 100 nm. *From Zhuang et al. (1999) with permission.*

sections, we will explore some of what is known about the structure/function relationships of these three main types of caecal cell. Additionally, as is true in most of the regions of the gut throughout larval development, smaller cells (typically adjacent to the basal side of the cell layer) that have been termed "regenerative" cells dot the epithelium. These are thought to be simple diploid precursor cells that will eventually either divide to produce new gut cells or undergo endoreplication of the chromosomal DNA to produce the typically polyploid mature gut cells (Ray et al., 2009). Finally, there are scattered and relatively small enteroendocrine cells within the caecal epithelium (as well as other regions of the gut) which are presumed to mediate chemical and neurological signalling in the gut (Brown and Lea, 1988).

The connection of the caecal diverticuli to the gut tube proper (the caecal neck) occurs via cells with specific structure and functional qualities. The cells of the caeca that are internal relative to the CM are distinguishable from those in this neck region (i.e. outside of the CM relative to the main lumen of the gut tube). Once we have completed a general description of gut cell structure, we will return to structure/function data as determined by recent investigations of the disposition of specific gene products and functionalities.

The posterior wall of the caecal neck adjoins the AMG and the architecture of the epithelial cells undergoes a significant change in character. The largest of the AMG cells, which are the majority of cells in his region of the gut tube, have few and small apical microvilli in stark contrast with the GC cells as well as the PMG cells (Clements, 1992; Zhuang et al., 1999). This of course suggests less of an absorptive function for these cells in relation to the other gut compartments, the GC or PMG cells. This also implies that the apical surface area of AMG cells is much lower than that of the GC and PMG cells, which possess extensive microvillar-based extension of the plasma membrane. The basal membranes of the AMG cells are similarly expansive to those of other gut cells. Numerous intracellular vesicles and labyrinthine extensions of the basal membranes fill much of the cytoplasm of the AMG cells (Volkman and Peters, 1989a,b; Zhuang et al., 1999). Most AMG cells possess large polyploid nuclei. As in the GC, there are scattered, smaller diploid cells that are either regenerative stem cells or neuroendocrine cells.

Figure 1.5 (Clark et al., 2005) shows an SEM image of a fourth instar *Aedes aegypti* larval gut. The alimentary canal from the GC (at top) to the malpighian tubules (MT) at the junction with the hindgut was dissected. The tube from AMG through PMG was slit anterior to posterior and then the epithelium curled upon itself such that the internal surface is now displayed as the outer surface of this preparation. Note that the upper half of

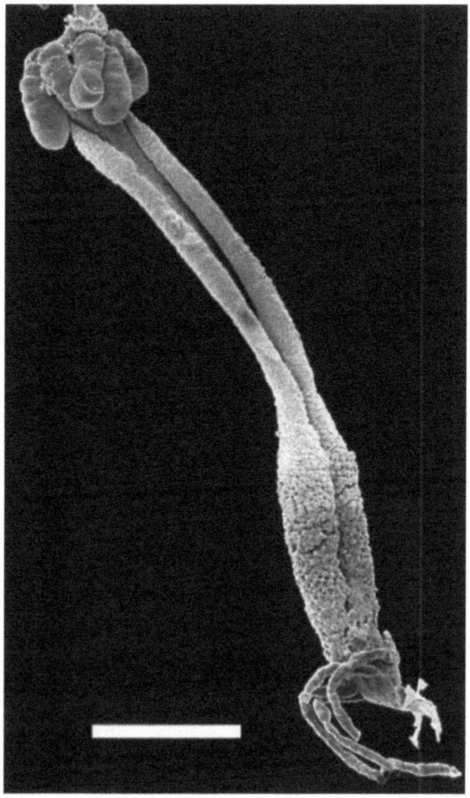

Figure 1.5 A scanning electronmicrograph of isolated larval *Aedes aegypti* midgut showing the different structurally distinct regions. The gut tube was cut along the length of the tube, and the tube then curled upon itself thus exposing the inner surfaces of the AMG and TR and the PMG. The GC are still intact and so the outer surface of the GC are shown at the top of the figure. The MTs are at the bottom. Note that, what the inner surface of the AMG, TR and PMG exhibit is a very different gross architecture. *From Clark et al. (2005) with permission. Scale bar represents 600 μm.*

the tube as shown here, which represents the AMG, has a very smooth appearance. The lower portion, the PMG, shows a more granular surface. At higher magnification or by viewing cross-sectioned material from the AMG and PMG it is evident that the difference in appearance of the everted tube is that the AMG cells have few and very short microvilli, whereas the major cells of the PMG have apical arrays of microvilli that are tightly packed and appear to be a solid cap when viewed by low magnification SEM (Clark et al., 2005). At the region where the AMG and PMG meet, there is a transitional region that is several cells broad from anterior to posterior along the long axis of the gut. The apical surfaces of cells in the transitional region show an increase in the numbers and dimensions of microvilli (Clark et al., 2005). Other distinctions have been described between the AMG, transitional region and the PMG and we will return to these shortly.

The PMG is broader than the AMG and the epithelial cells are also larger than those of the AMG. The apical surface is tremendously extended by microvilli that are tightly packed. Ultrastructural analyses show basal membrane infoldings, copious numbers of intracellular vesicles and extensive mitochondrial profiles, which are indicative of cells that are very metabolically active and involved in absorptive and secretory processes (Billingsley, 1990; Clements, 1992; Zhuang et al., 1999).

The hindgut is composed of the pyloris and MTs, ileum (anterior intestine), rectum and anal canal that are all of ectodermal origin in distinction with the endodermal midgut. The PM, which originates at the cardia at the posterior end of the foregut, begins to lose its integrity as it enters the pyloris of the hindgut (HG). The HG has a cuticular lining that encompasses the PM and the food bolus as it continues its journey toward excretion. The cells of the pyloris are thin epithelial cells and the funnel-shaped region of the HG has a posterior band of muscle forming a sphincter (the pyloric sphincter). At the most anterior extreme of the pyloris, the epithelial cells are quite small and contiguous with the stem cells of the posterior imaginal ring (Clements, 1992; Klowden, 2007). The five MTs are tubes with an opening into the pyloris and a closed terminus at the distal end of each. Two types of cells are described in the MTs: principal cells and stellate cells, which have differential embryological origins (Davies and Terhzaz, 2009; Dow, 2009). Principal cells are large with extensive apical microvilli and each cell can extend nearly around the circumference of the MT lumen, which is zigzag shaped due to the apical intrusion of the principal cells. Principal cells possess a large polytene nucleus and in later stages of larval development accumulations of membrane-bound inclusions or concretion bodies (Bradley and Snyder, 1989). The second type of cell in the MT is called the stellate cell and their shape is indeed star like. Stellate cells localise between some of the principal cells and can be difficult to detect with simple microscopy as they can appear to be the interconnections of the lateral membranes of principal cells. Recent physiological and immunohistochemical analyses have provided insights into the structural relationships between principal cell and stellate cells (see later discussion of this section). Stellate cells possess much smaller nuclei than principal cells and it is not clear whether or not they are diploid or polyploid. The function of the MTs is generally held to be similar to that of the vertebrate kidney in ion regulation and the formation of the primary urine (Beyenbach et al., 2009; Bradley, 1987).

The ileum or anterior intestine is a thin-walled epithelium within a prominent muscular tube. Relatively little is known of the functional specialties of the ileum other than it serves as the continuation of the pathway for the movement of the excreta. The muscular tube surrounding the ileum is at least partly responsible for pushing faecal material through the rectum and into the anal canal.

The rectum is a complex region of the gut and exhibits substantial architectural variations between genera of mosquitoes and insects in general. Simply stated it is an essential organ in the regulation of ionic balance in the animal and the retention or elimination of specific solutes. The cells of the insect rectum have at times been described as among the most complex cells in biology (Berridge and Oschman, 1972). The rectum of mosquito larvae varies in general architecture. The tubular nature of the alimentary canal continues into the ectodermal rectum. The length and architecture of the rectum varies between species but, generally, the rectum is a cuticle-lined epithelium that can exhibit longitudinal folds. The two subfamilies of mosquitoes, the Culicinae and the Anophelinae are somewhat different in terms of rectum structure (Bradley, 1987; Smith et al., 2007, 2008, 2010; White et al., 2013). Additionally, the osmolarity of the aquatic environment in which larvae develop can be reflected in sometimes-gross variations on the architectural details of the rectum. Culicinae species, such as *Aedes aegypti,* which select low ionic strength aquatic habitats exhibit a single compartment rectum. In contrast, Culicinae, such as *Aedes campestris,* that are tolerant of much higher osmolarities possess a rectum with distinct anterior and posterior compartments (Bradley, 1987; Smith et al., 2007, 2008, 2010; White et al., 2013). Anophelinae such as *Anopheles gambiae* (salt intolerant) and *Anopheles merus* (salt tolerant) possess rectal structures that are somewhat intermediate between the partitioned (anterior and posterior) rectum of salt-tolerant culicinae and the single-compartment rectum of salt intolerant species. Specifically, Anophelinae possess two distinct populations of cells but lack a clear compartmental divide between them (Bradley, 1987; Smith et al., 2007, 2008, 2010; White et al., 2013). The two populations are functionally and structurally distinct and have been named the dorsal anterior rectum (or DAR) cells and the remaining non-DAR cells (Smith et al., 2007). We will return to a discussion of the structure/function details of these cells later. In general, whether discussing Culicinae or Anophelinae larval rectum cells, the cells are characterised by extreme modifications of the plasma membrane on both the apical and basal sides. Meredith and Phillips (1973) described apical extension of the plasma membrane that takes the form of tightly packed parallel channels of membrane similar to tightly packed microvilli as described in the GC and the PMG. However, in the rectum, the parallel membrane channels do not end in free microvillar tips but rather in a surface that is embedded in a characteristic cuticular layer (Meredith and Phillips, 1973). The membrane channels or stacks are associated with numerous mitochondria and portasomes indicative of a very intense role in ion transport (Meredith and Phillips, 1973). The basal side of rectal cells can exhibit labyrinthine membrane infoldings that are much less organised into parallel arrays than the apical membranes but nonetheless very extensive. In Culicinae with divided anterior and posterior lobes, anterior cells look less extended in terms of plasma membrane amplification than do the posterior cells. In the Anophelinae, the DAR and non-DAR cells exhibit structural distinctions similar to those seen in the anterior and posterior cells of the Culicinae (Smith et al., 2008, 2010).

Throughout the epithelial tube that is the alimentary canal of insect larvae, we have described gross structural aspects and some details of the cell apical and basal plasma membranes. We have ignored the lateral membranes thus far. We will not dwell on the lateral membranes other than to state that as in all epithelia, cell–cell junctions exist in various arrangements, which of course hold the epithelia together as a sheet and also serve to influence the apical-to-basal movement or diffusion of molecules. A great deal of research has gone into the analyses of junctional complexes and the differences between vertebrate and insect model systems (e.g. Matter and Balda, 2003). Suffice it to say here that the lateral membranes of larval mosquito alimentary canal epithelial cells exhibit varied junctional complexes such that movement of molecules between cells is regulated but not always completely blocked (Neira Oviedo et al., 2009).

Figure 1.3 shows a cartoon rendering of the gross details of the larval mosquito alimentary canal. There are several highlights to recall as we continue into the next discussion of cell biology and cell polarity. The disposition of extracellular material that separates the food bolus from physical contact with the gut cells varies: in the foregut it is cuticular; from the cardia to the termination of the midgut at the pyloris there is a type II PM; from the beginning of the hindgut, the PM becomes surrounded by cuticular extracellular matrix; the gastric caeca possess a distinct additional barrier called the CM. The PM defines a fluid-filled environment called the ectoperitrophic space, which is in contact with the lumen of the gut and the lumen of the caeca. Various subdivisions of the gut are associated with characteristic structural specialisation of the several cell types located therein. In relation to the accessibility and effectiveness of orally administered biological toxins (a focus of this volume), gut structure is of obvious importance. In addition, certain details of the cell biology that have come to light in recent years have either already become demonstrably important to new control strategies or clearly have that potential. At this point, we are going to review a number of fairly recent investigations that have provided insight into the structure/function of the important cell types of the alimentary canal.

Material ingested by mosquito larvae will encounter the secretions of the SGs (i.e. saliva) as one of the earliest steps in digestion. In addition to aiding digestion, the saliva contains numerous gene products associated with immune surveillance and the inactivation of potential pathogens. Transcriptomic and proteomic analyses of the *Anopheles gambiae* larval SGs revealed the production and secretion of such immune effectors as defensins, lysozyme and TIL-domain proteins (Neira Oviedo et al., 2009). Hence, ingestion of materials that might inactivate such components of the saliva may well render the larvae more susceptible to biological toxins or organisms. Methods for generally inhibiting SG function would also be reasonable targets for the development of novel control strategies. The anterior and posterior lobes of the SGs are biochemically distinguishable but nothing is known about the actual compartmentalisation of specific salivary component synthesis and secretion. A clearer understanding of SG structure/function and cell biology will be a valuable pursuit in the future.

The glycocalyx-type extracellular matrix linings of the alimentary canal provide a range of functions such as facilitating the one-way movement of the food bolus from the mouth to the anus, physical protection of delicate cell surfaces from what can be abrasive particulates, barriers to full blown biological invasion from microbiota in the food bolus, size-exclusion barriers to macromolecular diffusion and even selective permeability and toxin sequestration. The lining of the foregut and the hindgut is cuticular and exhibits structural qualities of cuticle. The PM and the CM are also chitin-containing acellular barrier matrices with diverse functions (Hegedus et al., 2009; Lehane, 1997; Rudin and Hecker, 1989). The CM and PM are distinguishable from each other as well as the cuticular linings of the foregut and hindgut on several structural and biochemical bases. For example, lectin labelling shows that the CM and PM are readily labelled with several lectins including wheat germ agglutinin but the cuticular exoskeleton and the linings of the foregut and hindgut are distinguished from the PM and CM with ricinus communis 1 (RCA; Linser et al., 2008; Neira Oviedo et al., 2009). Figure 1.6 shows such a comparison between *Dolichus biflorus* Agglutinin (DBA) and RCA1 in a longitudinal section of a fourth instar *Anopheles gambiae* larva. The green fluorescence (DBA) highlights the PM and CM vividly and in contrast to the red signal (RCA1), which stains the exoskeleton cuticle and the linings of the foregut and hindgut. Indeed one can see that in the pyloris, the red cuticular lining of the hindgut surrounds the PM and that the PM begins to be compacted at the junction with the ileum. Furthermore, the origin of the PM (in green) is visible in the folded layering of the cardia while the foregut cuticle is visible within the innermost channel of the termination of the oesophagus (as colours are not presented in the print version of this volume, please visit the on-line version for full colour details). These chitin-containing extra cellular

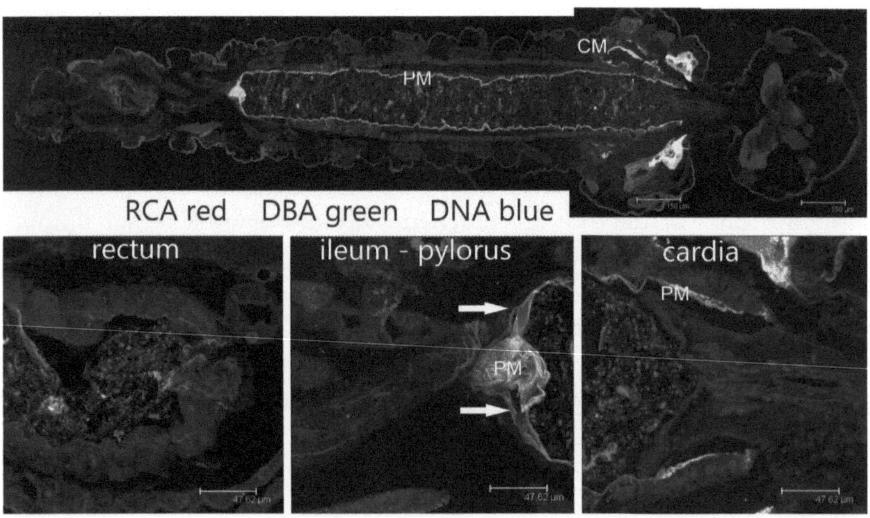

Figure 1.6 Figure shows longitudinal sections of paraffin embedded *Anopheles gambiae* fourth instar larvae at low (upper montage) and high (lower three panels) magnification with the anterior (head) to the right. Labelling was with TRITC-conjugated *Ricinus communis* Agglutinin I (red), FITC-conjugated *Dolichos biflorus* Agglutinin (green) and DRAQ5 for DNA (blue). For the purpose of this discussion, note that the green DBA staining labels the peritrophic membrane (PM) and the caecal membrane (CM), and the red RCA labels cuticular structures including the exoskeleton and the lining of the foregut adjacent to the beginnings of the PM in the cardia and at the posterior end of the PM at the beginning of the hindgut at the level of the ileum-pyloris junction (arrows). Also note that the rectum is lined by red, RCA+ cuticle. *From Linser et al. (2008) with permission.* (See the colour plate.)

matrices (ECMs) are structural barriers to any ingested material and hence need to be penetrated by any biological toxins or control organisms. Lectins, as used herein, are capable of discriminating specific details of glycoconjugate structure in terms of specific sugar structure, chemical linkages and other details. The analyses described here serve to show that the detailed glycobiology of the ECMs associated with gut possess distinct biochemical signatures. The impact of that varied biochemistry is largely untested.

The structure of specific cell types within the gut provides insight into their functional roles. As described earlier, the apical surface of the large, principal epithelial cell types of the GC and PMG possess extensive arrays of microvilli. This implies an absorptive role. AgAPN1, a GPI-anchored plasma membrane glycoprotein, first identified in adult *Anopheles gambiae* is a peptidase presumably involved in the final stages of protein digestion, liberating amino acids for absorption (Dinglasan et al., 2007). AgAPN1 is also a putative point of attachment for the malaria parasite *Plasmodium falciparum* (Armistead et al., 2014; Dinglasan et al., 2007; Mathias et al., 2013). Antibodies to this protein specifically label the microvillar arrays on the PMG cells and on a specific subset of the GC cells. Figure 1.7 shows AgAPN1 labelling of the PMG and the GC cells that form the neck region of each GC lobe. Only GC cells that are exterior to the CM label for this protein. In contrast, AgAPN2, another distinct cell surface aminopeptidase thought to be a binding site for the Cry11Ba toxin of *Bacillus thuringiensis* (Zhang et al., 2008) is also found on the microvilli of PMG cells and GC cells. But, in this case, only the GC cells that lie within the GC, internal to the CM, exhibit the protein. Additionally, the CAP cells (see earlier discussion) at the posterior extreme of each caecum contrast with the surrounding neighbouring GC cells by lacking AgAPN2 (Fig. 1.8; Harvey et al., 2010; Linser et al., 2007).

As mentioned earlier, one of the striking qualities of the larval mosquito alimentary canal is the extreme pH of portions of the gut lumen (Fig. 1.2). The evolutionary selective pressure for extremely high pH in the gut lumen is often associated with the high content of plant material in the diets of many insect larvae including caterpillars and mosquito larvae (Terra et al., 1996). This may have some truth but it should be noted that even in the Tsetse (Glossinidae) which derives all of its biological energy for its entire life cycle from blood meals, the gut pH can exceed 10 (Liniger et al., 2003). Regardless of the physiological ramifications of an alkaline digestive system, the mechanisms which drive the pH gradient along the length of the mosquito larva gut from nearly neutral at the level of the GC, to pH 10.5 or even

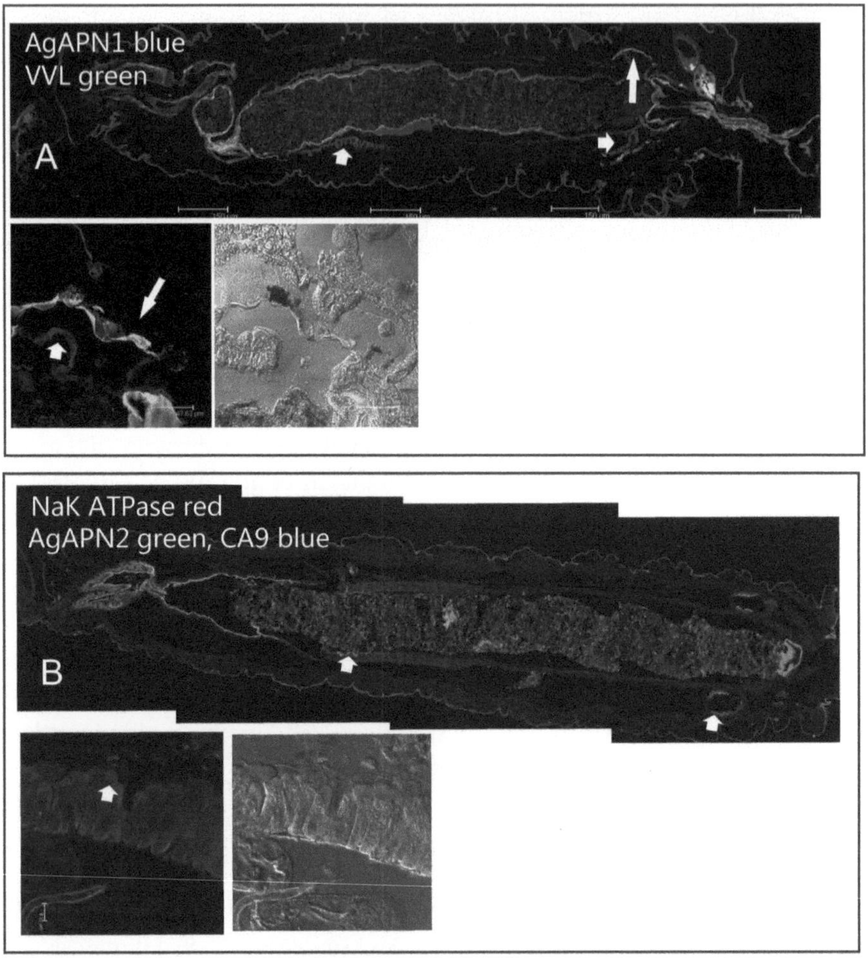

Figure 1.7 Figure shows the distribution of two integral membrane amino peptidases, AgAPN1 (panel A) and AgAPN2 (panel B) in larval *Anopheles gambiae* gut sections. In (A), AgAPN1 (blue) is seen on the brush border membranes (BBM) of the posterior midgut (PMG) cells (short arrow to left) and the BBM of the neck of the gastric caeca (GC) (short arrow to right and in high mag images at bottom of panel). Note that FITC-conjugated Vicia Villosa Lectin (green) was used to highlight the cuticular structures, the peritrophic membrane (PM) and the caecal membrane (CM) (long arrows) making the limitation of AgAPN1 to the neck cells of the GC evident. In (B), AgAPN2 (green) is compared to Na^+/K^+-ATPase (red) and the cytoplasmic marker carbonic anhydrase-9 (CA9). The short arrows indicate labelling for AgAPN2 on the BBM of the PMG and on the BBM of the GC cells internal to the CM. *From Linser et al. (2008) with permission.* (See the colour plate.)

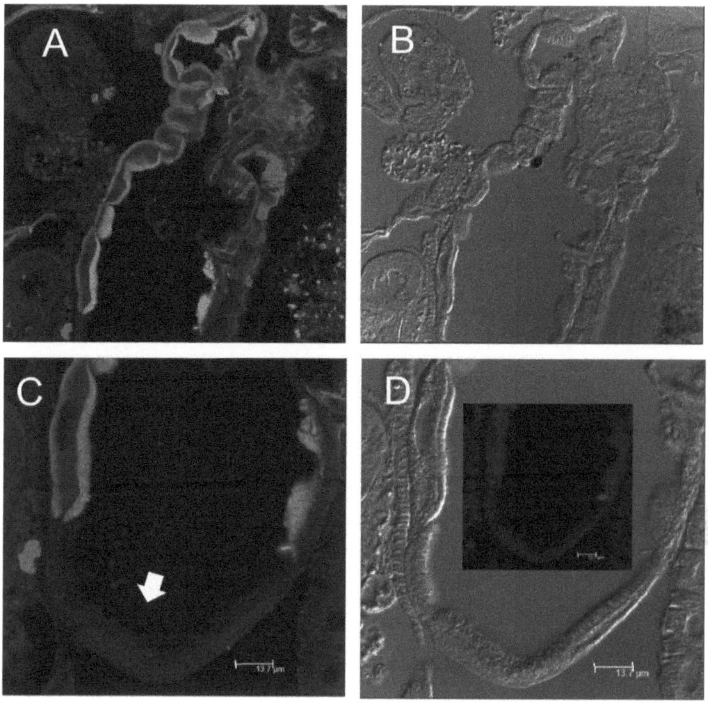

Figure 1.8 Figure shows the GC of fourth instar *Anopheles gambiae* at high magnifications to highlight the distribution of AgAPN2 (green) relative to the basal membrane marker Na^+/K^+-ATPase (red) and the cytoplasmic CA9 (blue). Panels A and B show the anterior portion of a caecum at the junction with the AMG. Note that AgAPN2 is prominently localised to the BBM of the anterior GC cells. Panels C and D show the posterior portion on the same caecum. Note the complete loss of staining for both Na^+/K^+-ATPase and AgAPN2 (arrow) at the posterior extreme of the caecum, which is the CAP cells. The inset in D shows only CA9 staining at the same magnification as shown in A and B to provide reference. *From Linser et al. (2008) and Harvey et al. (2010) with permission.* (See the colour plate.)

higher in the AMG, to pH 8 in the PMG and slightly acidic pH in the rectum are a study in functional cell polarity (Filippova et al., 1998; Zhuang et al., 1999). To clarify, the major process that drives the up and down pH gradient is the pumping of protons via a proton pump called vacuolar-ATPase (V-ATPase; Filippova et al., 1998; Zhuang et al., 1999). V-ATPase and its roles in epithelial physiology have been studied exhaustively in numerous model systems. For the purpose of our structural discussion here, pharmacological studies have demonstrated a central role in establishing and maintaining the highly alkaline environment within the gut. The V-ATPase is actually a macromolecular complex of several proteins and can be visualised with TEM as a sub-plasmalemmal knob termed a portasome. Typically, portasomes are associated with closely juxtaposed mitochondria and indeed the function of the proton pump is dependent on a substantial supply of ATP. Figure 1.4 shows TEMs of GC cells from larval *Aedes aegypti*. The two principal cell types of the GC are shown and the high-magnification inset reveals the studded plasma membrane of microvilli in which the central cytoplasmic domain of the microvillus is dominated by a mitochondrion (Meredith and Phillips, 1973; Volkman and Peters, 1989a; Zhuang et al., 1999). The internal pH of the GC and that of the adjacent AMG region is neutral to slightly alkaline. However, as the point of reference moves posteriorly in the AMG, the luminal pH rapidly rises to 10–11. The several log units of decrease in luminal proton concentration is accompanied by a shift in epithelial cell polarity, at least in reference to the position of the proton pumping V-ATPase. In the GC where the pH is near neutrality, the proton pump is situated on the apical side of the epithelial cell and hence is moving protons into the lumen. In the AMG, the proton pump is now localised to the basal side of the epithelial cells and hence involved in pulling protons out of the lumen in a transcellular pathway (Filippova et al., 1998; Zhuang et al., 1999). At the region of the PMG, the disposition of the V-ATPase reverses again such that the apical microvilli of the PMG cells possess high levels of V-ATPase in contrast to the basal plasma lemma, which shows little or none. This shifting pattern of V-ATPase (portasome) localisation from apical to basal and then back again in the course from anterior to posterior midgut has been verified both by immunochemical approaches (Fig. 1.9) as well as TEM (Filippova et al., 1998; Zhuang et al., 1999). In stark contrast to the V-ATPase paradigm, sodium–potassium ATPase (NaK-ATPase), which as the name implies pumps sodium and potassium (exchange) is expressed in opposite polarity to V-ATPase (Patrick et al., 2006). That is, in the GC NaK-ATPase is located on the basal side of most GC cells, in the AMG it is apical and in the PMG it returns to basal. The opposite cell polarity regarding these two ATPases is common in other areas of the gut but not universal (Patrick et al., 2006; Smith et al., 2007). Physiological modelling of the system has incorporated these important ion regulatory membrane proteins as well as numerous others in recent years. This complexity highlights the regional specialisation of the plasma membrane and such information is useful in developing novel targeting strategies for insect control.

The region of the gut epithelium where the cells undergo the remarkable flip-flop in functional polarisation of the plasma membrane has been termed

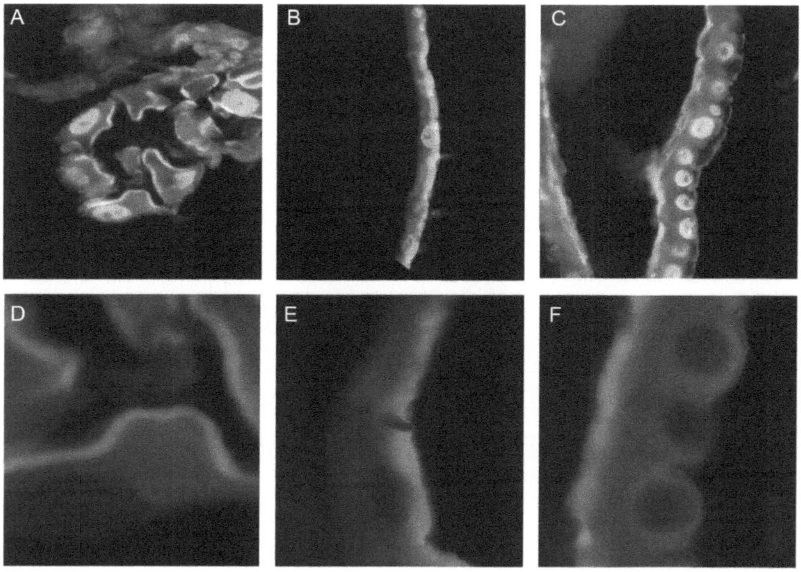

Figure 1.9 Figure demonstrates the apical to basal back to apical transitions in the polarised distribution of V-ATPase in the *Aedes aegypti* larval midgut. Panels A–C show regions of the GC, AMG and PMG, respectively. Green/yellow labelling shows the localisation of V-ATPase on the apical BBM of the GC (A) and PMG (C) and on the basal membrane of the AMG (B). Red labelling identifies the basal side of the epithelium in all cases as it depicts the external layer of muscle labelled with TRITC-conjugated Phalloidin. Blue is DAPI staining of cell nuclei. Panels D–F show the same series of gut regions at higher magnification. *From Zhuang et al. (1999) with permission.* (See the colour plate.)

the transitional region (TR; Clark et al., 2005; Smith et al., 2007). The cells of the TR exhibit a graded shift in several parameters including the numbers and size of microvilli, susceptibility to Cry4Ba toxin damage, the patterning of cell nuclei, and the flip-flop polarised location of the V-ATPase and NaK-ATPase (Clark et al., 2005; Smith et al., 2007). Figure 1.10 shows an analysis of this region of the *Anopheles gambiae* larval midgut with immunohistochemical markers for NaK-ATPase. Panel C provides a view of the shift in functional polarity with NaK-ATPase on the basal membrane infoldings in the PMG (to the right) and the shift of this protein to the apical side in the AMG cells (to the left; Smith et al., 2007). Also shown in this figure is the distribution of another important enzyme in the regulation of gut and cellular pH, carbonic anhydrase 9 (CA9) which is 1 of 12 mosquito carbonic anhydrase genes/proteins. This pH regulator is expressed by cells of the GC and the rectum as a cytoplasmic protein but is also secreted by the GC cells

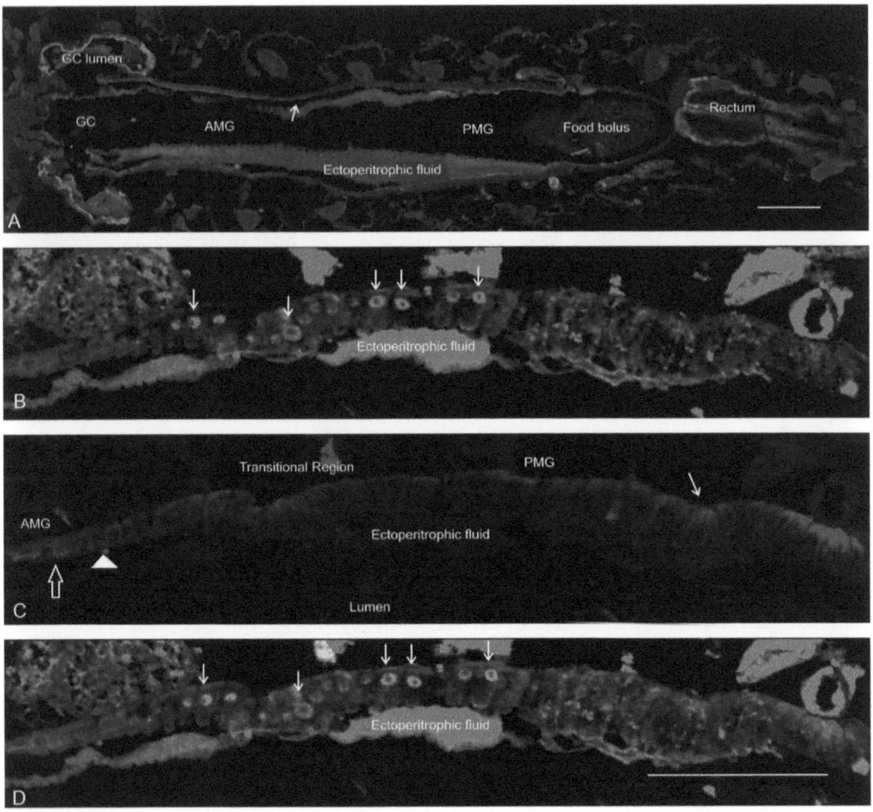

Figure 1.10 Figure shows the entire length of the larval *Anopheles gambiae* alimentary canal from the cardia at the extreme to the rectum at far right in panel A. CA9 is labelled in green and Na^+/K^+-ATPase in red. The arrow in (A) indicates the transition region at which Na^+/K^+-ATPase shifts from basal in the AMG to apical in the PMG. Panels B–D show the transition region at higher magnification. CA9 is expressed by several cell types but is secreted by GC cells into the ectoperitrophic space. In the transition region, CA9 is also found in cell nuclei (arrows, B and D). Panel C isolates the Na^+/K^+-ATPase signal and thus reveals the shift in epithelial cell polarity from basal (PMG, solid arrow) to apical (AMG, hollow arrow) of this very important physiological function. Panel D overlays the red and green images with a blue (DRAQ5, DNA) signal. *From Smith et al. (2007) with permission.* (See the colour plate.)

into the ectoperitrophic space and hence part of the digestive milieu and a component that will be encountered by ingested microbiota and toxins.

The PMG of larval mosquitoes is the largest region of the alimentary canal in terms of cell surface exposure to ingested materials. The apical brush border of the principal PMG cells is massive and difficult to estimate in terms of gross quantity. Its extensive nature has made it possible to engineer rather simple methodologies for isolating highly enriched membrane vesicle preparations termed the BBMV (brush border membrane vesicle; Harvey et al., 2009). This preparation has facilitated numerous analyses of the biochemical constituents of the apical plasma membrane from several perspectives including the purification of toxin receptor proteins (Zhang et al., 2008). Proteomic analyses have provided numerous potential targets for novel controls. Among proteins of the brush border membrane (BBM) are nutrient amino acid transporters, cadherins, alkaline phosphatase, sodium–proton antiporters and of course V-ATPase (Harvey et al., 2009 and elsewhere in this volume). The structure/function of the BBM of the PMG is yet to be explored to its fullest but to date many opportunities for physiological targeting or membrane binding of disruptive molecules has been realised with undoubtedly more to come.

The MTs of the insect alimentary canal are unique in structure and function and have received a great deal of attention over many years (Fig. 1.11). MTs represent a major component of the insect's machinery for regulating ionic homeostasis and have been loosely equated to the vertebrate kidney (Beyenbach, 2003; Dow, 2009). As mentioned earlier, there are five MTs extending from the HG. The odd number of biological structures in a bilaterian is in itself unusual. Additionally, the MTs are the exception to the rule that the alimentary canal is completely regenerated in the transition from larval to adult imago. The MTs of the adult are essentially the same cells as were present in the fourth instar larva. There are two main cell types in the MTs: the principal cells and the stellate cells. The principal cells make up the vast majority of the substance of the MTs and are large cells with large polytene nuclei. The tubular structure of the MTs is characterised by an irregular lumen that has a zigzag space defined by the apical microvillar arrays of the principal cells. The BBM of the principal cells is characterised by microvilli filled with mitochondria and plasmalemmal V-ATPase (Harvey et al., 2009; Patrick et al., 2006). Principal cells also contain large and variable numbers of membrane bound inclusions that contain organometallo concretions made up of complexed metal cations (Bradley and Snyder, 1989). Numerous transport proteins including the bicarbonate transporter NDAE1 are expressed on the basal/lateral membranes of principal cells. From the proximal to the distal extreme of a MT, the distribution of principal cell transport proteins and other markers is not uniform. A gradient of expression patterns has been described indicating that MTs of mosquitoes as well as other insects are regionally specialised along the long axis of each MT (Dow, 2009; Rheault et al., 2007). Stellate cells are far less numerous (~20% of the total

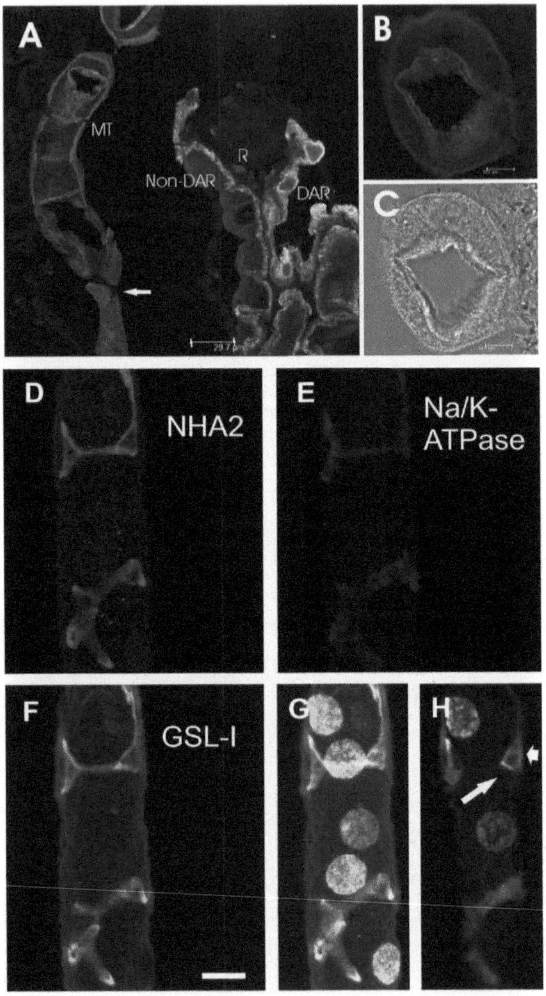

Figure 1.11 Figure reveals several details of MT cell biology in the hindgut region of *Anopheles gambiae* larva sections. Panel A shows immunolabelling for the bicarbonate transporter AgNDAE1 (red), V-ATPase (green) and CA9 (blue). The MT is characterised by basal membrane staining of the principal cells for AgNDAE1 and internal and apical staining for V-ATPase. Panel A also shows a partial view of the rectum and that CA9 is characteristic of the DAR cells whereas V-ATPase is also present in and on the non-DAR cells. Panels B and C show a high magnification view of a cross section through an MT displaying the basal labelling for AgNDAE1 and the apical labelling for V-ATPase. The arrow in (A) indicates a stellate cell which shows little signal in this labelling combination. Panels D–H show high magnification views of MT whole mounts labelled for proton antiporter NHA2 (red), Na^+/K^+-ATPase (blue) and Griffonia simplicifolia I lectin (green) and in D&H, DAPI for nuclei (aqua). NHA2, Na^+/K^+-ATPase and GSL-I all label the stellate cells of the MT. In the merge images G and H, the polarity of the markers is revealed. NHA2 and GSL1 co-localise (long arrow in H) to the apical aspect of the stellate cell whereas Na^+/K^+-ATPase is basal. *Figures A–C from Linser et al. (2012) and D–H from Xiang et al. (2012) with permission.* (See the colour plate.)

number of cells) and much smaller than principal cells. Stellate cells are first evident amongst the principal cells after the first (proximal to the hindgut) 10% of the length of each MT. Like the principal cells, stellate cells exhibit specific patterns of plasma lemmal proteins on both the apical and basal/lateral membranes (Beyenbach, 2003; Linser et al., 2012; Xiang et al., 2012). Models for the physiological roles of both principal cells and stellate cells have been proposed and supported by many varied studies (Beyenbach, 2003; Linser et al., 2012; Xiang et al., 2012). An interesting and controversial characteristic of MTs is that a substantial series of investigations supports the movement of certain ions such as Cl^- between cells, which indicates a certain specialisation of cell–cell junctions that can provide a regulated pericellular pathway (Beyenbach, 2003). In general, MTs monitor the makeup of the hemolymph and actively produce the primary urine, providing one of the key homeostatic functions within the larva.

The final component of the larval alimentary canal to be discussed herein is the rectum. As mentioned earlier, the rectum is contiguous with the anterior components of the alimentary canal by the ileum or posterior intestine. The rectum is a major regulator of excretion, elimination and retention of solutes critical for homeostasis. Mosquito larvae have adapted to many aquatic environments that can range in ionic strength from very low salinity fresh water to salt concentrations that exceed that of sea water (Smith et al., 2010; White et al., 2013). One rather obvious structural variation that correlates with salt tolerance is the structure of the rectum. Figure 1.12 shows the three major generalised structures of rectum from the simple bulb made up of seemingly a single type of cell found in culicine mosquitoes with low salt tolerance, to the bi-lobed rectum of salt-tolerant culicines to the regionally specialised two-cell type form in anopheline mosquito larvae (Smith et al., 2007, 2008). The salt tolerance of mosquito larvae has been investigated extensively and species that are capable of short term adaptation and those that have very narrow ranges of salinity are well documented in the classical literature (e.g. Bradley, 1987; Clements, 1992). Recent investigations have also shown that certain species of anopheline mosquitoes exhibit dynamic structural changes in the rectum when placed into changing conditions of salinity (Smith et al., 2008; White et al., 2013). It appears that the balance of functionalities compartmentalised within the two rectum cell types, DAR and non-DAR can change in response to ionic fluctuations. An interpretation of this dynamic is that the associations between transport mechanisms compartmentalised within the cell types can be modified to produce different potentials for vectorial and linked ion transport (Smith

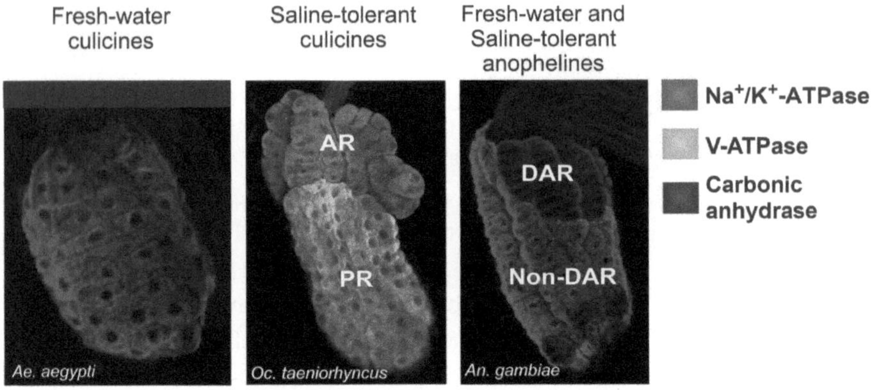

Figure 1.12 Figure depicts the gross structure of the three types of larval mosquito rectum as seen in fresh-water culicenes, salt-tolerant culicenes and all anophelines. Fresh-water culicenes such as *Aedes aegypti* have a uniform rectum with one primary cell type as seen in this whole mount immunolabelled for basal NaK-ATPase (left). Salt-tolerant culicenes such as *Ochlerotatus taeniorhynchus* have a rectum with distinct anterior and posterior regions. Anophelines possess a rectum that is distinguished by an anterior patch of distinct cells termed dorsal anterior rectum (DAR) cells and a posterior majority of non-DAR cells. AR, anterior rectum; PR, posterior rectum. *Image modified from Smith et al. (2008).* (See the colour plate.)

et al., 2008; White et al., 2013). Another aspect of rectum structure/function is that the PM that protects the midgut from microbial attack loses integrity in the hindgut. The cuticular lining of the rectum is very intimately associated with the epithelial cells and the gross structure of the rectum creates channels and crypts between folds. These pockets are typically teaming with bacteria in very close association with the rectum cuticle. The microbial flora that is resident in the insect gut has become an increasingly attention getting topic and may hold details of the biological balance in insect systems as it does in other complex metazoans (e.g. Engel and Moran, 2013).

The entire alimentary canal of the mosquito larva is surrounded by muscles in several forms from dense sheaths to glass-sponge-like baskets and networks. Figure 1.13 is representative of the circular and longitudinal musculature that produces waves of contraction in both anterior to posterior and posterior to anterior directions (Seron et al., 2004; Terra, 1990) The muscle contractions are influenced by neurochemical activity and involved in some aspects of food digestion (Krajniak, 2005). Although the anterior to posterior movement of the food bolus is largely driven by ingestion at one end and defecation at the other (Terra, 1990), the contraction of the various muscular investment of the gut plays several roles including the mixing of

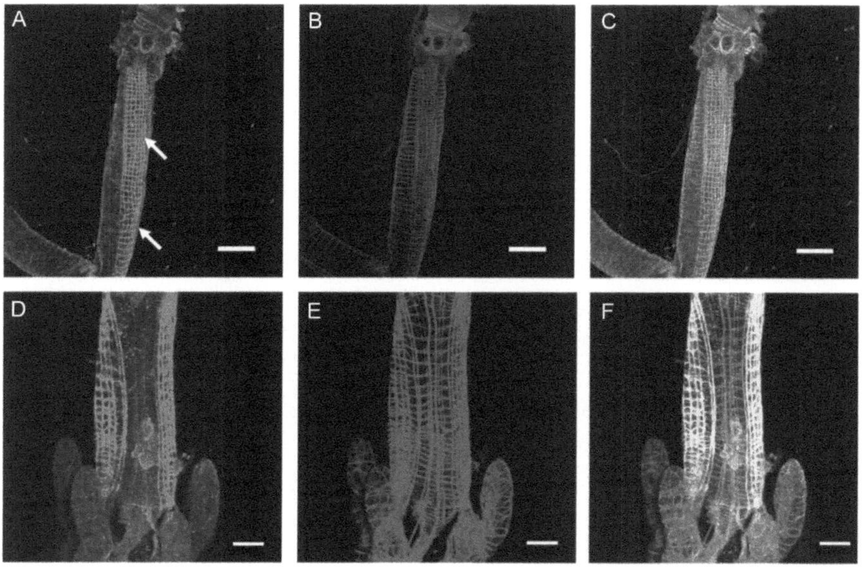

Figure 1.13 Figure shows whole mount preparations of isolated midgut from fourth instar larvae of *Aedes aegypti* (A–C) and *Anopheles gambiae* (D–F) immunolabelled for the peripheral membrane protein carbonic anhydrase CA-10 (green) and Phalloidin to reveal muscle (red). The arrows in (A) and the merged images show that a specific subset of the gut musculature labels for CA-10 whereas other muscles of the gut do not. Thus the musculature of the gut exhibits more structure/function variability than simple longitudinal versus circumferential. *From Seron et al. (2004) with permission.* (See the colour plate.)

components in the ectoperitrophic space and the GC compartment (Smith et al., 2008; White et al., 2013).

A final note on the mosquito larval gut epithelium; the epithelium in adult mosquitoes, by an unknown process, is either porous to or transports across whole protein molecules such as ingested vertebrate immunoglobulins (Beyenbach et al., 2009; Jeffers and Roe, 2008). Also, as noted above in the discussion on MTs, there is an apparent pathway between epithelial cells through which small molecules are selectively passed. Therefore, it seems possible that some molecules such as toxins that can pass through the type II PM of larvae might gain rapid access to the hemolymph and thus the basal aspect of the epithelial cells.

3. OTHER INSECTS

The general architecture of the alimentary canal of insects varies considerably depending on adaptive specialisations. A useful summary is presented by Engel and Moran (2013). The embryonic development of the insect gut has numerous common features and in both larvae of holometabolous insects and in adults of the same and nymphs and adults of hemimetabolous insects the macroscopic features may look very different but the same derivation of foregut and hindgut from ectoderm and the midgut from endoderm is the rule. Compartmentalisation of digestion, pH regulation, ion balance and the flow and recycling of water varies considerably and evolutionary opportunities have resulted in great diversity. The hypothetical pressures that have driven the evolutionary flow of compartmentalisation of function have been reviewed several times (e.g. Terra et al., 1996). Three groups of insects with agricultural impact that we will discuss are the lepidopterans (larvae), coleopterans (beetles and their larvae) and hemipterans (aphids).

3.1. Lepidopteran larvae (caterpillars)

Larvae of butterflies and moths have long been challenging pests to agriculture. Due to this fact and the additional fact that many are large and readily manipulated for biological research has led to a very substantial body of research for well over 100 years. Most texts on insect biology and physiology present caterpillars such as the larvae of the tobacco horn worm (*Manduca sexta*), and the silk worm (*Bombyx mori*), as favoured model systems. The gross architecture of lepidopteran larvae alimentary canal has of course the same general layout for most insects: foregut, midgut and, hindgut. Prominent diverticuli (gastric caeca) are absent and the midgut is by a large margin the most substantial part of the system. Individual caterpillars can yield gram quantities of dissected midgut tissue. The midgut is a pleated tube with cellular distinctions between the anterior, middle and posterior midgut (e.g. Cioffi, 1979; Fig. 1.14 [Dow, 1992]). The tube also contains a type II PM. Perhaps the most distinguishing characteristic of the lepidopteran larval midgut occurs at the cellular level with the existence of a specialised cell type known as a goblet cell (Dow, 1992; Harvey et al., 1983). Figure 1.14 also shows the gross cell biology of the lepidopteran gut epithelium. There are two principal cell types thought to be involved in digestion, ion balance, pH modulation and water flow. The first is a typical columnar epithelial cell with a very extensive apical microvillar array or BBM. These cells also exhibit varying degrees of basal membrane expansion, a large polyploid nucleus and varying vesicular content. The columnar cells are structurally similar to insect gut epithelial cells in general. In contrast, interspersed

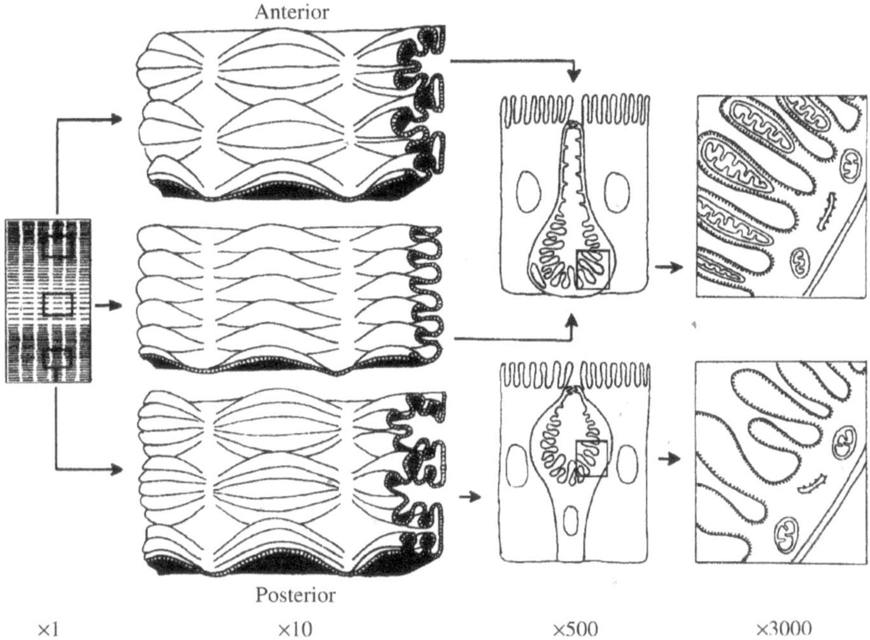

Figure 1.14 Structure of the lepidopteran midgut. Regional specialisation into three zones is visible at both coarse and progressively finer resolution. Approximate magnifications are shown. At 500 ×, the two forms of goblet cell are evident in the regions of the gut epithelium. *Taken from Dow (1992) with permission.*

among the columnar cells are the very distinct goblet cells. These cells are so named because of their architecture, which resembles a goblet with a prominent internal cavity. Goblet cells of the AMG are somewhat different from those of the PMG in that in the AMG goblet cells the goblet cavity is proximal to the basal aspect of the cell whereas in the PMG goblet cells the cavity is adjacent to the apical extreme of the cell (Fig. 1.14). The inner surface of the goblet cell plasma membrane adjacent to the cavity is studded with potasomes as described earlier in the mosquito system. The lepidopteran larval gut, due to its large size and its tractability as an experimental system, made it possible for numerous investigators to track down the nature of the portasome and to demonstrate the role that the V-ATPase plays in insect epithelial physiology (Harvey et al., 1983; Klein et al., 1991) and as a target for novel control strategies (Baum Chapter 5 this volume). The activity of this cell-surface proton pump is the driving force of another striking characteristic of the lepidopteran gut: a luminal pH that can be as high as 12 and transepithelial potential of nearly 240 mV (Harvey et al., 1983; Wieczorek,

1992). These are physiological extremes found nowhere else in nature. Certainly part of capacity for lepidopterans to generate these remarkable physiological conditions is contributed by the unique structure/function of the epithelium and in particular the goblet cells (Harvey et al., 1983; Wieczorek, 1992).

Lepidopterans do possess MTs although in the vicinity of the rectum they differ from mosquitoes. The distal ends of the MTs in lepidopterans (and coleopterans) are embedded in an extracellular matrix that holds them in a fixed juxtapositional relationship with the basal surface of the rectum (Azuma et al., 2012; Fermino et al., 2010; Ramsay, 1976). This association between the tissues is called the cryptonephric rectal complex (Azuma et al., 2012; Fermino et al., 2010; Ramsay, 1976). Most other insects have free distal ends that can actually be moved about the hemocoel by muscular contractions in the body wall. The physiological details of this fixed association with the rectum are poorly understood but is the subject of numerous investigations (Azuma et al., 2012; Fermino et al., 2010; Ramsay, 1976).

3.2. Coleopterans (beetles and their larvae)

As the most successful group of insects in terms of the numbers of species, coleopterans have physical and physiological adaptations for uncounted ecological niches and dietary specificities. Both larvae and adults can be agricultural pests and so control strategies can target the life stages differentially. The alimentary canal of coleopterans possesses the same basic elements as do other insects: foregut, midgut, hindgut with the same embryological origins (ectoderm, endoderm, ectoderm, respectively). The structural details vary greatly depending on food source and digestive strategies with tremendous variability in the presence, absence and/or nature of gastric caeca the numbers and extent of MTs and the relative size of the hindgut. Some exhibit alkaline extremes in regions of the gut and others do not (Terra et al., 1996). Some produce Type I peritrophic matices even in the larval stages (Ryerse et al., 1994).We will limit most of our comments to one species of very significant agricultural importance, the corn rootworm (*Diabrotica* sp.) The larvae of the western corn rootworm (*Dibrotica virgifera virgifera*) are of worldwide importance due to feeding damage to corn roots and the resulting losses in corn yield (Chu et al., 2013; Sayed et al., 2007; also refer to Chapters 4 and 5). Larvae feed on the roots of the corn plant whereas the adults feed on the reproductive components of the flowering plant. Many species of pestiferous beetles exhibit the capacity for rapid development of resistance to conventional pesticides as has been true for *Dibrotica* (Al-Deeb and Wilde, 2005; see also Chapter 6). Hence, control strategies are constantly evolving. A common strategy in use today is dependent on genetically modified crops, such as corn, which produce specific bacteriological toxins discussed in great length in Chapters 4 and 6. The corn rootworm is sensitive to certain Cry toxins of *Bacillus thuringiensis* (Bt) as discussed in Chapter 4. In 2003, genetically modified corn (maize) which produces Bt toxins was commercialised in the United States (EPA, 2003) for the primary purpose of controlling the western corn rootworm. The general pathobiology of Bt in a variety of susceptible insects will be discussed in Chapter 2 but suffice it to say here that the conditions in the gut of *Dibrotica v.v.* are appropriate for the cell-lytic action of Cry3Bb1. Ingestion of such genetically modified (GM) maize results in intoxication and death of *Dibrotica v.v.* The widespread use of Bt-maize has been successful in terms of reducing crop loss. But the selective forces brought in to play by the use of this and other toxic strategies have begun to create resistant rootworms (Frank et al., 2011; Petzold-Maxwell et al., 2012; Chapter 6). This has necessitated crop rotation and pesticide co-applications to support the continued utility of these GM crops. Additionally, the use of GM crops that can reduce the need for pesticides has generated a general reduction in such pesticide use. This in itself has produced circumstances in which other pestiferous insects that are not sensitive to the engineered GM toxins have been able to flourish and create new challenges.

3.3. Hemipterans (aphids)

Aphids are hemimetabolous insects. Therefore, they do not produce larval stages but rather nymphs that closely resemble the adult form. Aphids rarely reproduce sexually but rather undergo parthenogenic production of offspring by the females. Aphids feed on plant fluids typically by virtue of piercing mouth parts (proboscis with stylets) that enter the plant and suck phloem fluids from the plant (Pelton, 1938). The sugar-rich sap of the plant is very low in other essential nutrients for the aphid including amino acids and so it is necessary for aphids to intake large quantities of fluid and then excrete the excess sugary material in a fluid called honey dew. This sugary fluid on the surface of the plant can be a source of food for other insects or microbial plant pests (Dedryver et al., 2010). Aphids as agricultural pests have become a problem of greater proportions in recent years due to the various events that have reduced the use of chemical pesticides such as the use of GM crops, which in themselves do not impact aphid fluid-based feeding (Chougule and Bonning, 2012). Due to the feeding strategy of aphids, they have the potential to transmit other pathogens from one plant to another as they feed. Similar to the potential disease transmission that has been realised in hematophagus insects, sap-sucking aphids can transmit numerous viral diseases of plants (Brault et al., 2010; Ng and Perry, 2004). In some instances, a "dirty" proboscis and stylets is adequate to move a pathogen from one plant to another (non-circulative transmission). In many viral pathologies, the transmission event requires intermediate stages within the aphid (circulative transmission; reviewed in Bragard et al., 2013). In this form of transmission, the alimentary canal of the aphid plays a key role. As the fluid meal is taken into the gut, the virus must eventually exit the gut, enter the hemoceol, make its way to the SGs and then make its way into the saliva. This is paralleled by the processes of pathogen transmission in hematophagus insects. In some cases, the virus particles have evolved specific methods of access through the various cellular barriers in the path to the saliva that do not require actual viral replication within cells of the aphid (Bragard et al., 2013). Other virus/aphid relationships do in fact involve replication of the virus within the cells of the gut and eventual release into the hemoceol (Bragard et al., 2013). Some plant rhabdoviruses can be persistent in the aphid for life and can even be transmitted vertically to offspring (Hogenhout et al., 2008). The expansion of aphids as agricultural pests particularly for GM crops expressing Bt toxins has lead to efforts to engineer into the Bt transgene, peptide motifs that will render the toxins effective on specific aphids (Chougule and Bonning, 2012; Chougule et al., 2012). For Cyt toxins to be effective in any insect, there needs to be a specific binding interaction between the toxin and the BBM of insect gut epithelial cells. Recent manipulation of the Cyt2Aa toxin amino acid sequence in which a binding peptide was added seems to hold promise (Chougule and Bonning, 2012; Chougule et al., 2012).

4. CONCLUSIONS AND COMMENT

The insect alimentary canal is structurally complex and varied from species to species. As a target for developing arthropod control strategies, the gut is both an important and major barrier for insect control agents and hence an important target for intervention. The specific cell biology of the gut also provides novel targets for the development of gut-function-disrupting agents. With the recent and ongoing expansion of technological approaches to defining fundamental biological structure/function relationships, it is ever more possible to identify unique but vital functional molecular target sites. The era of "omic" biology in which we are currently immersed provides the investigator with unprecedented opportunities to discover and integrate knowledge of fundamental biology. Greater understanding of the cellular, biochemical and molecular structure and processes of insect gut will help develop rational strategies for intervening with target sites important for the survival of the pest stage, as well as exploiting existing macromolecules for delivering toxins, which may otherwise may not be effective.

REFERENCES

Al-Deeb, M., Wilde, G., 2005. Effect of Bt corn expressing Cry3Bb1 toxin on western corn rootworm (Coleoptera: Chrysomelidae) biology. J. Kansas Entomol. Soc. 78, 142–152.

Armistead, J.S., Morias, I., Mathias, D.K., Jardim, J.G., Joy, J., Fridman, A., Finnefrock, A.C., Baguchi, A., Plebanski, M., Scorpio, D.G., Churcher, T.S., Borg, N.A., Sattabongkot, J., Dinglasan, R.R., et al., 2014. Antibodies to a single, conserved epitope in Ahopheles APN1 inhibit universal transmission of Plasmodium falciparum and Plasmodium vivax malaria. Infect. Immun. 82, 818–829.

Azuma, M., Nagae, T., Maruyama, M., Kataoka, N., Miyake, S., 2012. Two water-specific aquaporins at the apical and basal plasma membranes of insect epithelia: molecular basis for water recycling through the cryptonephric complex of lepidopteran larvae. J. Insect Physiol. 58, 523–533.

Berridge, M.J., Oschman, J.L., 1972. Insect Epithelia. In: Berridge, M.J., Oschman, J.L. (Eds.), Transporting Epithelia. Academic Press, New York, NY, pp. 58–60.

Beyenbach, K.W., 2003. Transport mechanisms of diuresis in Malpighian tubules of insects. J. Exp. Biol. 206, 3845–3852.

Beyenbach, K.W., Baumgart, S., Lau, K., Piermarini, P.M., Zhang, S., 2009. Signalling to the apical membrane and to the paracellular pathway: changes in the cytosolic proteome of *Aedes* Malpighin tubules. J. Exp. Biol. 212, 329–340.

Billingsley, P.F., 1990. The midgut ultrastructure of hematophagus insects. Annu. Rev. Entomol. 35, 219–248.

Billingsley, P.F., Lehane, M.J., 1996. Structure and ultrastructure of the insect midgut. In: Lehane, M.J., Billingsley, P.F. (Eds.), Biology of the Insect Midgut. Chapman & Hall, London, pp. 3–30.

Black, W.C., Kondratieff, B.C., 2005. Evolution of arthropod disease vectors. In: Marquardt, W.C. (Ed.), Biology of Disease Vectors, second ed. Elsevier Academic Press, Burlington, MA, pp. 9–23.

Boudko, D.Y., Harvey, W.R., Linser, P.J., 2001. Alkalinization by chloride/bicarbonate pathway in larval mosquito midgut. Proc. Natl. Acad. Sci. U. S. A. 98, 15354–15359.

Bradley, T.J., 1987. Physiology of osmoregulation in mosquitoes. Annu. Rev. Entomol. 32, 439–462.

Bradley, T.J., Snyder, C., 1989. Fluid secretion and microvillar ultrastructure in mosquito Malpighian tubules. Am. J. Physiol. 257, R1096–R1102.

Bragard, C., Caciagli, P., Lemaire, O., Lopez-Moya, J.J., MacFarlane, S., Peters, D., Susi, P., Torrance, L., 2013. Status and prospects of plant virus control through interference with vector transmission. Annu. Rev. Phytopathol. 51, 177–201.

Brault, V., Uzest, M., Monsion, B., Jacquot, E., Blanc, S., 2010. Aphids as transport devices for plant viruses. C. R. Biol. 333, 524–538.
Brown, M.R., Lea, A.O., 1988. FMRFamide- and adipokinetic hormone-like immunoreactivity in the nervous system of the mosquito, *Aedes aegypti*. J. Comp. Neurol. 270, 606–614.
Chagas, A.C., Calvo, E., Rios-Velasquez, C.M., Pessoa, F.A., Madeiros, J.F., Ribeiro, J.M., 2013. A deep insight into the sialotranscriptome of the mosquito *Psorophora albipes*. BMC Genomics 14, 875–894.
Chougule, N.P., Bonning, B.C., 2012. Toxins for transgenic resistance to hemipteran pests. Toxins 4, 405–429.
Chougule, N.P., Li, H., Liu, S., Linz, L.B., Narva, K.E., Meade, T., Bonning, B.C., 2012. Retargeting of the *Bacillus thuringiensis* toxin Cyt2Aa against hemipteran insect pests. PNAS 110, 8465–8470.
Chu, C., Spencer, J.L., Curzi, M.J., Zavala, J.A., Seufferheld, M.J., 2013. Gut bacteria facilitate adaptation to crop rotation in the western corn rootworm. PNAS 110, 11917–11922.
Cioffi, M., 1979. The morphology and fine structure of the larval midgut of a moth (*Manduca sexta*) in relation to active ion transport. Tissue Cell 11, 467–479.
Clark, T.M., Hutchinson, M.J., Huegel, K.L., Moffett, S.B., Moffett, D.F., 2005. Additional morphological and physiological heterogeneity within the midgut of larval *Aedes aegypti* (Diptera: Culicidae) revealed by histology, electrophysiology, and effects of *Bacillus thuringiensis* endotoxin. Tissue Cell 37, 457–468.
Clements, A.N., 1992. The Biology of Mosquitoes. vol. 1. Chapman & Hall, London, pp. 100–149.
Corena, M.P., Fiedler, M.M., VanEkeris, L., Linser, P.J., 2004. A comparative study of carbonic anhydrase in the midgut of mosquito larvae. Comp. Biochem. Physiol. 137, 207–225.
Dadd, R.H., 1975. Alkalinity within the midgut of mosquito larvae with alkaline-active digestive enzymes. J. Insect Physiol. 21, 1847–1853.
Davies, S.A., Terhzaz, S., 2009. Organellar calcium signaling mechanisms in *Drosophila* epithelial function. J. Exp. Biol. 212, 387–400.
Dedryver, C., Le Ralec, A., Fabre, F., 2010. The conflicting relationships between aphids and men: a review of aphid damage and control strategies. C. R. Biol. 333, 539–553.
Dinglasan, R.R., Kalume, D.E., Kanzok, S.M., Ghosh, A.K., Muratova, O., Pandey, A., Jacobs-Lorena, M., 2007. Disruption of *Plasmodium falciparum* development by antibodies against a conserved mosquito midgut antigen. PNAS 104, 13461–13466.
Dow, J.T., 1992. pH gradients in lepidopteran midgut. J. Exp. Biol. 172, 355–375.
Dow, J.T., 2009. Insights into the Malpighian tubule from functional genomics. J. Exp. Biol. 212, 435–445.
Edwards, M.J., Jacobs-Lorena, M., 2000. Permeability and disruption of the peritrophic matrix and caecal membrane from *Aedes aegypti* and *Anopheles gambiae* mosquito larvae. J. Insect Physiol. 46, 1313–1320.
Engel, P., Moran, N.A., 2013. The gut microbiota of insects—diversity in structure and function. FEMS Microbiol. Rev. 37, 699–735.
EPA (Environmental Protection Agency), 2003. Event MON863 BtCry3Bb1 corn biopesticide registration action document. Available free at: http://www/epa.gov/pesticides/biopesticides/ingredients/tech_docs/brad_006484.htm.
Fermino, F., Conte, H., Falco, J.R.P., 2010. Analysis of nucleus activity in Malpighian tubules of *Diatrea saccharalis* (Fabricus) (Lepidoptera: Crambidae) larvae by critical electrolyte concentration. Neotrop. Entomol. 39, 568–571.
Filippova, M., Ross, L.S., Gill, S.S., 1998. Cloning of the V-ATPase B subunit cDNA from *Culex quinquefasciatus* and expression of the B and C subunits in mosquitoes. Insect Mol. Biol. 7, 223–232.

Frank, D.L., Bukowsky, R., French, B.W., Hibbard, B.E., 2011. Effect of MIR604 trans genic maize at different stages of development on western corn rootworm (Coleoptera: Chrysomelidae) in a central Missouri field environment. J. Econ. Entomol. 104, 2054–2061.
Gill, S.S., Cowles, E.A., Pietrantonio, P.V., 1992. The mode of action of *Bacillus thuringiensis* endotoxins. Annu. Rev. Entomol. 37, 615–636.
Goltsev, Y., Rezende, G.L., Vranizan, K., Lanzaro, G., Valle, D., Levine, M., 2009. Developmental and evolutionary basis for drought tolerance of the *Anopheles gambiae* embryo. Dev. Biol. 330, 462–470.
Harvey, W.R., Cioffi, M., Dow, J.A.T., Wolfersberger, M.G., 1983. Potassium ion transport ATPase in insect epithelia. J. Exp. Biol. 106, 91–117.
Harvey, W.R., Boudko, D.Y., Rheault, M.R., Okech, B.A., 2009. NHEvnat: an H+ V-ATPase electrically coupled to a Na+:nutrient amino acid transporter (NAT) forms a Na+/H+ exchanger (NHE). J. Exp. Biol. 212, 347–357.
Harvey, W.R., Okech, B.A., Linser, P.J., Becnel, J.J., Ahearn, G.A., Sterling, K.M., 2010. H^+-V-ATPase-energized transporters in brush border membrane vesicles from whole larvae of *Aedes aegypti*. J. Insect Physiol. 56, 1377–1389.
Hegedus, D., Erlandson, M., Gillott, C., Toprak, U., 2009. New insights into peritrophic matrix synthesis, architecture, and function. Annu. Rev. Entomol. 54, 285–302.
Hogenhout, S.A., Ammar, E.D., Whitfield, A.E., Redinbaugh, M.G., 2008. Insect vector interactions with persistently transmitted viruses. Annu. Rev. Phytopathol. 46, 327–359.
Jeffers, L.A., Roe, R.M., 2008. The movement of proteins across the insect and tick digestive system. J. Insect Physiol. 45, 319–332.
Johansen, O.A., Butt, F.H., 1941. Embryology of Insects and Myriopods. McGraw Hill, New York, NY.
Klein, U., Loffelmann, G., Wieczorek, H., 1991. The midgut as a model system for insect K+-transporting epithelia: immunocytochemical localization of a vacuolar-type H+ pump. J. Exp. Biol. 161, 61–75.
Klowden, M.J., 2007. Physiological Systems in Insects, second ed. Academic Press, Burlington, MA, pp. 137–179.
Krajniak, K.G., 2005. Annelid endocrine disruptors and a survey of invertebrate FMRFamide-related peptides. Integr. Comp. Biol. 45, 88–96.
Lehane, M.J., 1997. Peritrophoci matrix structure and function. Annu. Rev. Entomol. 42, 525–550.
Lehane, M.J., Billingsley, P.F., 1996. Biology of the Insect Midgut, first ed. Chapman & Hall, London, pp. 3–25.
Liniger, M., Acosta-Serrano, A., Van Den Abbeele, J., Renggli, C.K., Brun, R., Englund, P.T., Roditi, I., 2003. Cleavage of trypanosome surface glycoproteins by alkaline trypsin-like enzyme(s) in the midgut of *Glossina morsitans*. Int. J. Parasitol. 33, 1319–1328.
Linser, P.J., Boudko, D.Y., Corena, M.P., Harvey, W.R., Seron, T.J., 2007. The molecular genetics of larval mosquito biology. J. Am. Mosq. Control Assoc. 23 (2 Suppl.), 283–293.
Linser, P.J., Neira-Oviedo, M.V., Smith, K.E., 2008. Glycoconjugate analysis in *Anopheles gambiae*. Am. J. Trop. Med. Hyg. 79 (Suppl. 6), 150–199.
Linser, P.J., Neira Oviedo, M., Hirata, T., Seron, T.J., Smith, K.E., Piermarini, P.M., Romero, M.F., 2012. Slc4-like anion transporters of the larval mosquito alimentary canal. J. Insect Physiol. 58, 551–562.
Marinotti, O., Calvo, E., Nguyen, Q.K., Dissanayake, S., Ribiero, J.M., James, A.A., 2006. Genome-wide analysis of gene expression in adult *Anophleles gambiae*. Insect Mol. Biol. 15, 1–12.
Mathias, D.K., et al., 2013. A small molecule glycosaminoglycan mimetic blocks *Plasmodium* invasions of the mosquito midgut. PLoS Pathog. 9, e1003757.

Matter, K., Balda, M.S., 2003. Signalling to and from tight junctions. Mol. Cell. Biol. 4, 225–236.

Meredith, J., Phillips, J.E., 1973. Rectal ultrastructure in salt-and freshwater mosquito larvae in relation to physiological state. Z. Zellforsch. Mikrosk. Anat. 138, 1–22.

Neira Oviedo, M., Ribeiro, J., Heyland, A., VanEkeris, L., Moroz, T., Linser, P.J., 2009. The salivary transcriptome of *Anopheles gambiae* (Diptera: Culcidae) larvae: a microarray-based analysis. Insect Biochem. Mol. Biol. 39, 382–394.

Ng, J.C., Perry, K.L., 2004. Transmission of plant viruses by aphid vectors. Mol. Plant Pathol. 5, 505–511.

Patrick, M.L., Aimanova, K., Sanders, H.R., Gill, S.S., 2006. P-type Na^{+}/K^{+}-ATPase and V-type H^{+}-ATPase expression patterns in the osmoregulatory organs of larval and adult mosquito *Aedes aegypti*. J. Exp. Biol. 209, 4638–4651.

Pelton, J.Z., 1938. The alimentary canal of the aphid *Prociphilus Tesselata* Fitch. Ohio J. Sci. 38, 164–169.

Petzold-Maxwell, J.L., Cibile-Stewart, X., French, B.W., Gassmann, A.J., 2012. Adaptation by western corn rootwrom (Coleoptera: Chrysomelidae) to Bt maize: inheritance, fitness costs, and feeding preference. J. Econ. Entomol. 105, 1407–1418.

Ramsay, J.A., 1976. The rectal complex in the larvae of Lepidoptera. Philos. Trans. R. Soc. Lond. B Biol. Sci. 274, 203–226.

Ray, K., Mercedes, M., Chan, D., Choi, C.Y., Nishiura, J.T., 2009. Growth and differentiation of the larval mosquito midgut. J. Insect Sci. 9, 55–68.

Rheault, M.R., Okech, B.A., Keen, S.B.W., Miller, M.M., Meleshkevitch, E.A., Linser, P.J., Boubko, D.Y., Harvey, W.R., 2007. Molecular cloning, phylogeny and localization of AgNHA1: the first Na^{+}/H^{+} antiporter (NHA) from a metazoan, *Anopheles gambiae*. J. Exp. Biol. 210, 3848–3861.

Rudin, W., Hecker, H., 1989. Lectin-binding sites in the midgut of the mosquitoes *Anopheles stephensi* Liston and *Aedes aegypti* L. (Diptera: Culicidae). Parasitol. Res. 75, 268–279.

Ryerse, J.S., Purcell, J.P., Sammons, R.D., 1994. Structure and formation of the peritrophic membrane in the larva of the southern corn rootworm, *Diabrotica undecimpunctata*. Tissue Cell 26, 431–437.

Sayed, A., Nekl, E.R., Siqueira, A.A., Wang, H.C., Ffrenc-Constant, R.H., Bagley, M., Siegfried, B.D., 2007. A novel cadherin-like gene from western corn rootworm, *Diabrotica virgifera virgifera* (Coleoptera: Chrysomelidae), larval midgut tissue. Insect Mol. Biol. 16, 591–600.

Seron, T.J., Hill, J., Linser, P.J., 2004. A GPI-linked carbonic anhydrase in the larval mosquito midgut. J. Exp. Biol. 207, 4559–4572.

Smith, K.E., VanEkeris, L.A., Linser, P.J., 2007. Cloning and characterization of AgCA9, a novel α-carbonic anhydrase from *Anopheles gambiae* Giles *sensu stricto* larvae. J. Exp. Biol. 210, 3919–3930.

Smith, K.E., VanEkeris, L.A., Okech, B.A., Harvey, W.R., Linser, P.J., 2008. Larval anophleine mosquito recta exhibit a dramatic change in localization patterns of ion transport proteins in response to shifting salinity: a comparison between anopheline and culicine larvae. J. Exp. Biol. 211, 3067–3076.

Smith, K.E., VanEkeris, L., Valenti, M., Raymond, S., Smith, P., Linser, P.J., 2010. Physiological and pharmacological characterization of the anopheline rectum shows redistribution in function in response to varying salinity. Comp. Biochem. Physiol. C 157, 55–62.

Terra, W.R., 1990. Evolution of digestive systems of insects. Annu. Rev. Entomol. 35, 181–200.

Terra, W.R., Ferreira, C., Baker, J.E., 1996. Compartmentalization of digestion. In: Lehane, M.J., Billingsley, P.F. (Eds.), Biology of the Insect Midgut. Chapman & Hall, London, pp. 206–235.

Volkman, A., Peters, W., 1989a. Investigations on the midgut caeca of mosquito larvae-1. Fine structure. Tissue Cell 21, 243–251.
Volkman, A., Peters, W., 1989b. Investigations on the midgut caeca of mosquito larvae-II. Functional aspects. Tissue Cell 21, 253–261.
White, B.J., Kundert, P.N., Turissini, D.A., VanEkeris, L., Linser, P.J., Besansky, N.J., 2013. Dose and developmental responses of *Anopheles merus* larvae to salinity. J. Exp. Biol. 216, 3433–3441.
Wieczorek, H., 1992. The insect V-ATPase, a plasma membrane proton pump energizing secondary active transport: molecular analysis of electrogenic potassium transport in the tobacco hornworm midgut. J. Exp. Biol. 172, 335–343.
Xiang, M.A., Linser, P.J., Price, D.A., Harvey, W.R., 2012. Localization of two Na^+- or K^+–H^+ antiporters, AgNHA1 and AgNHA2, in Anopheles gambiae larval Malpighian tubules and the functional expression of AgNHA2 in yeast. J. Insect Physiol. 58, 570–579.
Zhang, R., Hua, G., Andacht, T.M., Adang, M.J., 2008. A 106-kDa aminopeptidase is a putative receptor for *Bacillus thuringiensis* Cry11Ba toxin in the mosquito *Anopheles gambiae*. Biochemistry 47, 11263–11272.
Zhuang, Z., Linser, P.J., Harvey, W.R., 1999. Antibody to H^+ V-ATPase subunit E colocalizes with portasomes in alkaline larval midgut of a freshwater mosquito (*Aedes aegypti*). J. Exp. Biol. 202, 2449–2460.

CHAPTER TWO

Diversity of *Bacillus thuringiensis* Crystal Toxins and Mechanism of Action

Michael J. Adang*,†, Neil Crickmore‡, Juan Luis Jurat-Fuentes§
*Department of Entomology, University of Georgia, Athens, Georgia, USA
†Department of Biochemistry and Molecular Biology, University of Georgia, Athens, Georgia, USA
‡School of Life Sciences, University of Sussex, Falmer, Brighton, United Kingdom
§Department of Entomology and Plant Pathology, University of Tennessee, Knoxville, Tennessee, USA

Contents

Advances in Insect Physiology, Volume 47
ISSN 0065-2806
http://dx.doi.org/10.1016/B978-0-12-800197-4.00002-6

Abstract

Parasporal crystals produced by *Bacillus thuringiensis* (*Bt*) bacteria are the main virulence factors underlying *Bt* toxicity to insects. Parasporal crystals are composed primarily of Cry and Cyt proteins that act on the midgut of susceptible insects. Cry proteins are an important component of *Bt* biopesticides and are vital tools for insect control via expression in transgenic crop plants. Some members of the Cry group are more distantly related including ETX/MTX and binary type toxins. Cry toxin structure and action involves critical steps in toxin activation, binding to receptors such as cadherin and then aminopeptidase or alkaline phosphatase probably in a 'sequential binding' manner. Specific Cry toxin–receptor interactions are a focus of this review. Recently, the importance of midgut ATP-binding cassette proteins to Cry intoxication of insects has been demonstrated. Mechanistic details involved in 'sequential binding' and 'pore formation' models are examined. The Cyt toxin of *Bt* subspecies *israelensis* is an important and interesting component in Cry–midgut interactions in mosquitoes. For some Cry toxins, Cyt serves as a receptor for docking to midgut membrane. Recent engineering work has demonstrated that Cyt can be re-targeted generating novel toxins for insect control. Overall, we review the remarkable progress made in the past 20 years in discovering novel Cry toxins and in elucidating complex mechanisms of Cry and Cyt toxin action; subjects relevant to the long-term control of insects that damage crops and vector human disease.

1. INTRODUCTION

The Gram-positive bacterium *Bacillus thuringiensis* (*Bt*) is characterized by the proteinaceous crystals that it synthesizes in the mother cell during sporulation (Aronson et al., 1986; Bulla et al., 1980). The history of the discovery and development of *Bt* has been extensively reviewed (Beegle and Yamamoto, 1992; Burges, 2001; Jurat-Fuentes and Jackson, 2012). Tens of thousands of *Bt* strains have been isolated and most *Bt* strains are active against larval stages of insects. Products based on *Bt* have been registered as pesticides in the United States since 1961. As with other biological pesticides, *Bt* offers a number of advantages over synthetic pesticides, including lack of polluting residues, high specificity to target insects and safety to non-target organisms. Consequently, the broadest uses of *Bt* are on food crops and in forestry where safety and specific action are desirable. Disadvantages of *Bt* are its high specificity and low persistence. While *Bt* is the most successful biopesticide for insect control it remains a small part, about 2%, of the total insecticide market. The widest usage of *Bt* for insect control is through transgenic plants producing *Bt* insecticidal proteins, particularly corn and cotton (James, 2009).

The presence of a parasporal crystal is the phenotypic trait of *Bt* and is used to separate this bacterium from other *Bacillus* species (Vilas-Bôas et al., 2007). The main components of the crystal are the delta-endotoxins, that act as the primary virulence factor for this pathogen (Raymond et al., 2010), although other proteins are also present in the crystal and have a role in toxin and/or crystal structure (Diaz-Mendoza et al., 2012; Staples et al., 2001). Toxins found in the crystal are classified into two families known as Cry and Cyt (Höfte and Whiteley, 1989). The Cry (from *cry*stal) toxins represent a large family currently consisting of around 300 different members (Crickmore et al., 2014). The Cyt toxins are characterized by possessing a general cytolytic activity *in vitro*, although they show a primarily dipteran-specific activity *in vivo* (Soberon et al., 2013a). A third family of protein toxins—the Vegetative Insecticidal Proteins (Vips)—are not classified as crystal toxins since they are secreted from vegetatively growing cells rather than included in the crystal during sporulation (de Maagd et al., 2003).

The characteristics of *Bt* crystal toxins will be discussed in Section 2 beginning with how Cry toxins are defined, named and classified into groups. We discuss conserved features of three-domain Cry toxins and the structurally distinct ETX/MTX and binary-like toxins that have Cry designations. We briefly describe Vip toxins with respect to their relationship to Cry toxins. Also in Section 2, we introduce parasporins, which typically have no known insect target, but have toxicity to specific human cell lines (Mizuki et al., 1999). Recent advances and innovations in toxin discovery are presented in Section 2. Recent reviews of mechanisms of three-domain Cry toxin action (Pardo-López et al., 2012; Vachon et al., 2012) have integrated complex events which occur in insect midgut after Cry protoxin processing and upon contact with the target midgut membrane. In Section 3, we discuss Cry toxin structure in the context of the respective role of each structural domain in Cry toxin action. We also discuss the importance of Cry solubilization from crystals and proteolytic processing to host specificity. Receptor molecules on insect midgut such as cadherin, aminopeptidase and alkaline phosphatase (ALP) have been extensively reviewed (Bravo et al., 2011; Pigott and Ellar, 2007), particularly in the context of models of Cry toxin action. In Section 4, we discuss those receptor molecules with an emphasis on combinations of Cry toxins and insect systems where receptor function has been established. We also include in the discussion evidence demonstrating a critical role in toxin action for

ATP-binding cassette (ABC) proteins (Heckel, 2012), and the relevance of glycolipids and other midgut molecules to toxin action. Models of Cry toxin action describing events occurring between Cry-binding midgut and pore formation are extant in recent literature (Bravo et al., 2004, 2011). In Section 5, we examine some of the mechanistic details involved in the 'sequential binding' and 'pore formation' models of toxin action. The Cyt toxin of *Bt* subsp. *israelensis* is a major crystal component and is critical to mosquitocidal toxicity (Ben-Dov, 2014). Since an excellent recent review summarizes the mechanism of Cyt toxin action (Soberon et al., 2013b), we will briefly describe Cyt pore formation, its role as a Cry receptor and recent engineering work that re-targets Cyt toxin resulting in a novel toxin for pest control (Chougule et al., 2013). An excellent review on mechanism of *Bt* Cry toxin action was previously published in Advances in Insect Physiology (Knowles, 1994) and we focus in this review to provide an update on the remarkable progress made in the past 20 years on the subject.

2. GENERAL CHARACTERISTICS OF *B. THURINGIENSIS* CRYSTAL TOXINS

2.1. Definition and classification of crystal toxins

Cry toxins are officially defined as proteins that have significant sequence similarity to existing toxins within the nomenclature or be a *B. thuringiensis* parasporal inclusion protein that exhibits pesticide activity, or some experimentally verifiable toxic effect, to a target organism (Crickmore et al., 1998). Naming of toxins is based solely on amino acid sequence identity and does not take into account their toxicity; thus, toxins that are active against the same order of insect will not necessarily share any similarity in their names. A toxin's name consists of four levels, e.g., Cry41Ab1, the first number is the primary level and all toxins sharing this first number (41 in the example above) will share significant sequence identity—at least 45%. Toxins sharing primary, secondary and tertiary level descriptors will have increasing sequence identity. Toxins that differ only in the quaternary level descriptor (e.g. Cry41Ab1 and Cry41Ab2) will have at least 95% sequence identity. A different quaternary level descriptor is given to all newly characterized toxins, and as a result some toxins are actually identical to others in the nomenclature, but have been assigned different names. The definition of a Cry toxin as given above does allow toxins that are neither produced by *B. thuringiensis* nor found in the crystal, to be given a Cry name. Examples include Cry16Aa that was cloned from

Clostridium bifermentans (Barloy et al., 1996) and Cry1Ia that is secreted rather than being part of a crystal (Varani et al., 2013). In both these cases, the proteins share significant sequence similarity to existing toxins and so are classified as Cry toxins. For proteins that are found in the crystal, but share only weak similarity to existing toxins, more evidence is required in order to classify them. Normally, this would be a demonstration that either the purified toxin or a recombinant form of the toxin has toxic activity to some target organism or cell. In the case of a recombinant toxin, the protein would not actually have to be produced by the parent strain. Indeed commercially important toxins, such as Cry2Ab, are often encoded by cryptic genes (Crickmore et al., 1994). The mere presence of a protein in the crystal of a *B. thuringiensis* strain is not sufficient for that protein to be classified as a Cry toxin. The same criteria are used when classifying Cyt toxins.

2.2. The diversity of Cry toxins

There are currently around 75 primary subgroups of Cry toxins—i.e., with different primary ranks in the nomenclature (Cry1, Cry2, Cry3, etc.). Figure 2.1 shows a cartoon illustrating the different lengths of these toxins. Their lengths vary from 369 (Cry34) to 1344 amino acids (Cry43). In their 1989 review, Höfte and Whiteley identified five conserved sequence blocks in all the Cry toxins, which are also shown in Fig. 2.1. Note that not all of the toxins contain these blocks and that some only contain a subset of them. Based on experimentally derived structures, and molecular modelling, it is believed that the toxins that contain all, or some, of these conserved blocks are likely to possess the same basic three-domain fold (Pardo-López et al., 2013). Figure 2.2 provides a more complete list of currently classified toxins, excluding those that share the same, quaternary rank. Of the 294 toxins in this list, 262 (89%) are predicted to have the three-domain fold and are coloured blue in this figure. It is likely that the majority of Cry toxins produced by *Bt* will be in this three-domain class, although the proportion of these toxins in Fig. 2.2 is likely to be inflated due to the fact that many of them were isolated by PCR techniques based on conserved sequences in existing toxins. Two other classes of Cry toxins have been previously identified: the ETX/MTX-like toxins and the Binary (Bin) like ones. There are 11 toxins in the former group (coloured orange in Fig. 2.2) which show sequence similarity to the *Clostridium perfringens* epsilon toxin (Bokori-Brown et al., 2011). The mosquitocidal MTX2 toxin from *Lysinibacillus sphaericus* is also related to this class (Berry, 2012). They are structurally distinct from the three-domain Cry toxins in that they adopt an elongated, and

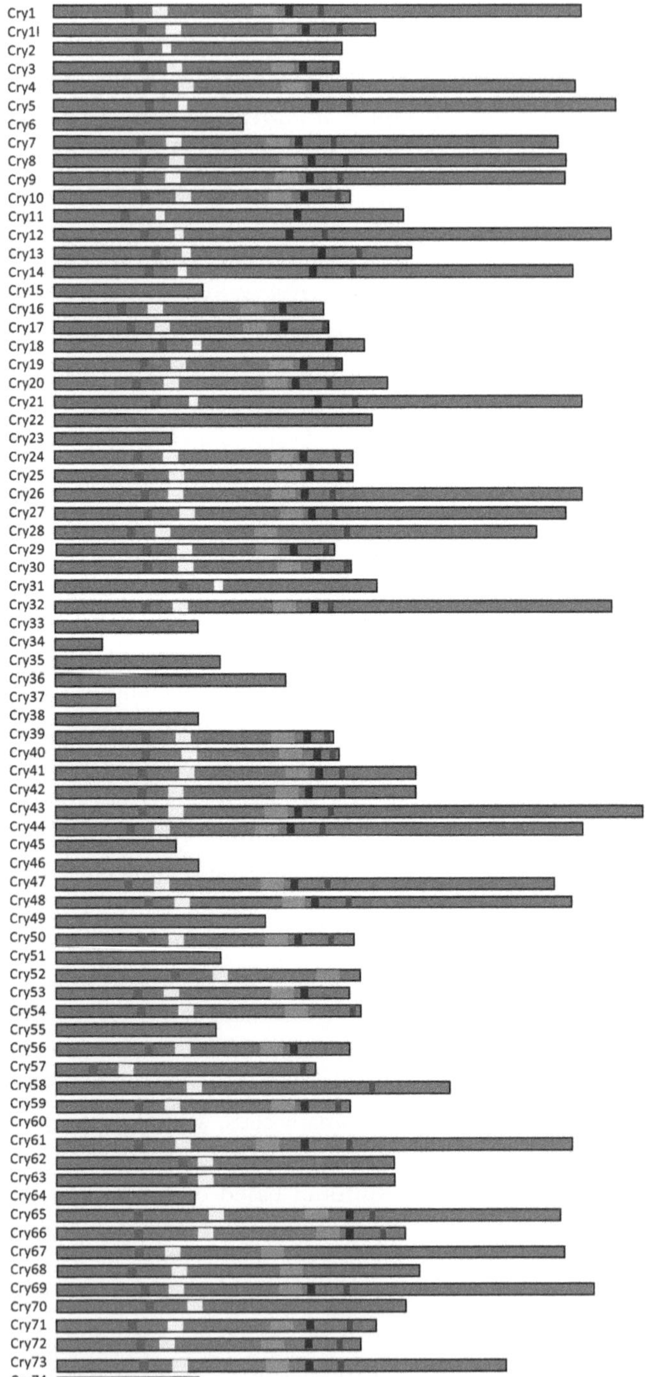

Figure 2.1 Graphical representation of the diversity of *Bt* Cry toxins. The length of each toxin is drawn to scale and the five conserved blocks described in Schnepf et al. (1998) are shown as coloured inserts. (See the colour plate.)

Three-domain	ETX–MTX	Bin	Parasporin

Cry1Aa	Cry2Aa	Cry8Aa	Cry15Aa	Cry31Aa	Cry40Aa	Cry58Aa	Cyt1Aa
Cry1Ab	Cry2Ab	Cry8Ab		Cry31Ab	Cry40Ba		Cyt1Ab
Cry1Ac	Cry2Ac	Cry8Ac	Cry16Aa	Cry31Ac	Cry40Ca	Cry59Aa	Cyt1Ba
Cry1Ad	Cry2Ad	Cry8Ad		Cry31Ad	Cry40Da	Cry59Ba	Cyt1Ca
Cry1Ae	Cry2Ae	Cry8Ba	Cry17Aa				Cyt1Da
Cry1Af	Cry2Af	Cry8Bb		Cry32Aa	Cry41Aa	Cry60Aa	Cyt2Aa
Cry1Ag	Cry2Ag	Cry8Bc	Cry18Aa	Cry32Ab	Cry41Ab	Cry60Ba	Cyt2Ba
Cry1Ah	Cry2Ah	Cry8Ca	Cry18Ba	Cry32Ba	Cry41Ba		Cyt2Bb
Cry1Ai	Cry2Ai	Cry8Da	Cry18Ca	Cry32Ca	Cry41Ca	Cry61Aa	Cyt2Bc
Cry1Ba	Cry2Aj	Cry8Db		Cry32Cb			Cyt2Ca
Cry1Bb	Cry2Ak	Cry8Ea	Cry19Aa	Cry32Da	Cry42Aa	Cry62Aa	Cyt3Aa
Cry1Bc	Cry2Ba	Cry8Fa	Cry19Ba	Cry32Ea			
Cry1Bd		Cry8Ga	Cry19Ca	Cry32Eb	Cry43Aa	Cry63Aa	Vip1Aa
Cry1Be	Cry3Aa	Cry8Ha		Cry32Fa	Cry43Ba		Vip1Ab
Cry1Bf	Cry3Ba	Cry8Ia	Cry20Aa	Cry32Ga	Cry43Ca	Cry64Aa	Vip1Ac
Cry1Bg	Cry3Bb	Cry8Ib	Cry20Ba	Cry32Ha	Cry43Cb		Vip1Ad
Cry1Bh	Cry3Ca	Cry8Ja		Cry32Hb	Cry43Cc	Cry65Aa	Vip1Ba
Cry1Bi		Cry8Ka	Cry21Aa	Cry32Ia			Vip1Bb
Cry1Ca	Cry4Aa	Cry8Kb	Cry21Ba	Cry32Ja	Cry44Aa	Cry66Aa	Vip1Bc
Cry1Cb	Cry4Ba	Cry8La	Cry21Ca	Cry32Ka			Vip1Ca
Cry1Da	Cry4Ca	Cry8Ma	Cry21Da	Cry32La	Cry45Aa	Cry67Aa	Vip1Da
Cry1Db	Cry4Cb	Cry8Na	Cry21Ea	Cry32Ma	Cry45Ba		
Cry1Dc	Cry4Cc	Cry8Pa	Cry21Fa	Cry32Mb		Cry68Aa	Vip2Aa
Cry1Ea		Cry8Qa	Cry21Ga	Cry32Na	Cry46Aa		Vip2Ab
Cry1Eb	Cry5Aa	Cry8Ra	Cry21Ha	Cry32Oa	Cry46Ab	Cry69Aa	Vip2Ac
Cry1Fa	Cry5Ab	Cry8Sa		Cry32Pa		Cry69Ab	Vip2Ad
Cry1Fb	Cry5Ac	Cry8Ta	Cry22Aa	Cry32Qa	Cry47Aa		Vip2Ae
Cry1Ga	Cry5Ad		Cry22Ab	Cry32Ra		Cry70Aa	Vip2Af
Cry1Gb	Cry5Ba	Cry9Aa	Cry22Ba	Cry32Sa	Cry48Aa	Cry70Ba	Vip2Ag
Cry1Gc	Cry5Ca	Cry9Ba	Cry22Bb	Cry32Ta	Cry48Ab	Cry70Bb	Vip2Ba
Cry1Ha	Cry5Da	Cry9Bb		Cry32Ua			Vip2Bb
Cry1Hb	Cry5Ea	Cry9Ca	Cry23Aa	Cry32Va	Cry49Aa	Cry71Aa	
Cry1Ia		Cry9Da		Cry32Wa	Cry49Ab		Vip3Aa
Cry1Ib	Cry6Aa	Cry9Db	Cry24Aa			Cry72Aa	Vip3Ab
Cry1Ic	Cry6Ba	Cry9Dc	Cry24Ba	Cry33Aa	Cry50Aa		Vip3Ac
Cry1Id		Cry9Ea	Cry24Ca		Cry50Ba	Cry73Aa	Vip3Ad
Cry1Ie	Cry7Aa	Cry9Eb		Cry34Aa			Vip3Ae
Cry1If	Cry7Ab	Cry9Ec	Cry25Aa	Cry34Ab	Cry51Aa	Cry74Aa	Vip3Af
Cry1Ig	Cry7Ba	Cry9Ed		Cry34Ac			Vip3Ag
Cry1Ja	Cry7Bb	Cry9Ee	Cry26Aa	Cry34Ba	Cry52Aa		Vip3Ah
Cry1Jb	Cry7Ca	Cry9Fa			Cry52Ba		Vip3Ai
Cry1Jc	Cry7Cb	Cry9Ga	Cry27Aa	Cry35Aa			Vip3Ba
Cry1Jd	Cry7Da			Cry35Ab	Cry53Aa		Vip3Bb
Cry1Ka	Cry7Ea	Cry10Aa	Cry28Aa	Cry35Ac	Cry53Ab		Vip3Ca
Cry1La	Cry7Fa			Cry35Ba			
Cry1Ma	Cry7Fb	Cry11Aa	Cry29Aa		Cry54Aa		Vip4Aa
Cry1Na	Cry7Ga	Cry11Ba	Cry29Ba	Cry36Aa	Cry54Ab		
Cry1Nb	Cry7Gb	Cry11Bb			Cry54Ba		Sip1Aa

Figure 2.2 List of *Bt* toxins. The toxins are grouped according to their primary ranking within the nomenclature. The background colour represents likely protein tertiary structural groupings and those toxins named with red text indicate the parasporin class. (See the colour plate.)

predominately β-sheet-based structure. However, like the three-domain toxins, they are believed to act via forming pores in the membranes of the target cells (Bokori-Brown et al., 2011). The Bin-like toxins are so named since they resemble the two components of the mosquitocidal binary toxin from *L. sphaericus*. While, neither the molecular structure nor the mechanism of action, of these toxins is particularly clear, there is some evidence that they form pores but also indications that one of the components acts intracellularly (Berry, 2012). There are 13 toxins in this class shown in Fig. 2.2 (coloured purple), although interestingly none appear to exist as homologous pairs as found with BinA and BinB from *L. sphaericus*. Cry49 has a binary partnership with the three-domain toxin Cry48 (Jones et al., 2007) for example, whilst Cry35 forms a binary partnership with Cry34 (Schnepf et al., 2005)—a toxin that does not fit into any of the three classes mentioned above but has similarities to the fungal aegerolysins (Berne et al., 2009). It is not known whether or not the Bin-like Cry36 has a binary partner. Other toxins appear to exist as binary pairs including the ETX/MTX-like Cry23 and Cry37 (Donovan et al., 2000) despite neither resembling the Bin toxins. Two further ETX/MTX-like toxins—Cry15 (Naimov et al., 2008) and Cry33 (Kim et al., 2003)—are believed to form binary partnerships with other proteins, but these partners have not been characterized sufficiently well enough to give them Cry designations. Other Cry toxins do not fit into these three main classes (three-domain, ETX/MTX and Bin-like) but like the aegerolysin-like Cry34 mentioned above, do have some resemblance to other known toxins or proteins. These relationships are shown in Fig. 2.3 in which the differently coloured filled circles represent the same groups of toxins identified in Fig. 2.2. The red open circles represent toxins with a shared characteristic such as the binary toxins discussed above, with the blue open circles representing related proteins found in other species. An example of such a relationship is Cry46 which shares similarities with the hydralysin (Sher et al., 2005) pore-forming toxins. Similarly, the Vip1, Vip4 and Cry37 toxins show similarity to the 1b component of the Clostridial iota toxin (Sakurai et al., 2009) with Vip2 resembling the 1a component. Two of the other toxins, Cry6 and Cry22 do not resemble any other known toxin, although they do show sequence similarity to a structural maintenance of chromosome and a cell wall anchor protein, respectively. No homologues have been identified for the Vip3 or Cry55 toxins. The Cyt toxins have a homologous fungal toxin, volva toxin (Weng et al., 2004).

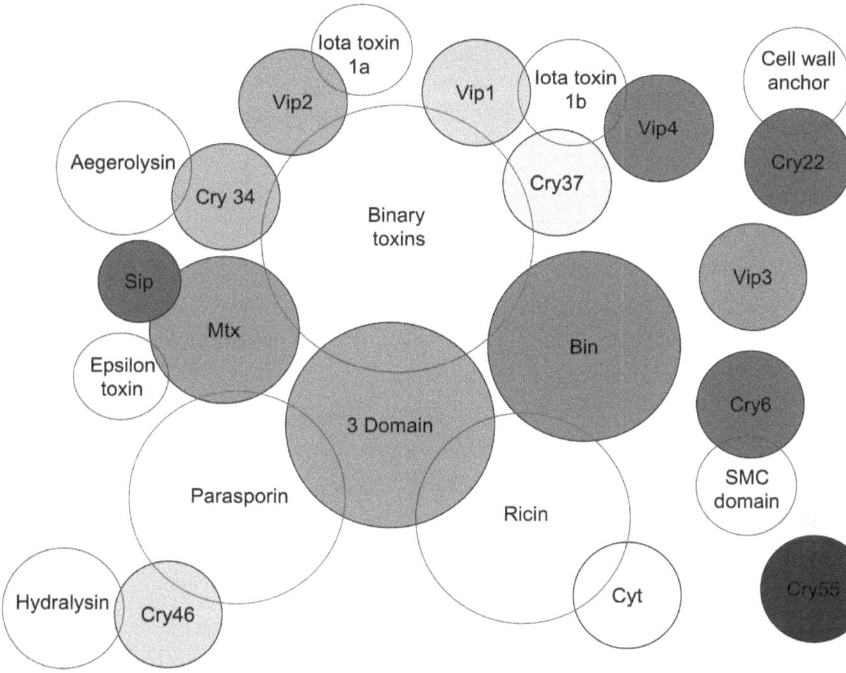

Figure 2.3 Venn diagram showing structural or functional relationships between toxins. The individually coloured circles represent the different structural groups identified in Fig. 2.2. Overlapping open blue circles indicate sequence similarity with non-*Bt* toxins or proteins. The open red circles indicate which of the structural groups contain members that share that particular characteristic namely binary toxin, parasporin or ricin domain containing. (See the colour plate.)

2.3. The parasporin toxins

In an attempt to understand, why so many *Bt* strains exist without a known insect target, a large screen was set up in the mid-1990s to look for activities against non-insect species. From this screen came a number of toxins which were non-haemolytic but had activity against one or more human cells (Mizuki et al., 1999). Many of these toxins were specifically active against cancer cell lines. Currently, there are six classes of toxin that have this activity; whilst they have been separately termed Parasporins they have also been allocated Cry toxin names (Akiba et al., 2009). In Fig. 2.1, the Parasporin toxins have been highlighted in red text. It can be seen from this figure that Cry31 (Parasporin 1), Cry41 (Parasporin 3) and Cry63 (Parasporin 6) are all three-domain toxins, whilst Cry45 (Parasporin 4) and Cry 64 (Parasporin 5) are both ETX/MTX-type toxins. The final toxin in this group Cry46 (Parasporin 2), as mentioned above, does not resemble any other Cry toxin but does have some similarity to hydralysin. Although Parasporin 1 appears to be a typical three-domain toxin, its mechanism of action is believed to be more complex than simple pore formation (Katayama et al., 2007). In particular, the toxin induces an increase in intracellular Ca^{2+} which is then associated with apoptosis as assessed by increases in Caspase-3 and PARP cleavage. Less is known about the two other three-domain toxins (Parasporins 3 and 6), although the cell swelling observed after toxin administration is consistent with pore formation (Nagamatsu et al., 2010; Yamashita et al., 2005). Parasporin 2 (Cry46) does not share significant sequence similarity with other known Cry toxins but its structure (Akiba et al., 2009) suggests that it ought to belong to the ETX/MTX family. Its mechanism of action is believed to be through pore formation after the toxin is targeted to lipid rafts on the cell surface where it interacts with glycosylphosphatidyl inositol (GPI)-anchored proteins (Kitada et al., 2009). Parasporin 4 appears to act as a cholesterol-independent β-pore-forming toxin that does not induce apoptosis (Okumura et al., 2011). Nothing is currently known about the mechanism of action of Parasporin 5.

2.4. The ricin domain

Figure 2.3 indicates that a number of toxins contain the so-called ricin domain (pfam00652, conserved domain 00161), a beta-trefoil-putative carbohydrate-binding, domain presumed to have arisen through gene triplication and containing the $(Q\text{-}X\text{-}W)_3$ conserved motif (Hazes, 1996). A number of the Parasporin toxins (Cry41A, Cry41B and putative parasporin Cry42) contain this domain leading to speculation that this lectin-like feature could be responsible for specific binding, and thus toxicity, to certain human cell lines. However, removal of the ricin domain either proteolytically or genetically did not appear to affect the activity of at least one of these Parasporins against human cells (Krishnan, 2013; Yamashita et al., 2005). Other toxins that contain this domain include Cry35A (Ellis et al., 2002) and Cyt1Ca (Manasherob et al., 2006), and in neither of these cases is there any evidence that the domain is involved in toxicity. Despite the fact that the beta-trefoil domain does have a role to play in the mechanism of action of other toxins such as ricin and pierisin (Matsushima-Hibiya et al., 2003), it has a ubiquitous presence in many genomes and may not always confer function.

2.5. Toxin discovery

The discovery of new toxin genes increased dramatically as innovations in molecular biology, most notably PCR and next-generation sequencing, became widespread. Originally, discovery started with the characterization of a toxin followed by the rather lengthy procedure of identifying and cloning its gene. As more genes were cloned similarities between them could be used to design probes, and later PCR primers, to pull out new genes (Höfte and Whiteley, 1989). The emphasis changed from characterizing interesting toxins to identifying strains with useful activities and then cloning and expressing putative toxin genes in order to find one with good toxicity. In recent years, the emphasis has changed once more towards the identification of novel genes through sequencing either the entire genome of the bacterium, or just the megaplasmids on which the toxin genes are normally found. To support this approach, computational pipelines have been developed to identify putative genes from genomic sequencing data (Ye et al., 2012). Whilst this mass sequencing approach has many advantages there is a need to be cautious, the fact that *Bt* strains often contain multiple copies of similar toxin genes can cause problems with both contig assembly and with database annotation (Guan et al., 2012). Whilst genomic sequencing can readily identify putative toxin genes that are related to known ones, it suffers from the same problem as previous PCR-based methods in being unable to identify completely novel toxin genes. One solution to this is to revert back to first analysing the protein and then finding the gene, but now making use of more advanced proteomic techniques to identify putative toxins from the *Bt* crystal (Lee et al., 2006) or via expression at particular growth phases (Huang et al., 2012).

3. CRY TOXIN STRUCTURE: FUNCTION

3.1. Overview of Cry structure

The majority of Cry toxins in the 73 subgroups have similar predicted three-domain structures despite differences in primary amino acid sequence and overall length of protoxin (Pardo-López et al., 2013). The presence of some of the five conserved blocks (Höfte and Whiteley, 1989) is a constant feature of the three-domain Cry toxins (Pardo-López et al., 2013). This conclusion is supported by three-dimensional structures resolved first for Cry3Aa (Li et al., 1991) by X-ray crystallography and then for Cry1Aa (Grochulski et al., 1995), Cry2Aa (Morse et al., 2001), Cry3Bb (Galitsky et al., 2001), Cry4Aa (Boonserm et al., 2006), Cry4Ba (Boonserm et al., 2005), Cry8Ea1 (Guo

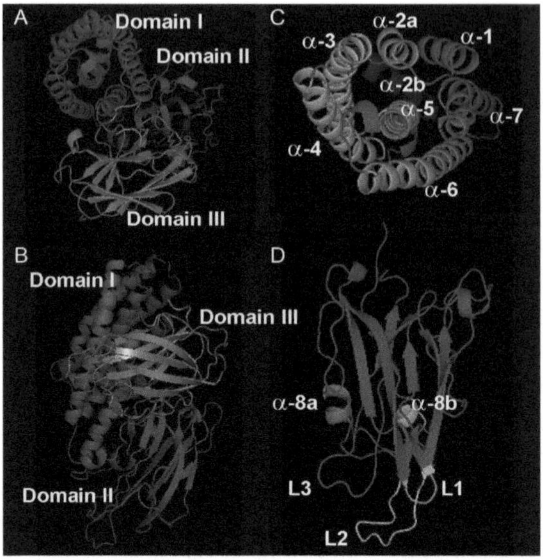

Figure 2.4 Views of Cry1Aa toxins from the Resource for Structural Bioinformatics Protein Data Bank (http://www.rcsb.org/pdb/) of (id: 1CIY) (Grochulski et al., 1995) modified using PyMOL Version 1.7 (http://www.pymol.org/). Domain amino acid sequence position information was obtained from NCBI data base (GenBank: AAP40639.1) and (Grochulski et al., 1995). Overall views of Cry1Aa toxin are shown (Panels A and B). Domain I is red, domain II is pink and domain III is orange. Domain I with the names of α-helices and domain II with named α-helices and loops are shown in Panels C and D, respectively. (See the colour plate.)

et al., 2009a), and nematicidal Cry5B toxin. Figure 2.4 shows views of the Cry1Aa molecule generated from three-dimensional structure data (Grochulski et al., 1995). The structure of domain I is a bundle of 7–8 α-helices with a centrally located hydrophobic α-helix 5. Based on observations that most of the α-helices of domain I are long enough to span a hydrophobic cellular membrane and that domain I has similarities to pore-forming domains in other bacterial toxins, it was hypothesized and later proven correct that domain I is involved in membrane insertion and pore formation (Li et al., 1991). Domain II, a three β-sheet structure, is involved in receptor binding, oligomerization, and membrane insertion. Domain III participates in receptor binding and possibly membrane insertion.

3.2. Cry domain I

The functional roles of the three Cry structural domains were implied by the details of Cry toxin structure (Li et al., 1991), comparisons with other bacterial toxins and then methodically investigated by functional analyses using wild-type and mutated Cry toxins. The importance of domain I α-helical bundles and specific amino acid residues to membrane insertion and pore formation were suggested by mutations such as A92D, R93D and Y153D (Chen et al., 1995; Wu and Aronson, 1992) in Cry1Ab that caused loss of toxicity without changing binding to brush border membrane. How Cry toxins transition from soluble to membrane-bound forms and the nature of the final form generating pores has been the subject of many functional analyses. Membrane partitioning of domain I involves intramolecular movements and insertion of a hydrophobic hairpin formed by α-helices 4 and 5. By restricting intramolecular movements of domain I helices, and Domains I and II relative to each other, researchers (Schwartz et al., 1997a) demonstrated that channel formation requires domain I swinging away from domain II and insertion of α-helices 4 and 5 into the membrane. According to the umbrella model based on the crystal structure of Cry3Aa (Li et al., 1991) toxin, helices lay on the surface of the cell membrane much like ribs of an umbrella with α4–5 helices inserting and forming the membrane pore. The hydrophobicity of the α4–5 pair, which is consistent with their insertion into membrane, has been compared to the pore-forming colicin toxins. Also, the central α4–5 helices alone in an antiparallel conformation have pore-forming ability in membranes (Gazit et al., 1998). Mutational alterations of α4 and the central α5 helix yield inactive Cry toxins, while mutations in other helices do not result in significant effects on insect toxicity (Aronson et al., 1995; Kumar and Aronson, 1999; Wu and Aronson, 1992). Additionally, mutated α4 residues that impair Cry1Aa pore formation in planar lipid bilayers (Masson et al., 1999) and in brush border membrane vesicles from midgut of larval tobacco hornworm, *Manduca sexta* (Girard et al., 2009) support a pore-lining function of helix α4. Similar types of mutations in helix 5 support its participation in channel formation (Florez et al., 2012). Interestingly, a fortuitous V171C exchange in α-helix 5 in Cry1Ab toxin had positive effects increasing translocation into midgut membranes and conferring a 25-fold increase in toxicity to the gypsy moth, *Lymantria dispar*, albeit a similar increase of toxicity to *M. sexta* was not observed (Alzate et al., 2010). Site-directed mutagenesis of residues involved in coiled-coil structures in helices α-3 and 6 of Cry1Ab resulted in mutant toxins capable of binding but unable to form oligomers or pores, supporting the relevance of these structures to oligomerization (Jiménez-Juárez et al., 2007).

3.3. Cry domain II

Domain II consists of three antiparallel β-sheets arranged in a triangular β-prism with projecting loop regions α8, loop 1, 2 and 3. Similar β-prism folding patterns have been noted in numerous carbohydrate-binding proteins (Sharma et al., 2007); including vitelline from hen's egg white and the plant lectins jacalin (Sankaranarayanan et al., 1996), KM^+ (Rosa et al., 1999), *Maclura pomiferan* agglutinin (Lee et al., 1989) and hundreds of others which are referred to as β-prism fold lectins (Sharma et al., 2007). With respect to Cry toxins, the three apical loops at the base of domain II are thought to be flexible and they vary considerably in length and amino acid sequence. The domain II loops are involved in recognition of midgut receptors (Pigott and Ellar, 2007; Smedley and Ellar, 1996), irreversible binding (Rajamohan et al., 1995) and toxicity to specific hosts. In general, toxins with high sequence similarity in the loops of domain II share at least some binding sites on midgut brush border (Hernández and Ferré, 2005; Jurat-Fuentes and Adang, 2001) and shared binding sites are a factor in cross-resistance to Cry toxins in pest insect species (Heckel et al., 2007).

The location of Cry toxin domain II loops deduced from solving a Cry crystal structure (Li et al., 1991), provided researchers critical information needed for designing experiments to investigate functional properties of Cry–midgut interactions. The following comments on structural–functional analyses will focus on the action of Cry1A toxins in lepidopteran larvae, with comparisons to Cry toxin action in other target insects. A primary approach to investigating the role of domain II loops in toxin function has been to mutagenize loops by deleting or substituting amino acid residues. Synthetic peptides corresponding to loop regions have also been valuable tools through their use as competitors to probe toxin–midgut interactions. Studies of Cry1A toxin action in lepidopteran larvae show that loops 2, 3 and α-8 of Cry1A domain II are involved in receptor recognition and toxicity (Lee et al., 2000, 2001; Lu et al., 1994; Rajamohan et al., 1995, 1996a,b). Loop 1 of Cry1A toxins has not been implicated in receptor recognition to the extent of other domain II loops. Toxicity assays showed that the G312 residue in Cry1Ac loop 1 is important for toxicity, but relatively non-toxic mutants are not significantly changed in binding affinity (Smedley and Ellar, 1996). In contrast, mutations in loop 2 residues of Cry1Ab have been particularly valuable for investigating toxin interaction with midgut membrane and isolated receptors due to the dominant role of loop 2 residues in toxin binding and toxicity. For example, mutations in Cry1Ab at Phe371 (deletion of Phe371 or F371A) and at Gly372 (G3724A) reduced toxicity to *M. sexta* larvae, yet while the toxins had similar reversible binding, irreversible binding to midgut brush border membrane vesicles (BBMV) was reduced (Rajamohan et al., 1995). These results suggested the involvement of loop 2 in insertion of toxin into brush border membrane. As cadherins, aminopeptidase and ALP were identified as Cry1A receptors in lepidopteran larvae those proteins were integrated into functional analyses of Cry toxins (Pigott and Ellar, 2007). The α-8 and loop 2 residues of Cry1Ab determine recognition of binding epitopes in cadherin BtR1 of *M. sexta* (Gomez et al., 2002; Gómez et al., 2003). Similarly, loop 2 of Cry1Aa is thought to recognize BtR175 cadherin of silkworm, *Bombyx mori* (Obata et al., 2009). According to a proposed model (Gomez et al., 2002), recognition of receptor epitopes by domain II loops involves hydropathic interactions between residues in the apical loops and epitopes on receptors. The apical residues of domain II loops have even been replaced with antibody complementarity determining regions creating toxins called Cry bodies that have insecticidal activity (Pigott et al., 2008). Recently, a model was proposed where there are no critical residues needed for functional binding sites on loops, but rather conformational flexibility of loops involved in binding provides for optimal toxin–cadherin receptor interactions and insect toxicity (Fujii et al., 2013).

3.4. Cry domain III

Domain III of Cry toxins has a complex role in toxin action that is less defined than the roles of domains I and II. The structure of domain III is a β-sandwich comprised of two antiparallel β-sheets compressed into a jelly roll topology (Li et al., 1991). Each β-sheet has five strands and the two long loops near the end of the β-sheets interface with helices α6 and α7 of domain I (Grochulski et al., 1995). The structure of Cry domain III is similar to the cellulose-binding domain of *Cellulomonas fimi* β-1,4-glucanase (Burton et al., 1999). Conserved blocks 3, 4 and 5 of Höfte and Whiteley (1989) are located in domain III. Block 4, which is composed of four alternating arginines in the second β-strand of the bottom sheet of domain III, is directed towards the other domains and has a suggested role in toxin stability (Grochulski et al., 1995) plus more specific functions. Mutations replacing Block 4 arginines had an effect on toxin-induced ion channel function with the suggestion that alternating arginines stabilize toxin organization within membrane (Chen et al., 1993; Masson et al., 2002b; Schwartz et al., 1997c). Domain III also is involved in receptor binding as revealed by analyses of Cry toxins mutated at predicted surface-exposed loops connecting β-strands and by analyses of hybrid Cry toxins that have domain III from another toxin. Aronson et al. (1995) targeted the unique loop region of Cry1Ac by mutagenizing residues S503 and S504 and attributed reduced *M. sexta* and tobacco budworm, *Heliothis virescens* toxicity as being due to reduced initial binding. How Cry1Ac binds molecules is unique among characterized toxins, as the loop forms a lectin-like pocket and binds carbohydrates (Burton et al., 1999; Jenkins et al., 1999). More specifically, this Cry1Ac loop region participates in recognition of *M. sexta* midgut and in binding to *N*-acetylgalactosamine (GalNAc) attached to aminopeptidase (de Maagd et al., 1999b). It is also notable that the pattern of Cry1Ac–GalNAc recognition applies to the midgut ALPs of *H. virescens* and cotton bollworm, *Helicoverpa armigera* (Jurat-Fuentes and Adang, 2004; Sarkar et al., 2009; Sengupta et al., 2013) and that mutations in the GalNAc binding region reduced toxicity. Recently, the GalNAc binding pocket on Cry1Ac was disrupted by amino acid replacements resulting in reduced binding to monomeric ALP and insecticidal activity. The results of Pardo-López et al. (2006) indicate that GalNAc binding to Cry1Ac induces a slight conformational change that enhances membrane insertion of an oligomeric prepore structure. The relevance of Cry1Ac domain III discussed above to binding and toxicity is also buttressed by studies showing that domain III of Cry1Ac mediates specificity in lepidopteran species including *L. dispar* (Lee et al., 1995), *H. virescens* and the cabbage looper, *Trichoplusia ni* (Ge et al., 1991). Construction of Cry1–Cry1Ac hybrid toxins where Cry1Ac is attached C-terminal to domains I–II from another Cry1 toxin in some cases resulted in Cry toxins more toxic to *H. virescens* (Karlova et al., 2005). Studies of Cry1C, an important toxin for its activity against armyworm, *Spodoptera* species, identified domain III as a critical determinant in insecticidal specificity. Similar to the Cry1Ac domain III loop discussed above, Cry1C has a predicted loop connecting two β-strands on the outer β-sheet (Herrero et al., 2004). The importance of this region in Cry1C was suggested by hybrid toxin G27 (domains I and II from Cry1Ea and domain III from Cry1Ca) when two residues were substituted and toxicity to *Spodoptera exigua*, but not *M. sexta* larvae was reduced (de Maagd et al., 1999a). Recently, Herrero et al. (2004) showed that mutation of residues in the Cry1Ca domain III loop reduced toxicity to beet armyworm, *S. exigua*, larvae and binding to BBMV, but did not alter oligomerization. Transfer of toxicity against *Spodoptera* species from Cry1C to Cry1Ac was achieved by transfer of Cry1C domain III (Bosch et al., 1994). A novel domain III exchange involved the transfer of Cry1Ab domain III onto Cry3Aa domains I–II in coleopteran activity that did not compete for binding with Cry3Aa-like proteins (Walters et al., 2010). This result supports an important binding role for the lepidopteran-active Cry1Ab domain III that is functional in the coleopteran western corn rootworm, *Diabrotica virgifera*.

3.5. Cry intoxication process

The production of Cry toxins represents a large commitment of resources, supporting the importance of these proteins for the evolutionary success of *B. thuringiensis* as an effective insect pathogen (Raymond et al., 2010). The conserved three-dimensional structure of most Cry toxins first described for Cry3Aa (Li et al., 1991) supports a common mode of action, with primary and secondary Cry protein structures being responsible for specificity (de Maagd et al., 2001). Figure 2.5 shows details of Cry toxin mode of

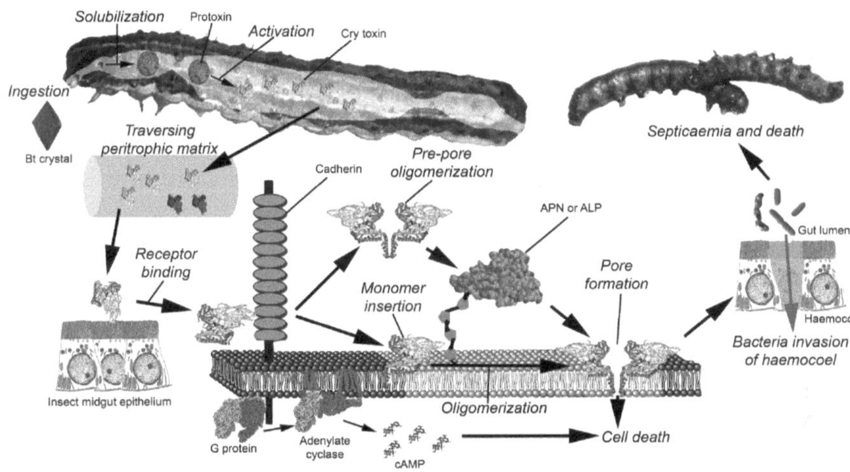

Figure 2.5 Representation of the current models of Cry toxin action in the insect midgut epithelium. After ingestion by susceptible larvae, toxin crystals are solubilized in the midgut fluids to yield the Cry protoxin form, which is processed to an activated Cry toxin form equivalent to the Cry toxin ingested by insects feeding on transgenic *Bt* crops. The Cry toxin core then traverses the peritrophic matrix, which is able to retain some of the toxin (darker toxin molecules in the figure). Once reaching the brush border membrane of the midgut epithelium, the Cry toxin binds with high affinity to cadherin, which results in activation of intracellular cell death pathways (represented by G protein, adenylate cyclase and cAMP in the figure), and/or according to the 'sequential binding' model, further proteolysis of the toxin monomer to result in formation of a pre-pore oligomer. This pre-pore oligomer is proposed to bind to alkaline phosphatase (ALP) or aminopeptidase (APN) to insert in the membrane forming a pore that leads to osmotic cell death. The insertion of toxin monomer and formation of a pore by oligomerization of inserted monomers is also presented as an alternative step. Disruption of the midgut epithelium barrier allows for bacterial invasion of the haemocoel, leading to septicaemia and death of the insect. (See the colour plate.)

action. It is well established that insecticidal Cry toxins target enterocytes in the insect host gut to compromise the gut epithelial barrier, which is key to facilitate access of *B. thuringiensis* to the haemocoel. The collapsing integrity of the gut epithelium during Cry intoxication also results in altered gut physiological conditions, which together with interaction with the midgut brush border membrane favours germination of the *B. thuringiensis* spore (Du and Nickerson, 1996). Ultimately, vegetative *B. thuringiensis* cells grow and multiply in the haemolymph (Johnston and Crickmore, 2009) resulting in septicaemia and insect death.

3.6. Cry toxin solubilization and proteolytic processing

The first step in the Cry intoxication process after ingestion by a susceptible host is the solubilization of the crystal by cleavage of interchain disulfide bonds to release Cry protoxins. Multiple factors affect solubility of Cry protoxins in the insect host gut, including gut physicochemical conditions (Jaquet et al., 1987); crystal composition (Aronson et al., 1991); intramolecular processing (Carroll et al., 1997); and in some Cry proteins, the action of 'helper' proteins (Naimov et al., 2011). The functional association between protoxin solubilization and host gut physicochemical conditions governs specificity in some cases, and it has been suggested to depend on the existence of typical disulfide bonds (Du et al., 1994) and predominant amino acids (Grochulski et al., 1995) in Cry proteins.

Solubilized protoxin molecules are then processed by endogenous *Bt* proteases and/or proteases in the host gut fluids to an active toxin core that is mostly resistant to further proteolysis. Typically, this toxin activation step is modelled *in vitro* by digesting Cry protoxins with trypsin or chymotrypsin (Andrews et al., 1985; Bietlot et al., 1989), the most common proteolytic enzymes in insect gut fluids (Terra and Ferreira, 1994). However, a number of reports suggest that Cry toxins activated with digestive enzymes from target insects may display distinct properties compared to trypsin or chymotrypsin-activated Cry toxins. For instance, increased toxin pore formation activity compared to activation with trypsin was reported when Cry3 toxins were activated in the presence of brush border membrane proteins of Colorado potato beetle, *Leptinotarsa decemlineata* (Rausell et al., 2004), or when Cry9Ca (Brunet et al., 2010b) or Cry1Ab (Gomez et al., 2014) were activated with digestive fluids from *M. sexta* larvae. In another example, Cry1Ie toxin oligomers formed after activation with trypsin were significantly less toxic than the monomeric toxin form (Guo et al., 2009b), which would be expected to oligomerize in the insect gut environment during the bioassays. The mechanisms explaining these differences are largely unknown, although a role for insect protease inhibitors in regulating Cry toxin activation (Brunet et al., 2010b) and insect lipids in promoting toxin oligomerization (Ma et al., 2012) have been proposed. Importantly, these observations may suggest that biologically relevant information may only be obtained when using Cry toxin activation with insect digestive fluids.

During the activation process, most well-studied Cry protoxins are sequentially digested to a 65- to 55-kDa toxin core by removal of about 500 and 43 amino acids from the C- and N-termini, respectively. In other cases, like the mosquitocidal Cry4A and Cry11Aa and the scarab-specific Cry8Da toxins, activation generates two peptide fragments that remain associated to form a toxin complex (Yamagiwa et al., 1999; Yamaguchi et al., 2010). The processing at the N-terminus has been shown to be crucial to subsequent steps in the Cry intoxication process (Bravo et al., 2002) and is the only region processed during activation of the smaller (67–75 kDa) Cry protoxins. Interestingly, in most Cry protoxins, this N-terminus is tightly associated to 20-kilobase (kb) DNA fragments (Bietlot et al., 1993) that may contain *cry* toxin genes (Xia et al., 2005). These DNA fragments not only appear to dictate sequential proteolysis of the Cry protoxin (Clairmont et al., 1998) but may also have a critical role in toxin binding specificity and membrane insertion (Ai et al., 2013; Guo et al., 2011).

The rate of Cry protoxin processing is important to determine susceptibility, as supported by enhanced activity in engineered Cry toxins with increased activation rates compared to wild-type toxins (Walters et al., 2008). Moreover, resistance to diverse Cry toxins has been associated with altered toxin processing or reduced rate of activation in a number of insects (Ferré and Van Rie, 2002). In this regard, endogenous *Bt* proteases may contribute to accelerate protoxin processing *in vivo* to increase insect susceptibility (Oppert, 1999). Activated Cry toxins then have to pass through the peritrophic matrix, a process that may reduce the amounts of active toxin interaction with the target tissue (midgut) and thus affecting susceptibility (Hayakawa et al., 2004; Rees et al., 2009). This protective role exerted by the peritrophic matrix can be overcome by *Bt* endogenous or exogenous chitinases (Kramer and Muthukrishnan, 1997).

4. MIDGUT CRY-BINDING PROTEINS AND RECEPTOR FUNCTION

Specific binding of Cry toxins to molecules located on the brush border membrane of the midgut cells is a major factor in determining the host range of Cry toxins. This feature of Cry–midgut interaction was revealed using BBMV prepared from midgut and labelled Cry toxins in binding assays. Major challenges of *Bt* researchers over the past two decades were the identification of molecules that specifically bind Cry toxins and then deducing functional relevance of this interaction to Cry toxin action. The proteins and glycoconjugates reported as involved in Cry toxin action are described in more detail in recent reviews (Bravo et al., 2011; Jurat-Fuentes and Jackson, 2012; Pigott and Ellar, 2007).

4.1. Aminopeptidase

Aminopeptidase-*N* (APN) (Knight et al., 1994, 1995; Sangadala et al., 1994) and cadherin (Vadlamudi et al., 1995) were the first proteins identified as putative Cry receptors in insects. Having two distinct proteins types as 'putative' receptor raised questions as to whether or not both proteins were involved in Cry toxin action; an issue that is resolved in some of the proposed intoxication models, as discussed below. APNs are tethered to the midgut brush border by GPI anchors (Garczynski and Adang, 1995) from where they cleave N-terminal amino acids from peptides, a step necessary for amino acid co-transport into epithelial cells (Terra et al., 1996). Phylogenetic analyses of lepidopteran APNs cluster these proteins into seven classes (Crava et al., 2010; Hughes, 2014). Among APNs for which expression data are available, members of the APN1 class are the most highly expressed in midgut tissue (Hughes, 2014). Interactions between Cry1 toxins and midgut APNs are known to involve recognition of epitopes on the primary protein structure, or in the case of Cry1Ac, an attached glycan with a terminal GalNAc moiety. APNs that bind Cry1Ac via GalNAc are identified in *M. sexta* (Masson et al., 1995), *H. virescens* (Gill et al., 1995; Luo et al., 1997), *L. dispar* (Jenkins et al., 2000) and *H. armigera* (Sarkar et al., 2009). Cry1Ac binding to APNs in *M. sexta* and *L. dispar* has sugar-dependent and -independent components (Jenkins et al., 2000; Masson et al., 1995) leading to proposal of a bivalent binding model involving sequential interactions with domain III and then domain II loop residues (Jenkins et al., 2000). Cry1Ac also recognizes a 106-kDa APN in *H. virescens* in a GalNAc-independent manner (Banks et al., 2001). Similarly, and although only known to recognize receptor proteins at amino acid epitopes, the Cry1Aa toxin binds through domains II and III to a 117-kDa APN in *B. mori* (Atsumi et al., 2005). A small patch of seven amino acids near the N-terminus of a cotton leafworm, *Spodoptera litura* APN serves as epitope for recognition by loops 2 and 3 of Cry1C domain II (Kauer et al., 2014).

In the 1990s, evidence for APN function as Cry1 receptors was provided by the ability of an APN preparation to enhance Cry1-induced pore formation in membrane vesicles (Luo et al., 1997; Sangadala et al., 1994) and ion channels in membrane bilayers (Schwartz et al., 1997c). This approach was particularly challenging due to the difficulty of purifying midgut APNs with the lipid moiety remaining on the GPI anchor (Garczynski and Adang, 1995). *In vivo* evidence of Cry receptor was suggested by work of Gill and Ellar (2002) who expressed *M. sexta* APN in transgenic *Drosophila* rendering larvae susceptible to Cry1Ac. Silencing midgut APNs in *S. litura*, sugarcane borer (*Diatraea saccharalis*) and *H. armigera* by RNA interference (RNAi) established their role as Cry1C (Rajagopal et al., 2002), Cry1Ab (Yang et al., 2010), and Cry1Ac (Sivakumar et al., 2007) receptors, respectively. Mutation or reduced expression of specific APNs has been correlated with resistance to Cry1 toxins in *D. saccharalis* (Yang et al., 2010), *S. exigua* (Herrero et al., 2005) and *H. armigera* (Zhang et al., 2009). In a greenhouse-derived strain of *T. ni*, Cry1Ac resistance was associated with differential alteration of two APNs; with APN1 being downregulated and APN6 being upregulated in resistant larvae (Tiewsiri and Wang, 2011). Also, the lack of 110-kDa APN1 protein in resistant larvae correlated with reduced transcript levels and was conferred by a trans-regulatory mechanism (Tiewsiri and Wang, 2011).

The role of APNs in the action of mosquitocidal Cry toxins has received considerable attention. Diverse APNs have been identified as putative receptors of Cry11Aa and Cry11Ba in yellow fever mosquito, *Aedes aegypti* (Chen et al., 2009b, 2013; Likitvivatanavong et al., 2011), and Cry11Ba in malaria mosquitoes, *Anopheles albimanus* and *An. gambiae* (Abdullah et al., 2006; Zhang et al., 2008). The specific APNs in *Aedes* and *Anopheles* that bind Cry11Aa and Cry11Ba, respectively, do so with high affinity (nM range). Interestingly, in bioassays, the presence of partial APN fragments enhanced Cry larval mortality (Chen et al., 2013; Zhang et al., 2010), an unexpected observation from a functional receptor that may suggest interactions between APN and toxin are reversible. Silencing of APN expression in *Ae. aegypti* by RNAi resulted in increased tolerance to Cry4Ba toxin (Saengwiman et al., 2011), supporting a functional toxin receptor role for this APN.

4.2. Cadherin

Cadherin-like proteins are widely accepted as functional Cry toxin receptors. Contrary to the generally observed localization for cadherin proteins to regions of cell–cell interactions, cadherin proteins binding Cry toxins localize mostly to the brush border membrane of midgut cells (Aimanova et al., 2006; Chen et al., 2005; Hara et al., 2003; Valaitis, 2011). Cadherins have also been found to be present in the basement membrane anchoring midgut epithelial cells of *M. sexta* and *L. dispar* larvae (Chen et al., 2005; Valaitis, 2011). A cadherin called BT-R1 (*Bt* receptor 1) was first identified as a Cry toxin receptor in *M. sexta* larvae (Vadlamudi et al., 1995). Cry1A toxins were shown to bind Bt-R1 with high affinity and expression of Bt-R1 in cultured insect cells conferred sensitivity to Cry1A toxins (Hua et al., 2004; Keeton and Bulla, 1997). Functional evidence of a Cry1A receptor role for cadherins has been reported for *B. mori* (Nagamatsu et al., 1999), *O. nubilalis* (Flannagan et al., 2005) and *H. virescens* (Jurat-Fuentes and Adang, 2006). Cadherin mutations associated with resistance to Cry1A proteins are well documented for a number of insects species, including pink bollworm, *Pectinophora gossypiella* (Morin et al., 2003), *H. armigera* (Xu et al., 2005) and *H. virescens* (Gahan et al., 2001). Gene silencing of a cadherin from *S. exigua* identified that protein as a putative receptor of Cry1Ca toxin, extending the role of cadherin as a Cry receptor in lepidopteran larvae outside the Cry1A class of toxins (Park and Kim, 2013; Ren et al., 2013). Cadherin proteins are also reported as functional receptors of Cry3Aa and Cry3Bb toxins in Coleoptera (Fabrick et al., 2009; Hua et al., 2014). In Diptera, cadherins are identified as putative receptors of Cry11Aa and Cry11Ba in *Ae. aegypti* and Cry11Ba in *An. gambiae* (Chen et al., 2009a; Hua et al., 2013; Likitvivatanavong et al., 2011). Although Cry4Ba binds to cadherins in *Ae. aegypti* (Bayyareddy et al., 2009; Hua et al., 2008) and *An. gambiae* (Hua et al., 2008), the binding affinities are lower than for Cry11 binding, which may explain why silencing of cadherin expression by RNAi in *Aedes* larvae increased larval tolerance to Cry11Aa but had no effect on Cry4Ba toxicity (Rodriguez-Almazan et al., 2012). Cadherin has been associated with resistance to *Bt* subsp. *israelensis* in a laboratory-selected strain of *Ae. aegypti*, but the cadherin is not a known Cry-binding protein (Bonin et al., 2009).

4.3. Alkaline phosphatase

Membrane-bound midgut ALPs are a major group of Cry-binding proteins identified in Lepidoptera, Coleoptera and Diptera larvae; and in many cases, receptor function has been established. Interactions between Cry1Ac and ALP were recognized when incubation with Cry1Ac reduced ALP enzymatic activity in midgut proteins from *H. virescens* and *M. sexta* larvae (English and Readdy, 1989; Sangadala et al., 1994). Binding of Cry proteins to ALPs in BBMVs from *M. sexta* (McNall and Adang, 2003), *H. virescens* (Krishnamoorthy et al., 2007) and *Ae. aegypti* (Bayyareddy et al., 2009) were detected by proteomic analyses. As noted above for APN, binding of Cry1Ac to ALP from *H. virescens* and *H. armigera* larvae involves interactions with GalNAc (Jurat-Fuentes and Adang, 2004; Ning et al., 2010; Sengupta et al., 2013). In the case of Cry1Ab, initial interaction with ALP in *M. sexta* through β-16 in domain III is critical to binding to ALP and toxicity to larvae (Arenas et al., 2010). RNAi silencing of APN and ALP genes in *M. sexta* larvae showed that binding to ALP is more important for Cry1Ab toxicity than to APN; in contrast to Cry1Ac which relies more on APN (Flores-Escobar et al., 2013). Correlations between reduced ALP expression and resistance to Cry1 toxins in strains of *H. virescens, H. armigera* and the fall armyworm (*S. frugiperda*) further support an *in vivo* role of ALPs in Cry intoxication of lepidopteran larvae (Jurat-Fuentes et al., 2011). In Diptera, mosquitocidal Cry11Aa and Cry11Ba toxins bind ALP in *Aedes* and *Anopheles* BBMV, and this interaction is relevant to *in vivo* toxicity, as feeding toxin-binding regions of ALP with Cry toxins to larvae reduced toxicity (Fernandez et al., 2006; Hua et al., 2009). Evidence for receptor functionality *in vitro* has been provided from expression of an *Aedes* ALP on the surface of insect cells conferring susceptibility to Cry4Ba toxin (Dechklar et al., 2011). Recently, ALP1 in *Ae. aegypti* was established as a functional receptor for Cry11Aa and Cry4Ba by RNAi silencing (Jimenez et al., 2012). Receptor ALPs were under expressed in a *Bt* var. *israelensis*-resistant strain of *Ae. aegypti* (Tetreau et al., 2012). In Coleoptera, midgut ALPs binding Cry3Aa in the yellow mealworm, *Tenebrio molitor* (Zuniga-Navarrete et al., 2013), or Cry1Ba6 in cotton boll weevil, *Anthonomus grandis* (Martins et al., 2010), have been identified, with confirmation of receptor function yet to be determined (Martins et al., 2010).

4.4. ABC transporter

The ABC family of proteins in insects are related to the multi-drug resistance proteins in animals (Heckel, 2012). The comparative genomics of the ABC gene family in arthropods and the relationship of ABC proteins to xenobiotics and insecticide transport were recently reviewed (Dermauw and Van Leeuwen, 2014). Initial evidence for a role of ABCC2 proteins in the Cry1 mode of action was provided by genetic linkage between an *ABCC2* mutation and resistance to Cry1A toxins in strains of *H. virescens* (Gahan et al., 2010). Isolation of resistant loci into distinct strains allowed for the demonstration that mutations in cadherin explained lack of Cry1Aa binding (Jurat-Fuentes et al., 2004) in these *H. virescens* strains, an inactivating mutation in the *ABCC2* gene was associated to lack of Cry1Ab and Cry1Ac binding (Gahan et al., 2010). Resistance to Cry1Ac was also mapped to the *ABCC2* locus in diamondback moth, *Plutella xylostella*, and *T. ni* providing further support for the function of this protein in Cry1A toxin action (Baxter et al., 2011). Direct confirmation of the importance of ABC transporters to Cry1A toxin action was obtained from studies of the silkworm, *B. mori*. In this insect, a single tyrosine insertion on an outer lumen-facing loop of an ABCC2 protein was linked to resistance against the Cry1Ab toxin, and germ line transformation of this insertion conferred resistance to this toxin in susceptible silkworms (Atsumi et al., 2012). Moreover, the importance of the tyrosine insertion and functional role of the ABCC2 protein in susceptibility to Cry1Ab, Cry1Ac, Cry1Fa and even relatively unrelated Cry8Ca in *B. mori* was demonstrated by comparing expression of mutated and wild-type ABCC2 genes in cultured insect cells (Tanaka et al., 2013). Although these studies illustrate a critical role for ABCC2 proteins in Cry toxin action, there is no experimentally tested mechanistic description of ABCC2-Cry toxin interactions available.

4.5. Other Cry-binding (receptor) proteins and molecules

Glycolipids from *M. sexta*, a glycoconjugate from *L. dispar*, members of the polycalin protein family, an ADAM metalloprotease in *L. decemlineata*, a sodium solute transporter in flour beetle (*Tribolium castaneum*), amylases in *An. albimanus* and *An. gambiae* and an α-glucosidase in *An. gambiae* have been reported as Cry toxin-binding molecules and putative receptors. Initial evidence for interactions between lipids and Cry toxins was provided by studies of Cry1A protoxin and activated toxin binding to glycolipids from pupae of the blowfly *Calliphora vicina* (Dennis et al., 1986). Although toxin binding to glycolipids was detected, this evidence was not widely considered, probably because fly larvae are not usually susceptible to Cry1A toxins. Functional evidence for glycolipids as Cry toxin receptors emerged from studies of resistance to Cry5Ba in the nematode, *Caenorhabditis elegans* (Griffitts et al., 2005). The same study included evidence for specific Cry1A toxin binding to glycolipids extracted from *M. sexta* midgut tissue, supporting a role for lipids in Cry intoxication. A role for glycolipids in formation of Cry toxin oligomers through interaction with domain II of the toxin was proposed (Ma et al., 2012). However, the functional relevance of lipid–Cry toxin interactions to toxicity has not been experimentally established.

In the gypsy moth (*L. dispar*), a 270 kDa glycoconjugate binds Cry1Aa, Cry1Ab and Cry1Ba toxins (Valaitis et al., 2001), although the biological relevance of this interactions is unknown. A 252 kDa member of the polycalin family in *B. mori* (P252) was shown to bind Cry1A toxins in a GalNAc-independent manner (Hossain et al., 2004; Pandian et al., 2008). This protein exists in the insect as an oligomer that forms a complex with Cry1A toxins, but it does not interfere with activity against *B. mori* larvae (Pandian et al., 2010). Interactions between Cry toxins and members of the polycalin protein family have also been reported in *H. armigera* (Angelucci et al., 2008; Ma et al., 2012). Taken together, these observations may reflect the importance of interactions between Cry toxins and lipids or proteins in the midgut fluids to subsequent binding to receptors on midgut cells, a possibility that needs further experimental study.

Diverse proteins have been reported to participate in Cry3A toxicity in *L. decemlineata*. In this system, proteolysis of Cry3Aa by BBMV proteins reduces toxin pore formation (Rausell et al., 2007), yet interactions of Cry3Aa domain II loop 1 with an ADAM metalloprotease are important for effective toxin pore formation (Ochoa-Campuzano et al., 2007). More recently, a prohibitin protein has been proposed to participate in recruiting of the toxin to the ADAM protease (Ochoa-Campuzano et al., 2013). Unfortunately, the critical function of prohibitin prevented the use of silencing approaches to test its relevance for Cry intoxication. Interestingly, prohibitin has been also described to bind Cry4Ba in *Ae. Aegypti* (Bayyareddy et al., 2009), although both Cry4Ba and Cry11Aa toxins were also reported to bind a 70-kDa α-amylase in *An. albimanus* (Fernandez-Luna et al., 2010). The functional role of these *in vitro* interactions for *in vivo* toxicity needs to be experimentally tested.

5. MODELS OF CRY TOXIN ACTION

Although extensively investigated and reviewed (Jurat-Fuentes and Jackson, 2012), mechanistic details for interactions between midgut brush border proteins and Cry toxins that are conducive to toxicity remain controversial (Vachon et al., 2012). It is well established that activated Cry toxins must recognize sites on insect midgut cell proteins to exert toxicity and that this binding step determines specificity, yet it is not sufficient to predict susceptibility (Garczynski et al., 1991; Wolfersberger, 1990). Since Cry toxin binding is mostly localized to the brush border membrane of the midgut epithelium where binding proteins are located (Bravo et al., 1992; Chen et al., 2005), vesicle preparations representing an enriched fraction for this tissue (BBMV) have been used as *in vitro* model to study interactions between Cry toxins and midgut cells. While diverse proteins, lipids and glycoconjugates have been proposed to contain Cry toxin-binding sites and represent functional receptors (Pigott and Ellar, 2007), most data available are focused on cadherin, APN, and ALP proteins. Interestingly, Cry1A and Cry3Aa protoxins bind specifically to midgut cadherin proteins in target insects (Fabrick and Tabashnik, 2007; Fabrick et al., 2009; Gomez et al., 2014), although the relevance of this binding toxicity *in vivo* is unclear considering the fast rate of activation of protoxins in the midgut lumen after solubilization.

The 'sequential binding' model developed for Cry1A toxin–receptor interactions (Pardo-López et al., 2013) suggests initial reversible binding of Cry1A-activated toxins to abundant APN and ALP proteins, favouring higher affinity toxin interactions with less abundant cadherin proteins (Pacheco et al., 2009a). This binding of toxin to the extracellular cadherin region most proximal to the cell membrane (Pigott and Ellar, 2007) is proposed to promote proteolytic removal of helix $\alpha 1$ and result in formation of a pre-pore toxin oligomer that displays high affinity for APN and ALP proteins (Bravo et al., 2004; Gómez et al., 2002). Binding to APN and ALP proteins facilitates concentration of toxin oligomers on specialized membrane regions called lipid rafts (Zhuang et al., 2002), facilitating insertion and formation of a toxin pore that leads to cell death by osmotic shock. In agreement with this model, reversible (initial binding to APN and ALP, binding to cadherin, and toxin oligomer binding to APN and ALP) and irreversible (toxin insertion on the membrane) binding components have been described for Cry toxins, with irreversible binding being directly correlated to toxicity (Liang et al., 1995). Further evidence supporting this model comes from reports of Cry toxin oligomer formation *in vitro* in the presence of cadherin peptides (Fabrick et al., 2009; Pacheco et al., 2009b; Peng et al., 2010) and augmented toxicity associated with increased production of Cry toxin oligomers (Gao et al., 2011; Pacheco et al., 2009b). Lack or low levels of susceptibility to Cry1A toxins in heterologous systems expressing cadherin (Aimanova et al., 2006; Hua et al., 2004; Jurat-Fuentes and Adang, 2006) proteins are in agreement with the dependence on sequential receptor interactions for Cry toxicity. Unfortunately, alternative reports of heterologous cadherin or APN expression previously suggested as evidence against sequential binding (Vachon et al., 2012), did not report quantitative data or monitored only changes in cell morphology, which complicates accurate assessment of cytotoxicity. Only a report of heterologous expression of APN from *M. sexta* in *Drosophila melanogaster* larvae resulted in high susceptibility to Cry1Ac could be considered as evidence for the sequential binding model, yet in this case, the existence of cadherin and ALP *Drosophila* orthologs capable of interacting with the toxin was not tested.

Nevertheless, there are some reported observations that are difficult to explain with the 'sequential binding' model (Vachon et al., 2012), some of which may be a consequence of the Cry1A–*M. sexta* model system used for its development, potentially limiting extended applicability. For instance, the model does not consider the role of membrane lipids or ABC transporters, which have been proposed as relevant Cry toxin receptors (Gahan et al., 2010; Griffitts et al., 2005; Ma et al., 2012). Moreover, modified Cry1A toxins with cadherin-independent oligomerization (Soberón et al., 2007) did not overcome resistance due to alterations in cadherin receptors and were less potent against susceptible insects than wild-type toxins (Tabashnik et al., 2011). This reduced potency compared to wild-type toxin may be explained by the pre-pore oligomer formed by modified Cry1A toxins differing from the oligomer formed by wild-type toxin (Gomez et al., 2014). However, lack of direct correlation between amounts of oligomer formed *in vitro* and toxicity *in vivo* in some cases (Gomez et al., 2014; Gómez et al., 2002) question the relevance of these pre-pore oligomers for toxicity. In this regard, Cry toxin oligomeric structures are also formed in solution (Masson et al., 2002a; Walters et al., 1994) and form functional pores in model lipid membranes in the absence of receptor proteins (Gomez et al., 2014; Vié et al., 2001). The functional relevance of pre-pore Cry toxin oligomers for toxin insertion in the membrane is further challenged by toxin mutants unable to form oligomers but capable of binding irreversibly, suggesting insertion, in BBMV from susceptible insects (Tigue et al., 2001). The observation that these mutants could insert as monomers but not cause toxicity supports the possibility that oligomerization may occur after insertion of toxin monomers in the membrane (Vachon et al., 2012).

Two alternative models for Cry toxin pore formation on the midgut cell membrane have been proposed (Knowles, 1994). In the 'penknife' model (Hodgman and Ellar, 1990), only the hydrophobic α5 and α6 helices in domain I insert into the membrane, while the rest of the domain lays flat on the membrane surface. Alternatively, most available evidence supports the 'umbrella' model, in which a conformational change in the toxin molecule upon binding to receptors results in insertion of a hydrophobic hairpin composed of α4 and α5 helices (Li et al., 2001; Schwartz et al., 1997c). The position of domains II and III of the Cry toxins during insertion is still a matter of debate. While protease protection studies (Aronson, 2000; Aronson et al., 1999) and mutagenesis (Dean et al., 1996; Nair and Dean, 2008) support that the whole toxin molecule is confined to the membrane, there is also evidence suggesting that the toxin remains associated to membrane receptors after pore formation (Fortier et al., 2007) and at least residues of domain III may be exposed to the solvent (Pardo-López et al., 2006). In contrast to the pre-pore oligomers presented by the 'sequential binding' model, proposals modelling pore formation support that association between inserted toxin monomers results in a tetrameric ion channel with a diameter of approximately 15 Å (Groulx et al., 2010, 2011). However, trimeric pores have been proposed from sequence analysis (Torres et al., 2008) and lipid membrane experiments (Ounjai et al., 2007). Evidence supporting the possibility that Cry toxin pores may arrange from association of diverse Cry toxin monomers (heterooligomers) has been presented (Carmona et al., 2011), which may help explain synergistic and inhibitory effects among Cry toxins (Ibargutxi et al., 2008). More importantly, the significance of heterooligomeric pore formation for binding competition studies needs to be considered and tested experimentally.

The α4 helix lining the Cry toxin pore seems to control the passage of ions through the pore (Kumar and Aronson, 1999; Masson et al., 1999), although there are also reports suggesting a role in controlling pore permeability by alternative domain I helices (Alcantara et al., 2001; Arnold et al., 2001). Pore properties as well as formation are also dependent on midgut epithelium components (Peyronnet et al., 2001; Schwartz et al., 1997b) and ionic composition (Brunet et al., 2010a; Fortier et al., 2005). Variable pore properties depending on the environment may reflect adaptation of Cry toxins to diverse functional environments (Schnepf et al., 1998).

An alternative model for Cry intoxication disregards pore formation and highlights the activation of intracellular oncotic cell death pathways as responsible for enterocyte death (Zhang et al., 2008). However, evidence supporting this model is limited to Cry1Ab and studies with insect cell cultures expressing a cadherin from *M. sexta* (Zhang et al., 2005). In these cells, binding of Cry1Ab toxin was associated with cell death and increased cAMP production. It is important to note that the ovarian Hi5 cell line used for these analyses is susceptible to the Cry1Ac toxin in the absence of expression of proposed midgut toxin receptors, suggesting that they may display a mechanism of susceptibility that differs from cells in the insect midgut epithelium. Evidence supporting a role for intracellular pathways in Cry intoxication of midgut cells was provided by a deletion of 55 amino acids in the cytoplasmic domain of a cadherin being genetically linked with non-recessive resistance to Cry1Ac in *H. armigera* (Zhang et al., 2012). However, only a 20% reduction in susceptibility to Cry1Ac was detected in Sf9 cells expressing the wild type versus the cadherin with the cytoplasmic deletion, challenging the relative importance of cadherin-mediated intracellular signalling in Cry susceptibility.

Although speculative, a third possibility would consider effects of both pore formation and intracellular cell death pathways (Pigott and Ellar, 2007). Thus, there are examples of bacterial pore-forming toxins that induce host cell death through oncosis (Zhou et al., 2009). The observation that modified Cry toxins that do not depend on binding to cadherin for oligomerization display lower activity against susceptible insects (Soberón et al., 2007), may suggest that intracellular signalling activated by wild type, but not by modified Cry toxins binding to cadherin may contribute to toxicity. Consequently, it is plausible that intracellular cell death may occur in the presence of lower Cry toxin concentrations, while higher toxin concentrations would promote increased toxin insertion and formation of pores. Further work to establish functional connections between pore formation and intracellular signalling for cytotoxicity are needed to clarify the molecular events resulting in midgut cell death by Cry toxins.

In the case of binary Cry toxins, low sequence homology suggests differences in the intoxication model when compared with the three-domain Cry toxins. Most data available are focused on the Cry34/35 complex (Ellis et al., 2002). In this case, while the Cry35 protein displays lectin folds suggestive of receptor binding and the Cry34 protein has homology to proteins involved in intracellular signalling (Schnepf et al., 2005), they probably exert their toxicity through pore formation (Masson et al., 2004). Data from binding assays support that Cry34 toxin greatly enhances Cry35 binding to *D. virgifera* BBMV proteins (Li et al., 2013). Interestingly, comparisons between results from homologous competition tests in the presence or absence of Cry34 suggest lower binding affinity for the Cry34/35 complex. Binding of Cry35 to *D. virgifera* BBMV proteins in the absence or presence of Cry34 was to binding sites not recognized by Cry3Aa, Cry8Ba or Cry6Aa toxins (Li et al., 2013). The identity of the specific Cry34/35 or Cry35 receptors is unknown.

6. CYTOLYTIC TOXINS

The Cyt (cytolytic) toxins were originally discovered in *Bt* subspecies *israelensis* (Goldberg and Margalit, 1977) and have been recently reviewed (Ben-Dov, 2014; Soberon et al., 2013b). These proteins seem to be specific to some *Bt* subspecies (Tyrell et al., 1981). The first feature observed for the 28 kDa protein from *Bt* subsp. *israelensis* was its cytolytic activity against cultured mammalian and insects cells and toxicity when injected in mice, while it did not display activity against larvae of the cabbage butterfly, *Pieris brassicae* (Thomas and Ellar, 1983a). Against mosquitoes, the 28-kDa protein was less active than native *Bt* subsp. *israelensis* crystals (Chilcott and Ellar, 1988; Davidson and Yamamoto, 1984; Yamamoto et al., 1983), and it synergized activity of other *Bt* subsp. *israelensis* crystal proteins (Chilcott and Ellar, 1988; Wu and Chang, 1985). Currently, there are also *cyt* genes that have been described in *Bt* strains targeting lepidopteran or coleopteran insects (Guerchicoff et al., 1997), adding to, three *cyt* toxin gene families (*cyt1*, *cyt2* and *cyt3*) that include 11 holotype toxins in the current nomenclature (Crickmore et al., 2014).

The three-dimensional Cyt structures resolved to date support a conserved α–β structural model including two α-helix hairpins flanking a β-sheet core containing seven to eight β-strands (Cohen et al., 2008, 2011; Li et al., 1996). This highly conserved structure is also revealed in sequence alignments, which identify highly conserved blocks (Butko, 2003). Mutagenic studies identified β-sheet residues as critical for toxicity, while mutations of residues on the helical domains did not affect toxicity, suggesting a critical role for the β-sheet core.

Ingested Cyt1A and Cyt2A protoxins are processed by digestive proteases at the same sites in the N- and C-termini to a stable toxin core of 25 and 23 kDa, respectively (Koni and Ellar, 1994). Activated Cyt toxins display high affinity for membrane lipids containing unsaturated acyl chains (Gill et al., 1987; Thomas and Ellar, 1983b), which are abundant in midgut brush border membrane of Diptera (Li et al., 1996). A putative phospholipid-binding site pocket homologous to *Erwinia* virulence factor (Evf) was described for Cyt2Ba (Rigden, 2009). Two non-exclusive mechanisms for Cyt1A-induced cytolysis have been proposed: pore formation and detergent-like membrane disruption (Butko, 2003). Pore formation by

Cyt toxins is supported by evidence from studies with planar bilayers (Knowles et al., 1989) and erythrocytes (Promdonkoy and Ellar, 2003), and by the importance of pre-pore oligomerization for toxicity (López-Diaz et al., 2013). According to this model, binding to membrane lipids leads to a conformational change in the toxin resulting in movement of the amphiphilic helices to expose the hydrophobic face of the β-sheet for membrane insertion (Li et al., 1996). Three of the β-strands are long enough to insert spanning the cell membrane (Promdonkoy and Ellar, 2005), which is followed by oligomerization to form a β-barrel pore conducive to cell death by osmotic shock (Du et al., 1999; Li et al., 1996). Recent reports suggest the existence of pre-pore oligomers (López-Diaz et al., 2013), yet the relevance of these oligomers for toxicity has been questioned (Cantón et al., 2014). In contrast to pore formation, the detergent model advocates that cell death occurs through localization of the toxin to the membrane surface to induce detergent-like defects in lipid packing, which result in leakage of intracellular molecules (Butko et al., 1996). In support of this model, large Cyt toxin aggregates, rather than smaller oligomers capable of forming pores, are observed during binding to cell membranes (Chow et al., 1989), membrane permeation (Rodriguez-Almazan et al., 2011) and lipid vesicles exposed to Cyt1A fragment into smaller forms as it would be expected from a detergent-like action (Manceva et al., 2005). It is generally accepted that both models may occur *in vivo* depending on toxin concentration, with lower concentrations favouring oligomeric pores and higher concentrations leading to membrane breaks (Butko, 2003).

Synergistic effects between Cyt and other bacterial toxins have frequently been reported, even to overcome resistance. For instance, a Cyt1A overcame resistance to Cry4 toxins (Wirth et al., 1997) and delayed evolution of resistance to Cry11Aa (Wirth et al., 2005) in *Culex quinquefasciatus*. Synergism was also observed for Bin toxins in strains of *Cx. pipiens* and *Cx. quinquefasciatus* (Thiery and Hamon, 1998) and for Mtx1 toxicity in *Cx. quinquefasciatus* (Zhang et al., 2006). The Cyt synergistic effect for Cry toxins is proposed to involve Cyt1A binding to domain II of Cry toxins (Lailak et al., 2013) in solution or on the membrane plane to promote formation of a Cry toxin pre-pore oligomer (Pérez et al., 2007). Formation of this Cry oligomer is independent of the Cyt oligomerization, binding or insertion (López-Diaz et al., 2013), which may help explain why synergism between Cyt1A and Cry3A toxin against the cottonwood leaf beetle, *Chrysomela scripta*, does not correspond with binding of Cry3A to membrane-bound Cyt1A (Federici and Bauer, 1998). Because binding to lipids limits specificity of Cyt toxins for insecticidal applications, Cyt2Aa toxin has been engineered to contain a pea aphid gut-binding peptide resulting in a novel toxin with specific activity against major aphid pests (Chougule et al., 2013). Lack of competition for midgut sites between the gut-binding peptide of the engineered Cyt toxin and Cry toxins suggests the potential use of both toxins for sustainable pest control.

ACKNOWLEDGEMENTS

This work was supported by USDA NIFA award number 2010-65105-20590 to M. J. A. (University of Georgia) and Biotechnology Risk Assessment Grant Program competitive grant no. 2010-33522-21700 from USDA NIFA to J. L. J-F. (University of Tennessee). We thank Dr. Leara Rhodes (University of Georgia) for editing an earlier version of this manuscript. Ruchir Mishra is thanked for generating and assembling the three-dimensional Cry images.

REFERENCES

Abdullah, M.A., Valaitis, A.P., Dean, D.H., 2006. Identification of a *Bacillus thuringiensis* Cry11Ba toxin-binding aminopeptidase from the mosquito, *Anopheles quadrimaculatus*. BMC Biochem. 7, 16.

Ai, B., Li, J., Feng, D., Li, F., Guo, S., 2013. The elimination of DNA from the Cry toxin-DNA complex is a necessary step in the mode of action of the Cry8 toxin. PLoS One 8.

Aimanova, K.G., Zhuang, M., Gill, S.S., 2006. Expression of Cry1Ac cadherin receptors in insect midgut and cell lines. J. Invertebr. Pathol. 92, 178–187.

Akiba, T., Abe, Y., Kitada, S., Kusaka, Y., Ito, A., Ichimatsu, T., Katayama, H., Akao, T., Higuchi, K., Mizuki, E., Ohba, M., Kanai, R., Harata, K., 2009. Crystal structure of the parasporin-2 *Bacillus thuringiensis* toxin that recognizes cancer cells. J. Mol. Biol. 386, 121–133.

Alcantara, E.P., Alzate, O., Lee, M.K., Curtiss, A., Dean, D.H., 2001. Role of alpha-helix seven of *Bacillus thuringiensis* Cry1Ab delta-endotoxin in membrane insertion, structural stability, and ion channel activity. Biochemistry 40, 2540–2547.

Alzate, O., Osorio, C., Florez, A.M., Dean, D.H., 2010. Participation of valine 171 in alpha-Helix 5 of *Bacillus thuringiensis* Cry1Ab delta-endotoxin in translocation of toxin into *Lymantria dispar* midgut membranes. Appl. Environ. Microbiol. 76, 7878–7880.

Andrews Jr., R.E., Bibilos, M.M., Bulla Jr., L.A., 1985. Protease activation of the entomocidal protoxin of *Bacillus thuringiensis* ssp. *kurstaki*. Appl. Environ. Microbiol. 50, 737–742.

Angelucci, C., Barrett-Wilt, G.A., Hunt, D.F., Akhurst, R.J., East, P.D., Gordon, K.H., Campbell, P.M., 2008. Diversity of aminopeptidases, derived from four lepidopteran gene duplications, and polycalins expressed in the midgut of *Helicoverpa armigera*: identification of proteins binding the delta-endotoxin, Cry1Ac of *Bacillus thuringiensis*. Insect Biochem. Mol. Biol. 38, 685–696.

Arenas, I., Bravo, A., Soberon, M., Gomez, I., 2010. Role of alkaline phosphatase from *Manduca sexta* in the mechanism of action of *Bacillus thuringiensis* Cry1Ab toxin. J. Biol. Chem. 285, 12497–12503.

Arnold, S., Curtiss, A., Dean, D.H., Alzate, O., 2001. The role of a proline-induced broken-helix motif in alpha-helix 2 of *Bacillus thuringiensis* delta-endotoxins. FEBS Lett. 490, 70–74.

Aronson, A., 2000. Incorporation of protease K into larval insect membrane vesicles does not result in disruption of integrity or function of the pore-forming *Bacillus thuringiensis* delta-endotoxin. Appl. Environ. Microbiol. 66, 4568–4570.
Aronson, A.I., Beckman, W., Dunn, P., 1986. *Bacillus thuringiensis* and related insect pathogens. Microbiol. Rev. 50, 1–24.
Aronson, A.I., Geng, C., Wu, L., 1999. Aggregation of *Bacillus thuringiensis* Cry1A toxins upon binding to target insect larval midgut vesicles. Appl. Environ. Microbiol. 65, 2503–2507.
Aronson, A.I., Han, E.S., McGaughey, W., Johnson, D., 1991. The solubility of inclusion proteins from *Bacillus thuringiensis* is dependent upon protoxin composition and is a factor in toxicity to insects. Appl. Environ. Microbiol. 57, 981–986.
Aronson, A.I., Wu, D., Zhang, C., 1995. Mutagenesis of specificity and toxicity regions of a *Bacillus thuringiensis* protoxin gene. J. Bacteriol. 177, 4059–4065.
Atsumi, S., Miyamoto, K., Yamamoto, K., Narukawa, J., Kawai, S., Sezutsu, H., Kobayashi, I., Uchino, K., Tamura, T., Mita, K., Kadono-Okuda, K., Wada, S., Kanda, K., Goldsmith, M.R., Noda, H., 2012. Single amino acid mutation in an ATP-binding cassette transporter gene causes resistance to Bt toxin Cry1Ab in the silkworm, *Bombyx mori*. Proc. Natl. Acad. Sci. U.S.A. 109, E1591–E1598.
Atsumi, S., Mizuno, E., Hara, H., Nakanishi, K., Kitami, M., Miura, N., Tabunoki, H., Watanabe, A., Sato, R., 2005. Location of the *Bombyx mori* aminopeptidase N type 1 binding site on Bacillus thuringiensis Cry1Aa toxin. Appl. Environ. Microbiol. 71, 3966–3977.
Banks, D.J., Jurat-Fuentes, J.L., Dean, D.H., Adang, M.J., 2001. *Bacillus thuringiensis* Cry1Ac and Cry1Fa delta-endotoxin binding to a novel 110 kDa aminopeptidase in *Heliothis virescens* is not N-acetylgalactosamine mediated. Insect Biochem. Mol. Biol. 31, 909–918.
Barloy, F., Delecluse, A., Nicolas, L., Lecadet, M.M., 1996. Cloning and expression of the first anaerobic toxin gene from *Clostridium bifermentans* subsp. *malaysia*, encoding a new mosquitocidal protein with homologies to *Bacillus thuringiensis* delta-endotoxins. J. Bacteriol. 178, 3099–3105.
Baxter, S.W., Badenes-Perez, F.R., Morrison, A., Vogel, H., Crickmore, N., Kain, W., Wang, P., Heckel, D.G., Jiggins, C.D., 2011. Parallel evolution of *Bacillus thuringiensis* toxin resistance in Lepidoptera. Genetics 189, 675–679.
Bayyareddy, K., Andacht, T.M., Abdullah, M.A., Adang, M.J., 2009. Proteomic identification of *Bacillus thuringiensis* subsp. *israelensis* toxin Cry4Ba binding proteins in midgut membranes from *Aedes* (*Stegomyia*) *aegypti* Linnaeus (Diptera, Culicidae) larvae. Insect Biochem. Mol. Biol. 39, 279–286.
Beegle, C.C., Yamamoto, T., 1992. History of *Bacillus thuringiensis* Berliner research and development. Can. Entomol. 124, 587–616.
Ben-Dov, E., 2014. *Bacillus thuringiensis* subsp. *israelensis* and its dipteran-specific toxins. Toxins (Basel) 6, 1222–1243.
Berne, S., Lah, L., Sepcic, K., 2009. Aegerolysins: structure, function, and putative biological role. Protein Sci. 18, 694–706.
Berry, C., 2012. The bacterium, *Lysinibacillus sphaericus*, as an insect pathogen. J. Invertebr. Pathol. 109, 1–10.
Bietlot, H.P., Carey, P.R., Choma, C., Kaplan, H., Lessard, T., Pozsgay, M., 1989. Facile preparation and characterization of the toxin from *Bacillus thuringiensis* var. *kurstaki*. Biochem. J. 260, 87–91.
Bietlot, H.P., Schernthaner, J.P., Milne, R.E., Clairmont, F.R., Bhella, R.S., Kaplan, H., 1993. Evidence that the CryIA crystal protein from *Bacillus thuringiensis* is associated with DNA. J. Biol. Chem. 268, 8240–8245.
Bokori-Brown, M., Savva, C.G., Fernandes da Costa, S.P., Naylor, C.E., Basak, A.K., Titball, R.W., 2011. Molecular basis of toxicity of *Clostridium perfringens* epsilon toxin. FEBS J. 278, 4589–4601.

Bonin, A., Paris, M., Tetreau, G., David, J.P., Despres, L., 2009. Candidate genes revealed by a genome scan for mosquito resistance to a bacterial insecticide: sequence and gene expression variations. BMC Genomics 10, 551.

Boonserm, P., Davis, P., Ellar, D.J., Li, J., 2005. Crystal structure of the mosquito-larvicidal toxin Cry4Ba and its biological implications. J. Mol. Biol. 348, 363–382.

Boonserm, P., Mo, M., Angsuthanasombat, C., Lescar, J., 2006. Structure of the functional form of the mosquito larvicidal Cry4Aa toxin from *Bacillus thuringiensis* at a 2.8-angstrom resolution. J. Bacteriol. 188, 3391–3401.

Bosch, D., Schipper, B., van der Kleij, H., de Maagd, R.A., Stiekema, W.J., 1994. Recombinant *Bacillus thuringiensis* crystal proteins with new properties: possibilities for resistance management. Nat. Biotechnol. 12, 915–918.

Bravo, A., Gómez, I., Conde, J., Muñoz-Garay, C., Sánchez, J., Miranda, R., Zhuang, M., Gill, S.S., Soberón, M., 2004. Oligomerization triggers binding of a *Bacillus thuringiensis* Cry1Ab pore-forming toxin to aminopeptidase N receptor leading to insertion into membrane microdomains. Biochim. Biophys. Acta 1667, 38–46.

Bravo, A., Hendrickx, K., Jansens, S., Peferoen, M., 1992. Immunocytochemical analysis of specific binding of *Bacillus thuringiensis* insecticidal crystal proteins to lepidopteran and coleopteran midgut membranes. J. Invertebr. Pathol. 60, 247–253.

Bravo, A., Likitvivatanavong, S., Gill, S.S., Soberón, M., 2011. *Bacillus thuringiensis*: a story of a successful bioinsecticide. Insect Biochem. Mol. Biol. 41, 423–431.

Bravo, A., Sanchez, J., Kouskoura, T., Crickmore, N., 2002. N-terminal activation is an essential early step in the mechanism of action of the *Bacillus thuringiensis* Cry1Ac insecticidal toxin. J. Biol. Chem. 277, 23985–23987.

Brunet, J.F., Vachon, V., Juteau, M., Van Rie, J., Larouche, G., Vincent, C., Schwartz, J.L., Laprade, R., 2010a. Pore-forming properties of the *Bacillus thuringiensis* toxin Cry9Ca in *Manduca sexta* brush border membrane vesicles. Biochim. Biophys. Acta Biomembr. 1798, 1111–1118.

Brunet, J.F., Vachon, V., Marsolais, M., Van Rie, J., Schwartz, J.L., Laprade, R., 2010b. Midgut juice components affect pore formation by the *Bacillus thuringiensis* insecticidal toxin Cry9Ca. J. Invertebr. Pathol. 104, 203–208.

Bulla Jr., L.A., Bechtel, D.B., Kramer, K.J., Shethna, Y.I., Aronson, A.I., Fitz-James, P.C., 1980. Ultrastructure, physiology, and biochemistry of *Bacillus thuringiensis*. Crit. Rev. Microbiol. 8, 147–204.

Burges, H.D., 2001. *Bacillus thuringiensis* in pest control. Pestic. Outlook 12, 90–98.

Burton, S.L., Ellar, D.J., Li, J., Derbyshire, D.J., 1999. N-acetylgalactosamine on the putative insect receptor aminopeptidase N is recognised by a site on the domain III lectin-like fold of a *Bacillus thuringiensis* insecticidal toxin. J. Mol. Biol. 287, 1011–1022.

Butko, P., 2003. Cytolytic toxin Cyt1A and its mechanism of membrane damage: data and hypotheses. Appl. Environ. Microbiol. 69, 2415–2422.

Butko, P., Huang, F., PusztaiCarey, M., Surewicz, W.K., 1996. Membrane permeabilization induced by cytolytic delta-endotoxin CytA from *Bacillus thuringiensis* var *israelensis*. Biochemistry 35, 11355–11360.

Cantón, P.E., López-Díaz, J.A., Gill, S.S., Bravo, A., Soberón, M., 2014. Membrane binding and oligomer membrane insertion are necessary but insufficient for *Bacillus thuringiensis* Cyt1Aa toxicity. Peptides 53, 286–291.

Carmona, D., Rodríguez-Almazán, C., Muñoz-Garay, C., Portugal, L., Pérez, C., de Maagd, R.A., Bakker, P., Soberón, M., Bravo, A., 2011. Dominant negative phenotype of *Bacillus thuringiensis* Cry1Ab, Cry11Aa and Cry4Ba mutants suggest hetero-oligomer formation among different Cry toxins. PLoS One 6.

Carroll, J., Convents, D., Van Damme, J., Boets, A., Van Rie, J., Ellar, D.J., 1997. Intramolecular proteolytic cleavage of *Bacillus thuringiensis* Cry3A delta-endotoxin may facilitate its coleopteran toxicity. J. Invertebr. Pathol. 70, 41–49.

Chen, J., Aimanova, K.G., Fernandez, L.E., Bravo, A., Soberón, M., Gill, S.S., 2009a. *Aedes aegypti* cadherin serves as a putative receptor of the Cry11Aa toxin from *Bacillus thuringiensis* subsp. *israelensis*. Biochem. J. 424, 191–200.
Chen, J., Aimanova, K.G., Pan, S., Gill, S.S., 2009b. Identification and characterization of *Aedes aegypti* aminopeptidase N as a putative receptor of *Bacillus thuringiensis* Cry11A toxin. Insect Biochem. Mol. Biol. 39, 688–696.
Chen, J., Brown, M.R., Hua, G., Adang, M.J., 2005. Comparison of the localization of *Bacillus thuringiensis* Cry1A delta-endotoxins and their binding proteins in larval midgut of tobacco hornworm, *Manduca sexta*. Cell Tissue Res. 321, 123–129.
Chen, J., Likitvivatanavong, S., Aimanova, K.G., Gill, S.S., 2013. A 104 kDa *Aedes aegypti* aminopeptidase N is a putative receptor for the Cry11Aa toxin from *Bacillus thuringiensis* subsp. *israelensis*. Insect Biochem. Mol. Biol. 43, 1201–1208.
Chen, X.J., Curtiss, A., Alcantara, E., Dean, D.H., 1995. Mutations in domain I of *Bacillus thuringiensis* d-endotoxin CryIAb reduce the irreversible binding of toxin to *Manduca sexta* brush border membrane vesicles. J. Biol. Chem. 270, 6412–6419.
Chen, X.J., Lee, M.K., Dean, D.H., 1993. Site-directed mutations in a highly conserved region of *Bacillus thuringiensis* δ-endotoxin affect inhibition of short circuit current across *Bombyx mori* midguts. Proc. Natl. Acad. Sci. U.S.A. 90, 9041–9045.
Chilcott, C.N., Ellar, D.J., 1988. Comparative toxicity of *Bacillus thuringiensis* var. *israelensis* crystal proteins *in-vivo* and *in-vitro*. J. Gen. Microbiol. 134, 2551–2558.
Chougule, N.P., Li, H., Liu, S., Linz, L.B., Narva, K.E., Meade, T., Bonning, B.C., 2013. Retargeting of the *Bacillus thuringiensis* toxin Cyt2Aa against hemipteran insect pests. Proc. Natl. Acad. Sci. U.S.A. 110, 8465–8470.
Chow, E., Singh, G.J., Gill, S.S., 1989. Binding and aggregation of the 25-kilodalton toxin of *Bacillus thuringiensis* subsp. *israelensis* to cell membranes and alteration by monoclonal antibodies and amino acid modifiers. Appl. Environ. Microbiol. 55, 2779–2788.
Clairmont, F.R., Milne, R.E., Pham, V.T., Carrière, M.B., Kaplan, H., 1998. Role of DNA in the activation of the Cry1A insecticidal crystal protein from *Bacillus thuringiensis*. J. Biol. Chem. 273, 9292–9296.
Cohen, S., Albeck, S., Ben-Dov, E., Cahan, R., Firer, M., Zaritsky, A., Dym, O., 2011. Cyt1Aa toxin: crystal structure reveals implications for its membrane-perforating function. J. Mol. Biol. 413, 804–814.
Cohen, S., Dym, O., Albeck, S., Ben-Dov, E., Cahan, R., Firer, M., Zaritsky, A., 2008. High-resolution crystal structure of activated Cyt2Ba monomer from *Bacillus thuringiensis* subsp. *israelensis*. J. Mol. Biol. 380, 820–827.
Crava, C.M., Bel, Y., Lee, S.F., Manachini, B., Heckel, D.G., Escriche, B., 2010. Study of the aminopeptidase N gene family in the lepidopterans *Ostrinia nubilalis* (Hubner) and *Bombyx mori* (L.): sequences, mapping and expression. Insect Biochem. Mol. Biol. 40, 506–515.
Crickmore, N., Baum, J., Bravo, A., Lereclus, D., Narva, K., Sampson, K., Schnepf, E., Sun, M., Zeigler, D.R., 2014. *Bacillus thuringiensis* toxin nomenclature. http://www.lifesci.sussex.ac.uk/home/Neil_Crickmore/Bt/.
Crickmore, N., Wheeler, V.C., Ellar, D.J., 1994. Use of an operon fusion to induce expression and crystallisation of a *Bacillus thuringiensis* delta-endotoxin encoded by a cryptic gene. Mol. Gen. Genet. 242, 365–368.
Crickmore, N., Zeigler, D.R., Feitelson, J., Schnepf, E., Van Rie, J., Lereclus, D., Baum, J., Dean, D.H., 1998. Revision of the nomenclature for the *Bacillus thuringiensis* pesticidal crystal proteins. Microbiol. Mol. Biol. Rev. 62, 807–813.
Davidson, E.W., Yamamoto, T., 1984. Isolation and assay of the toxic component from the crystals of *Bacillus thuringiensis* var. *israelensis*. Curr. Microbiol. 11, 171–174.
de Maagd, R.A., Bakker, P., Staykov, N., Dukiandjiev, S., Stiekema, W., Bosch, D., 1999a. Identification of *Bacillus thuringiensis* delta-endotoxin Cry1C domain III amino acid residues involved in insect specificity. Appl. Environ. Microbiol. 65, 4369–4374.

de Maagd, R.A., Bakker, P.L., Masson, L., Adang, M.J., Sangadala, S., Stiekema, W., Bosch, D., 1999b. Domain III of the *Bacillus thuringiensis* delta-endotoxin Cry1Ac is involved in binding to *Manduca sexta* brush border membranes and to its purified aminopeptidase N. Mol. Microbiol. 31, 463–471.

de Maagd, R.A., Bravo, A., Berry, C., Crickmore, N., Schnepf, H.E., 2003. Structure, diversity, and evolution of protein toxins from spore-forming entomopathogenic bacteria. Annu. Rev. Genet. 37, 409–433.

de Maagd, R.A., Bravo, A., Crickmore, N., 2001. How *Bacillus thuringiensis* has evolved specific toxins to colonize the insect world. Trends Genet. 17, 193–199.

Dean, D.H., Rajamohan, F., Lee, M.K., Wu, S.J., Chen, X.J., Alcantara, E., Hussain, S.R., 1996. Probing the mechanism of action of *Bacillus thuringiensis* insecticidal proteins by site-directed mutagenesis—a mini review. Gene 179, 111–117.

Dechklar, M., Tiewsiri, K., Angsuthanasombat, C., Pootanakit, K., 2011. Functional expression in insect cells of glycosylphosphatidylinositol-linked alkaline phosphatase from *Aedes aegypti* larval midgut: a *Bacillus thuringiensis* Cry4Ba toxin receptor. Insect Biochem. Mol. Biol. 41, 159–166.

Dennis, R.D., Wiegandt, H., Haustein, D., Knowles, B., Ellar, D.J., 1986. Thin layer chromatography overlay technique in the analysis of the binding of the solubilized protoxin of *Bacillus thuringiensis* var. *kurstaki* to an insect glycosphingolipid of known structure. Biomed. Chromatogr. 1, 31–37.

Dermauw, W., Van Leeuwen, T., 2014. The ABC gene family in arthropods: comparative genomics and role in insecticide transport and resistance. Insect Biochem. Mol. Biol. 45, 89–110.

Diaz-Mendoza, M., Bideshi, D.K., Federici, B.A., 2012. A 54-kilodalton protein encoded by pBtoxis is required for parasporal body structural integrity in *Bacillus thuringiensis* subsp. *israelensis*. J. Bacteriol. 194, 1562–1571.

Donovan, W.P., Donovan, J.C., Slaney, A.C., 2000. *Bacillus thuringiensis* cryET33 and cryET34 compositions and uses thereof US Patent 6,063,756.

Du, C., Martin, P.A.W., Nickerson, K.W., 1994. Comparison of disulfide contents and solubility at alkaline pH of insecticidal and noninsecticidal *Bacillus thuringiensis* protein crystals. Appl. Environ. Microbiol. 60, 3847–3853.

Du, C., Nickerson, K.W., 1996. *Bacillus thuringiensis* HD-73 spores have surface-localized Cry1Ac toxin: physiological and pathogenic consequences. Appl. Environ. Microbiol. 62, 3722–3726.

Du, J., Knowles, B.H., Li, J., Ellar, D.J., 1999. Biochemical characterization of *Bacillus thuringiensis* cytolytic toxins in association with a phospholipid bilayer. Biochem. J. 338, 185–193.

Ellis, R.T., Stockhoff, B.A., Stamp, L., Schnepf, H.E., Schwab, G.E., Knuth, M., Russell, J., Cardineau, G.A., Narva, K.E., 2002. Novel *Bacillus thuringiensis* binary insecticidal crystal proteins active on western corn rootworm, *Diabrotica virgifera virgifera* LeConte. Appl. Environ. Microbiol. 68, 1137–1145.

English, L., Readdy, T.L., 1989. Delta endotoxin inhibits a phosphatase in midgut epithelial membranes of *Heliothis virescens*. Insect Biochem. 19, 145–152.

Fabrick, J., Oppert, C., Lorenzen, M.D., Morris, K., Oppert, B., Jurat-Fuentes, J.L., 2009. A novel *Tenebrio molitor* cadherin is a functional receptor for *Bacillus thuringiensis* Cry3Aa toxin. J. Biol. Chem. 284, 18401–18410.

Fabrick, J.A., Tabashnik, B.E., 2007. Binding of *Bacillus thuringiensis* toxin Cry1Ac to multiple sites of cadherin in pink bollworm. Insect Biochem. Mol. Biol. 37, 97–106.

Federici, B.A., Bauer, L.S., 1998. Cyt1Aa protein of *Bacillus thuringiensis* is toxic to the cottonwood leaf beetle, *Chrysomela scripta*, and suppresses high levels of resistance to Cry3Aa. Appl. Environ. Microbiol. 64, 4368–4371.

Fernandez-Luna, M.T., Lanz-Mendoza, H., Gill, S.S., Bravo, A., Soberon, M., Miranda-Rios, J., 2010. An alpha-amylase is a novel receptor for *Bacillus thuringiensis* ssp. *israelensis* Cry4Ba and Cry11Aa toxins in the malaria vector mosquito *Anopheles albimanus* (Diptera: Culicidae). Environ. Microbiol. 12, 746–757.

Fernandez, L.E., Aimanova, K.G., Gill, S.S., Bravo, A., Soberón, M., 2006. A GPI-anchored alkaline phosphatase is a functional midgut receptor of Cry11Aa toxin in *Aedes aegypti* larvae. Biochem. J. 394, 77–84.

Ferré, J., Van Rie, J., 2002. Biochemistry and genetics of insect resistance to *Bacillus thuringiensis*. Annu. Rev. Entomol. 47, 501–533.

Flannagan, R.D., Yu, C.G., Mathis, J.P., Meyer, T.E., Shi, X., Siqueira, H.A., Siegfried, B.-D., 2005. Identification, cloning and expression of a Cry1Ab cadherin receptor from European corn borer, *Ostrinia nubilalis* (Hubner) (Lepidoptera: Crambidae). Insect Biochem. Mol. Biol. 35, 33–40.

Flores-Escobar, B., Rodriguez-Magadan, H., Bravo, A., Soberon, M., Gomez, I., 2013. Differential role of *Manduca sexta* aminopeptidase-N and alkaline phosphatase in the mode of action of Cry1Aa, Cry1Ab, and Cry1Ac toxins from *Bacillus thuringiensis*. Appl. Environ. Microbiol. 79, 4543–4550.

Florez, A.M., Osorio, C., Alzate, O., 2012. Protein engineering of *Bacillus thuringiensis* δ-endotoxins. In: Sansinenea, E. (Ed.), *Bacillus thuringiensis* Biotechnology. Springer Netherlands, Netherlands, pp. 93–113.

Fortier, M., Vachon, V., Kirouac, M., Schwartz, J.L., Laprade, R., 2005. Differential effects of ionic strength, divalent cations and pH on the pore-forming activity of *Bacillus thuringiensis* insecticidal toxins. J. Membr. Biol. 208, 77–87.

Fortier, M., Vachon, V., Marceau, L., Schwartz, J.L., Laprade, R., 2007. Kinetics of pore formation by the *Bacillus thuringiensis* toxin Cry1Ac. Biochim. Biophys. Acta Biomembr. 1768, 1291–1298.

Fujii, U., Tanaka, S., Stsuki, M., Hoshino, Y., Morimoto, C., Kotani, T., Harashima, Y., Endo, H., Yoshizawa, Y., Sato, R., 2013. Cry1Aa binding to the cadherin receptor does not require conserved amino acid sequences in the domain II loops. Biosci. Rep. 33 (1), e00010. http://dx.doi.org/10.01042/BSR20120113.

Gahan, L.J., Gould, F., Heckel, D.G., 2001. Identification of a gene associated with Bt resistance in *Heliothis virescens*. Science 293, 857–860.

Gahan, L.J., Pauchet, Y., Vogel, H., Heckel, D.G., 2010. An ABC transporter mutation is correlated with insect resistance to *Bacillus thuringiensis* Cry1Ac toxin. PLoS Genet. 6, e1001248.

Galitsky, N., Cody, V., Wojtczak, A., Ghosh, D., Luft, J.R., Pangborn, W., English, L., 2001. Structure of the insecticidal bacterial delta-endotoxin Cry3Bb1 of *Bacillus thuringiensis*. Acta Crystallogr. Sec. D Biol. Crystallogr. 57, 1101–1109.

Gao, Y., Jurat-Fuentes, J.L., Oppert, B., Fabrick, J.A., Liu, C., Gao, J., Lei, Z., 2011. Increased toxicity of *Bacillus thuringiensis* Cry3Aa against *Crioceris quatuordecimpunctata*, *Phaedon brassicae* and *Colaphellus bowringi* by a *Tenebrio molitor* cadherin fragment. Pest Manage. Sci. 67, 1076–1081.

Garczynski, S.F., Adang, M.J., 1995. *Bacillus thuringiensis* CryIA(c) d-endotoxin binding aminopeptidase in the *Manduca sexta* midgut has a glycosyl-phosphatidylinositol anchor. Insect Biochem. Mol. Biol. 25, 409–415.

Garczynski, S.F., Crim, J.W., Adang, M.J., 1991. Identification of putative insect brush border membrane-binding molecules specific to *Bacillus thuringiensis* delta-endotoxin by protein blot analysis. Appl. Environ. Microbiol. 57, 2816–2820.

Gazit, E., La Rocca, P., Sansom, M.S.P., Shai, Y., 1998. The structure and organization within the membrane of the helices composing the pore-forming domain of *Bacillus thuringiensis* d-endotoxin are consistent with an "umbrella-like" structure of the pore. Proc. Natl. Acad. Sci. U.S.A. 95, 12289–12294.

Ge, A.Z., Pfister, R.M., Dean, D.H., 1991. Functional domains of *Bacillus thuringiensis* insecticidal crystal proteins: refinement of *Heliothis virescens* and *Trichoplusia ni* specificity domains on CryIA(c). J. Biol. Chem. 266, 17954–17958.

Gill, M., Ellar, D., 2002. Transgenic *Drosophila* reveals a functional in vivo receptor for the *Bacillus thuringiensis* toxin Cry1Ac1. Insect Mol. Biol. 11, 619–625.

Gill, S., Cowles, E.A., Francis, V., 1995. Identification, isolation, and cloning of a *Bacillus thuringiensis* CryIAc toxin-binding protein from the midgut of the lepidopteran insect *Heliothis virescens*. J. Biol. Chem. 270, 27277–27282.

Gill, S.S., Singh, G.J., Hornung, J.M., 1987. Cell membrane interaction of *Bacillus thuringiensis* subsp. *israelensis* cytolytic toxins. Infect. Immun. 55, 1300–1308.

Girard, F., Vachon, V., Prefontaine, G., Marceau, L., Schwartz, J.L., Masson, L., Laprade, R., 2009. Helix alpha 4 of the *Bacillus thuringiensis* Cry1Aa toxin plays a critical role in the postbinding steps of pore formation. Appl. Environ. Microbiol. 75, 359–365.

Goldberg, L.J., Margalit, J., 1977. A bacterial spore demonstrating rapid larvicidal activity against *Anopheles sergentii, Uranotaenia unguicalata, Culex univitattus, Aedes aegyptii* and *Culex pipiens*. Mosq. News 37, 355–358.

Gómez, I., Dean, D.H., Bravo, A., Soberâon, M., 2003. Molecular basis for *Bacillus thuringiensis* Cry1Ab toxin specificity: two structural determinants in the *Manduca sexta* Bt-R1 receptor interact with loops alpha-8 and 2 in domain II of Cy1Ab toxin. Biochemistry 42, 10482–10489.

Gomez, I., Sanchez, J., Miranda, R., Bravo, A., Soberon, M., 2002. Cadherin-like receptor binding facilitates proteolytic cleavage of helix α-1 in domain I and oligomer pre-pore formation of *Bacillus thuringiensis* Cry1Ab toxin. FEBS Lett. 513, 242–246.

Gómez, I., Sánchez, J., Miranda, R., Bravo, A., Soberón, M., 2002. Cadherin-like receptor binding facilitates proteolytic cleavage of helix alpha-1 in domain I and oligomer prepore formation of *Bacillus thuringiensis* Cry1Ab toxin. FEBS Lett. 513, 242–246.

Gomez, I., Sanchez, J., Munoz-Garay, C., Matus, V., Gill, S.S., Soberon, M., Bravo, A., 2014. *Bacillus thuringiensis* Cry1A toxins are versatile proteins with multiple modes of action: two distinct pre-pores are involved in toxicity. Biochem. J. 459, 383–396.

Griffitts, J.S., Haslam, S.M., Yang, T., Garczynski, S.F., Mulloy, B., Morris, H., Cremer, P.S., Dell, A., Adang, M.J., Aroian, R.V., 2005. Glycolipids as receptors for *Bacillus thuringiensis* crystal toxin. Science 307, 922–925.

Grochulski, P., Masson, L., Borisova, S., Pusztai-Carey, M., Schwartz, J.L., Brousseau, R., Cygler, M., 1995. *Bacillus thuringiensis* CryIA(a) insecticidal toxin: crystal structure and channel formation. J. Mol. Biol. 54, 447–464.

Groulx, N., Juteau, M., Blunck, R., 2010. Rapid topology probing using fluorescence spectroscopy in planar lipid bilayer: the pore-forming mechanism of the toxin Cry1Aa of *Bacillus thuringiensis*. J. Gen. Physiol. 136, 497–513.

Groulx, N., McGuire, H., Laprade, R., Schwartz, J.L., Blunck, R., 2011. Single molecule fluorescence study of the *Bacillus thuringiensis* toxin Cry1Aa reveals tetramerization. J. Biol. Chem. 286, 42274–42282.

Guan, P., Ai, P., Dai, X., Zhang, J., Xu, L., Zhu, J., Li, Q., Deng, Q., Li, S., Wang, S., Liu, H., Wang, L., Li, P., Zheng, A., 2012. Complete genome sequence of *Bacillus thuringiensis* serovar *Sichuansis* strain MC28. J. Bacteriol. 194, 6975.

Guerchicoff, A., Ugalde, R.A., Rubinstein, C.P., 1997. Identification and characterization of a previously undescribed cyt gene in *Bacillus thuringiensis* subsp. *israelensis*. Appl. Environ. Microbiol. 63, 2716–2721.

Guo, S., Li, J., Liu, Y., Song, F., Zhang, J., 2011. The role of DNA binding with the Cry8Ea1 toxin of *Bacillus thuringiensis*. FEMS Microbiol. Lett. 317, 203–210.

Guo, S., Ye, S., Liu, Y., Wei, L., Xue, J., Wu, H., Song, F., Zhang, J., Wu, X., Huang, D., Rao, Z., 2009a. Crystal structure of *Bacillus thuringiensis* Cry8Ea1: an insecticidal toxin toxic to underground pests, the larvae of *Holotrichia parallela*. J. Struct. Biol. 168, 259–266.

Guo, S., Zhang, Y., Song, F., Zhang, J., Huang, D., 2009b. Protease-resistant core form of *Bacillus thuringiensis* Cry1Ie: monomeric and oligomeric forms in solution. Biotechnol. Lett. 31, 1769–1774.

Hara, H., Atsumi, S., Yaoi, K., Nakanishi, K., Higurashi, S., Miura, N., Tabunoki, H., Sato, R., 2003. A cadherin-like protein functions as a receptor for *Bacillus thuringiensis* Cry1Aa and Cry1Ac toxins on midgut epithelial cells of *Bombyx mori* larvae. FEBS Lett. 538, 29–34.

Hayakawa, T., Shitomi, Y., Miyamoto, K., Hori, H., 2004. GalNAc pretreatment inhibits trapping of *Bacillus thuringiensis* Cry1Ac on the peritrophic membrane of *Bombyx mori*. FEBS Lett. 576, 331–335.

Hazes, B., 1996. The (QxW)3 domain: a flexible lectin scaffold. Protein Sci. 5, 1490–1501.

Heckel, D.G., 2012. Learning the ABCs of Bt: ABC transporters and insect resistance to *Bacillus thuringiensis* provide clues to a crucial step in toxin mode of action. Pestic. Biochem. Physiol. 104, 103–110.

Heckel, D.G., Gahan, L.J., Baxter, S.W., Zhao, J.Z., Shelton, A.M., Gould, F., Tabashnik, B.E., 2007. The diversity of Bt resistance genes in species of Lepidoptera. J. Invertebr. Pathol. 95, 192–197.

Hernández, C.S., Ferré, J., 2005. Common receptor for *Bacillus thuringiensis* toxins Cry1Ac, Cry1Fa, and Cry1Ja in *Helicoverpa armigera*, *Helicoverpa zea*, and *Spodoptera exigua*. Appl. Environ. Microbiol. 71, 5627–5629.

Herrero, S., Gechev, T., Bakker, P.L., Moar, W.J., de Maagd, R.A., 2005. *Bacillus thuringiensis* Cry1Ca-resistant *Spodoptera exigua* lacks expression of one of four aminopeptidase N genes. BMC Genomics 6, 96.

Herrero, S., Gonzalez-Cabrera, J., Ferre, J., Bakker, P.L., de Maagd, R.A., 2004. Mutations in the *Bacillus thuringiensis* Cry1Ca toxin demonstrate the role of domains II and III in specificity towards *Spodoptera exigua* larvae. Biochem. J. 384, 507–513.

Hodgman, T.C., Ellar, D.J., 1990. Models for the structure and function of the *Bacillus thuringiensis* d-endotoxins determined by compilation analysis. J. DNA Seq. Map. 1, 97–106.

Höfte, H., Whiteley, H.R., 1989. Insecticidal crystal proteins of *Bacillus thuringiensis*. Microbiol. Rev. 53, 242–255.

Hossain, D.M., Shitomi, Y., Moriyama, K., Higuchi, M., Hayakawa, T., Mitsui, T., Sato, R., Hori, H., 2004. Characterization of a novel plasma membrane protein, expressed in the midgut epithelia of *Bombyx mori*, that binds to Cry1A toxins. Appl. Environ. Microbiol. 70, 4604–4612.

Hua, G., Jurat-Fuentes, J.L., Adang, M.J., 2004. Fluorescent-based assays establish *Manduca sexta* Bt-R_{1a} cadherin as a receptor for multiple *Bacillus thuringiensis* Cry1A toxins in *Drosophila* S2 cells. Insect Biochem. Mol. Biol. 34, 193–202.

Hua, G., Park, Y., Adang, M.J., 2014. Cadherin AdCad1 in *Alphitobius diaperinus* larvae is a receptor of Cry3Bb toxin from *Bacillus thuringiensis*. Insect Biochem. Mol. Biol. 45, 11–17.

Hua, G., Zhang, Q., Zhang, R., Abdullah, A.M., Linser, P.J., Adang, M.J., 2013. AgCad2 cadherin in *Anopheles gambiae* larvae is a putative receptor of Cry11Ba toxin of *Bacillus thuringiensis* subsp. *jegathesan*. Insect Biochem. Mol. Biol. 43, 153–161.

Hua, G., Zhang, R., Abdullah, M.A., Adang, M.J., 2008. *Anopheles gambiae* cadherin AgCad1 binds the Cry4Ba toxin of *Bacillus thuringiensis israelensis* and a fragment of AgCad1 synergizes toxicity. Biochemistry 47, 5101–5110.

Hua, G., Zhang, R., Bayyareddy, K., Adang, M.J., 2009. *Anopheles gambiae* alkaline phosphatase is a functional receptor of *Bacillus thuringiensis jegathesan* Cry11Ba toxin. Biochemistry 48, 9785–9793.

Huang, S., Ding, X., Sun, Y., Yang, Q., Xiao, X., Cao, Z., Xia, L., 2012. Proteomic analysis of *Bacillus thuringiensis* at different growth phases by using an automated online two-dimensional liquid chromatography-tandem mass spectrometry strategy. Appl. Environ. Microbiol. 78, 5270–5279.

Hughes, A.L., 2014. Evolutionary diversification of aminopeptidase N in Lepidoptera by conserved clade-specific amino acid residues. Mol. Phylogenet. Evol. 76C, 127–133.

Ibargutxi, M.A., Muñoz, D., Escudero, I.R.d., Caballero, P., 2008. Interactions between Cry1Ac, Cry2Ab, and Cry1Fa *Bacillus thuringiensis* toxins in the cotton pests *Helicoverpa armigera* (Hübner) and *Earias insulana* (Boisduval). Biol. Control 47, 89–96.

James, C., 2009. Global Status of Commercialized Biotech/GM Crops, ISAAA Brief. ISAAA, Ithaca, NY.

Jaquet, F., Hutter, R., Luthy, P., 1987. Specificity of *Bacillus thuringiensis* δ-endotoxin. Appl. Environ. Microbiol. 53, 500–504.

Jenkins, J.L., Lee, M.K., Sandagala, S., Adang, M.J., Dean, D.H., 1999. Binding of *Bacillus thuringiensis* Cry1Ac toxin to *Manduca sexta* aminopeptidase-N receptor is not directly related to toxicity. FEBS Lett. 462, 373–376.

Jenkins, J.L., Lee, M.K., Valaitis, A.P., Curtiss, A., Dean, D.H., 2000. Bivalent sequential binding model of a *Bacillus thuringiensis* toxin to gypsy moth aminopeptidase N receptor. J. Biol. Chem. 275, 14423–14431.

Jiménez-Juárez, N., Muñoz-Garay, C., Gomez, I., Saab-Rincon, G., Damian-Almazo, J.Y., Gill, S.S., Soberón, M., Bravo, A., 2007. Bacillus thuringiensis Cry1Ab mutants affecting oligomer formation are non-toxic to *Manduca sexta* larvae. J. Biol. Chem. 282, 21222–21229.

Jimenez, A.I., Reyes, E.Z., Cancino-Rodezno, A., Bedoya-Perez, L.P., Caballero-Flores, G.G., Muriel-Millan, L.F., Likitvivatanavong, S., Gill, S.S., Bravo, A., Soberon, M., 2012. *Aedes aegypti* alkaline phosphatase ALP1 is a functional receptor of *Bacillus thuringiensis* Cry4Ba and Cry11Aa toxins. Insect Biochem. Mol. Biol. 42, 683–689.

Johnston, P.R., Crickmore, N., 2009. Gut bacteria are not required for the insecticidal activity of *Bacillus thuringiensis* toward the tobacco hornworm, *Manduca sexta*. Appl. Environ. Microbiol. 75, 5094–5099.

Jones, G.W., Nielsen-Leroux, C., Yang, Y., Yuan, Z., Dumas, V.F., Monnerat, R.G., Berry, C., 2007. A new Cry toxin with a unique two-component dependency from *Bacillus sphaericus*. FASEB J. 21, 4112–4120.

Jurat-Fuentes, J.L., Adang, M.J., 2001. Importance of Cry1 delta-endotoxin domain II loops for binding specificity in *Heliothis virescens* (L.). Appl. Environ. Microbiol. 67, 323–329.

Jurat-Fuentes, J.L., Adang, M.J., 2004. Characterization of a Cry1Ac-receptor alkaline phosphatase in susceptible and resistant *Heliothis virescens* larvae. Eur. J. Biochem. 271, 3127–3135.

Jurat-Fuentes, J.L., Adang, M.J., 2006. The *Heliothis virescens* cadherin protein expressed in *Drosophila* S2 cells functions as a receptor for *Bacillus thuringiensis* Cry1A but not Cry1Fa toxins. Biochemistry 45, 9688–9695.

Jurat-Fuentes, J.L., Gahan, L.J., Gould, F.L., Heckel, D.G., Adang, M.J., 2004. The HevCaLP protein mediates binding specificity of the Cry1A class of *Bacillus thuringiensis* toxins in *Heliothis virescens*. Biochemistry 43, 14299–14305.

Jurat-Fuentes, J.L., Jackson, T.A., 2012. Bacterial entomopathogens. In: Vega, F.E., Kaya, H.K. (Eds.), Insect Pathology, second ed. Elsevier, San Diego, pp. 265–349.

Jurat-Fuentes, J.L., Karumbaiah, L., Jakka, S.R., Ning, C., Liu, C., Wu, K., Jackson, J., Gould, F., Blanco, C., Portilla, M., Perera, O., Adang, M., 2011. Reduced levels of membrane-bound alkaline phosphatase are common to lepidopteran strains resistant to Cry toxins from *Bacillus thuringiensis*. PLoS One 6, e17606.

Karlova, R., Weemen-Hendriks, M., Naimov, S., Ceron, J., Dukiandjiev, S., de Maagd, R.A., 2005. *Bacillus thuringiensis* delta-endotoxin Cry1Ac domain III enhances activity against *Heliothis virescens* in some, but not all Cry1-Cry1Ac hybrids. J. Invertebr. Pathol. 88, 169–172.

Katayama, H., Kusaka, Y., Yokota, H., Akao, T., Kojima, M., Nakamura, O., Mekada, E., Mizuki, E., 2007. Parasporin-1, a novel cytotoxic protein from *Bacillus thuringiensis*, induces Ca2+ influx and a sustained elevation of the cytoplasmic Ca2+ concentration in toxin-sensitive cells. J. Biol. Chem. 282, 7742–7752.

Kauer, R., Sharma, A., Gupta, D., Kalita, M., Bhatnagar, R.K., 2014. *Bacillus thuringiensis toxin*, Cry1C interacts with 128HLHFHLP134 region of aminopeptidase N of agricultural pest, *Spodoptera litura*. Process Biochem. 49, 688–696.

Keeton, T.P., Bulla Jr., L.A., 1997. Ligand specificity and affinity of BT-R1, the *Bacillus thuringiensis* Cry1A toxin receptor from *Manduca sexta*, expressed in mammalian and insect cell cultures. Appl. Environ. Microbiol. 63, 3419–3425.

Kim, H.S., Saitoh, H., Yamashita, S., Akao, T., Park, Y.S., Maeda, M., Tanaka, R., Mizuki, E., Ohba, M., 2003. Cloning and characterization of two novel crystal protein genes from a *Bacillus thuringiensis* serovar *dakota* strain. Curr. Microbiol. 46, 33–38.

Kitada, S., Abe, Y., Maeda, T., Shimada, H., 2009. Parasporin-2 requires GPI-anchored proteins for the efficient cytocidal action to human hepatoma cells. Toxicology 264, 80–88.

Knight, P.J., Knowles, B.H., Ellar, D.J., 1995. Molecular cloning of an insect aminopeptidase N that serves as a receptor for *Bacillus thuringiensis* CryIA(c) toxin. J. Biol. Chem. 270, 17765–17770.

Knight, P.J.K., Knowles, B.H., Ellar, D.J., 1994. The receptor for *Bacillus thuringiensis* CryIA(c) delta-endotoxin in the brush border membrane is aminopepidase N. Mol. Microbiol. 11, 429–436.

Knowles, B.H., 1994. Mechanism of action of *Bacillus thuringiensis* insecticidal δ-endotoxins. Adv. Insect Physiol. 24, 275–308.

Knowles, B.H., Blatt, M.R., Tester, M., Horsnell, J.M., Carroll, J., Menestrina, G., Ellar, D.J., 1989. A cytolytic delta-endotoxin from *Bacillus thuringiensis* var. *israelensis* forms cation-selective channels in planar lipid bilayers. FEBS Lett. 244, 259–262.

Koni, P.A., Ellar, D.J., 1994. Biochemical characterization of *Bacillus thuringiensis* cytolytic delta-endotoxins. Microbiology 140, 1869–1880.

Kramer, K.J., Muthukrishnan, S., 1997. Insect chitinases: molecular biology and potential use as biopesticides. Insect Biochem. Mol. Biol. 27, 887–900.

Krishnamoorthy, M., Jurat-Fuentes, J.L., McNall, R.J., Andacht, T., Adang, M.J., 2007. Identification of novel Cry1Ac binding proteins in midgut membranes from *Heliothis virescens* using proteomic analyses. Insect Biochem. Mol. Biol. 37, 189–201.

Krishnan, V., 2013. Investigation of Parasporins, the Cytotoxic Proteins from the Bacterium *Bacillus thuringiensis*. University of Sussex, United Kingdom.

Kumar, A.S., Aronson, A.I., 1999. Analysis of mutations in the pore-forming region essential for insecticidal activity of a *Bacillus thuringiensis* delta-endotoxin. J. Bacteriol. 181, 6103–6107.

Lailak, C., Khaokhiew, T., Promptmas, C., Promdonkoy, B., Pootanakit, K., Angsuthanasombat, C., 2013. *Bacillus thuringiensis* Cry4Ba toxin employs two receptor-binding loops for synergistic interactions with Cyt2Aa2. Biochem. Biophys. Res. Commun. 435, 216–221.

Lee, K.Y., Kang, E.Y., Park, S., Ahn, S.K., Yoo, K.H., Kim, J.Y., Lee, H.H., 2006. Mass spectrometric sequencing of endotoxin proteins of *Bacillus thuringiensis* ssp. *konkukian* extracted from polyacrylamide gels. Proteomics 6, 1512–1517.

Lee, M.K., Jenkins, J.L., You, T.H., Curtiss, A., Son, J.J., Adang, M.J., Dean, D.H., 2001. Mutations at the arginine residues in alpha8 loop of *Bacillus thuringiensis* delta-endotoxin Cry1Ac affect toxicity and binding to *Manduca sexta* and *Lymantria dispar* aminopeptidase N. FEBS Lett. 497, 108–112.

Lee, M.K., Rajamohan, F., Jenkins, J.L., Curtiss, A.S., Dean, D.H., 2000. Role of two arginine residues in domain II, loop 2 of Cry1Ab and Cry1Ac *Bacillus thuringiensis* delta-endotoxin in toxicity and binding to *Manduca sexta* and *Lymantria dispar* aminopeptidase N. Mol. Microbiol. 38, 289–298.

Lee, M.K., Young, B.A., Dean, D.H., 1995. Domain III exchanges of *Bacillus thuringiensis* Cry1A toxins affect binding to different gypsy moth midgut receptors. Biochem. Biophys. Res. Commun. 216, 306–312.

Lee, X., Johnston, R.A., Rose, D.R., Young, N.M., 1989. Crystallization and preliminary X-ray diffraction studies of the complex of *Maclura pomifera* agglutinin with the disaccharide Gal beta 1-3GalNAc. J. Mol. Biol. 210, 685–686.

Li, H., Olson, M., Lin, G., Hey, T., Tan, S.Y., Narva, K.E., 2013. *Bacillus thuringiensis* Cry34Ab1/Cry35Ab1 interactions with western corn rootworm midgut membrane binding sites. PLoS One 8, e53079.

Li, J., Derbyshire, D.J., Promdonkoy, B., Ellar, D.J., 2001. Structural implications for the transformation of the *Bacillus thuringiensis* delta-endotoxins from water-soluble to membrane-inserted forms. Biochem. Soc. Trans. 29, 571–577.

Li, J., Koni, P.A., Ellar, D.J., 1996. Structure of the mosquitocidal delta-endotoxin CytB from *Bacillus thuringiensis* sp *kyushuensis* and implications for membrane pore formation. J. Mol. Biol. 257, 129–152.

Li, J.D., Carroll, J., Ellar, D.J., 1991. Crystal structure of insecticidal delta-endotoxin from *Bacillus thuringiensis* at 2.5 A resolution. Nature 353 (6347), 815–821.

Liang, Y., Patel, S.S., Dean, D.H., 1995. Irreversible binding kinetics of *Bacillus thuringiensis* CryIA delta-endotoxins to gypsy moth brush border membrane vesicles is directly correlated to toxicity. J. Biol. Chem. 270, 24719–24724.

Likitvivatanavong, S., Chen, J., Bravo, A., Soberon, M., Gill, S.S., 2011. Cadherin, alkaline phosphatase, and aminopeptidase N as receptors of Cry11Ba toxin from *Bacillus thuringiensis* subsp. jegathesan in *Aedes aegypti*. Appl. Environ. Microbiol. 77, 24–31.

López-Diaz, J.A., Cantón, P.E., Gill, S.S., Soberón, M., Bravo, A., 2013. Oligomerization is a key step in Cyt1Aa membrane insertion and toxicity but not necessary to synergize Cry11Aa toxicity in *Aedes aegypti* larvae. Environ. Microbiol. 15, 3030–3039.

Lu, H., Rajamohan, F., Dean, D.H., 1994. Identification of amino acid residues of *Bacillus thuringiensis* δ-endotoxin CryIAa associated with membrane binding and toxicity to *Bombyx mori*. J. Bacteriol. 176, 5554–5559.

Luo, K., Sangadala, S., Masson, L., Mazza, A., Brousseau, R., Adang, M.J., 1997. The *Heliothis virescens* 170-kDa aminopeptidase functions as 'Receptor A' by mediating specific *Bacillus thuringiensis* Cry1A δ-endotoxin binding and pore formation. Insect Biochem. Mol. Biol. 27, 735–743.

Ma, G., Rahman, M.M., Grant, W., Schmidt, O., Asgari, S., 2012. Insect tolerance to the crystal toxins Cry1Ac and Cry2Ab is mediated by the binding of monomeric toxin to lipophorin glycolipids causing oligomerization and sequestration reactions. Dev. Comp. Immunol. 37, 184–192.

Manasherob, R., Itsko, M., Sela-Baranes, N., Ben-Dov, E., Berry, C., Cohen, S., Zaritsky, A., 2006. Cyt1Ca from Bacillus thuringiensis subsp. israelensis: production in *Escherichia coli* and comparison of its biological activities with those of other Cyt-like proteins. Microbiology 152, 2651–2659.

Manceva, S.D., Pusztai-Carey, M., Russo, P.S., Butko, P., 2005. A detergent-like mechanism of action of the cytolytic toxin Cyt1A from *Bacillus thuringiensis* var. *israelensis*. Biochemistry 44, 589–597.

Martins, E.S., Monnerat, R.G., Queiroz, P.R., Dumas, V.F., Braz, S.V., de Souza Aguiar, R.W., Gomes, A.C., Sanchez, J., Bravo, A., Ribeiro, B.M., 2010. Midgut GPI-anchored proteins with alkaline phosphatase activity from the cotton boll weevil

(*Anthonomus grandis*) are putative receptors for the Cry1B protein of *Bacillus thuringiensis*. Insect Biochem. Mol. Biol. 40, 138–145.

Masson, L., Lu, Y.J., Mazza, A., Brousseau, R., Adang, M.J., 1995. The CryIA(c) receptor purified from *Manduca sexta* displays multiple specificities. J. Biol. Chem. 270, 20309–20315.

Masson, L., Mazza, A., Sangadala, S., Adang, M.J., Brousseau, R., 2002a. Polydispersity of *Bacillus thuringiensis* Cry1 toxins in solution and its effect on receptor binding kinetics. Biochim. Biophys. Acta 1594, 266–275.

Masson, L., Schwab, G., Mazza, A., Brousseau, R., Potvin, L., Schwartz, J.L., 2004. A novel *Bacillus thuringiensis* (PS149B1) containing a Cry34Ab1/Cry35Ab1 binary toxin specific for the western corn rootworm *Diabrotica virgifera virgifera* LeConte forms ion channels in lipid membranes. Biochemistry 43, 12349–12357.

Masson, L., Tabashnik, B.E., Liu, Y.B., Schwartz, J.L., 1999. Helix 4 of the *Bacillus thuringiensis* Cry1Aa toxin lines the lumen of the ion channel. J. Biol. Chem. 274, 31996–32000.

Masson, L., Tabashnik, B.E., Mazza, A., Prâefontaine, G., Potvin, L., Brousseau, R., Schwartz, J.L., 2002b. Mutagenic analysis of a conserved region of domain III in the Cry1Ac toxin of *Bacillus thuringiensis*. Appl. Environ. Microbiol. 68, 194–200.

Matsushima-Hibiya, Y., Watanabe, M., Hidari, K.I., Miyamoto, D., Suzuki, Y., Kasama, T., Koyama, K., Sugimura, T., Wakabayashi, K., 2003. Identification of glycosphingolipid receptors for pierisin-1, a guanine-specific ADP-ribosylating toxin from the cabbage butterfly. J. Biol. Chem. 278, 9972–9978.

McNall, R.J., Adang, M.J., 2003. Identification of novel *Bacillus thuringiensis* Cry1Ac binding proteins in *Manduca sexta* midgut through proteomic analysis. Insect Biochem. Mol. Biol. 33, 999–1010.

Mizuki, E., Ohba, M., Akao, T., Yamashita, S., Saitoh, H., Park, Y.S., 1999. Unique activity associated with non-insecticidal *Bacillus thuringiensis* parasporal inclusions: in vitro cell-killing action on human cancer cells. J. Appl. Microbiol. 86, 477–486.

Morin, S., Biggs, R.W., Sisteron, M.S., Shriver, L., Ellers-Kirk, C., Higginson, D., Holley, D., Gahan, L.J., Heckel, D.G., Carriere, Y., Dennehy, T.J., Brown, J.K., Tabashnik, B.E., 2003. Three cadherin alleles associated with resistance to *Bacillus thuringiensis* in pink bollworm. Proc. Natl. Acad. Sci. U.S.A. 100, 5004–5009.

Morse, R.J., Yamamoto, T., Stroud, R.M., 2001. Structure of Cry2Aa suggests an unexpected receptor binding epitope. Structure (Camb.) 9, 409–417.

Nagamatsu, Y., Koike, T., Sasaki, K., Yoshimoto, A., Furukawa, Y., 1999. The cadherin-like protein is essential to specificity determination and cytotoxic action of the *Bacillus thuringiensis* insecticidal CryIAa toxin. FEBS Lett. 460, 385–390.

Nagamatsu, Y., Okamura, S., Saitou, H., Akao, T., Mizuki, E., 2010. Three Cry toxins in two types from *Bacillus thuringiensis* strain M019 preferentially kill human hepatocyte cancer and uterus cervix cancer cells. Biosci. Biotechnol. Biochem. 74, 494–498.

Naimov, S., Boncheva, R., Karlova, R., Dukiandjiev, S., Minkov, I., de Maagd, R.A., 2008. Solubilization, activation, and insecticidal activity of *Bacillus thuringiensis* serovar *thompsoni* HD542 crystal proteins. Appl. Environ. Microbiol. 74, 7145–7151.

Naimov, S., Valkova, R., Dukiandjiev, S., Minkov, I., de Maagd, R.A., 2011. Carboxy-terminal extension effects on crystal formation and insecticidal properties of Cry15Aa. J. Invertebr. Pathol. 108, 56–58.

Nair, M.S., Dean, D.H., 2008. All domains of Cry1A toxins insert into insect brush border membranes. J. Biol. Chem. 283, 26324–26331.

Ning, C., Wu, K., Liu, C., Gao, Y., Jurat-Fuentes, J.L., Gao, X., 2010. Characterization of a Cry1Ac toxin-binding alkaline phosphatase in the midgut from *Helicoverpa armigera* (Hubner) larvae. J. Insect Physiol. 56, 666–672.

Obata, F., Kitami, M., Inoue, Y., Atsumi, S., Yoshizawa, Y., Sato, R., 2009. Analysis of the region for receptor binding and triggering of oligomerization on *Bacillus thuringiensis* Cry1Aa toxin. FEBS J. 276, 5949–5959.

Ochoa-Campuzano, C., Martínez-Ramírez, A.C., Contreras, E., Rausell, C., Real, M.D., 2013. Prohibitin, an essential protein for Colorado potato beetle larval viability, is relevant to *Bacillus thuringiensis* Cry3Aa toxicity. Pestic. Biochem. Physiol. 107, 299–308.

Ochoa-Campuzano, C., Real, M.D., Martínez-Ramírez, A.C., Bravo, A., Rausell, C., 2007. An ADAM metalloprotease is a Cry3Aa *Bacillus thuringiensis* toxin receptor. Biochem. Biophys. Res. Commun. 362, 437–442.

Okumura, S., Saitoh, H., Ishikawa, T., Inouye, K., Mizuki, E., 2011. Mode of action of parasporin-4, a cytocidal protein from *Bacillus thuringiensis*. Biochim. Biophys. Acta 1808, 1476–1482.

Oppert, B., 1999. Protease interactions with *Bacillus thuringiensis* insecticidal toxins. Arch. Insect Biochem. Physiol. 42, 1–12.

Ounjai, P., Unger, V.M., Sigworth, F.J., Angsuthanasombat, C., 2007. Two conformational states of the membrane-associated *Bacillus thuringiensis* Cry4Ba δ-endotoxin complex revealed by electron crystallography: implications for toxin-pore formation. Biochem. Biophys. Res. Commun. 361, 890–895.

Pacheco, S., Gómez, I., Arenas, I., Saab-Rincon, G., Rodríguez-Almazán, C., Gill, S.S., Bravo, A., Soberón, M., 2009a. Domain II loop 3 of *Bacillus thuringiensis* Cry1Ab toxin is involved in a "ping pong" binding mechanism with *Manduca sexta* aminopeptidase-N and cadherin receptors. J. Biol. Chem. 284, 32750–32757.

Pacheco, S., Gómez, I., Gill, S.S., Bravo, A., Soberón, M., 2009b. Enhancement of insecticidal activity of *Bacillus thuringiensis* Cry1A toxins by fragments of a toxin-binding cadherin correlates with oligomer formation. Peptides 30, 583–588.

Pandian, G.N., Ishikawa, T., Togashi, M., Shitomi, Y., Haginoya, K., Yamamoto, S., Nishiumi, T., Hori, H., 2008. *Bombyx mori* midgut membrane protein P252, which binds to *Bacillus thuringiensis* Cry1A, is a chlorophyllide-binding protein, and the resulting complex has antimicrobial activity. Appl. Environ. Microbiol. 74, 1324–1331.

Pandian, G.N., Ishikawa, T., Vaijayanthi, T., Hossain, D.M., Yamamoto, S., Nishiumi, T., Angsuthanasombat, C., Haginoya, K., Mitsui, T., Hori, H., 2010. Formation of macromolecule complex with *Bacillus thuringiensis* Cry1A toxins and chlorophyllide binding 252-kDa lipocalin-like protein locating on *Bombyx mori* midgut membrane. J. Membr. Biol. 237, 125–136.

Pardo-López, L., Gómez, I., Rausell, C., Sanchez, J., Soberón, M., Bravo, A., 2006. Structural changes of the Cry1Ac oligomeric pre-pore from *Bacillus thuringiensis* induced by N-acetylgalactosamine facilitates toxin membrane insertion. Biochemistry 45, 10329–10336.

Pardo-López, L., Soberón, M., Bravo, A., 2013. *Bacillus thuringiensis* insecticidal three-domain Cry toxins: mode of action, insect resistance and consequences for crop protection. FEMS Microbiol. Rev. 37, 3–22.

Pardo-López, L., Soberón, M., Bravo, A., 2012. *Bacillus thuringiensis* insecticidal three-domain Cry toxins: mode of action, insect resistance and consequences for crop protection. FEMS Microbiol. Rev. 37, 3–22.

Park, Y., Kim, Y., 2013. RNA interference of cadherin gene expression in *Spodoptera exigua* reveals its significance as a specific Bt target. J. Invertebr. Pathol. 114, 285–291.

Peng, D., Xu, X., Ye, W., Yu, Z., Sun, M., 2010. *Helicoverpa armigera* cadherin fragment enhances Cry1Ac insecticidal activity by facilitating toxin-oligomer formation. Appl. Microbiol. Biotechnol. 85, 1033–1040.

Pérez, C., Muñoz-Garay, C., Portugal, L.C., Sánchez, J., Gill, S.S., Soberón, S., Bravo, A., 2007. *Bacillus thuringiensis* ssp. *israelensis* Cyt1Aa enhances activity of Cry11Aa toxin by facilitating the formation of a pre-pore oligomeric structure. Cell. Microbiol. 9, 2931–2937.

Peyronnet, O., Vachon, V., Schwartz, J.L., Laprade, R., 2001. Ion channels induced in planar lipid bilayers by the *Bacillus thuringiensis* toxin Cry1Aa in the presence of gypsy moth (*Lymantria dispar*) brush border membrane. J. Membr. Biol. 184, 45–54.

Pigott, C.R., Ellar, D.J., 2007. Role of receptors in *Bacillus thuringiensis* crystal toxin activity. Microbiol. Mol. Biol. Rev. 71, 255–281.

Pigott, C.R., King, M.S., Ellar, D.J., 2008. Investigating the properties of *Bacillus thuringiensis* Cry proteins with novel loop replacements created using combinatorial molecular biology. Appl. Environ. Microbiol. 74, 3497–34511.

Promdonkoy, B., Ellar, D.J., 2003. Investigation of the pore-forming mechanism of a cytolytic δ-endotoxin from *Bacillus thuringiensis*. Biochem. J. 374, 255–259.

Promdonkoy, B., Ellar, D.J., 2005. Structure-function relationships of a membrane pore forming toxin revealed by reversion mutagenesis. Mol. Membr. Biol. 22, 327–337.

Rajagopal, R., Sivakumar, S., Agrawal, N., Malhotra, P., Bhatnagar, R.K., 2002. Silencing of midgut aminopeptidase N of *Spodoptera litura* by double-stranded RNA establishes its role as *Bacillus thuringiensis* toxin receptor. J. Biol. Chem. 277, 46849–46851.

Rajamohan, F., Alcantara, E., Lee, M.K., Chen, X.J., Curtiss, A., Dean, D.H., 1995. Single amino acid changes in domain II of *Bacillus thuringiensis* CryIAb δ-endotoxin affect irreversible binding to *Manduca sexta* midgut membrane vesicles. J. Bacteriol. 177, 2276–2282.

Rajamohan, F., Cotrill, J.A., Gould, F., Dean, D.H., 1996a. Role of domain II, loop 2 residues of *Bacillus thuringiensis* CryIAb δ-endotoxin in reversible and irreversible binding to *Manduca sexta* and *Heliothis virescens*. J. Biol. Chem. 271, 2390–2396.

Rajamohan, F., Hussain, S.R., Cotrill, J.A., Gould, F., Dean, D.H., 1996b. Mutations at domain II, loop 3, of *Bacillus thuringiensis* CryIAa and CryIAb delta-endotoxins suggest loop 3 is involved in initial binding to lepidopteran midguts. J. Biol. Chem. 271, 25220–25226.

Rausell, C., García-Robles, I., Sánchez, J., Muñoz-Garay, C., Martínez-Ramírez, A.C., Real, M.D., Bravo, A., 2004. Role of toxin activation on binding and pore formation activity of the *Bacillus thuringiensis* Cry3 toxins in membranes of *Leptinotarsa decemlineata* (Say). Biochim. Biophys. Acta 1660, 99–105.

Rausell, C., Ochoa-Campuzano, C., Martínez-Ramírez, A.C., Bravo, A., Real, M.D., 2007. A membrane associated metalloprotease cleaves Cry3Aa *Bacillus thuringiensis* toxin reducing pore formation in Colorado potato beetle brush border membrane vesicles. Biochim. Biophys. Acta 1768, 2293–2299.

Raymond, B., Johnston, P.R., Nielsen-LeRoux, C., Lereclus, D., Crickmore, N., 2010. *Bacillus thuringiensis*: an impotent pathogen? Trends Microbiol. 18, 189–194.

Rees, J.S., Jarrett, P., Ellar, D.J., 2009. Peritrophic membrane contribution to Bt Cry delta-endotoxin susceptibility in Lepidoptera and the effect of calcofluor. J. Invertebr. Pathol. 100, 139–146.

Ren, X.L., Chen, R.R., Zhang, Y., Ma, Y., Cui, J.J., Han, Z.J., Mu, L.L., Li, G.Q., 2013. A *Spodoptera exigua* cadherin serves as a putative receptor for *Bacillus thuringiensis* Cry1Ca toxin and shows differential enhancement of Cry1Ca and Cry1Ac toxicity. Appl. Environ. Microbiol. 79, 5576–5583.

Rigden, D.J., 2009. Does distant homology with Evf reveal a lipid binding site in *Bacillus thuringiensis* cytolytic toxins? FEBS Lett. 583, 1555–1560.

Rodriguez-Almazan, C., Reyes, E.Z., Zuniga-Navarrete, F., Munoz-Garay, C., Gomez, I., Evans, A.M., Likitvivatanavong, S., Bravo, A., Gill, S.S., Soberon, M., 2012. Cadherin binding is not a limiting step for *Bacillus thuringiensis* subsp. israelensis Cry4Ba toxicity to *Aedes aegypti* larvae. Biochem. J. 443, 711–717.

Rodriguez-Almazan, C., Ruiz de Escudero, I., Canton, P.E., Munoz-Garay, C., Perez, C., Gill, S.S., Soberon, M., Bravo, A., 2011. The amino- and carboxyl-terminal fragments of the *Bacillus thuringiensis* Cyt1Aa toxin have differential roles in toxin oligomerization and pore formation. Biochemistry 50, 388–396.

Rosa, J.C., De Oliveira, P.S., Garratt, R., Beltramini, L., Resing, K., Roque-Barreira, M.C., Greene, L.J., 1999. KM+, a mannose-binding lectin from *Artocarpus integrifolia*: amino acid sequence, predicted tertiary structure, carbohydrate recognition, and analysis of the beta-prism fold. Protein Sci. 8, 13–24.

Saengwiman, S., Aroonkesorn, A., Dedvisitsakul, P., Sakdee, S., Leetachewa, S., Angsuthanasombat, C., Pootanakit, K., 2011. *In vivo* identification of *Bacillus thuringiensis* Cry4Ba toxin receptors by RNA interference knockdown of glycosylphosphatidylinositol-linked aminopeptidase N transcripts in *Aedes aegypti* larvae. Biochem. Biophys. Res. Commun. 407, 708–713.

Sakurai, J., Nagahama, M., Oda, M., Tsuge, H., Kobayashi, K., 2009. *Clostridium perfringens* iota-toxin: structure and function. Toxins 1, 208–228.

Sangadala, S., Walters, F.S., English, L.H., Adang, M.J., 1994. A mixture of *Manduca sexta* aminopeptidase and phosphatase enhances *Bacillus thuringiensis* insecticidal CryIA(c) toxin binding and $^{86}Rb^{(+)}$-K^{+} efflux *in vitro*. J. Biol. Chem. 269, 10088–10092.

Sankaranarayanan, R., Sekar, K., Banerjee, R., Sharma, V., Surolia, A., Vijayan, M., 1996. A novel mode of carbohydrate recognition in jacalin, a Moraceae plant lectin with a beta-prism fold. Nat. Struct. Biol. 3, 596–603.

Sarkar, A., Hess, D., Mondal, H.A., Banerjee, S., Sharma, H.C., Das, S., 2009. Homodimeric alkaline phosphatase located at *Helicoverpa armigera* midgut, a putative receptor of Cry1Ac contains alpha-GalNAc in terminal glycan structure as interactive epitope. J. Proteome Res. 8, 1838–1848.

Schnepf, E., Crickmore, N., Van Rie, J., Lereclus, D., Baum, J., Feitelson, J., Zeigler, D.R., Dean, D.H., 1998. Bacillus thuringiensis and its pesticidal crystal proteins. Microbiol. Mol. Biol. Rev. 62, 775–806.

Schnepf, H.E., Lee, S., Dojillo, J., Burmeister, P., Fencil, K., Morera, L., Nygaard, L., Narva, K.E., Wolt, J.D., 2005. Characterization of Cry34/Cry35 binary insecticidal proteins from diverse *Bacillus thuringiensis* strain collections. Appl. Environ. Microbiol. 71, 1765–1774.

Schwartz, J.L., Juteau, M., Grochulski, P., Cygler, M., Prâefontaine, G., Brousseau, R., Masson, L., 1997a. Restriction of intramolecular movements within the Cry1Aa toxin molecule of *Bacillus thuringiensis* through disulfide bond engineering. FEBS Lett. 410, 397–402.

Schwartz, J.L., Lu, Y.J., Soehnlein, P., Brousseau, R., Masson, L., Laprade, R., Adang, M.J., 1997b. Ion channels formed in planar lipid bilayers by *Bacillus thuringiensis* toxins in the presence of *Manduca sexta* midgut receptors. FEBS Lett. 412, 270–276.

Schwartz, J.L., Potvin, L., Chen, X.J., Brousseau, R., Laprade, R., Dean, D.H., 1997c. Single-site mutations in the conserved alternating-arginine region affect ionic channels formed by CryIAa, a *Bacillus thuringiensis* toxin. Appl. Environ. Microbiol. 63, 3978–3984.

Sengupta, A., Sarkar, A., Priya, P., Ghosh Dastidar, S., Das, S., 2013. New insight to structure-function relationship of GalNAc mediated primary interaction between insecticidal Cry1Ac toxin and HaALP receptor of Helicoverpa armigera. PLoS One 8, e78249.

Sharma, A., Chandran, D., Singh, D.D., Vijayan, M., 2007. Multiplicity of carbohydrate-binding sites in beta-prism fold lectins: occurrence and possible evolutionary implications. J. Biosci. 32, 1089–1110.

Sher, D., Fishman, Y., Zhang, M., Lebendiker, M., Gaathon, A., Mancheno, J.M., Zlotkin, E., 2005. Hydralysins, a new category of beta-pore-forming toxins in cnidaria. J. Biol. Chem. 280, 22847–22855.

Sivakumar, S., Rajagopal, R., Venkatesh, G.R., Srivastava, A., Bhatnagar, R.K., 2007. Knockdown of aminopeptidase-N from *Helicoverpa armigera* larvae and in transfected Sf21 cells by RNA interference reveals its functional interaction with *Bacillus thuringiensis* insecticidal protein Cry1Ac. J. Biol. Chem. 282, 7312–7319.

Smedley, D.P., Ellar, D.J., 1996. Mutagenesis of three surface-exposed loops of a *Bacillus thuringiensis* insecticidal toxin reveals residues important for toxicity, receptor recognition and possibly membrane insertion. Microbiology 142, 1617–1624.

Soberon, M., Lopez-Diaz, J.A., Bravo, A., 2013a. Cyt toxins produced by *Bacillus thuringiensis*: a protein fold conserved in several pathogenic microorganisms. Peptides 41, 87–93.

Soberón, M., Pardo-López, L., López, I., Gómez, I., Tabashnik, B.E., Bravo, A., 2007. Engineering modified Bt toxins to counter insect resistance. Science 318, 1640–1642.

Staples, N., Ellar, D., Crickmore, N., 2001. Cellular localization and characterization of the *Bacillus thuringiensis* Orf2 crystallization factor. Curr. Microbiol. 42, 388–392.

Tabashnik, B.E., Huang, F., Ghimire, M.N., Leonard, B.R., Siegfried, B.D., Rangasamy, M., Yang, Y., Wu, Y., Gahan, L.J., Heckel, D.G., Bravo, A., Soberon, M., 2011. Efficacy of genetically modified Bt toxins against insects with different genetic mechanisms of resistance. Nat. Biotechnol. 29, 1128–1131.

Tanaka, S., Miyamoto, K., Noda, H., Jurat-Fuentes, J.L., Yoshizawa, Y., Endo, H., Sato, R., 2013. The ATP-binding cassette transporter subfamily C member 2 in *Bombyx mori* larvae is a functional receptor for Cry toxins from *Bacillus thuringiensis*. FEBS J. 280, 1782–1794.

Terra, W.R., Ferreira, C., 1994. Insect digestive enzymes: properties, compartmentalization and function. Comp. Biochem. Physiol. 109B, 1–62.

Terra, W.R., Ferreira, C., Jordao, B.P., Dillon, R.J., 1996. Digestive enzymes. In: Lehane, M.J., Billingsley, P.F. (Eds.), Biology of the Insect Midgut. Chapman & Hall, London, pp. 153–194.

Tetreau, G., Bayyareddy, K., Jones, C.M., Stalinski, R., Riaz, M.A., Paris, M., David, J.P., Adang, M.J., Despres, L., 2012. Larval midgut modifications associated with Bti resistance in the yellow fever mosquito using proteomic and transcriptomic approaches. BMC Genomics 13, 248.

Thiery, I., Hamon, S., 1998. Bacterial control of mosquito larvae: investigation of stability of *Bacillus thuringiensis* var. *israelensis* and *Bacillus sphaericus* standard powders. J. Am. Mosq. Control Assoc. 14, 472–476.

Thomas, W.E., Ellar, D.J., 1983a. *Bacillus thuringiensis* var *israelensis* crystal delta-endotoxin: effects on insect and mammalian cells in vitro and in vivo. J. Cell Sci. 60, 181–197.

Thomas, W.E., Ellar, D.J., 1983b. Mechanism of action of *Bacillus thuringiensis* var *israelensis* insecticidal delta-endotoxin. FEBS Lett. 154, 362–368.

Tiewsiri, K., Wang, P., 2011. Differential alteration of two aminopeptidases N associated with resistance to *Bacillus thuringiensis* toxin Cry1Ac in cabbage looper. Proc. Natl. Acad. Sci. U.S.A. 108, 14037–14042.

Tigue, N.J., Jacoby, J., Ellar, D.J., 2001. The alpha-helix 4 residue, Asn135, is involved in the oligomerization of Cry1Ac1 and Cry1Ab5 *Bacillus thuringiensis* toxins. Appl. Environ. Microbiol. 67, 5715–5720.

Torres, J., Lin, X., Boonserm, P., 2008. A trimeric building block model for Cry toxins in vitro ion channel formation. Biochim. Biophys. Acta Biomembr. 1778, 392–397.

Tyrell, D.J., Bulla Jr., L.A., Andrews Jr., R.E., Kramer, K.J., Davidson, L.I., Nordin, P., 1981. Comparative biochemistry of entomocidal parasporal crystals of selected *Bacillus thuringiensis* strains. J. Bacteriol. 145, 1052–1062.

Vachon, V., Laprade, R., Schwartz, J.L., 2012. Current models of the mode of action of *Bacillus thuringiensis* insecticidal crystal proteins: a critical review. J. Invertebr. Pathol. 111, 1–12.

Vadlamudi, R.K., Weber, E., Ji, I., Ji, T.H., Bulla Jr., L.A., 1995. Cloning and expression of a receptor for an insecticidal toxin of *Bacillus thuringiensis*. J. Biol. Chem. 270, 5490–5494.

Valaitis, A.P., 2011. Localization of Bacillus thuringiensis Cry1A toxin-binding molecules in gypsy moth larval gut sections using fluorescence microscopy. J. Invertebr. Pathol. 108, 69–75.

Valaitis, A.P., Jenkins, J.L., Lee, M.K., Dean, D.H., Garner, K.J., 2001. Isolation and partial characterization of gypsy moth BTR-270, an anionic brush border membrane glycoconjugate that binds *Bacillus thuringiensis* Cry1A toxins with high affinity. Arch. Insect Biochem. Physiol. 46, 186–200.

Varani, A.M., Lemos, M.V., Fernandes, C.C., Lemos, E.G., Alves, E.C., Desiderio, J.A., 2013. Draft genome sequence of *Bacillus thuringiensis* var. *thuringiensis* strain T01-328, a Brazilian isolate that produces a soluble pesticide protein, Cry1Ia. Genome Announc. 1(5), e00817–13.

Vié, V., Van Mau, N., Pomarède, P., Dance, C., Schwartz, J.L., Laprade, R., Frutos, R., Rang, C., Masson, L., Heitz, F., Le Grimellec, C., 2001. Lipid-induced pore formation of the *Bacillus thuringiensis* Cry1Aa insecticidal toxin. J. Membr. Biol. 180, 195–203.

Vilas-Bôas, G.T., Peruca, A.P., Arantes, O.M., 2007. Biology and taxonomy of *Bacillus cereus*, *Bacillus anthracis*, and *Bacillus thuringiensis*. Can. J. Microbiol. 53, 673–687.

Walters, F.S., deFontes, C.M., Hart, H., Warren, G.W., Chen, J.S., 2010. Lepidopteran-active variable-region sequence imparts coleopteran activity in eCry3.1Ab, an engineered *Bacillus thuringiensis* hybrid insecticidal protein. Appl. Environ. Microbiol. 76, 3082–3088.

Walters, F.S., Kulesza, C.A., Phillips, A.T., English, L.H., 1994. A stable oligomer of *Bacillus thuringiensis* delta-endotoxin, CryIIIA. Insect Biochem. Mol. Biol. 24, 963–968.

Walters, F.S., Stacy, C.M., Lee, M.K., Palekar, N., Chen, J.S., 2008. An engineered chymotrypsin/cathepsin G site in domain I renders *Bacillus thuringiensis* Cry3A active against Western corn rootworm larvae. Appl. Environ. Microbiol. 74, 367–374.

Weng, Y.P., Lin, Y.P., Hsu, C.I., Lin, J.Y., 2004. Functional domains of a pore-forming cardiotoxic protein, volvatoxin A2. J. Biol. Chem. 279, 6805–6814.

Wirth, M.C., Georghiou, G.P., Federici, B.A., 1997. CytA enables CryIV endotoxins of *Bacillus thuringiensis* to overcome high levels of CryIV resistance in the mosquito, *Culex quinquefasciatus*. Proc. Natl. Acad. Sci. U.S.A. 94, 10536–10540.

Wirth, M.C., Park, H.W., Walton, W.E., Federici, B.A., 2005. Cyt1A of *Bacillus thuringiensis* delays evolution of resistance to Cry11A in the mosquito *Culex quinquefasciatus*. Appl. Environ. Microbiol. 71, 185–189.

Wolfersberger, M.G., 1990. The toxicity of two *Bacillus thuringiensis* delta-endotoxins to gypsy moth larvae is inversely related to the affinity of binding sites on midgut brush border membranes for the toxins. Experientia 46, 475–477.

Wu, D., Aronson, A.I., 1992. Localized mutagenesis defines regions of the *Bacillus thuringiensis* delta-endotoxin involved in toxicity and specificity. J. Biol. Chem. 267, 2311–2317.

Wu, D., Chang, F.N., 1985. Synergism in mosquitocidal activity of 26 and 65 kDa proteins from *Bacillus thuringiensis* subsp. *israelensis* crystal. FEBS Lett. 190, 232–236.

Xia, L., Sun, Y., Ding, X., Fu, Z., Mo, X., Zhang, H., Yuan, Z., 2005. Identification of *cry*-type genes on 20-kb DNA associated with Cry1 crystal proteins from *Bacillus thuringiensis*. Curr. Microbiol. 51, 53–58.

Xu, X., Yu, L., Wu, Y., 2005. Disruption of a cadherin gene associated with resistance to Cry1Ac δ-endotoxin of *Bacillus thuringiensis* in *Helicoverpa armigera*. Appl. Environ. Microbiol. 71, 948–954.

Yamagiwa, M., Esaki, M., Otake, K., Inagaki, M., Komano, T., Amachi, T., Sakai, H., 1999. Activation process of dipteran-specific insecticidal protein produced by *Bacillus thuringiensis* subsp. *israelensis*. Appl. Environ. Microbiol. 65, 3464–3469.

Yamaguchi, T., Sahara, K., Bando, H., Asano, S.-I., 2010. Intramolecular proteolytic nicking of *Bacillus thuringiensis* Cry8Da toxin in BBMVs of Japanese beetle. J. Invertebr. Pathol. 105, 243–247.

Yamamoto, T., Iizuka, T., Aronson, J.N., 1983. Mosquitocidal protein of *Bacillus thuringiensis* subsp. *israelensis*: identification and partial isolation of the protein. Curr. Microbiol. 9, 279–284.

Yamashita, S., Katayama, H., Saitoh, H., Akao, T., Park, Y.S., Mizuki, E., Ohba, M., Ito, A., 2005. Typical three-domain Cry proteins of *Bacillus thuringiensis* strain A1462 exhibit cytocidal activity on limited human cancer cells. J. Biochem. 138, 663–672.

Yang, Y., Zhu, Y.C., Ottea, J., Husseneder, C., Leonard, B.R., Abel, C., Huang, F., 2010. Molecular characterization and RNA interference of three midgut aminopeptidase N isozymes from *Bacillus thuringiensis*-susceptible and -resistant strains of sugarcane borer, *Diatraea saccharalis*. Insect Biochem. Mol. Biol. 40, 592–603.

Ye, W., Zhu, L., Liu, Y., Crickmore, N., Peng, D., Ruan, L., Sun, M., 2012. Mining new crystal protein genes from *Bacillus thuringiensis* on the basis of mixed plasmid-enriched genome sequencing and a computational pipeline. Appl. Environ. Microbiol. 78, 4795–4801.

Zhang, B., Liu, M., Yang, Y., Yuan, Z., 2006. Cytolytic toxin Cyt1Aa of *Bacillus thuringiensis* synergizes the mosquitocidal toxin Mtx1 of *Bacillus sphaericus*. Biosci. Biotechnol. Biochem. 70, 2199–2204.

Zhang, H., Wu, S., Yang, Y., Tabashnik, B.E., Wu, Y., 2012. Non-recessive Bt toxin resistance conferred by an intracellular cadherin mutation in field-selected populations of cotton bollworm. PLoS One 7, e53418.

Zhang, R., Hua, G., Andacht, T.M., Adang, M.J., 2008. A 106-kDa aminopeptidase is a putative receptor for *Bacillus thuringiensis* Cry11Ba toxin in the mosquito *Anopheles gambiae*. Biochemistry 47, 11263–11272.

Zhang, R., Hua, G., Urbauer, J.L., Adang, M.J., 2010. Synergistic and inhibitory effects of aminopeptidase peptides on *Bacillus thuringiensis* Cry11Ba toxicity in the mosquito *Anopheles gambiae*. Biochemistry 49, 8512–8519.

Zhang, S., Cheng, H., Gao, Y., Wang, G., Liang, G., Wu, K., 2009. Mutation of an aminopeptidase N gene is associated with *Helicoverpa armigera* resistance to *Bacillus thuringiensis* Cry1Ac toxin. Insect Biochem. Mol. Biol. 39, 421–429.

Zhang, X., Candas, M., Griko, N.B., Rose-Young, L., Bulla, L.A., 2005. Cytotoxicity of *Bacillus thuringiensis* Cry1Ab toxin depends on specific binding of the toxin to the cadherin receptor BT-R(1) expressed in insect cells. Cell Death Differ. 12, 1407–1416.

Zhou, X., Konkel, M.E., Call, D.R., 2009. Type III secretion system 1 of *Vibrio parahaemolyticus* induces oncosis in both epithelial and monocytic cell lines. Microbiology 155, 837–851.

Zhuang, M., Oltean, D.I., Gómez, I., Pullikuth, A.K., Soberón, M., Bravo, A., Gill, S.S., 2002. *Heliothis virescens* and *Manduca sexta* lipid rafts are involved in Cry1A toxin binding to the midgut epithelium and subsequent pore formation. J. Biol. Chem. 277, 13863–13872.

Zuniga-Navarrete, F., Gomez, I., Pena, G., Bravo, A., Soberon, M., 2013. A *Tenebrio molitor* GPI-anchored alkaline phosphatase is involved in binding of *Bacillus thuringiensis* Cry3Aa to brush border membrane vesicles. Peptides 41, 81–86.

CHAPTER THREE

Lysinibacillus sphaericus: Toxins and Mode of Action, Applications for Mosquito Control and Resistance Management

Maria Helena Neves Lobo Silva Filha*, Colin Berry†, Lêda Regis*
*Centro de Pesquisas Aggeu Magalhães-Fiocruz, Recife-Pernambuco, Brazil
†Cardiff School of Biosciences, Cardiff University, Cardiff, United Kingdom

Contents

Advances in Insect Physiology, Volume 47
ISSN 0065-2806
http://dx.doi.org/10.1016/B978-0-12-800197-4.00003-8

Abstract

Lysinibacillus sphaericus (*Ls*) strains that produce insecticidal proteins show high activity against mosquito larvae. The most active of these is the binary (Bin) toxin that acts following ingestion and, after midgut processing and binding to specific receptors, provokes cytopathological effects and leads to larval death. Bin toxin displays specific action against some species of medical importance (e.g. *Culex* and *Anopheles*) and it is safe to non-target organisms. These features have led to the production of biolarvicides based on this bacterium and its effectiveness to control mosquito larvae has been widely related in the literature. The field utilisation of *Ls* has also shown that resistance could be selected among exposed populations and the mechanisms and genes involved in this process have been described. Management strategies can be successfully employed to avoid resistance and *Ls* can be used within integrated programmes as a selective and efficient agent to control mosquitoes.

1. INTRODUCTION

1.1. Background

The utilisation of entomopathogenic bacteria for insect control started in the 1960s with the discovery and development of *Bacillus thuringiensis* (*Bt*) varieties that produced insecticidal proteins active against agricultural insect pests. The *B. thuringiensis* serovar. *israelensis* (*Bti*) discovered by Goldberg and Margalit (1978) was the first serotype identified as active against Diptera larvae (de Barjac, 1978). This entomopathogenic bacterium enjoyed a rapid development from the characterisation of its properties to field utilisation (Becker, 1997; Guillet et al., 1990; Margalit and Dean, 1985), mainly because of the serious resistance problems encountered by synthetic insecticides in vector-control programmes during that period. The second mosquitocidal bacterium *Lysinibacillus sphaericus* (*Ls*), previously designated as *B. sphaericus*, was identified by Neide in 1904 (Neide, 1904). Characterisation of this species as a mosquito pathogen was initiated by Kellen, much later, when a toxic strain was isolated from cadavers of *Culiseta incidens* larvae (Kellen et al., 1965). The Kellen (K) strain displayed a low level of toxicity and did not attract much interest for its development as a control agent. The discovery by Singer (1973), of the SSII-1 strain, that displayed a higher activity than the K strain renewed the interest in this bacterium and motivated the search for new strains. Later, strains with high activity were discovered (Singer, 1977; Weiser, 1984; Wickremesingue and Mendis, 1980) that led to the development of the use of *Ls* as a mosquito-control agent.

Insecticidal factors produced by *Ls* were identified in strains isolated worldwide, and these isolates were classified according to their toxicity to mosquitoes. Early studies showed that the high activity of some strains was associated with the production of crystalline inclusions during the bacterial sporulation (Fig. 3.1) (de Barjac and Charles, 1983; Kalfon et al., 1984; Payne and Davidson, 1984; Yousten and Davidson, 1982). The crystals are synthesised during stage III of sporulation, and once formed, they remain associated with the spore within the exosporium (Kalfon et al., 1984; Yousten and Davidson, 1982). A study showed that mutant strains that were blocked from the early stages of sporulation did not produce crystals and lost their toxicity toward larvae, which confirmed the essential role played by the crystals for the mosquitocidal activity of these strains (Charles et al., 1988). The active crystals contain the binary (Bin) protoxin, which is the major insecticidal protein produced by *Ls* (Baumann et al., 1985).

1.2. General features and strains

Ls is a Gram positive, aerobic, sporulating and cosmopolitan bacterium. This species is a saprophytic organism that occurs in several habitats, including soil and aquatic environments. The most important morphological characteristic is the presence of spherical terminal spores, and the most marked phenotypic features are its inability to grow under anaerobic conditions or to use sugars as source of energy (White and Lotay, 1980). Instead, this species uses carbon compounds, such as organic acids and other amino acids. The reclassification of this organism as *Ls* was due to some differential phenotypic features, including the capacity to produce peptidoglycans containing lysine and aspartate (Ahmed et al., 2007).

In terms of classification, *Ls* is quite a heterogeneous species (Nakamura, 2000), and in view of its application to mosquito control, the strains can be divided into toxic and non-toxic strains that share 79% DNA homology (Krych et al., 1980). This chapter addresses *Ls* strains classified in the

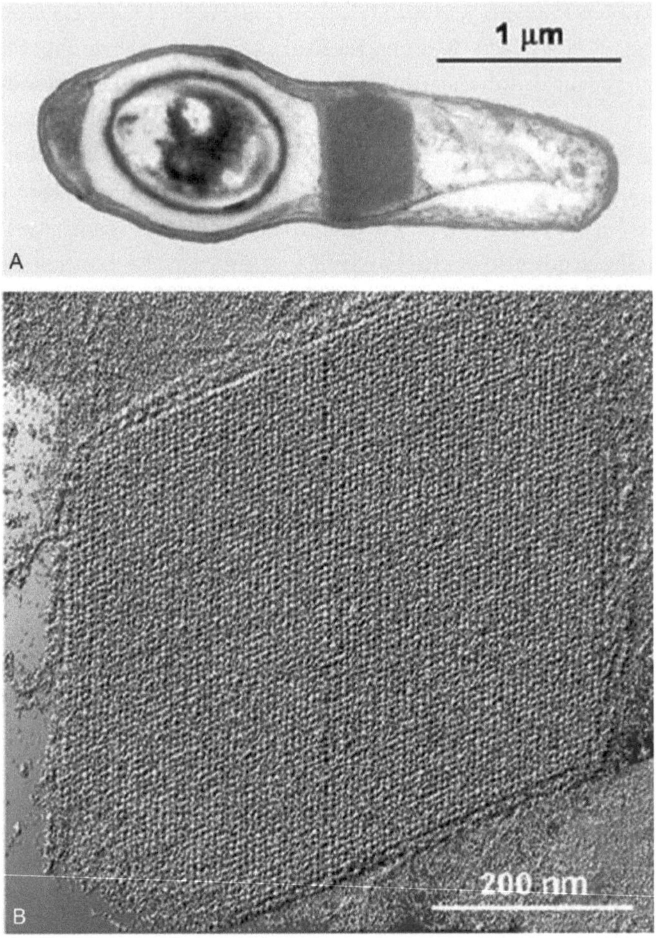

Figure 3.1 Micrography of *Lysinibacillus sphaericus* strain 2297 at the end of sporulation. (A) Spore and crystal are in the left- and right side of the exosporium, respectively. (B) Crystal lattice. *Taken from Charles et al. (2010).*

DNA homology group IIA, as will be described below. The highly toxic strains isolated to date, can produce different insecticidal factors as the Bin protoxin from crystalline inclusions and other toxins, such as the mosquitocidal toxins (Mtxs) and the Cry48/Cry49 toxin, as will be described in Section 2.

General classifications of *Ls* strains have been performed based on different systems. The DNA homology of the strains (Krych et al., 1980) and serotyping based on the flagellar antigen (de Barjac, 1990; de Barjac et al., 1985) are the most commonly used methods. Other approaches have also been proposed to classify this diverse group, such as bacteriophage typing (Yousten, 1984b; Yousten et al., 1980), numerical classification using the taxonomy of phenotypic features (Alexander and Priest, 1990), analysis of the cellular fatty acids (Frachon et al., 1991), ribotyping (Aquino de Muro et al., 1992), and the profiling of randomly amplified polymorphic DNA (Woodburn et al., 1995). The toxic strains were found to belong to few groups according to all the classifications used. DNA homology analysis placed the strains into five groups (I–V) that probably represent distinct species, and those that displayed some level of mosquito activity were clustered in subgroup IIA. The flagellar antigen analysis generated approximately 49 serotypes, and nine serotypes host the highly toxic strains. The serotype 5a5b contains active strains that produce the Bin toxin and includes 1593 and 2362, which are the most commonly strains used to date for the production of larvicides. Other high-toxicity strains producing the Bin toxin from DNA group IIA include the IAB881, IAB59, 2297 and IAB872 strains, belonging to serotypes 3, 6, 25 and 48, respectively. Table 3.1 shows some representative examples of strains and their classification based on previously published data (Charles et al., 1996). Other highly toxic strains have also been isolated and employed for the production of biolarvicides. The C3-41 strain (serotype 5a5b) isolated in China, for instance, has been extensively used in control programmes in that country (Yuan et al., 1999). The screening of new strains for improved larvicidal activity and optimal performance for large-scale production continues to sustain the development of new products (Hire et al., 2010; Poophati et al., 2013; Prabhu et al., 2013; Sun et al., 1996).

1.3. The relevance of *L. sphaericus* as a mosquito-control agent

Ls occupies an important position among the set of insecticidal agents that are currently available for mosquito control. This bacterium produces insecticidal factors that display high activity against culicids of medical relevance that include *Culex* and *Anopheles* species. The production of biolarvicides based on *Ls* has been successfully achieved since the 1980s (Lacey, 2007; Lecadet, 1996), and its high insecticidal potency allied to a selective spectrum (Section 2) are important features that allowed the development of commercial products. In addition, the development of *Ls* biolarvicides at the industrial level was supported by the successful production and application of *Bt*.

Table 3.1 Examples of *Lysinibacillus sphaericus* strains and their larvicidal properties to mosquito larvae

Strain	Origin	Flagellar serotype	DNA group	Larvicidal activity[a]	Genes encoding mosquitocidal proteins[b]		
					Crystal	Mtxs	Cry48/Cry49
Kellen K	USA	1a	IIA	Low	–	1, 2, 3	Nd
SSII-1	India	2a2b	IIA	Medium	–	1, 2, 3	Nd
IAB881	Ghana	2a2b	Nd[c]	Medium	+	1, (2 and 3 Nd)	+
LP1-G	China	3	IIA	Medium	+	1, (2 and 3 Nd)	+
1593M	Indonesia	5a5b	IIA	High	+	1, 2 (3 Nd)	–
2362	Nigeria	5a5b	IIA	High	+	1, 2, 3	–
1691	El Salvador	5a5b	IIA	High	+	1, 2, 3	Nd
IAB59	Ghana	6	Nd	High	+	1, 2, 3	+
2297	Sri Lanka	25	IIA	High	+	1, 2, 3	–
IAB872	Ghana	48	Nd	High	+	1, (2 and 3 Nd)	Nd
C3-41	China	5a5b	IIA	High	+	1 (pseudo) 2, 3, 4	–

[a]Based on criteria defined by Charles et al. (1996).
[b]+ Presence, – absence.
[c]Not determined.
Modified and extended from Charles et al. (1996)

The experiences of *Ls* larvicide utilisation provided a solid background for its adoption in mosquito-control programmes worldwide (Section 4). These larvicides were first introduced for *Cx. pipiens* control in many areas of France as early as 1987 to reduce the nuisance caused by this species (Thiery et al., 1996). An initiative by the World Health Organisation (WHO) created a multi-centre study for the evaluation of the field effectiveness of *Ls* to control *Cx. quinquefasciatus* larvae in urban areas of tropical countries, with particular attention to the role of this mosquito as the vector of *Wuchereria bancrofti* the causative agent of filariasis (WHO, 1993). These pioneer trials, along with others, supported the broad utilisation of these larvicides. The control of *Culex* spp. has gained more importance recently in regard to their role as vectors of emergent arboviruses, such as the West Nile Virus that has provoked important epidemics in human populations (Kramer et al., 2008; Petersen and Fischer, 2012). The control of anophelines is a challenge, and previous studies have shown that some relevant species involved in *Plasmodium* transmission are susceptible to *Ls*. The control of *Anopheles* (*An.*) *stephensi* and *An. sinensis* in India and China, respectively, showed the operational viability of *Ls* to control this group of mosquitoes (Kumar et al., 1994; Thiery et al., 1996; Yuan et al., 2000). In addition to its application for vector control in urban areas, the selective activity and biocompatibility of *Ls* is of great utility when the target species breed in environmentally sensitive areas. The utilisation of *Ls* also raises concerns about the selection of resistance. High levels of resistance achieved due to its utilisation have been reported, and the major findings on this issue are presented in Section 5 along with the strategies that can be introduced for the management of resistance in Section 6. These strategies can ensure the effectiveness of *Ls* when used in the scope of integrated control programmes and can overcome the potential onset of resistance. Different aspects of *Ls* and its properties as an entomopathogenic bacterium have been covered by previous reviews and book chapters that can provide additional information (Baumann et al., 1991; Becker, 2000; Berry, 2012; Charles and Nielsen-LeRoux, 2000; Charles et al., 1996, 2010; Delecluse et al., 2000; Lacey, 2007; Porter et al., 1993; Regis and Nielsen-LeRoux, 2000).

2. TOXINS AND MODE OF ACTION

2.1. Spectrum of action

Ls strains in DNA group IIA may produce a range of toxins as detailed in this section, including those with activity against mosquitoes, which are the Bin, the Mtx and the Cry48Aa/49Aa toxins. The profile of toxins produced by individual strains is varied and contributes to the designation of bacteria as either high- or low-toxicity strains with respect to their activity against mosquito larvae. The strains with the highest activity are characterised by the presence of the Bin protoxin that is produced as parasporal crystals during sporulation (de Barjac et al., 1980; Payne and Davidson, 1984; Yousten, 1984a). Strains lacking Bin crystals display low toxicity and some of them produce Mtx toxins, however, the latter undergo degradation during their production in the vegetative phase and they do not contribute to provide a high activity to the strains. The performance of strain 1593 that produces Bin and Mtx toxins to larvae is, for instance, 3000-fold superior compared to the SSII-1 strain that produces Mtx toxins only (Myers et al., 1979). Additional studies on different strains have shown that Bin accounts for most activity recorded for the sporulated cultures and this is the main active ingredient of biolarvicides based on *Ls*, as reviewed by Charles et al. (1996). According to the Insect Resistance Action Committee (www.irac-online.org), the insecticidal toxins from *Ls* are classified into the mode of action group 11 (Moa11), along with *Bti*, and those agents are defined as "bacterial disruptors of insect midgut membranes". The midgut of mosquito larvae is the central site for the action of these toxins, since they act following ingestion, are processed under specific conditions in this environment and they act on specific receptors located on the epithelium, to cause mortality of larvae. More details of insect midgut are provided in Chapter 1.

Mosquitoes are the principal targets of the *Ls* toxins and this is reflected in the activity spectrum of the individual toxins. However, *Ls* toxicity to *Phlebotomus* sandflies has been reported for high-toxicity strains 1593 and 2362 that may result in larval death and reduced fecundity of surviving insects (Penner and Wilamowski, 1996; Robert et al., 1997, 1998; Wahba, 2000). Strain 2362 also showed low toxicity against *Lutzomyia* sandflies (Wermelinger et al., 2000). In addition, larvicidal effects of *Ls* extracts against the nematode *Trichostrongylus colubriformis* have also been reported (Bone and Tinelli, 1987) and some toxicity was seen against the crustacean *Palaemonetes pugio* (Key and Scott, 1992).

Within the mosquitoes, there is differential toxicity to the species studied. The most susceptible are *Culex* spp, in particular, those from the *Cx. pipiens* complex, but one exception in this genus is *Cx. cinereus* larvae (Nicolas and Dossou-Yovo, 1987). Anophelines including species of medical importance such as *An. gambiae*, *An. stephensi*, *An. albimanus*, *An. quadrimaculatus*, *An. darlingi* and *An. nuneztovari* are also susceptible to the Bin toxin (Arredondo-Jimenez et al., 1990; Davidson, 1989; Karch et al., 1992; Lacey et al., 1988b; Rodrigues et al., 1998, 1999; Young et al., 1990). *Aedes* or *Ochlerotatus* show a variable scenario including susceptible species such as *Oc. atropalpus*, *Ae. vexans* and *Oc. trivittatus*, as well as *Ae. aegypti* larvae that are refractory to Bin toxin (Berry et al., 1993; Delecluse et al., 2000; Nielsen-Leroux and Charles, 1992). The lethal concentration (LC) of *Ls* for these larvae is between 100- and 1000-fold higher than the respective LC for *Cx. pipiens* larvae (Thiery and de Barjac, 1989). The screening of *Ls* activity has also demonstrated susceptible larvae from *Psorophora* and *Mansonia* species. On the other hand, *Simulium* larvae that are susceptible to *Bti* cannot be targeted by the Bin toxin. Table 3.2 presents a non-exhaustive list of mosquito susceptibilities to *Ls* strains. The most common species targeted by *Ls* in field-control trials or programmes are described in Section 4.

2.2. Binary toxin

The Bin toxin, comprising the BinA and BinB proteins, is the best characterised of the toxins from *Ls*. It is produced during early sporulation by highly mosquitocidal strains (El-Bendary et al., 2005; Kalfon et al., 1984) and is deposited as a parasporal crystalline inclusion within the exosporium (Davidson and Myers, 1981; Kalfon et al., 1984; Yousten and Davidson, 1982). In these strains, Bin contributes to the majority of the toxicity and this fact, in turn, is a factor in the relative ease with which mosquitoes can develop resistance to *Ls* (see Section 5). Bin toxin acts in the midgut and the major steps of its mode of action in culicid larvae are: ingestion of crystals; dissolution of the crystal matrix under the alkaline pH conditions and release of Bin protoxin in the midgut lumen; processing of the Bin protoxin into active toxin; binding of the active toxin to specific receptors available on the midgut epithelium; occurrence of cytopathological effects on the midgut are followed by larval death, which is provoked by

Table 3.2 Sensitivity of mosquito species to *Lysinibacillus sphaericus*

Species	Susceptibility (strain tested)	References
Culex pipiens	High (1593, 2013-4); moderate (SSII-1)	Thiery and de Barjac (1989), Wraight et al. (1987)
Cx. quinquefasciatus	High (1593); moderate (SSII-1)	Cheong and Yap (1985), Mulligan et al. (1978), Wraight et al. (1987)
Cx. nigripalpus	High (1593, 1404, SSII-1)	Ramoska et al. (1977)
Cx. salinarius	High (1593)	Wraight et al. (1987)
Cx. restuans	High (2013-4)	Wraight et al. (1987)
Cx. tarsalis	High (1593)	Mulligan et al. (1978)
Cx. cinereus	Refractory (2362)	Nicolas and Dossou-Yovo (1987)
Culiseta melanura	High (1593)	Wraight et al. (1987)
Mansonia uniformis	High	Cheong and Yap (1985)

Continued

Table 3.2 Sensitivity of mosquito species to *Lysinibacillus sphaericus*—cont'd

Species	Susceptibility (strain tested)	References
Psorophora columbiae	High (1593); moderate (SSII-1)	Ramoska et al. (1977)
Anopheles stephensi	High (multiple strains)	Davidson (1989), Thiery and de Barjac (1989)
An. gambiae	High (2362)	Davidson (1989), Nicolas et al. (1987)
An. albimanus	High (2362)	Davidson (1989)
An. quadrimaculatus	High (2362)	Davidson (1989), Young et al. (1990)
An. darlingi	High (2362)	Rodrigues et al. (1998, 1999)
An. nuneztovari	High (2362)	Rodrigues et al. (1998, 1999)
An. braziliensis	High (2362)	Rodrigues et al. (1998, 1999)
An. balabacensis	Moderate	Cheong and Yap (1985)
Ochlerotatus nigromaculis	High	Mulligan et al. (1978)
Oc. atropalpus	High	Berry et al. (1993)
Oc. intrudens	High (1593)	Wraight et al. (1987)
Oc. triseriatus	Moderate (SSII-1); low (1593)	Wraight et al. (1987)
Oc. canadensis	Moderate (SSII-1)	Wraight et al. (1987)
Oc. fitchii	Moderate (1593)	Wraight et al. (1987)
Oc. stimulans	Moderate (1593); low (SSII-1)	Wraight et al. (1987)
Oc. taeniorhynchus	Moderate (1593, 1404, SSII-1)	Ramoska et al. (1977)
Aedes vexans	Moderate	Wraight et al. (1987)
Ae. aegypti	Refractory (1593); low (SSII-1 vegetative cells)	Thiery and de Barjac (1989), Wraight et al. (1987)
Toxorhynchites rutilus	Refractory (2362, 2297)	Lacey et al. (1988b)

mechanisms that are still under investigation. Recently, activity of Bin toxins against human cancer cells has also been reported (Luo et al., 2014).

The analysis of Bin proteins isolated from parasporal crystals was reported in the mid-1980s (Baumann et al., 1985; Narasu and Gopinathan, 1986) and initially it appeared that they were derived from a larger precursor protein (Broadwell and Baumann, 1986). Subsequent cloning of the gene encoding BinA (Berry and Hindley, 1987; Hindley and Berry, 1987) and BinB (Baumann et al., 1987, 1988) showed that, in fact, the two components were produced from a single operon as independent proteins of approximately 42 and 51 kDa, respectively. The proteins are produced in approximately equimolar amounts and form a co-crystal in sporulating *Ls* whereas the individual components expressed in recombinant *Ls* did not form crystals (Charles et al., 1993). The combination of BinA + BinB forms crystals in *Ls* and *Bt* strains but not in recombinant *B. subtilis* (Baumann and Baumann, 1991; Broadwell et al., 1990a; Charles et al., 1993; Yuan et al., 1999) suggesting that the former, insect pathogenic bacteria, encode a factor that facilitates crystallisation and that is absent from *B. subtilis*. Bin protein synthesis is enhanced by recombinant co-expression of the P20 protein from *Bt* (Park et al., 2007) but a region downstream of the *bin* operon in *Ls* strain 2297 reduces Bin synthesis (Park et al., 2009). The activity of the Bin toxin appears to be synergistic with the Cyt1Aa protein from *Bt* when the two are co-expressed in acrystaliferous *Bt* (Li et al., 2000) but expression of Cyt1Ab in *Ls* did not show synergy although it did help to overcome Bin resistance in *Culex* larvae (Thiery et al., 1998). Other studies have shown synergy of *Bt* Cyt and Cry toxins with *Ls* against wild-type or Bin-resistant *Culex*, which may indicate synergy with Bin toxins although the use of *Ls* cells in these assays may also indicate synergy with other toxins that they produce (see below) (Wirth et al., 2000a,c, 2001a, 2004).

Circular dichroism analysis has suggested that BinA and BinB are predominantly composed of beta sheet (Hire et al., 2009; Kale et al., 2013; Srisucharitpanit et al., 2012) although BinB in wild-type and truncated forms has also been reported to contain considerable alpha helix (Tangsongcharoen et al., 2011). Crystallization of BinB protein (Chiou et al., 1999; Srisucharitpanit et al., 2013) and BinA/BinB co-crystals (Smith et al., 2004) have been described and the structure of BinB has recently been published (Srisucharitpanit et al., 2014). This protein has an N-terminal domain with a beta-trefoil architecture found in lectins and a C-terminal region rich in extended beta sheets that shows structural similarity to aerolysin beta pore forming toxins. These results may suggest a role for the N-terminal region in receptor binding and roles for the C-terminal region and the related BinA structure in formation of a beta pore. Association of the two proteins with each other and the membrane may result in conformational changes to the structures (Boonserm et al., 2006; Kale et al., 2013).

As described, the ingestion of the Bin proteins by mosquito larvae results in their solubilisation in the alkaline environment of the gut (Charles, 1987) and activation of the protoxin forms by proteolytic cleavage mediated by gut proteinases (Aly et al., 1989; Broadwell and Baumann, 1987; Brownbridge and Margalit, 1987; Davidson et al., 1987, 1990). These will activate both subunits BinA (51 kDa) and BinB (42 kDa) into smaller polypeptides of 43 and 39 kDa, respectively. After proteolysis, the 43 kDa BinB derivative results from the removal of 21 and 53 residues from the N- and C-termini, respectively (Clark and Baumann, 1990). For the BinA active fragment of 39 kDa, cleavage of 10 and 17 amino acids occur in these respective positions (Broadwell et al., 1990c). The correct processing of the Bin subunits and their presence in equimolar amounts are essential conditions that assure the optimal activity of this Bin toxin (Broadwell et al., 1990b; Davidson et al., 1990; Nicolas et al., 1993; Oei et al., 1990). Similar patterns of protoxin cleavage occur on exposure to digestive enzymes from non-susceptible larvae, indicating that the protoxin processing is not the origin of insect specificity (Nicolas et al., 1990). In solution, NaOH solubilised BinA/BinB crystal proteins may associate into a $BinA_2BinB_2$ heterotetramer but this association may be lost on trypsin activation (Smith et al., 2005) with activated proteins showing weak interactions between the two proteins (Kale et al., 2013), although formation of oligomeric complexes between activated toxins has also been suggested (Srisucharitpanit et al., 2012).

The solubilised and active toxin binds regionally to the larval midgut in the gastric caecum and posterior midgut (Davidson, 1988, 1989; Mulla et al., 1984a; Oei et al., 1992) leading to toxicity. In *Culex* larvae, the BinB component of the toxin is responsible for receptor binding and the BinA component subsequently binds to BinB or the BinB/receptor complex (Charles et al., 1997; Oei et al., 1992). The situation in *An. gambiae* appears a little more complex with a possible role for BinA in binding as well (Charles et al., 1997). The receptor for Bin binding has been identified in *Cx. pipiens* as a midgut-bound α-glucosidase (Silva-Filha et al., 1999). Orthologs of this protein have been identified in other mosquito species including *An. gambiae* (Opota et al., 2008) and the refractory *Ae. aegypti* (Ferreira et al., 2010) and differences in this receptor are believed to be the crucial factor in determining sensitivity (Section 3). Changes in the receptor are also known to cause resistance to the Bin toxin (Section 5).

Toxin binding to the midgut receptors is essential, and soon after the ingestion of Bin crystals, cytopathological alterations can be observed. The pathogenesis associated with *Ls* treatment was first described based on the study on the action of strain SSII-1, which produces only Mtx toxins, on *Cx. quinquefasciatus* larvae (Davidson, 1979). Subsequently, studies showed the cytopathological alterations in midgut cells of larvae treated with different strains all of which produced the Bin toxin as their major toxin (Charles, 1987; de Melo et al., 2008; Silva Filha and Peixoto, 2003; Singh and Gill, 1988). For *Cx. pipiens*, the major alterations observed in the midgut cells are the intense disruption of microvilli, intense cytoplasmic vacuolisations (or cytolysosomes) with broken membranes, pronounced swelling of mitochondria and break-down of the endoplasmatic reticulum. Ultra-structural effects investigated using an *in vitro* processed form of Bin toxin to treat *Cx. pipiens*-cultured cells, showed similar effects as those seen in midgut cells from larvae (Davidson and Titus, 1987). The study of Singh and Gill (1988) also recorded damage in neural and muscles tissues that were detected later than the major effects that were primarily observed in the midgut cells. Treatment of resistant *Cx. pipiens* larvae that lacked the midgut receptors with high concentration of Bin toxin showed minor alterations that were comparable with those observed for *Ae. aegypti*, a refractory species that does not have functional receptors in their midgut (Charles, 1987; de Melo et al., 2008). Physiological studies of the Bin action are not available except for one that shows an inhibition of the oxygen uptake of mitochondria and in the activity of the enzyme choline acetyl transferase in larvae treated with the Bin toxin (Narasu and Gopinathan, 1988).

The mode of action of the Bin toxin, following receptor binding, remains somewhat unclear. Many reports have suggested that to exhibit toxicity, both BinA and BinB components are absolutely required (Broadwell et al., 1990b; Charles et al., 1993; Nicolas et al., 1993; Oei et al., 1990), with optimal activity reported when components are present in approximately equimolar amounts (Davidson et al., 1990). Nevertheless, toxicity of BinA in purified form (Hire et al., 2009) or produced in *B. subtilis* (de la Torre et al., 1989) or in *Bt* (Nicolas et al., 1993) has been reported by some authors. When the *Cx. pipiens* Bin receptor Cpm1 was expressed in Madin Darby canine kidney cells, patch-clamp experiments showed that toxin binding is followed by the induction of currents that are likely to be due to the opening of pores (Pauchet et al., 2005). Experiments with *Culex* cells in culture (Cokmus et al., 1997) and artificial membranes (Schwartz et al., 2001), also suggest that the toxin may be able to form pores and indicated that BinA was better able to form pores than BinB, consistent with the model whereby BinB is the receptor-binding component and BinA forms a pore. In contrast, a separate report described the ability of BinB to interact with artificial membranes and form pores in the absence of BinA (Boonserm et al., 2006). This pore formation was proposed to be through membrane insertion of beta sheet rather than alpha helical structures. Thus, an alternative model for Bin toxicity may involve receptor binding and pore formation by BinB coupled with an unknown role for BinA or by a BinA/BinB complex and this may be supported by the crystal structure of BinB (Srisucharitpanit et al., 2014).

In addition to the possibility of pore formation, a further effect is characteristic of Bin intoxication. The vacuolisation of target cells is seem (Charles, 1987; Davidson, 1988; Pauchet et al., 2005) accompanied by the uptake of labelled toxins into vesicles (Davidson, 1988), a phenomenon that only occurs when both BinA and BinB components are present together (Oei et al., 1992). A detailed study of this phenomenon was carried out using Madin Darby cells expressing the Bin receptor Cpm1 (Opota et al., 2011). This investigation showed the opening of cationic pores in the membrane and demonstrated that the large vacuoles formed in target cells were autophagic. These structures were transient but, having disappeared from the cells, these vacuoles then reappeared following cell division: a novel phenomenon termed post-mitotic vacuolation. The uptake of Bin into the cells, along with their receptor, was shown to be via recycling endosomes; structures that are distinct from the large transient autophagic vesicles. Thus, Bin intoxication induces autophagy, while Bin uptake in separate structures protects it from degradation by targeting to recycling pathways. The overall significance of these events for toxicity remains to be clarified but Bin trafficking may allow it access to tissues beyond the midgut.

The BinA and BinB proteins are related to each other and to a family of Bin-like proteins including Cry49 from *Ls* (see below), Cry35 and Cry36 from *Bt*, and sequences of unknown function from *B. cereus* group strains (e.g. accession number ZP_17404242) and *Chlorobium phaeobacteroides* (accession number Y_911930) (Baumann et al., 1988; Jones et al., 2007). Cry36 acts alone to cause insect mortality, whereas Cry35 requires the 14 kDa Cry34 protein for toxicity and Cry49 requires the three-domain family toxin Cry48 for its function. The various interactions that this family of proteins may require for toxicity, further complicates our understanding of their modes of action.

The Bin toxin proteins themselves are highly conserved. Strains isolated from around the world produce Bin toxins (Priest et al., 1997) but only six variants of BinA (in which nine amino acids are altered) and four variants of BinB (in which six amino acids are altered) have been described (Hire et al., 2009; Humphreys and Berry, 1998; Priest et al., 1997). Two variants, Bin1 and Bin2 have been shown to share the same receptor (Silva-Filha et al., 2004) and cross-resistance between variants is seen (Yuan et al., 2003). Nevertheless, Bin toxin variants can show differential activity against mosquito targets. BinA variants were shown to alter the activity against the marginal target *Ae. aegypti* and to alter the progression of growth and mortality for *Cx. quinquefasciatus* and these effects were localised to amino acids 99 and 104 in this protein (Berry et al., 1993). Reciprocal exchange of the amino acid at position 93 of the BinA protein between the BinA2 variant, which is highly active against *Cx. pipiens* larvae and the BinA4 variant, which is non-toxic to this insect, showed that this residue was also a key determinant of activity (Yuan et al., 2001). Deletion experiments have defined the essential core regions of the Bin toxins. BinA can be truncated by 17 residues at both the N- and C-termini without loss of toxicity. BinB can be truncated by 34 residues at the N-terminus and 53 residues at the C-terminus without loss of toxicity (Broadwell et al., 1990c; Clark and Baumann, 1990, 1991; Limpanawat et al., 2009; Oei et al., 1990; Sebo et al., 1990). Analysis of the binding of non-toxic variants suggested that the N-terminal region of BinB may have a role in interaction with the receptor and its C-terminus, along with both the N- and C-termini of BinA may be involved in the interaction of the two proteins (Oei et al., 1992). Predictions of structural disorder within the BinA and BinB proteins have suggested that the N- and C-termini may be flexible, consistent with a role in protein–protein interactions (Kale et al., 2013). More detailed analysis of the N-terminal region of BinB and receptor binding, confirmed the importance of residues 33–158 in this interaction and, particularly, the sequences Ile-Arg-Phe (residues 85–87) and Phe-Gln-Phe (residues 147–149) (Romão et al., 2011). Mutagenesis studies on BinB indicated that individual substitution of Pro35, Glu36, Phe41 and Tyr42, resulted in reduced activity but all were able to interact with BinA (Singkhamanan et al., 2013). Pro35 and Phe41 to alanine substitutions could bind to the larval midgut at a comparable level to the wild-type BinB but binding of the Tyr42 to alanine mutant was reduced and the Pro35Ala replacement decreased penetration of the membrane. Block mutation of BinB from residues 113–150 showed the protein to have some tolerance to mutation in this region (Singkhamanan et al., 2010). Alanine replacement of individual amino acids Phe149 and Tyr150 resulted in loss of toxicity and loss of midgut binding for the latter mutant. Toxicity could be rescued by replacement of these two residues with other aromatic residues. In BinB, the substitution of Cys67 or Cys161 reduced BinB interaction with BinA and eliminated toxicity while replacement of Cys241 had no effect (Boonyos et al., 2010). In similar experiments with BinA, alanine or serine replacement drastically reduces (Cys195) or abolishes (Cys31, Cys47) activity (Promdonkoy et al., 2008). These three cysteine residues in each protein were shown not to be involved in disulphide bonding and the nature of their roles is not clear at present.

Substitution of charged residues with alanine in BinA produced a loss (Arg97) or reduction in activity (Glu98, Arg101, Glu114) but no change in the ability of BinA to interact with BinB (Sanitt et al., 2008). Substitution of Arg312 has also been shown to eliminate toxicity (Elangovan et al., 2000). Replacement of tryptophan residues at positions 222 and 226 in BinA produces proteins that are still able to permeabilise liposomes but which have lost biological activity (Kunthic et al., 2011). It has been shown that alanine mutants in both BinA and BinB can eliminate toxicity and that, for each protein, a non-toxic variant mutated close to the N-terminus and a non-toxic variant mutated close to the C-terminus, can complement each other to produce a toxic combination (Shanmugavelu et al., 1998).

2.3. Cry48/Cry49

While the Bin toxin is the major sporulation-associated toxin in most *Ls* strains, the ability of the spores of a small number of strains to overcome Bin resistance in mosquitoes, indicated the presence of another toxin in these strains and an approximately 49 kDa protein was identified as a candidate (Pei et al., 2002; Shi et al., 2001). The genes encoding the Cry49 protein and a second crystal protein (Cry48) related to the three-domain toxins of *Bt* has been cloned and expressed (Jones et al., 2007) and show a very narrow target range, so far active only against *Culex* mosquitoes (Jones et al., 2008). It is of interest that the two proteins form a novel type of Bin toxin since both components are necessary for activity and no other three-domain protein has ever been shown to have a requirement for another protein for its activity (Jones et al., 2007). Consistent with the presence of a Bin-type toxin and a three-domain toxin, the cytopathology of *Culex* cells exposed to the Cry48/Cry49 toxin shows features of both toxin types, including the vacuolation observed on Bin intoxication (de Melo et al., 2009) (Fig. 3.2). Similar effects were seen when Bin susceptible larvae were treated with the synergistic combination of Bin and Cry11Aa toxins.

2.4. Mosquitocidal toxin 1

The mosquitocidal toxin 1 (Mtx1) was first discovered in *Ls* strain SSII-1, a strain that shows low toxicity and lacks the Bin toxin (Thanabalu et al.,

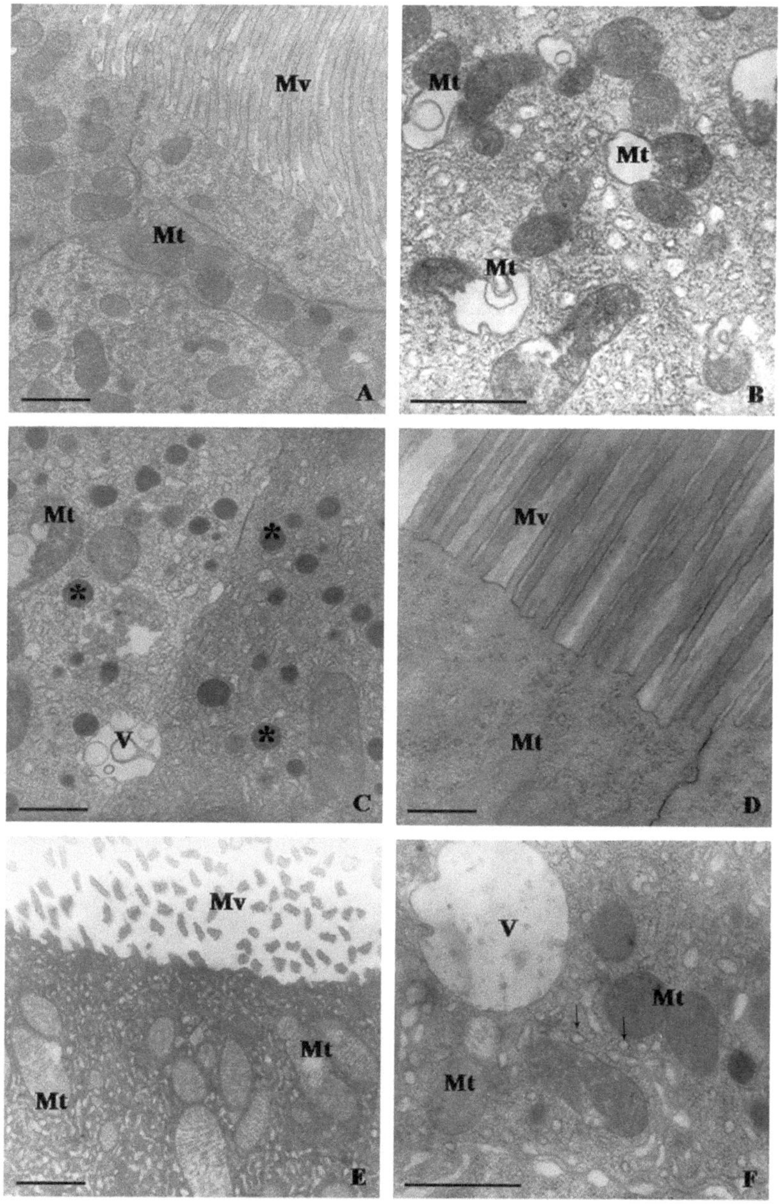

Figure 3.2 Cytopathological effects of Cry48Aa/Cry49Aa toxin to *Culex quinquefasciatus*. Transverse ultrathin sections of the posterior midgut from fourth instar larvae treated with Cry48Aa/Cry49Aa toxin from *Lysinibacillus sphaericus* IAB59 strain. (A) Cells from a non-treated larva rich in microvilli (Mv) and mitochondria (Mt); (B) cells from 1 h-treated larva showing mitochondrial vacuolation (Mt); (C) cells from 1 h-treated larva presenting electron-dense granules (*), cytoplasmic vacuoles (V) and small vesicles from endoplasmic reticulum breakdown; mitochondria (Mt); (D) cells from 1 h-treated larva with preserved microvilli (Mv) and mitochondrial swelling (Mt) at the apical side of the cell; (E) cells from 6 h-treated larva showing evident microvilli damage (Mv) and mitochondrial vacuolation (Mt). (F) Cells from 6 h-treated larva presenting large cytoplasmic vacuoles (V) and small ribosome-coated vesicles (arrows); mitochondria (Mt). Scale bars = 1 μm. *Taken from de Melo et al. (2009).* (See the colour plate.)

1991). The toxin is present in many high- and low-toxicity strains but is lacking in some strains that produce Bin toxin such as LP1-G (Liu et al., 1993; Priest et al., 1997) and is present as a disrupted pseudogene in others (e.g. strain C3-41) (Hu et al., 2008). The toxin is active against both *Cx. quinquefasciatus* and *Ae. aegypti* larvae but has no effect on the mosquito *Toxorhynchites splendens* (Thanabalu, 1992; Thanabalu et al., 1992a). It also has low-level toxicity against *Chironomus riparus* (Partridge and Berry, 2002). Mtx1 acts synergistically with Mtx2 from *Ls* (Rungrod et al., 2009) and Cry11 but is antagonistic with Cyt1Aa (Wirth et al., 2007). In combination with other toxins from mosquitocidal *Bt*, synergy can also be seen and mosquitoes resistant to individual *Bt* toxins may show some cross-resistance to Mtx1 (Wirth et al., 2014).

The gene encoding Mtx1 is preceded by an inverted repeat region, characteristic of a binding site for a regulatory protein (Thanabalu, 1992). The production of a reporter protein driven by the *mtx1* promoter was found to be higher in *B. subtilis* than in *Ls*, suggesting regulation in the latter species (Ahmed et al., 1995) and it has been speculated that the gene upstream of *mtx1* may encode a BglG family regulator of Mtx1 production (Berry, 2012). The toxicity of purified Mtx1 is high, showing equivalent potency to the Bin toxin (Thanabalu et al., 1992a) and the low-level toxicity of *Ls* strain SSII-1 is thought to be due to low levels of Mtx1 production and to the degradation of Mtx1 by the producing bacterium (Thanabalu and Porter, 1995). The proteinase responsible for this degradation has been identified (Wati et al., 1997; Yang et al., 2007b) and is known as sphericase/sfericase (Almog et al., 2003; Yoshida et al., 1977). Some success in protecting the toxin by expression in proteinase negative strains of *Ls* has been reported (Thanabalu and Porter, 1995) but the broad specificity of sfericase has blocked attempts to stabilise the toxin by mutagenesis (Yang et al., 2007b).

Produced as a 100 kDa protein, Mtx1 is processed by gut enzymes to an approximately 27 kDa moiety, with regional sequence similarity to ADP-ribosylating toxins and a C-terminal, 70 kDa moiety containing lectin-like beta-trefoil repeat sequences (Hazes and Read, 1995; Thanabalu et al., 1992a). This cleavage produces the two subunits that enable Mtx1 to function as a two component, AB toxin in which the 70 kDa protein is expected to act as the receptor-binding unit, probably mediated by the lectin-like motifs. This is expected to allow the enzymatic subunit (27 kDa) to enter the susceptible cells where it will modify target proteins by ADP-ribosylation. The C-terminal 70 kDa protein is able to cause morphological changes in *Cx. quinquefasciatus* and *Ae. aegypti* cells in culture but not to *An. gambiae* cells or human HeLa cells (Thanabalu et al., 1993). However, both components are required for toxicity to insects. The ADP-ribosyl transferase subunit has been shown to promote ADP-ribosylation of Mtx1 components and proteins of 38 and 42 kDa from *Cx. quinquefasciatus* cell extracts (Thanabalu et al., 1993). Following activation with chymotrypsin, the kinetics of the Mtx1 ADP-ribosyl transferase have been elucidated with respect to the NAD^+ substrate using soybean trypsin inhibitor as an artificial substrate and ribosylation is seen to occur at arginine residues (Carpusca et al., 2006; Schirmer et al., 2002a). The ADP-ribosylation activity of the catalytic subunit can be inhibited by the 70 kDa protein, which binds non-covalently, mediated by residues Asp273 and Asp275, and a C-terminal region (Carpusca et al., 2004). Furthermore, transfection of the catalytically active subunit into HeLa cells produced cytotoxic effects (Schirmer et al., 2002a), suggesting that specificity of Mtx1 is due to the 70 kDa component. Expression of the enzymatic subunit alone in *Escherichia coli* proved problematic due to toxicity of the protein to host cells. This was shown to be due to ADP-ribosylation of the bacterial elongation factor Tu by the toxin (Schirmer et al., 2002b).

The structure of the catalytic subunit (with a small sequence from the N-terminus of the 70 kDa protein) has been solved (Reinert et al., 2006) as has the structure of the full-length protoxin (Treiber et al., 2008). In the latter form, the putative receptor-binding moiety (70 kDa subunit) can be seen with the four lectin domains curled around the catalytic domain and the authors propose a mechanism for the toxin whereby proteolytic activation occurs at the exposed site between the two subunits; the binding subunit then interacts with glycolipid via the lectin-like domains and is endocytosed with the catalytic unit non-covalently attached. At low pH, the Mtx1 forms a heptamer, membrane insertion of N-terminal segments occurs, leading to translocation of the catalytic core into the cytosol. This translocation also separates the enzyme from the 70 kDa protein, thereby freeing it from inhibition.

2.5. Other Mtx toxins

A further family of related proteins, synthesised during the vegetative phase, has also been identified in *Ls* strains. The best studied member of this family, Mtx2, was the second insecticidal toxin to be isolated from the low-toxicity strain SSII-1 (Thanabalu and Porter, 1996). It is unrelated in structure or function to Mtx1 but is related to Mtx3, later identified in SSII-1 (Liu et al., 1996), and Mtx4 identified in the genome of the high-toxicity strain C3-41 (Hu et al., 2008), a strain that also encodes Mtx2 and Mtx3 along with a pseudogene related to this family of proteins, indicating probable gene duplication in this family of genes (Berry, 2012). In addition, the proteins are related to *Clostridium* epsilon toxin and Cry15, Cry23, Cry33 and Cry38 from *Bt* (de Maagd et al., 2003) in the Pfam ETX_MTX2 superfamily of proteins. The similarity to epsilon toxin suggests that these Mtx toxins act by pore formation. Mtx3 is active against *Cx. quinquefasciatus* larvae and is highly conserved, with only one conservative amino acid replacement reported (Liu et al., 1996). Most studies on this family have focused on Mtx2. This protein shows less conservation than Mtx3. Although a series of individual mutations in the *mtx2* genes did not alter their toxicity to *Cx. quinquefasciatus*, one mutant, from lysine to threonine at residue 224, abolished activity. Double mutants at positions 224 and 279 significantly affected activity against this insect and *Ae. aegypti* so that specificity can, effectively, be switched between the two species (Chan et al., 1996). The Mtx2 family of proteins features a putative N-terminal signal sequence, although their secretion has not been confirmed. Mtx2 can be truncated by up to 23 amino acids at its N-terminus (including this signal sequence) without loss of activity but deletion of only five residues from the C-terminus produced inactive toxin (Phannachet et al., 2010). Mtx2 shows instability on exposure to conditioned media or the *Ls* proteinase sfericase (Yang et al., 2007b) so that, like Mtx1, it appears to be a rather short-lived toxin. Mtx2 is synergistic with Cyt1Aa and Cry11Aa (Wirth et al., 2007) while interactions with other *Bt* toxins is somewhat complex with some cross-resistance exhibited (Wirth et al., 2014).

2.6. Sphaericolysin

Sphaericolysin is a cholesterol dependent cytolysin that was discovered in *Ls* strain A3-2 that was isolated from the crop of *Myrmeleon bore* (ant lion) larvae (Nishiwaki et al., 2007). The protein shows injection toxicity against the cockroach, *Blattella germanica*, and, to a lesser extent, against the common cutworm, *Spodoptera litura*. Sphaericolysin induces pores in erythrocyte membranes and inserts to form a complex with an external diameter of around 35 nm. The residue Tyr187, equivalent to Tyr159 in perfringolysin is important in pore formation and the Tyr to alanine mutant markedly reduced haemolytic activity as did dosing with cholesterol. A cholesterol binding motif is located close to the C-terminus of the protein and is assumed to have a role in membrane interaction. The sequence of sphaericolysin includes an N-terminal signal peptide and secretion of the protein has been demonstrated *in vivo* in *S. litura* larvae (Nishiwaki et al., 2007).

The sphaericolysin family of proteins appears to be well conserved in *Ls* strains, even outside the DNA group 11A that is normally associated with insecticidal activity. A gene with 100% sequence identity to that from strain A3-2 has been found in the group IIB type-strain 7055 (C. Berry, unpublished) and a closely related (91.4% identical) protein has been identified in the *Ls* strain B354, belonging to DNA homology group V (From et al., 2008). Evidence for the presence of related genes in DNA group IV was also shown. The strain B354 protein was shown to cause haemolysis and osmotic protection assays suggested formation of a pore of 5–7 nm.

Very closely related proteins are common amongst other Gram positive bacteria including the *B. cereus* group and *Paenibacillus alvei* (Fig. 3.3). The crystal structure of one of these homologs, anthrolysin, has been solved (PDB accession number 3CQF) and, based on this, a model of the sphaericolysin structure can be built (Fig. 3.4).

2.7. S-layer proteins

A recent report has indicated that the SlpC S-layer proteins, produced during the vegetative stage of growth of *Ls* strains OT4b25, OT4b26, III(3)7 and 2362, can show insecticidal activity against *Cx. quinquefasciatus* mosquitoes (Lozano et al., 2011). This is in contrast to a previous report that the S-layer protein of strain 2362 is non-toxic (Bowditch et al., 1989). SlpC from *Ls* strain C3-41 also shows no toxicity and has been the subject of extensive study (Hu et al., 2008; Li et al., 2013). Previously, the insecticidal activity of a *Bt* S-layer protein has been reported (Pena et al., 2006). Although the toxic variants of the *Ls* SlpC were compared to other SlpC proteins from *Ls* in terms of % identities (>94%), specific comparisons and nucleotide accession numbers were not published (Lozano et al., 2011) so that identification of potential residues responsible for toxicity is not possible at this time.

2.8. Safety issues

The safety of *Ls* to non-target organisms is one of the most prominent features of this entomopathogenic bacterium. It is innocuous for most organisms and this brings the advantage that the lack of negative effects on natural predators allows the possibility of co-introducing other biological agents in

```
s A3-2      MGIK---KTIKFILCLSISLCILNYPSISFAETLDTNSSSVKSK-SDIDTGIANLNY
Ls B354   MVYLRIRNYFKVTKVLVCLSAILCIINYPIISFAATLDTNVANEKSKSNDIDTGIANLNY
Bcer      MIFLNIKKNTKRRKFLTCLLVSLCTINYSSISFAETQASNVTDVTKNASGINTGIANLKY
BtkAlv    MIFLNIKKNTKRRKFLACLLVSLCTINYSSISFAETQASNATDVTKNASGIDTGIANLKY
P alv        MKKKSNHLKGRKVLVSLLVSLQVFAFASISSAAPTE----------PNDIDMGIAGLNY
B14905       MGIR---KTIKFLLCLSISLCLLTYPSISLAETLDTNSS------SDIDTGIANLNY

Ls A3-2   NNREVLAVNGDRVDSFVPKEGLNSNDKFIVVERNKKSLTTSPVDISIIDSMANRTYPGAL
Ls B354   NNSEVLAVNGDKINSFVPKEGINSNDKFIVVERNKKSLTTSPVDISIIDSMVNRTYPGAL
Bcer      NNQEVLAVNGDKVESFVPKESINSNGKFVVVEREKKSLTTSPVDISIIDSVVNRTYPGAV
BtkAlv    NNQEVLAVNGDKVESFVPKESINSNGKFVVVEREKKSLTTSPVDISIIDSVVNRTYPGAV
P alv     NRNEVLAIQGDQISSFVPKEGIQSNGKFIVVERDKKSLTTSPVDISIVDSITNRTYPGAI
B14905    NNREILAVNGDRVDSFVPKEGVSSNDKFIVVERNKKSLTTSPVDISIIDSMANRTYPGAL

Ls A3-2   QLANQAFVDNQPNLLVAKRKPLNISIDLPGMKRENTLTVDNPTYGNVSGAVDELVSTWSE
Ls B354   QLANKAFVDNQPSLLVAKRKPLNISIDLPGMKRENTTTVDNPTYGNVSGAVDELVSTWSE
Bcer      QLANKAFADNQPSLLVAKRKPLNISIDLPGMRKENTITVQNPTYGNVAGAVDDLVSTWNE
BtkAlv    QLANKAFADNQPSLLVAKRKPLNISIDLPGMRKENTITVQNPTYGNVAGAVDDLVSTWNE
P alv     QLANKDFADNQPSLVMAARKPLDISIDLPGLKNENTISVQNPNYGTVSSAIDQLVSTWGE
B14905    QLANQAFVDNQPNLLVAKRKPLNISIDLPGMKRENTLTVDNPTYGNVSGAVDELVSTWSE

Ls A3-2   KYSSTHTLPARLQYSESMVYSKSQIASALNVNAQVLDNSLGIDFNAIANGEKKVMVAAYK
Ls B354   KYSTTHTLPARLQYSESMVYSKSQIASALNVNAQVLDNSLGIDFNAIANGEKKVMVAAYK
Bcer      KYSTTHTLPARMQYTESMVYSKSQIASALNVNAKYLDNSLNIDFNAVANGEKKVMVAAYK
BtkAlv    KYSTTHTLPARMQYTESMVYSKSQIASALNVNAKYLDNSLNIDFNAVANGEKKVMVAAYK
P alv     KYSSTHTLPARLQYAESMVYSQNQISSALNVNAKVLNGTLGIDFNAVANGEKKVMVAAYK
B14905    KYSTTHTLPARLQYSESMVYSKSQIASALNVNAQVLDNSLGIDFNAIANGEKKVMVAAYK

Ls A3-2   QIFYTVSAELPNNPSDLFDDSVDFAELTRKGVSNDAPPVMVSNVAYGRTIYVKLETSSKS
Ls B354   QIFYTVSAEQPNNPSDLFDDSVHFEELTRKGVSNDAPPVMVSNVAYGRTIYVKLETSSKS
Bcer      QIFYTVSAELPNNPSDLFDNSVTFDELTRKGVSNSAPPVMVSNVAYGRTIYVKLETTSKS
BtkAlv    QIFYTVSAELPNNPSDLFDNSVTFDELTRKGVSNSAPPVMVSNVAYGRTIYVKLETTSKS
P alv     QIFYTVSAGLPNNPSDLFDDSVTFAELARKGVSNEAPPLMVSNVAYGRTIYVKLETTSKS
B14905    QIFYTVSAELPNNPSDLFDDSVNFAELTRKGVNNDAPPVMVSNVAYGRTIYVKLETNSKS

Ls A3-2   KDVQAAFKALLKNVNTNVETSAQYKDIFEESSFTAVVLGGDSQKHNQIVSKDFNDIREVI
Ls B354   KDVQAAFKALLKDVNTNIETSGQYKDIFEESSFTAVVLGGDSQKHNQVVSKDFNDIREII
Bcer      KDVQAAFKALLK--NNSVETSGQYKDIFEESTFTAVVLGGDAKEHNKVVTKDFNEIRNII
BtkAlv    KDVQAAFKALLK--NNSVETSGQYKDIFEESTFTAVVLGGDAKEHNKVVTKDFNEIRNII
P alv     NDVQTAFKLLLN--NPSIQASGQYKDIYENSSFTAVVLGGDAQTHNQVVTKDFNVIQSVI
B14905    KDVQAAFKALLKNANTNVETSAQYKDIFEESSFTAVVLGGDSQKHNQIVSKDFNDIREVI

Ls A3-2   KDNGEFSLKNPAYPISYTSVFLKDHSIAAVHNNTDYIETTATEYSKGKIILDHYGAYVAQ
Ls B354   KDNGEFSLKNPAYPISYTSVFLKDNSIAAVHNNTDYIETTATEYSKGKIILDHYGAYVAQ
Bcer      KDNAELSLKNPAYPISYTSTFLKDNATAAVHNNTDYIETTTTEYSSAKMTLDHYGAYVAQ
BtkAlv    KDNAELSLKNPAYPISYTSTFLKDNSTAAVHNNTDYIETTTTEYSSAKMTLDHYGAYVAQ
P alv     KDNAQFSSKNPAYPISYTSVFLKDNSIAAVHNNTEYIETKTTEYSKGKIKLDHSGAYVAQ
B14905    KDNGEFSLKNPGYPISYTSVFLKDHSIAAVHNNTDYIETTATEYSKGKIILDHYGAYVAQ

Ls A3-2   FEIAWDEFSYDENGNEVLTHKTWDGNWRDKTAHFSTVIPLPANSKNIRIYARECTGLAWE
Ls B354   FEVAWDEISYDENGNEVLSHKTWDGNWKDKTAHFSTVIPLPPNAKNIRIYARECTGLAWE
Bcer      FDVSWDEFTFDQKGNEVLTHKTWDGSGKDKTAHYSTVIPLPPNSKNIKIVARECTGLAWE
BtkAlv    FDVSWDEFTFDQKGNEVLTHKTWDGSGKDKTAHYSTVIPLPPNSKNIKIVARECTGLAWE
P alv     FEVYWDEFSYDADGQEIVTRKSWDGNWRDRSAHFSTEIPLPPNAKNIRIFARECTGLAWE
B14905    FEVAWDEFSYDENGNEVLTHKTWDGNWKDKTAHFSTVIPLPANSKNIRIYARECTGLAWE

Ls A3-2   WWRTVIDEYNVPLSNEIKVSIGGTTLYPTGSIE
Ls B354   WWRTIIDEYNVPISNEIKVSIGGTTLYPTGNIN
Bcer      WWRTIINEQNVPLTNEIKVSIGGTTLYPTASISH
BtkAlv    WWRTIINEQNVPLTNEIKVSIGGTTLYPTASISH
P alv     WWRTVVDEYNVPLASDINVSIWGTTLYPKSSITH
B14905    WWRTVIDEYNVPMSNEIKVSIGGTTLYPTGSIE
```

breeding sites to achieve an additional level of control. The known insecticidal toxins from *Ls* strains have narrow spectra of action, and screening of other possible target species for these toxins confirmed the high specificity within the Diptera group, which is limited to culicid species, further restricted with respect to some *Aedes* and *Ochlerotatus* species (Table 3.2). The safety for other invertebrate species, especially those insects and other organisms in co-habitation with mosquito larvae in the aquatic environment, has been recorded through laboratory evaluations as well as field trials. Among the groups of organisms that are not affected by standard doses of *Ls* are associated fauna of invertebrates and vertebrates of breeding sites, which includes macroinvertebrates, insects belonging to Odonata, Coleoptera, Diptera orders, crustaceans, amphibians and fish (Brown et al., 2004; Davidson et al., 1977, 1984; Lacey and Mulla, 1990; Lacey and Siegel, 2000; Mulla et al., 1984b; Mulligan et al., 1978; Rodcharoen et al., 1991; Sternberg et al., 2012; Yousten, 1984a; Yousten et al., 1991, 1992). A more extended field study on wetlands, evaluated the possible impact of *Ls* treatments for mosquito control on invertebrates associated with this environment (Merritt et al., 2005). Quantitative parameters to measure abundance and diversity throughout a 3-year period indicated the lack of effects due to exposure in the groups analysed. The effects of *Ls* on mammals were assessed in the early days of its development and studies based on specific procedures confirmed the safety of this microbial agent for the group (Shadduck et al., 1980; Siegel and Shadduck, 1990).

Another safety concern is the fate of *Ls* spores in the environment. Laboratory and field studies showed that spores can recycle in specific conditions, for instance, in the larval cadavers but this feature has been seen rather as an advantage, because it can provide extended control in some breeding sites (Charles and Nicolas, 1986; Matanmi et al., 1990; Nicolas et al., 1987). It has been also demonstrated that recycling does not occur

Figure 3.3 Alignment of sphaericolysin and related proteins. Identical residues are shaded green or cyan, spaces introduced to optimise the alignment are shaded grey. The sequences shown are *Ls* A3-2 (sphaericolysin from *Lysinibacillus sphaericus* strain A3-2, accession number BAF62176), *Ls* B354 (sphaericolysin from *L. sphaericus* strain B354, accession number EU043116), Bcer (anthrolysin from *Bacillus cereus* accession number ZP_03231124), BtkAlv (alveolysin from *B. thuringiensis* subsp. *kurstaki* accession number ZP_04117355), P alv (*P. alvei* alveolysin accession number ZP_10866820), B14905 (perfringolysin O precursor from *Bacillus* sp B14905 accession number ZP_01724867). Alignment generated using ClustalX (Thompson et al., 1994). (See the colour plate.)

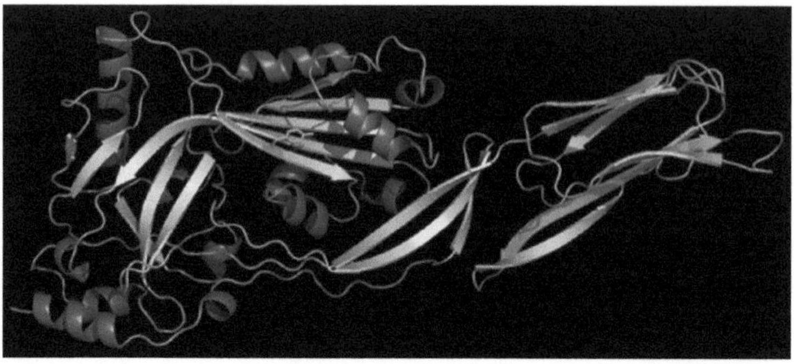

Figure 3.4 Modelled structure of sphaericolysin. The structure of sphaericolysin from *Lysinibacillus sphaericus* strain A3-2 was generated using Swiss model (Schwede et al., 2003) based on the structure of anthrolysin O from *Bacillus anthracis* (sequence identical to the *B. cereus* anthrolysin shown in Fig. 3.3, PDB accession number 3CQF, Bourdeau et al., 2009). Model coloured according to secondary structure: alpha helix (brown), beta sheet (yellow), other (grey). The putative cholesterol binding loop is shown in green and residue Tyr 187 in blue. (See the colour plate.)

in a number of environments such as treated water and dried environments that are re-flooded, which do not ensure the conditions needed for the spore germination (Davidson et al., 1984). Investigation of the fate of spores in non-target invertebrates shows that they are eliminated and suggested that their growth was not maintained in these organisms (Yousten et al., 1991, 1992, 1995). The accumulation of spores in the soil and aquatic environments can also be a concern associated with the intensive use of these biolarvicides. (Guidi et al., 2011) performed a quantitative study of the fate of spores after *Ls* application in catch basins and concluded that the number of spores did not increase significantly during the trial and the indigenous microbiota was not affected.

Safety studies of the Mtx and Cry48/Cry49 toxins have been limited since these toxin factors have not been effectively used as major active principles in the commercial preparations. However, laboratory studies have shown their narrow spectra of action, in particular, Cry48/Cry49, and it is likely that there will be no safety concerns related to their potential utilisation.

3. RECEPTORS OF THE BINARY TOXIN

3.1. Binding of the binary toxin to larvae midgut

The Bin toxin displays the highest activity among the set of *Ls* toxins. After the ingestion of Bin crystals and successful midgut processing as described in

Section 2, the activated form of the toxin recognises and binds to the midgut epithelium of susceptible larvae. The cytopathological effects and the lethal action observed in Bin-treated larvae depend, essentially, on the ability of this toxin to bind to the midgut cells. Treatment of larvae with labelled-toxin and further analysis of the midgut showed the interaction of the Bin toxin along the epithelium, and different patterns were described (Davidson, 1988, 1989; Oei et al., 1992). For *Cx. pipiens*, the most susceptible species, Bin displays a marked and regionalised binding to the gastric caeca and posterior midgut (Fig. 3.5). For anopheline larvae such as *An. gambiae*, *An. stephensi*, *An. albimanus* and *An. quadrimaculatus*, the binding to their midguts was variable and less defined than that found for *Cx. pipiens*. For *Ae. aegypti*, Bin refractory larvae, this interaction was very weak, compared to the Bin binding for the other species. The midgut binding patterns observed are directly correlated to the *in vivo* susceptibility for the species studied.

Quantitative assays that were performed to measure the Bin binding affinity for the proteins from the midgut, confirmed this association. *In vitro* binding assays between radiolabelled toxin and midgut brush-border membrane fractions (BBMF) enriched with proteins from the microvilli of apical cells, showed that Bin toxin binds with high affinity to BBMF from *Cx. pipiens* and *Cx. quinquefasciatus* larvae (Nielsen-Leroux and Charles, 1992; Nielsen-Leroux et al., 1995, 1997, 2002; Oliveira et al., 2004; Silva-Filha et al., 1997, 2004, 2008). Binding is saturable, reversible and

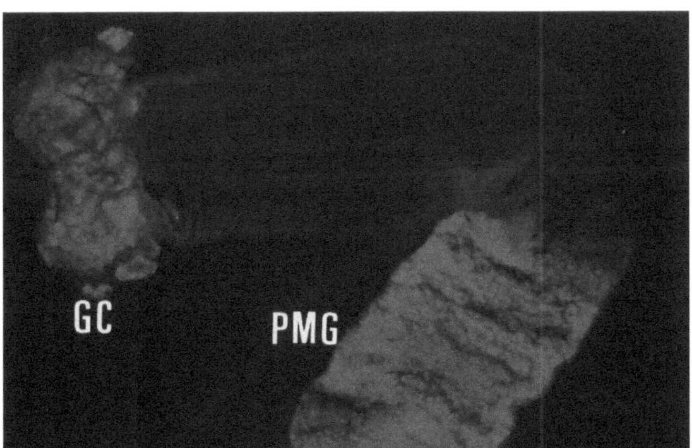

Figure 3.5 Binding of fluorescent-labelled binary toxin regionalized in the gastric caeca (GC) and posterior midgut (PMG) of *Culex quinquefasciatus* larvae. *Figure kindly provided by Colin Berry, Cardiff University (UK).* (See the colour plate.)

the values of the dissociation constants recorded in different studies (K_d = 5–20 nM) indicated the high affinity of the complex formed by the toxin receptor in these species (Fig. 3.6). Specific binding of Bin toxin was also detected for the BBMF of *An. gambiae* and *An. stephensi*, however, the plateau of saturation and affinity (K_d = 30–110 nM) did not reach the same levels detected for BBMF samples from the *Cx. pipiens* complex (Silva-Filha et al., 1997). The LC of Bin toxin for these species is, overall, 10-fold higher than that for *Cx. pipiens* (Davidson, 1989). The level of specific binding detected for the BBMF of *Ae. aegypti* is equivalent to the non-specific binding detected, and the data obtained do not fit to the ligand–receptor model (Nielsen-Leroux and Charles, 1992). The Bin LC for this species is between 100 and 1000-fold higher, compared to that for *Cx. pipiens*. The availability and binding capacity of midgut molecules that act as receptors for the Bin toxin is likely to be the main factor that determines the status of susceptibility among related mosquito species (Table 3.3). The findings regarding the mechanism of resistance to the Bin toxin have reinforced the central role of the midgut receptors for the Bin toxin mode of action (see Section 5).

3.2. Receptors

Midgut-bound proteins from insect larvae play an essential role as receptors for insecticidal proteins from entomopathogenic bacteria. Aminopeptidases (APNs), cadherins (CADs), alkaline phosphatases (ALPs) and α-amylase have been characterised as receptors to the Cry toxins from *Bt* strains in lepidopteran and mosquito larvae (Abdullah et al., 2006; Chen et al., 2009a,b; Fernández et al., 2006; Fernandez-Luna et al., 2010; Hua et al., 2008, 2009; Jurat-Fuentes and Adang, 2004; Knight et al., 1994; Luo et al., 1997; Vadlamudi et al., 1995; Valaitis et al., 2001). More information on receptors of *Bt* toxins is presented in Chapter 2. The receptors of the Bin toxin from *Ls* are α-glucosidases and they were characterised in three species as follows: Cpm1 for *Cx. pipiens* maltase 1 (Darboux et al., 2001; Silva-Filha et al., 1999), Cqm1 for *Cx. quinquefasciatus* maltase 1 (Romão et al., 2006) and Agm3 for *An. gambiae* maltase 3 (Opota et al., 2008). *Aedes aegypti* has also a gene that encodes an orthologue of the Bin toxin receptor, Aam1 protein (*Ae. aegypti* maltase 1) that displays 74% identity to the Cqm1 α-glucosidase; however, Aam1 is not able to bind to Bin toxin (Ferreira et al., 2010; Nene et al., 2007; Opota et al., 2008; Romão et al., 2006).

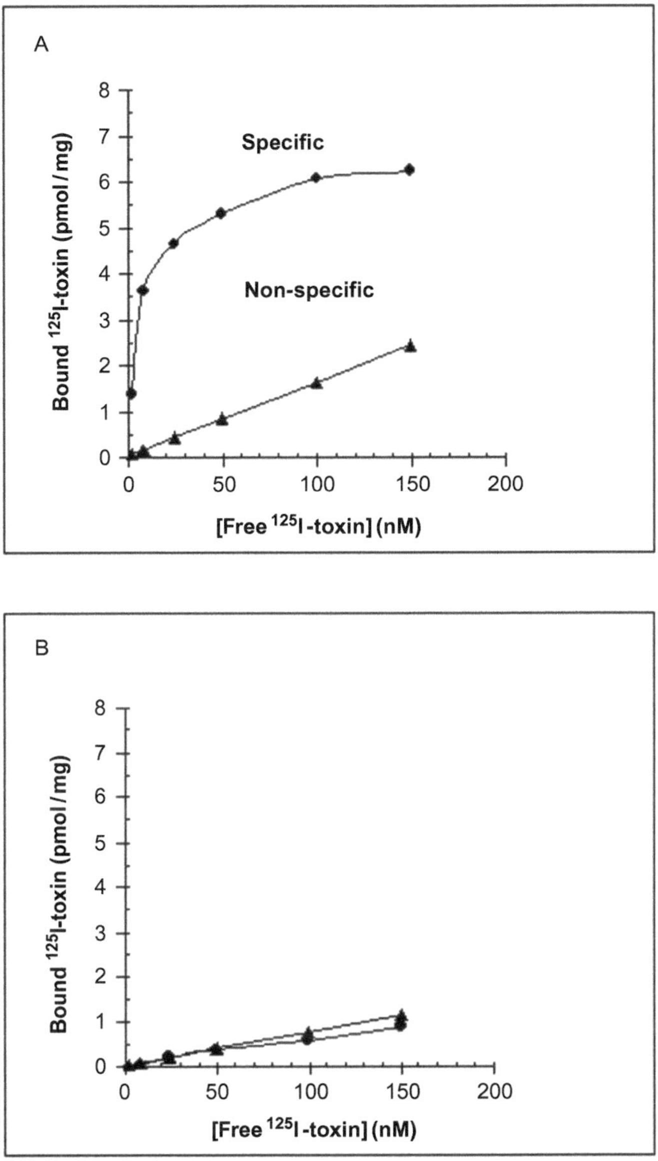

Figure 3.6 *In vitro* saturation binding assays between binary toxin and brush-border membrane fractions (BBMF) from fourth instar *Culex quinquefasciatus* larvae from a susceptible colony (A) and from CqRL1/2362, a *Lysinibacillus sphaericus*-resistant colony (B). *Taken from Oliveira et al. (2004).*

Table 3.3 Mosquito susceptibility to *Lysinibacillus sphaericus* and the capacity of the Binary (Bin) toxin to interact with larval midgut

Characteristic	*Culex pipiens* [a]	*Cx. quinquefasciatus* [a]	*Anopheles gambiae* [b]	*An. stephensi* [b]	*Aedes aegypti* [c]
Larval susceptibility	High	High	Moderate	Moderate	Refractory
Bin binding to midgut	Regionalized	ND[d]	Diffuse	Diffuse	Non-specific
Bin-specific binding to BBMF[e]	Yes	Yes	Yes	Yes	No
Bin binding affinity or K_d (nM)	6–20	5–15	30	110	Nd
No. of receptors B_{max} (pmol/mg)	3–8	6–10	5	4	Nd
Cytopathological effects	Yes	Yes	Yes	Yes	Minor

[a]Charles (1987), Davidson (1988, 1989), de Melo et al. (2008), Nielsen-Leroux and Charles (1992), Nielsen-Leroux et al. (1995), Nielsen-Leroux et al. (1997), Oei et al. (1992), Silva-Filha et al. (2004), Silva-Filha et al. (2008), Silva Filha and Peixoto (2003) and Singh and Gill (1988).
[b]Charles (1987), Davidson (1988, 1989) and Silva-Filha et al. (1997).
[c]Charles (1987), Davidson (1988, 1989), Ferreira et al. (2010) and Nielsen-Leroux and Charles (1992).
[d]Not determined.
[e]Brush-border membrane fractions.

Those α-glucosidases (EC 3.2.1.20) belong to the α-amylase family that is composed of several enzymes that have the ability to hydrolase starch (Janecek, 1997). This was the first amylase *sensu lato* identified as a receptor of insect toxins, followed by the report of an α-amylase identified as the receptor of Cry4Ba and Cry11Aa toxins in *An. albimanus* larvae (Fernandez-Luna et al., 2010). The primary role of the α-glucosidases is the digestion of carbohydrates through their ability to hydrolyse α-1-4 links between glucose residues of carbohydrates (Krasikov et al., 2001). The studies of these enzymes in mosquitoes are still scarce despite their essential role for the digestion of carbohydrates. A recent study showed that the genes coding for α-glucosidases from *Cx. quinquefasciatus*, *An. gambiae* and *Ae. aegypti* are organised in two clusters, and each one contains three to five genes (Gabrisko, 2013). Some α-glucosidases have been characterised in the salivary glands, since these enzymes are fundamental for the hydrolysis of sucrose, a main component of the adult diet (Marinotti and James, 1990; Marinotti et al., 1996). The α-glucosidases have also importance in the digestion of carbohydrates the activity of which is mainly located in the posterior midgut as characterised in *An. aquasalis* and *An. gambiae* (Souza-Neto et al., 2007; Zheng et al., 1995). In whole extracts and BBMF from *Cx. quinquefasciatus* larvae, four α-glucosidases were detected through in-gel catalytic assays, using 4-methylumbelliferyl α-D-glucopyranoside as a substrate (Romão et al., 2006). Among them, only Cqm1 is recognised by a polyclonal antibody raised against this protein and this suggests that the remaining α-glucosidases did not share a high level of identity with Cqm1 (Romão et al., 2006).

The Cpm1 α-glucosidase was the first receptor characterised for the Bin toxin in *Cx. pipiens* larvae and it shares 97% and 66% identity with the Cqm1 and Agm3 orthologs, respectively. The genes encoding these proteins are organised in a similar manner in three exons and two introns that are positioned in the same regions (Darboux et al., 2001; Opota et al., 2008; Romão et al., 2006). The open reading frames of the Cpm1/Cqm1 and Agm3 genes encode proteins of 580 and 588 residues, respectively (Table 3.4). Among the conserved features observed in the protein sequences, there are four conserved α-glucosidase motifs, some predicted glycosylation sites and a conserved signalling sequence for a glycosylphosphatidylinositol (GPI) anchor at the C-terminal end.

Most studies of these proteins were based on their expression using a *Spodoptera frugiperda* (Sf9) cell line that proved to be an adequate system for their characterisation (Darboux et al., 2002, 2007; Opota et al., 2008).

Table 3.4 Features of four ortholog α-glucosidases from mosquito larvae

	Culex pipiens	***Cx. quinquefasciatus***	***Anopheles gambiae***	***Aedes aegypti***
	Cpm1[a]	**Cqm1[b]**	**Agm3[c]**	**Aam1[d]**
Gene ORF (bp)	1740	1740	1764	1746
Protein no. of residues	580	580	588	582
Protein identity to Cpm1 (%)	100	97	66.4	74.2
Protein-predicted mass (kDa)	66	66	67	66
Protein observed mass (≈kDa)	66	64–66	67	73
No. of predicted glycosylation sites	3	3	3	6
In vivo glycosylation[e]	Nd[f]	No	Nd	Yes
Location along the midgut	Caeca and posterior cells	Midgut	Posterior cells	Midgut
α-Glucosidase activity	Yes	Yes	Nd	Yes
Binding to the Bin toxin	Yes	Yes	Yes	No

[a]Darboux et al. (2002, 2007) and Silva-Filha et al. (1999).
[b]Romão et al. (2006).
[c]Opota et al. (2008).
[d]Ferreira et al. (2010), Opota et al. (2008) and Romão et al. (2006).
[e]Glycosylation was verified using PNGaseF.
[f]Not determined.

The expression of Cpm1 in Madin and Darby kidney canine (MDKC) cells was also successfully achieved. Both catalytic activity and Bin binding capacity were observed, in addition, this recombinant receptor was able to mediate the cytopathological effects derived from the treatment with Bin toxin (Pauchet et al., 2005). The expression of Cpm1/Cqm1 as membrane-bound proteins is essential for the activity of Bin toxin on the larva and they are bound to the epithelium through a GPI anchor. These proteins can be solubilized from the midgut through the action of the enzyme phosphatidylinositol phospholipase C that specifically releases GPI-bound proteins from the cell surface (Darboux et al., 2002; Ferreira et al., 2010; Silva-Filha et al., 1999). Mutations in the *cpm1/cqm1* genes that prevent the expression of these molecules as GPI-bound proteins are the most important resistance mechanism shown in *Cx. pipiens* larvae (Section 5).

A second essential feature of Cpm1/Cqm1 proteins as receptors of the Bin toxin is their conformational state. The characterisation of these receptors was performed using proteins from larvae or recombinant proteins, under their native state, since denatured samples failed to bind the Bin toxin (Romão et al., 2006; Silva-Filha et al., 1999). The analysis of Cqm1 showed that its apparent molecular mass differs between 66 and 85 kDa under denaturing and semi-denaturing conditions, respectively, suggesting that the native protein assumes a different conformation (Ferreira et al., 2010).

The direct binding of the Bin toxin to one class of receptors, also contrasts with the mode of action of Cry toxins that seem to rely on the sequential binding to two midgut receptors (Arenas et al., 2010; Bravo et al., 2004, 2007). In this model, the Cry toxins bind primarily to CADs, and this step is essential to confer a conformational change to oligomers that are then able to bind with high affinity to GPI-anchored receptors such as APNs and/or ALPs. Taking into account the complexity of the interactions between the Cry toxins and their receptors, the Bin toxin binding to its single class of receptors can be considered a vulnerable step in mode of action of *Ls*.

3.3. Comparative analysis of the Cqm1 and Aam1 α-glucosidases

The comparative studies of the orthologs Cqm1 and Aam1 provided data to understand the molecular basis of Bin toxin action on culicid larvae and, in particular, the refractoriness of *Ae. aegypti* larvae. Aam1 has 74% identity and 83% similarity to Cqm1, higher levels than those found for the Agm3 protein that is the receptor for the Bin toxin in *An. gambiae*, a moderately susceptible species. The open reading frame of the *aam1* gene is organised in a similar fashion, however, the entire gene is significantly bigger compared to the ≈1.8 kb observed for the *cpm1/cqm1/agm3* orthologs, due to the great size of the two introns found in the *aam1* sequence (Nene et al., 2007). Previous studies ruled out some factors that could be the reason for the refractoriness of *Ae. aegypti*. The processing of Bin toxin in *Ae. aegypti* midgut occurs in a similar manner as in *Cx. pipiens* and this is not the reason for the refractoriness of the latter species (Nielsen-Leroux and Charles, 1992). Subsequently, it was demonstrated that Aam1 is produced as a GPI-anchored protein and this protein is expressed during larval and adult stages. In midgut brush-border membranes from *Ae. aegypti* larvae, Aam1 is

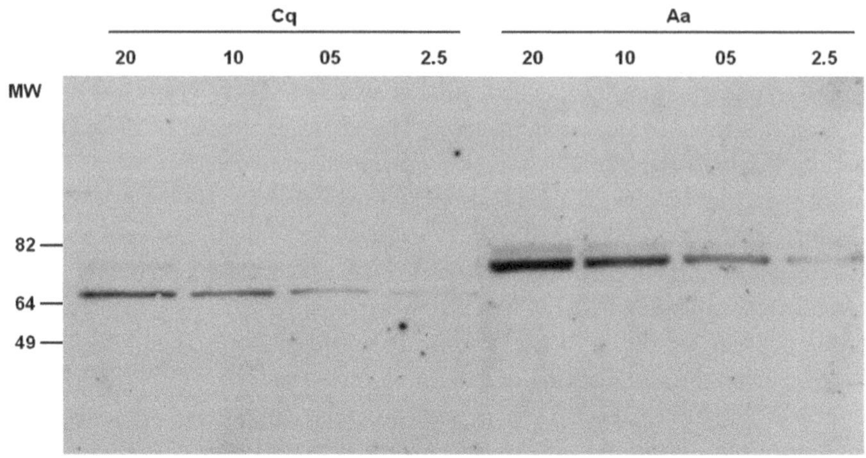

Figure 3.7 Immunoblotting of midgut brush-border proteins from *Culex quinquefasciatus* (Cq) and *Aedes aegypti* (Aa) larvae. Protein samples (2.5–20 μg) were separated on SDS-PAGE, transferred to nitrocellulose membrane and submitted to immunodetection with an anti-Cqm1 antibody. MW, molecular weight markers in kDa. *Taken from Ferreira et al. (2010).*

expressed in a comparable level to that observed for Cqm1 (Fig. 3.7), however, this protein does not display capacity to bind the Bin toxin (Ferreira et al., 2010). One major difference between Cqm1 and Aam1 is N-glycosylation and, although both sequences present predicted sites (Table 3.4), only Aam1 was found to have carbohydrates added that significantly increase its size, while the glycosylation of Cqm1, if present, could not be detected (Ferreira et al., 2010). Cqm1 treatment with the endoglycosidase PNGase F does not abolish its capacity to bind to BinB, suggesting that this interaction does not involve glycans. On the other hand the hypothesis that glycans from Aam1 could hide or prevent the access of the Bin toxin to the binding site was refuted because the removal of carbohydrates by deglycosylation with PNGase F did not provoke any effect in this aspect (Ferreira et al., 2010). The differential role of this major post-translational modification in Aam1 remains unknown. Further investigation using wild and mutant Cqm1 and Aam1 proteins expressed in Sf9 cells showed that non-conserved residues of Cqm1 are needed for the BinB subunit binding (Ferreira et al., 2014). This study showed that a segment of Cqm1 N-terminus is responsible for binding to BinB subunit and the replacement of the doublet 159GG160, located in a loop region, by the respective residues (KL) from Aam1 protein abolished binding, showing that they are required for this interaction. This loop from the Aam1 ortholog shows other divergent residues and the insertion of five amino acids that could be implicated in the lack of capacity of this protein to bind BinB. Those residues showed to be essential for binding, probably by making possible the access of Bin toxin to amino acids that may act as binding sites themselves. Additional investigation of these proteins will certainly be important to understand the molecular basis of Bin toxin selectivity in the insect midgut.

The studies of the molecules involved in the mode of action and selectivity of the Bin toxin has been restricted to the midgut receptors. Despite the role of α-glucosidases as the major target site of the Bin toxin, it is possible that novel molecules and pathways could be involved in the mode of action and selectivity of *Ls*, as demonstrated by recent studies in the action of Cry toxins to mosquitoes and lepidopterans. The immune defence involving the mitogen-activated protein kinase p38 (MAPK p38) pathway seems to participate in the response to pore-forming toxins and it is responsible for protecting *Ae. aegypti* larvae from the action of Cry11Aa toxin (Cancino-Rodezno et al., 2010; Porta et al., 2011). Silencing of genes from this pathway rendered larvae hypersensitive to that toxin indicating its importance to modulate the Cry toxin action. Mutations in the ABC transporters were directly linked to the resistance of lepidopteran larvae to *Bt* Cry1A toxins also proving their involvement in the mode of action (Atsumi et al., 2012; Gahan et al., 2010). The role of other molecules in the mode of action of Bin toxin, therefore, remains a field to be explored.

4. APPLICATIONS FOR MOSQUITO CONTROL

4.1. Field trials

Following the discovery and characterisation of potent *Ls* strains, such as 1593, 2362 and 2297, field trials against several mosquito species have been conducted in countries located at different latitudes, including India, Malaysia, Thailand, USA, Australia, Ivory Coast, Burkina Faso, Tanzania, Mali and other African countries, as reviewed by Lacey (2007). Small- and medium-scale tests targeting mosquito larvae in the genera *Culex*, *Anopheles*, *Aedes*, *Ochlerotatus*, *Mansonia* and *Psorophora* breeding in a large variety of site types provided evidence of persistent larvicidal activity and the potential for spore recycling in polluted aquatic habitats. These findings attracted the interest of the scientific community in areas like entomology, microbiology and tropical health, stimulating further field studies, especially against those mosquito species breeding in environments characterised by organically enriched water. As a result, fundamental information about the activity of this bacterium, its efficacy and its persistence in *Culex* breeding sites under natural conditions has been produced. In the following years, further field experiments against the vectors of lymphatic filariasis and malaria were conducted in many places in Africa (Barbazan et al., 1997; Hougard et al., 1993; Karch et al., 1991; Ragoonanansingh et al., 1992; Seyoum and Abate, 1997; Skovmand and Sanogo, 1999) and in the South and Central Americas (Arredondo-Jimenez et al., 1990; Berrocal et al., 2000; Berti and Gonzales, 2004; Blanco-Castro et al., 2000; Consoli et al., 1997; Montero Lago et al., 1991; Regis et al., 1995, 1996, 2000a,b; Rivera et al., 1997; Rodrigues et al., 2008; Silva-Filha et al., 2001; Suarez and Morales, 1999).

These small- and medium-scale field trials to evaluate *Ls*-based products in aquatic habitats under different climatic conditions helped to build a significant body of knowledge on appropriate formulations, doses, methods for product application and for measuring the impact of treatments on mosquito populations. Table 3.5 presents a set of field evaluations of experimental or commercial products against mosquito larvae in a wide range of breeding sites. *Cx. quinquefasciatus* and *Cx. pipiens* are undoubtedly the most important target species because *Ls* displays the highest activity to this group and is able to show a prolonged activity in their breeding sites (Davidson et al., 1984; Des Rochers and Garcia, 1984; Mulla et al., 1984a; Nicolas et al., 1987). The abundance of these species in urban areas is due to the availability of breeding sites with organically polluted water, which offer optimal conditions for their development. Waste water lagoons resulting from human activities (e.g. dairies) can also be excellent breeding sites for some *Culex* spp., and such environments can be responsible for a huge production of mosquitoes (Jones et al., 1990; Matanmi et al., 1990; Mulla et al., 1984a, 1988a). *Anopheles* is the second most important group that has been successfully targeted by *Ls*. Control strategies to fight different species that are *Plasmodium* vectors have adopted *Ls*-based larvicides (Table 3.5). Some important anopheline vector species (e.g. *An. gambiae*, *An. stephensi*, *An. darlingi* and *An. nuneztovari*) are susceptible to *Ls*, and field trials have been conducted in a diversity of scenarios in malaria-endemic regions. The control of anophelines is often complex due to factors such as their ecological features, the diversity and extension of breeding sites, and constraints to establishing larvicide applications in those areas. Nevertheless, *Ls* has been an important ally in *Anopheles* control programmes, especially when it is integrated with other strategies. *Psorophora*, *Mansonia*, *Aedes* and *Ochlerotatus* are also targets

Table 3.5 Field trials using *Lysinibacillus sphaericus*-based products to control mosquito larvae

Targeted species	Environment	References
Culex quinquefasciatus	Containers	Kuppusamy et al. (1987), Lacey et al. (1984)
	Sod-lined potholes	Lacey et al. (1988b)
	Polluted sites, polluted ponds, cesspools, cesspits, catch basins, settling basins	Hougard (1990), Karch et al. (1991, 1990), Kumar et al. (1996), Mulla et al. (1997, 1999), Nicolas et al. (1987), Paing et al. (1987), Siegel and Novak (1997, 1999), Skovmand and Bauduin (1997), Skovmand and Sanogo (1999), Su and Mulla (1999), Su (2008)
	Diverse urban habitats	Andrade et al. (2007), Barbazan et al. (1997), Hougard et al. (1993), Mulla et al. (2001), Regis et al. (1995, 1996, 2000b), Silva-Filha et al. (2001), Skovmand et al. (2009)
	Dairy wastewater	Jones et al. (1990), Mulla et al. (1984a)
Cx. quinquefasciatus, *Cx. annulirostris*, *Anopheles funestus*, *An. albimanus*	Polluted pools	Barbazan et al. (1998), Brown et al. (2004), Montero Lago et al. (1991), Ragoonanansingh et al. (1992)
Cx. quinquefasciatus, *Cx. stigmatosoma*, *Cx. tarsalis*	Experimental containers	Su and Mulla (1999)
Cx. quinquefasciatus, *Cx. tritaeniorhynchus*, *Cx. vishnui*	Diverse urban habitats	Yadav et al. (1997)
Cx. pipiens	Artificial pools, settling basins	Berry et al. (1987), Karch et al. (1990)
Cx. pipiens, *Cx. restuans*	Catch basins	Siegel and Novak (1997)

Continued

Table 3.5 Field trials using *Lysinibacillus sphaericus*-based products to control mosquito larvae—cont'd

Targeted species	Environment	References
Cx. pipiens, Ochlerotatus trivittatus,	Waste water lagoons	Berry et al. (1987)
Cx. peus, Cx. stigmatosoma	Dairy wastewater	Matanmi et al. (1990), Mulla et al. (1988a)
Cx. nigripalpus	Wastewater	Lacey et al. (1988a)
Cx. restuans	Woodland pools	Lacey et al. (1988b)
Culex spp	Sod-line ponds	Lord (1991)
An. gambiae	Irrigation ponds, sunlit ponds, rain puddles	Fillinger et al. (2003), Karch et al. (1991), Skovmand and Bauduin (1997), Skovmand and Sanogo (1999)
An. stephensis	Artificial breeding sites	Kumar et al. (1994)
An. arabiensis	Pools, rice fields water ditches	Romi et al. (2003), Shililu et al. (2003)
An. quadrimaculatus, Psorophora columbiae, An. crucians	Rice fields	Dennett et al. (2001), Dennett and Meisch (2000), Lacey et al. (1986, 1988b)
An. albimanus	Lake habitats	Rivera et al. (1997)
An. albimanus, Culex spp	Experimental potholes	Arredondo-Jimenez et al. (1990)
An. culicifacies	Rice fields	Sundararaj and Reuben (1991)
An. darlingi	Artificial breeding sites	Berrocal et al. (2000), Galardo et al. (2013), Rodrigues et al. (2008)
An. nuneztovari	Field pools	Rojas et al. (2001)
An. marajoara, An. triannulatus, An. braziliensis	Field pools	Moreno et al. (2010)
An. aquasalis	Field ponds	Berti and Gonzales (2004), Moser et al. (2002, 2012)
Mansonia spp, *Ma. indiana, Ma. uniformis, Ma. dyari*	Rice ditches and rice ponds	Floore and Wardz (2009), Pradeepkumar et al. (1988), Yap et al. (1991)

Table 3.5 Field trials using *Lysinibacillus sphaericus*-based products to control mosquito larvae—cont'd

Targeted species	Environment	References
Ps. columbiae	Rice field plots, flood-water habitats	Bowles et al. (1990), Groves and Meisch (1996), Lacey et al. (1986), Lord (1991)
Ps. columbiae, Oc. nigromaculis	Irrigated fields	Mulla et al. (1985), Mulla et al. (1988b)
Oc. nigromaculis	Irrigated pastures	Mulla et al. (1988b)
Oc. triseriatus, Cx. restuans, Cx. pipiens	Tires	Siegel and Novak (1997, 1999), Siegel et al. (2001)
Aedes vexans	Riparian woodland	Becker (2003)
Culiseta incidens	Tires	Kramer (1990)

of *Ls*-based larvicides, and there are successful examples of trials for controlling these mosquitoes in their typical habitats, which are commonly situated in rural areas, such as irrigated fields, pastures, rice fields and other flooded environments (Table 3.5). Numerous trials to control mosquitoes based on *Ls* larvicides have been conducted; however, it is not possible to establish a common methodology to employ this agent, and each programme should be designed according to the target's biology and ecology, the biotic and abiotic conditions, and other factors that play a role in the proliferation of that species.

4.2. Factors affecting field performance

Since the first few years of studies, a cast of biotic and abiotic factors influencing the larvicidal activity and its persistence in the environment was identified, and this knowledge was essential to plan strategies for the operational use of microbial larvicides. Relevant factors limiting or facilitating the toxic action of bacteria are the intrinsic susceptibility of the mosquito species, the feeding behaviour of larvae, climatic variables (e.g. temperature, rainfall and solar radiation), habitat conditions (e.g. water temperature, organic content, larval density, density of non-target organisms, water flow and sunlight exposure), features of the formulation used (e.g. presentation, solar protection and dispersion) and methods of application. Among the numerous factors that can negatively affect the performance of *Ls* products in the field, exposure to solar radiation might be the most critical. Although

Ls is less susceptible to this factor than *Bti* (Silapanuntakul et al., 1983), both produce insecticidal proteins and, thus, can suffer degradation due to solar radiation (Burke et al., 1983; Karch and Charles, 1987). Prolonged *Ls* persistence is often observed in septic tanks and other breeding sites that are protected from the sunlight (Silva-Filha et al., 2001). In a similar manner, *Bti* can achieve enhanced performance under shadowed conditions (Araújo et al., 2007; Melo-Santos et al., 2009). Sedimentation of the crystal-spore complex also reduces the activity of *Ls* because *Culex* and *Anopheles* larvae feed at the surface of the breeding sites. Some studies showed a correlation between the crystal spores settling and the loss of toxicity in experimental sites (Davidson et al., 1984; Lacey and Lacey, 1990; Skovmand and Guillet, 2000). Water flow, typically found in breeding sites such as catch basins or water ditches in urban areas, has a negative impact on the persistence of *Ls* because the current carries away the active ingredient.

Ls persistence for long periods is most likely the major feature that contributes to its field effectiveness, particularly in urban areas (Davidson et al., 1984; Mulla et al., 1984a; Mulligan et al., 1980; Silapanuntakul et al., 1983). This persistence occurs because *Ls* has the ability to recycle in mosquito cadavers or in the soil of breeding sites. This process involves a new cycle of spore germination, vegetative development and the production of new batches of spores and crystals that can be released in the environment and sustain the larvicidal activity. The recycling has been demonstrated under laboratory and field conditions and can take place in some mosquito species (Becker et al., 1995; Correa and Yousten, 1995; Des Rochers and Garcia, 1984; Hertlein et al., 1979; Karch and Coz, 1986; Menon et al., 1982; Nicolas et al., 1987). Laboratory studies have shown that *de novo* production of spores is approximately 20-fold greater than the amount that was originally ingested by larvae (Charles and Nicolas, 1986). Recycling can provide effective results if the favourable conditions required for bacterial germination occur (mainly in the mosquito cadavers) and if the newly produced spore crystals are located in the feeding zone of larvae. The availability of spore crystals on the bottom of breeding sites cannot contribute to activity against larvae (Singer, 1980).

4.3. Trials against the vectors of lymphatic filariasis

In a context of the control of lymphatic filariasis transmission, the World Health Organization-Special Programme for Research and Training in Tropical Diseases (WHO-TDR) organised in 1990 an informal consultation to develop a strategy for large-scale field trials aiming to evaluate the efficacy of *Ls* against *Culex* populations under different ecological conditions in Brazil, Cameroun, India, Sri Lanka and Tanzania (WHO, 1993). The projects were conducted in four steps, according to the protocol developed by TDR, which also considered the effects of vector control on the transmission of *Wuchereria bancrofti*: a baseline collection of data (i), a preparation phase (ii), the implementation of an 18 month-larviciding period (iii), and a 6-month follow-up phase after larviciding stopped (iv). Entomological and parasitological data gathered from all the phases showed a remarkable impact of *Ls* treatment of breeding sites on reducing the vector-biting rate and generating a significant decline in human exposure to filarial infective larvae (Barbazan et al., 1997; Becker, 2000; Maxwell et al., 1999; Regis et al., 2000b). The output of these projects, together with previous knowledge and experience from field trials, contributed to constructing a solid basis for the design of large-scale implementation of *Ls*-based larvicides under different ecological conditions. The large-scale use of *Ls* also resulted in the emergence of resistance in larvae of the *Cx. pipiens* complex (see Section 5). However, given the known advantages of *Ls* over conventional larvicides, that discovery promptly led to a research effort to elucidate the resistance process and mechanisms. As a consequence, solutions to avoid, delay or manage resistance were soon made available (see Section 6). Since then, *Ls* has been applied in association with *Bti* in rotation or as a mixture, or associated with other control agents, and the susceptibility of mosquito larvae to the Bin toxin should be duly monitored.

4.4. Recent large-scale trials

The discovery of resistance to *Ls* does not seem to have lessened the interest in the use of this entomopathogen for the control of mosquito larvae, as suggested by the many large-scale field tests reported in recent years (Table 3.6). It is important to emphasise that the use of *Ls* and *Bti* mixture, recommended as a resistance-management strategy, gives more advantages to the use of these biolarvicides in vector-control programmes: this mixture greatly reduces the risk of developing resistance to *Ls*, broadens the spectrum of target species, including *Ae. aegypti*, and takes advantage of the synergy between the toxins of the two species, beyond the separate performance of each one (Wirth et al., 2004). Recent field trials have been conducted, mostly against malaria vectors, as in Tanzania, Kenya, Ivory Coast, Burkina Faso, Benin and Brazil, but also against *Culex* and *Aedes* species in other

Table 3.6 Examples of field trials using combined *Lysinibacillus sphaericus* and *Bacillus thuringiensis* serovar. *israelensis* strategies for mosquito control

Country	Targeted species	References
Kenya, Mbita	*Anopheles gambiae*, *An. funestus*	Fillinger and Lindsay (2006)
Gambia, Farafeni Town	*An. gambiae*, *An. arabiensis*, *An. melas*	Majambere et al. (2007)
Turkey, Antalya	*Culex pipiens*	Cetin et al. (2007)
Colombia, Cali	*Cx. quinquefasciatus*, *Aedes aegypti*	Giraldo-Calderon et al. (2008)
Cote d'Ivoire, Tiémélekro	*An. funestus*, *An. gambiae*	Tchicaya et al. (2009)
Tanzania, Dar es Salaam	*An. gambiae*, *An. funestus*, *An. coustani*, *Cx. quinquefasciatus*	Geissbuhler et al. (2009)
Kenya, Highland Valley	*An. gambiae*, *An. funestus*, *An. arabiensis*	Fillinger et al. (2009)
Poland, Wroclaw	*Anopheles* spp.	Rydzanicz et al. (2009)
Benin	*Cx. quinquefasciatus* and other mosquito species	Lingenfelser et al. (2010)
Kenya, Malindi	*An. gambiae*, *Cx. quinquefasciatus*	Mwangangi et al. (2011)
USA, Stratford CN	*Cx. pipiens*, *Cx. restuans*, *Ae. japonicus*	Anderson et al. (2011)
USA, California	*Cx. tarsalis*, *Ochlerotatus melanimon*	Dritz et al. (2011)
Switzerland, Chiasso, Ticino	*Cx. pipiens*, *Ae. albopictus*	Guidi et al. (2013a)
Tanzania, Dar es Salaam	*An. gambiae*, *An. funestus*	Maheu-Giroux and Castro (2013)
Brazil, Amapá	*An. darlingi*	Galardo et al. (2013)

countries (Table 3.6). A common finding in these field trials is that all the tested species of malaria vectors were highly susceptible to *Ls*, in each location that testing was conducted. Furthermore, an evaluation of the impact of the biolarvicide on the vector population and malaria transmission was performed in most of these works. As an example, Fillinger and Lindsay (2006) have demonstrated that microbial larvicides (*Bti* and *Ls*) reduced the malarial vector mosquito larvae and adult females by more than 90% in West Kenya. In a location in Tanzania, one year of community-based larvicide application reduced transmission by the primary malaria vector, *An. gambiae*, by 31% (Fillinger et al., 2008). As described, the first field trials of *Ls* against mosquitoes were conducted, mainly in Africa, against *Culex* and *Anopheles* species in the 1980s (Table 3.5). Currently, there appears to be a renewed interest in the use of entomopathogenic bacteria to control malaria vectors, which can be attributed to factors related to the fast urbanisation of the African population and relatively easy access to most breeding sites in urban areas (Fillinger et al., 2008), to the exophagic behaviour that has been exhibited by *An. gambiae* in such areas (Fillinger and Lindsay, 2011; Geissbuhler et al., 2007; Killeen et al., 2007), and to the insecticide resistance, mainly to pyrethroids used in insecticide-treated nets, that has emerged for the primary malaria vectors in many regions of Africa (Cuamba et al., 2010; Munhenga et al., 2008; Ranson et al., 2011). In view of this scenario, there is a strong belief that integrated vector management targeting both larval and adult mosquitoes is the future for malaria control (Townson et al., 2005).

4.5. Operational use in mosquito-control programmes

Presently, *Ls* is almost always used in association with *Bti* as a mixture or in rotation and integrated with other environmentally friendly control measures. These treatments are used in some countries to combat vector-borne diseases or the nuisance caused by mosquitoes. Some examples of operational use will be presented below. In Germany, the KABS (German Mosquito Control Organisation) has conducted a comprehensive programme in a combined control strategy to control *Ae. vexans* and other species of mosquitoes in the Rhine Valley, that has remained active since 1985 (Becker, 1997). The KABS was the first to adopt the use of *Ls*, as a complement to the routine use of *Bti* to fight the nuisance caused by mosquitoes, especially *Culex*, during the tourist season. There was no record of resistance to these entomopathogens or of negative impacts on non-target fauna, which were carefully monitored (Becker, 1998). In Tanzania, the microbial larvicides *Bti* and *Ls* have been used operationally since 2006 in the context of the Urban Malaria Control Programme in Dar es Salaam (Fillinger et al., 2008; Geissbuhler et al., 2009). In Brazil, Recife city, with 1.5 million inhabitants, is an endemic area for Bancroftian filariasis transmitted by *Cx. quinquefasciatus*, with a 6% prevalence estimated two decades ago (Maciel et al., 1996). The Recife City Hall's Department of Health has been running a Filariasis Control Programme, integrating vector control and mass treatment of humans with diethylcarbamazine since 2003. Vector-control actions, including *Ls* treatments, were established in some critical areas (Cartaxo et al., 2011; Silva-Filha et al., 2008) and were progressively expanded; since 2006, the whole city (94 districts) has been treated. A conjugate product containing *Ls* and *Bti* crystals was also applied in a city district in 2010–2011, in an attempt to develop a suitable approach to control both *Cx. quinquefasciatus* and *Ae. aegypti* in urban areas using a single product that assures effectiveness and low potential for the development of resistance (C. M. F. Oliveira, personal communication). Another programme has been conducted in the Pinheiros River, which crosses the metropolitan area of São Paulo (Brazil). This river's banks are a major source of *Cx. quinquefasciatus* in the city, and larvicidal treatment to reduce mosquito populations is conducted on 22.4 km of the river. From July 2003 to July 2006, 30 treatments were applied, using an *Ls*-based product in rotation with a *Bti*-based larvicide. The larvicides were applied using a boat and the frequency of treatments is based on surveillance of mosquito immature stages. No resistance was found in the *Cx. quinquefasciatus* local population (Andrade et al., 2007; Silva-Filha et al., 2008). In the United States, *Ls* and *Bti*-based products, as well as conjugate products, have been used in many counties to fight several species of mosquitoes since the introduction of these products for vector control.

5. RESISTANCE

5.1. Factors involved in the selection of resistance

The production of biolarvicides based on *Ls* strains with high larvicidal action supported their large-scale utilisation in many countries, especially to fight larvae from the *Cx. pipiens* complex and anophelines. Field utilisation showed the effectiveness of this agent but also revealed the onset of resistance. Some aspects that have contributed to this phenomenon are mentioned below. The environmental conditions of subtropical and tropical countries promote mosquito proliferation throughout the year and require continuous treatment cycles, which increase the selection pressure imposed on these populations (Barbazan et al., 1997, 1998; Hougard and Back, 1992; Hougard et al., 1993; Mulla et al., 2001; Regis et al., 1995, 2000b; Silva-Filha et al., 2001; Skovmand and Bauduin, 1997; Skovmand et al., 2009; Yuan et al., 2000). The persistence of *Ls* in some breeding sites (e.g. septic tanks) due to its capacity for recycling (Becker et al., 1995; Charles and Nicolas, 1986; Karch and Coz, 1986) also contributes to a prolonged exposure of larvae to the insecticidal crystals. The adoption of larvicides as the major tool in control programmes is a strategy that increases the selection pressure, while the adoption of integrated strategies can significantly reduce the breeding site area that must be covered by larvicide application. Finally, the intrinsic mode of action of *Ls*, based on the action of one major toxin that targets a single class of receptors (Nielsen-Leroux and Charles, 1992), is critical for the selection of resistance as reviewed by Wirth (2010) and Ferreira and Silva-Filha (2013).

5.2. Laboratory and field reports

Resistance associated with the Bin toxin from *Ls* biolarvicides has been reported in mosquito colonies artificially selected in the laboratory as well as in field populations exposed to this agent (Table 3.7). The first resistant laboratory colony (GEO) was established from *Cx. pipiens* field samples from California (EUA), and it achieved a high level of resistance ratio (RR $\approx$ 100,000) after 12 generations of selection pressure (Wirth et al., 2000b). The samples employed for the foundation of this colony had already showed a reduced susceptibility to *Ls*, which could be the reason for the high resistance achieved in such a short period. Another laboratory selection performed with populations from California, the L-SEL colony, showed a RR of 37-fold after 80 generations (Rodcharoen and Mulla, 1994), a much slower rate. Two laboratory colonies of *Cx. quinquefasciatus* from Brazil (CqRL1/2362) and China (RLCq1/C3-41) were selected on a similar basis with *Ls* strains 2362 and C3-41, respectively. A high-resistance level (RR > 100,000) was achieved in both cases after 46 and 13 generations of selection, respectively (Pei et al., 2002). In a second colony (CqRL2/IAB59) selected with the IAB59 strain in Brazil, the resistance to Bin evolved more slowly (Amorim et al., 2007), likely due to the presence of the Cry48Aa/Cry49Aa toxin produced by this strain (Jones et al., 2007). This colony showed an RR $\approx$ 46,000 towards the IAB59 strain after 70 generations, and a strong cross-resistance (RR $\approx$ 70,000) to the Bin toxin was detected, as expected. The selection of a colony with the IAB59 strain in China, RLCq2/IAB59, showed a high level of resistance (RR > 100,000) that was detected after 18 generations (Pei et al., 2002).

Field resistance was first detected in a *Cx. pipiens* population from Southern France that was submitted to *Ls* exposure for approximately

Table 3.7 Reports of *Culex pipiens* and *Culex quinquefasciatus* resistance to *Lysinibacillus sphaericus* selected under laboratory (Lab) conditions or after field exposure

Sample	Country	Origin	RR[a]	Binding to receptors[b]	*r* alleles[c]	Inheritance[d]	References
GEO	USA	Lab	>100,000	No	$cpm1_{GEO}$	R/A	Darboux et al. (2002), Nielsen-Leroux et al. (1995), Wirth et al. (2000b)
L-SEL	USA	Lab	37	Nd	Nd	Nd	Rodcharoen and Mulla (1994)
CqRL1/2362	Brazil	Lab	>100,000	No	$cqm1_{REC}$	R/A	Oliveira et al. (2004), Pei et al. (2002), Romão et al. (2006)
CqRL2/IAB59	Brazil	Lab	≈40,000	No	$cqm1_{REC}$	R/A	Amorim et al. (2007), Oliveira et al. (2004), and Pei et al. (2002)
RLCq1/C3-41	China	Lab	>100,000	No	*cqm1R*	R/A	Amorim et al. (2007), Guo et al. (2013), Oliveira et al. (2004), Pei et al. (2002)
RLCq2/IAB59	China	Lab	>100,000	Nd	Nd	Nd	Pei et al. (2002)
SPHAE	France	Field	>20,000	Yes	Nd	R/S	Nielsen-Leroux et al. (1997), Nielsen-Leroux et al. (2002)
Kochi	India	Field	≈150*	No	Nd	Nd	Rao et al. (1995)
Coque	Brazil	Field	≈10	Yes	Nd	Nd	Silva-Filha et al. (1995)

RFCq1	China	Field	>20,000	Nd	Nd	Nd	Yuan et al. (2000)
BP	France	Field	>10,000	No	$cpm1_{BP}$, $cpm1_{BP}$-*del*	R/S	Darboux et al. (2007), Nielsen-Leroux et al. (2002)
TUNIS	Tunisia	Field	≈750*	Yes	Nd	R/S	Nielsen-Leroux et al. (2002)
Nonthaburi	Thailand	Field	>125,000	Nd	Nd	Nd	Mulla et al. (2003)

[a]Resistance ratio between the lethal concentration of *L. sphaericus* for the test sample and that obtained for a susceptible reference colony. *These field samples were submitted to further laboratory selection and achieved higher levels of resistance (>10,000).
[b]Detection of binary toxin binding to midgut microvilli proteins from larvae.
[c]Resistance alleles.
[d]Inheritance of resistance: R/A (recessive and autosomal), R/S (recessive and sex-linked), Nd (not determined).

five years (Sinègre et al., 1994). This finding, although based on larval samples from a few breeding sites, revealed for the first time high levels of resistance (RR > 20,000) in field populations. This result raised concerns about the intensive use of *Ls* that were confirmed by subsequent resistance reports. *Cx. quinquefasciatus* or *Cx. pipiens* populations from India (Rao et al., 1995), China (Yuan et al., 2000), Tunisia (Nielsen-Leroux et al., 2002), Thailand (Mulla et al., 2003), and a second population (BP) from France (Chevillon et al., 2001; Nielsen-Leroux et al., 2002), exposed to treatments within control programmes, showed RRs that were, in most cases, higher than 10,000-fold (Table 3.7). The *Ls* strains employed in these field-control programmes were 2362 and 1593, in addition to the C3-41 strain isolated and widely used for the production of this biolarvicide in China (Yuan et al., 1999). In Brazil, a *Cx. quinquefasciatus* population from Coque (Recife) exposed to *Ls* during a 2-year period showed a low resistance level (RR ≈ 10-fold) (Silva-Filha et al., 1995). The utilisation of *Ls* for the control of *Cx. quinquefasciatus* in two other urban areas of Água Fria (Recife) and Rio Pinheiros (São Paulo) in Brazil did not result in the selection of resistance (Silva-Filha et al., 2008). Factors recorded in those areas, such as the interruption of treatments for some periods, migration of individuals and rotations with *Bti* treatment, might have contributed to the decrease of the selection pressure. Nevertheless, the other reports have undoubtedly indicated that high levels of resistance can be achieved in the field, and this resistance remains a central issue concerning *Ls* utilisation.

The reports described above indicate that different factors may play an important role in the development of resistance. The initial susceptibility of the population to the larvicides used and the frequency of resistance alleles can be crucial to the evolution of resistance (Andow et al., 2000; Gould et al., 1997; Tabashnik et al., 2003, 2006). Laboratory Bin-resistant colonies showed heterogeneous responses when they were subjected to selection pressure, suggesting that their genetic background could be a determining factor. Baseline studies of the susceptibility of *Cx. pipiens* larvae to *Ls* are limited; however, those performed with *Bti* showed variation. The evaluation of approximately 50 *Cx. pipiens* populations without historic *Bti* exposure have shown RRs to this agent from less than 3 to 12.5-fold and indicates pre-existing natural variations (Vasquez et al., 2009; Wirth et al., 2001b). Generally, the susceptibility of the target populations is not assessed before the introduction of insecticides in a control programme, although such data can be critical to detect the development of resistance. For this reason, the evaluation of larval susceptibility through bioassays and the screening of alleles associated with *Ls* resistance in *Culex* larvae should be routinely performed.

5.3. Mechanisms and inheritance of resistance

Laboratory-selected and field-derived colonies have been used to characterise resistance mechanisms to the Bin toxin. Most of the studies have focused on the detection of functional receptors in the larval midgut because these molecules are essential for the Bin toxin's mode of action (Section 2). Quantitative *in vitro* binding assays showed that the Bin toxin fails to interact with BBMF from larvae of the laboratory-selected colonies GEO (EUA), CqRL1/2362 (Brazil) and CqRL/C3-41 (China) (Darboux et al., 2002; Guo et al., 2013; Oliveira et al., 2004). For the field-derived colonies BP from France (Darboux et al., 2007) and Kochi from India (S. Poophathi and C. Nielsen-LeRoux, unpublished data), this mechanism was also behind the resistance. In contrast, larvae from the SPHAE (France) and TUNIS (Tunisia) field-derived colonies showed abundant and functional receptors (Nielsen-Leroux et al., 2002), as was recorded for the susceptible colonies that were used as references (Nielsen-Leroux et al., 1997, 2002). It is likely that another mechanism related to the post-binding step is responsible for the high-resistance levels observed in those larvae, but it remains unknown. Larvae from the field population of Coque (Brazil) displayed capacity to bind the Bin toxin; however, the resistance level ($RR \approx 10$) was most likely quite low to be correlated to a loss in the capacity of binding (Silva-Filha et al., 1995). The available data show that the lack of receptors for the Bin toxin in the larval midgut is the major mechanism reported for resistance to *Ls*. The availability of those resistant colonies allowed studies of the inheritance of resistance to be performed through crosses and back-crosses, and the data are summarised in Table 3.7. Monofactorial and recessive inheritance was detected in all the cases studied, and some alleles of the *cpm1/cqm1* genes that presented mutations associated with resistance due to failures in Bin binding to receptors have been identified in three laboratory-selected colonies (GEO, CqRL1/2362 and CqRL/C3-41) and in one field population (BP). For the SPHAE and TUNIS populations, it has been demonstrated that one gene, sex-linked and also recessively inherited, is responsible for the high level of resistance displayed by these larvae (Nielsen-Leroux et al., 1997, 2002). Further studies of these colonies might reveal new pathways involved in the mode of action of *Ls*.

5.4. Resistance alleles of the *cpm1/cqm1* gene

The availability of laboratory colonies with high levels of resistance provided a robust model for the characterisation of mechanisms and genes involved in the resistance to the Bin toxin. The identification of the gene coding for the

Cpm1 receptor in *Cx. pipiens* (Darboux et al., 2001) opened perspectives for the investigation of the molecular basis of resistance. To date, seven alleles from *cpm1/cqm1* genes associated with Bin resistance were characterised and the first allele was described in larvae from the GEO colony (Darboux et al., 2002). The allele $cpm1_{GEO}$ showed a mutation that provoked the loss of the GPI anchor and prevented the expression of a full Cpm1 midgut-bound protein. As a consequence, this truncated protein is absent in the epithelial cell membrane of the midgut and, therefore, cannot act as a receptor for the Bin toxin. This mechanism consistently explains the high level of resistance observed for GEO larvae, which is associated with the lack of functional receptors on the midgut (Nielsen-Leroux et al., 1995; Wirth et al., 2000b). This pioneer study was followed by the identification of six other resistance alleles of the *cpm1/cqm1* gene in populations from different countries. Each allele is characterised by a particular mutation and all of them code for transcripts of truncated or non-functional proteins, which disrupt the Bin toxin binding to the midgut cells. The mutation in the *r* alleles associated to Bin resistance that have been identified to date are shown in Fig. 3.8 and their features are described below.

The $cpm1_{GEO}$ allele is characterised by a nonsense mutation T1706A that changes the Leu-569 to a premature stop codon, in the proximity of the putative GPI signature located in the C-terminus of the Cpm1 protein. This mutation prevents GPI anchoring, while the α-glucosidase activity and the toxin binding capacity of this molecule remains intact, as demonstrated by the respective recombinant proteins expressed in Sf9 cells (Darboux et al., 2002). This mutant protein lacks only eleven amino acid residues compared to the wild-type Cpm1 protein that is composed of 580 amino acids. A second independent allele was identified in the CqRL1/2362 *Cx. quinquefasciatus* colony from Brazil that was also characterised by a high level of resistance and lack of midgut receptors for Bin toxin (Oliveira et al., 2004; Pei et al., 2002). In larvae from this colony the $cqm1_{REC}$ allele had a deletion of 19 nucleotides (nt) located from nucleotides 1276–1294 of its sequence (Romão et al., 2006), resulting in the reading frame shift that creates a premature stop codon at position 1362. The potential truncated protein coded by this transcript would have only 437 residues and it remains unknown if this protein is produced in these larvae, since immunodetection assays failed to detect its presence (Romão et al., 2006). These authors hypothesised that the premature stop codon could be recognised by the ubiquitous nonsense-mediated decay pathway of mRNA degradation (Gonzalez et al., 2001; Wagner and Lykke-Andersen, 2002; Wilkinson and Shyui, 2002) causing the removal of this transcript.

```
atgcgaccgctgggagctttgagcctagtcgctctgttggcgatgaccgtcaacgggctagcgattcgcgaaccggactcgaaggactgg   90
 M  R  P  L  G  A  L  S  L  V  A  L  L  A  M  T  V  N  G  L  A  I  R  E  P  D  S  K  D  W    30

tatcagcacgcgacgttctaccagatctatccgaggtcgttttggacagcaacggcgatgggattggtgatttggcggggatcacctcc  180
 Y  Q  H  A  T  F  Y  Q  I  Y  P  R  S  F  L  D  S  N  G  D  G  I  G  D  L  A  G  I  T  S    60

aagatgaagtacttggcggatattgggattgacgcgacgtggttgagcccgccgttcaagtcaccgctgaaagactttgggtatgatgtg  270
 K  M  K  Y  L  A  D  I  G  I  D  A  T  W  L  S  P  P  F  K  S  P  L  K  D  F  G  Y  D  V    90

tcggatttctacgctatccagccggagtacgggaatttgacggatttgacaagttggtggaggaatcgcacaagaatgggatcaagctg  360
 S  D  F  Y  A  I  Q  P  E  Y  G  N  L  T  D  F  D  K  L  V  E  E  S  H  K  N  G  I  K  L   120
                                                                                    *
atgctggatttattccgaaccactcgagtgaccagcacgagtggtttgtgaagtctgtggcgcgggatccggagtacagtgactttat  450
 M  L  D  F  I  P  N  H  S  S  D  Q  H  E  W  F  V  K  S  V  A  R  D  P  E  Y  S  D  F  Y   150

gtgtggagacccccggcgacaggggtggtccgccgaataattggatctcggtatttggagggtcggcgtggacatataatacagcgaga  540
 V  W  R  P  P  A  T  G  G  G  P  P  N  N  W  I  S  V  F  G  G  S  A  W  T  Y  N  T  A  R   180

ggggagtactatctgcaccagtttacgcctcagcagccggatttgaactaccggaatccgaaggtgctggcggagatgaccaagatgttg  630
 G  E  Y  Y  L  H  Q  F  T  P  Q  Q  P  D  L  N  Y  R  N  P  K  V  L  A  E  M  T  K  M  L   210

ttcttctggttggaccgtggagtggatgggttccggttggacgctatcaaccacatgttcgaggatgagcagttccgggatgagccgttg  720
 F  F  W  L  D  R  G  V  D  G  F  R  L  D  A  I  N  H  M  F  E  D  E  Q  F  R  D  E  P  L   240

tctgggtggggacagtcgggagagtacgattctttggatcacatttataccaaggacatcccggatgtgtatgatgtggtttacaactgg  810
 S  G  W  G  Q  S  G  E  Y  D  S  L  D  H  I  Y  T  K  D  I  P  D  V  Y  D  V  V  Y  N  W   270

cgggatcagatggataagtattccgcggagaagggccgtactatcattctgatgacggaagcgtattcgagcatcgagggaacgatgctg  900
 R  D  Q  M  D  K  Y  S  A  E  K  G  R  T  I  I  L  M  T  E  A  Y  S  S  I  E  G  T  M  L   300

tactacgagagtgcggaccggaagcgccagggagctcacatgccgttcaacttccagctgatctatgactttaagaaggagcagaacgcc  990
 Y  Y  E  S  A  D  R  K  R  Q  G  A  H  M  P  F  N  F  Q  L  I  Y  D  F  K  K  E  Q  N  A   330

gttgggctgaagagttcgatcgactggtggatgaacaacatgccggctcgacacactccaagctgggtagctggatcgcatgatcactcg 1080
 V  G  L  K  S  S  I  D  W  W  M  N  N  M  P  A  R  H  T  P  S  W  V  A  G  S  H  D  H  S   360

cgggtagcgtcacgagttggactagatcgggttgatcaggtcatgacgttgctgcataccctgccggggactagtattacgtattacggt 1170
 R  V  A  S  R  V  G  L  D  R  V  D  Q  V  M  T  L  L  H  T  L  P  G  T  S  I  T  Y  Y  G   390

gaagaggtagccatgcaggacttcaaagaggctcagcagttcgacaaccgagatccgaaccggacgccgatgcagtgggactcgtcgacc 1260
 E  E  V  A  M  Q  D  F  K  E  A  Q  Q  F  D  N  R  D  P  N  R  T  P  M  Q  W  D  S  S  T   420

agtgccgggttcagtaccaacaccaacacctggctccgagttcatccggactacgctcggtacaacgtagacgtgatgcagaagaatcca 1350
 S  A  G  F  S  T  N  T  N  T  W  L  R  V  H  P  D  Y  A  R  Y  N  V  D  V  M  Q  K  N  P   450

caaagtaccttccaccacttccagcatctgaccaagcttcgacggcaccggacgatgcagagtggtgagtacgtgcacaagacggtcgga 1440
 Q  S  T  F  H  H  F  Q  H  L  T  K  L  R  R  H  R  T  M  Q  S  G  E  Y  V  H  K  T  V  G   480

accaaggtgtacgcattgctgagggaactccgtggtgaggattcgttcctgacggtgctgaacatggccggagcagaggataccgtggat 1530
 T  K  V  Y  A  L  L  R  E  L  R  G  E  D  S  F  L  T  V  L  N  M  A  G  A  E  D  T  V  D   510

ctgggagatttcgtgaatcttccgcagaagatgcgagtcgaggtggcgcaaccgaactccaagtcgaaggcgggcaatgaggtcgacatc 1620
 L  G  D  F  V  N  L  P  Q  K  M  R  V  E  V  A  Q  P  N  S  K  S  K  A  G  N  E  V  D  I   540

ggcaagttgacgctgggaccgtacgattcggttgtgctgagagcaacggtttcgtcggcagctgctatcaacctttcgattggattgctg 1710
 G  K  L  T  L  G  P  Y  D  S  V  V  L  R  A  T  V  S  S  A  A  A  I  N  L  S  I  G  L  L   570

ctagcgattatggccaggtatatttcgtgtag                                                           1743
 L  A  I  M  A  R  Y  I  F  V  .                                                              580
```

Figure 3.8 Nucleotide and deduced amino acid sequence of the open reading frame from the *cqm1* gene, which codes the Cqm1 receptor in *Culex quinquefasciatus* larvae (GenBank accession number DQ333335). Seven polymorphisms associated to resistance to binary toxin that were independently identified in different alleles of the *cpm1/cqm1* genes, are represented in this sequence for illustration purpose only. (1) *cqm1R* allele shows a deletion of a cytosine (asterisk, boxed) at position 445 that creates a stop codon downstream (tga, boxed); (2) *cpm1*$_{BP}$ shows a nonsense mutation that creates a stop codon (Gln396Stop) at this position (arrow); (3) *cpm1*$_{BP}$*-del* has an alternative splicing event responsible for the deletion of 66 residues (bold and underlined); (4) *cqm1*$_{REC}$ has a 19-nt deletion (bold and underlined); (5) *cqm1*$_{REC-D25}$ has a 25-nt deletion encompassing the 19-nt from the previous deletion (bold and underlined) and six subsequent bases (bold and boxed); (6) *cqm1*$_{REC-D16}$ has a 16-nt deletion (bold and italics); the three last deletions (19-, 25-, 16-nt) create the stop codon located at the same position 1362 (tga, boxed); (7) *cqm1*$_{GEO}$ has a nonsense mutation (T1706A) that creates a stop codon (Leu569Stop) at this position (arrow).

During a screening of the *cqm1*$_{REC}$ in field populations from Recife city (Brazil), two new alleles containing mutations potentially associated with resistance were found (Chalegre et al., 2012). The *cqm1*$_{REC\text{-}D16}$ allele showed a 16-nt deletion (1306–1321) that was found in a few individuals from non-treated populations. The second allele named *cqm1*$_{REC\text{-}D25}$ had a 25-nt deletion, which encompassed the 19-nt deletion from the *cqm1*$_{REC}$ and six additional nucleotides downstream of this sequence (1276–1300), and it was found in one larva from a *Ls*-treated population. The 19-, 16- and 25-nt deletions are closely located in the sequence and their changes in the sequence reading frame generated the same premature stop codon located at position 1362.

Two *r* alleles detected in the BP population from Southern France that showed a high level of resistance to *Ls*, were also associated with a lack of functional midgut receptors (Chevillon et al., 2001; Nielsen-Leroux et al., 2002). Darboux et al. (2007) confirmed the co-existence of these alleles in that population and identified their mutations. The *cpm1*$_{BP}$ allele shows a nonsense mutation Gln396Stop that produces a protein without the last 184 residues including the signature for the GPI anchor. Similar to other cases, it is expected that these truncated proteins are secreted as soluble proteins, instead of being produced as midgut membrane proteins. The second allele found among BP larvae, the *cpm1*$_{BP}$*-del*, has a 198-nt internal deletion provoked by the insertion of a transposon that is responsible for generating an alternative splicing event. This allele codes for a transcript of a protein with a deletion of 66 amino acids located between the Val422 and Gln487, and it retains the predicted GPI-anchor in its C-terminus. The respective recombinant Cpm1$_{BP}$-del protein expressed in Sf9 cells showed that it correct locates to the cell membrane but this protein is unable to bind the Bin toxin. It is likely that the loss of these amino acids residues provokes an alteration of the protein conformation and prevents its binding to the Bin toxin. The frequency of these alleles verified in a sample of 108 larvae was similar: 50% carried a copy of each allele, 25% were homozygous for the *cpm1*$_{BP}$, and the remaining individuals were homozygous for the *cpm1*$_{BP}$*-del*.

The molecular basis of the resistance behind the CqRL/C3-41 laboratory-selected colony from China was recently identified (Guo et al., 2013). Cloning and analysis of the *cqm1* gene from the CqRL/C3-41 larvae revealed the *cqm1R* allele characterised by the deletion of a cytosine at position 445. This single-deletion changes the reading frame of the subsequent 39 residues and a stop codon is created at position 582 in this sequence. The transcript codes a truncated protein composed of 194 residues and homozygous larvae for this allele are resistant due to absence of full-length Cqm1 midgut-bound molecules.

The characterisation of these alleles indicates that *cpm1/cqm1* is a highly polymorphic gene and five ($cqm1_{REC}$, $cqm1_{REC\text{-}D16}$, $cqm1_{REC\text{-}D25}$ from Brazil; $cpm1_{BP}$ and $cpm1_{BP}$*-del* from France), among the seven mutations recorded, are located in the same region of the gene indicating that this seems to be a hot-spot for such events. In addition, these mutations can have a high impact because they are responsible for provoking full refractoriness due to the lack of membrane-bound receptors. Mechanisms of target site alteration to synthetic insecticides, for instance, are often associated to mutations in their genes that provoke a reduction in their capacity for binding the active principle (Du et al., 2013; Rinkevich et al., 2013), and they do not provoke the total loss of activity. This has serious implications because the Cpm1/Cqm1 receptors are essential for the action of the Bin toxin, and the selection of individuals with such polymorphisms in their genes can lead to operational failures.

5.5. Diagnosis and field survey of resistance

The investigation of *Ls* resistance is commonly performed through susceptibility bioassays that are employed to compare the dose–response of larvae from unknown samples to the response of larvae from a susceptible sample, used as a reference. Bioassays allow the determination of the LCs of the given *Ls* sample tested for 50% (LC_{50}) and 90% (LC_{90}) of the larvae, exposed for 48 h (WHO, 1985). The ratio between the LC for the test sample and that for the reference colony, named RR, informs the level of resistance detected. It is difficult to establish the RR value that truly corresponds to initial stages of resistance and, often, when the RR of a given sample to *Ls* and *Bt* toxins is equal or greater than 10-fold, it is referred to as a low level of resistance. This parameter could be greatly improved if baseline data for the susceptibility of the populations under analysis were available. Populations can exhibit natural variations to the larvicides prior to their utilisation that are not necessarily a consequence of the selection pressure (Robertson et al., 1995).

Bioassay is also the major tool used to evaluate larval susceptibility to *Ls*, however, studies on the genetics of resistance to this agent have shown that this trait is recessively inherited, at least for the samples analysed to date (Table 3.7). In this case, heterozygous individuals carrying *r* alleles can hardly be detected in the bioassays because they will display a susceptible phenotype. Nevertheless, in most cases the *r* alleles have not been identified and their screening is performed using techniques based on crosses and followed by the analysis of the progeny susceptibility, such as the F2 and F1 screens that have been employed to characterise *r* alleles to *Bt* toxins in pest insects (Andow et al., 2000; Gould et al., 1997; Xu et al., 2009; Yang et al., 2006, 2007a; Zhao et al., 2010). These approaches although efficient, are indirect methods to detect *r* alleles. On the other hand, the identification of such alleles allows the development of molecular methods for their direct detection. Polymerase chain reaction (PCR) methods have been specifically designed to target specific alleles or to amplify strategic gene segments that host mutations associated to resistance. Molecular methods for the detection of mutations in the voltage-gated sodium channel genes that confer knock-down resistance to pyrethroids have been employed for the screening of resistance in mosquito populations (Saavedra-Rodriguez et al., 2007; Sarkar et al., 2009; Singh et al., 2009; Tripet et al., 2006).

For *Ls*, the identification of alleles of the *cpm1/cqm1* gene associated with resistance opened perspectives for their molecular detection in field populations. The $cqm1_{REC}$ primarily identified in the CqRL1/2362 *Cx. quinquefasciatus* laboratory-selected colony, was detected by allele specific PCR in all field populations surveyed from Recife city (Brazil), including six non-treated and one *Ls*-treated populations. Its frequency in non-treated samples ranged from 0.001 to 0.017 and the average of five populations was 0.003 (Chalegre et al., 2009, 2012). In three evaluations performed in the treated area, $cqm1_{REC}$ showed an average frequency of 0.048, which was significantly higher than those recorded for the non-treated ones (Chalegre et al., 2009, 2012). Bioassay analysis performed with these populations did not reveal RR values significantly between the treated and the non-treated areas but the early detection of an increased number of individuals carrying this allele in the treated area is likely to be correlated to the selection pressure imposed by the *Ls* exposure. This method can be strategic for the surveillance of resistance, especially for recessive *r* alleles. The two other alleles $cqm1_{REC\text{-}D16}$ and $cqm1_{REC\text{-}D25}$, that were identified in the *Cx. quinquefasciatus* field samples from Recife city, showed a frequency between 0.001 and 0.006 in five populations and a more limited distribution in the samples compared to the $cqm1_{REC}$ (Chalegre et al., 2012). The study suggests that the $cqm1_{REC}$ is the most important allele associated with resistance in these populations and it can be used as a marker for the surveillance in that specific area. To date, the other *r* alleles to *Ls* recorded in the literature have not been tracked in the *Cx. pipiens* populations.

5.6. Biological cost of resistance

The biological performance of individuals from three laboratory-selected resistant colonies has been evaluated. The L-SEL *Cx. quinquefasciatus* colony from California with a RR of 37-fold (Rodcharoen and Mulla, 1994) showed lower fecundity and fertility compared with their susceptible counterparts (Rodcharoen and Mulla, 1997). A marked reduction was observed in those parameters and, for instance, the number of larvae hatched per raft from the resistant females was less than 50%, compared to the susceptible colony. The evaluation of the CqRL1/2362 colony from Brazil with a high-resistance level (RR > 100,000) showed alteration in those parameters, and, although statistically significant, those were minor reductions. The fecundity, which was found to be the most affected parameter for instance, was only 8% lower than that recorded for the susceptible colony (de Oliveira et al., 2003). Individuals from the CqRL1/2362 colony also exhibit a longer developmental time (from egg to egg) and females required more time to lay the first batch of eggs. A third study performed with the CqRL2/IAB59 colony, selected with the IAB59 strain (Amorim et al., 2007), did not show significant differences in the fertility, fecundity or pupal weight of the resistant individuals compared to their susceptible counterparts (Amorim et al., 2010).

Data from these studies indicate that *Ls* resistance is likely to be associated with discrete biological costs rather than to a dramatic impact on the fitness of the resistant individuals from these colonies, at least under laboratory conditions. It is worth noting that, in some cases, the resistance to *Bt* toxins might be associated to high biological costs that can prevent the maintenance of the insect colonies in the laboratory (Anilkumar et al., 2008; Tabashnik et al., 1994). This does not seem to be case of the *Ls*-selected colonies that were maintained under laboratory conditions after a considerable number of generations (M.H. Silva-Filha, personal communication).

One direct consequence of *Ls* resistance in these insects seems to be the replacement of the midgut-bound Cqm1 α-glucosidase by a soluble form of this protein that could display, or not, its catalytic activity. In some cases, it is possible that the soluble form of the protein is not expressed neither, as described bellow. The mutant protein from GEO larvae, if expressed for instance, would retain its α-glucosidase activity (Darboux et al., 2002). On the other hand, the Cqm1 mutant proteins encoded by the resistance alleles from BP and CqRL/C3-41 larvae did not display α-glucosidase activity when they were expressed in heterologous hosts (Darboux et al., 2007;

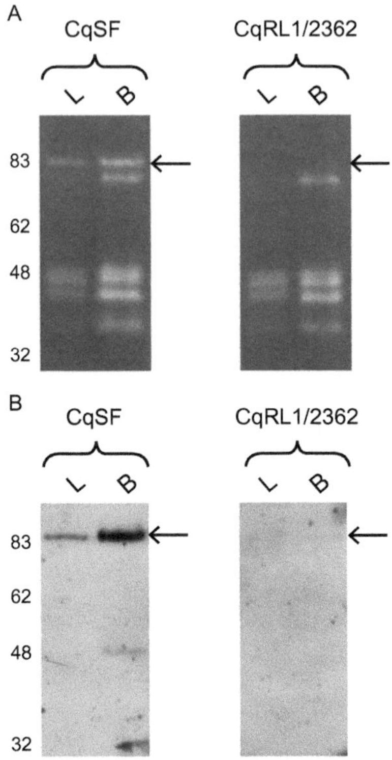

Figure 3.9 Alpha-glucosidases from susceptible (CqSF) and *Lysinibacillus sphaericus*-resistant (CqRL1/2362) *Culex quinquefasciatus* larvae. (A) In-gel α-glucosidase assay performed with whole crude extracts (L) and midgut *microvilli* proteins (B) from larvae. Bands indicating cleavage of the substrate were visualised with a UV transilluminator. (B) Immunoblotting of the samples shown in (A) with the anti-Cqm1 antiserum showing the Cqm1 protein indicated by arrows. Molecular weight markers (kDa) are shown on the left. *Taken from Romão et al. (2006).*

Guo et al., 2013). The analysis of larvae from the CqRL1/2362 resistant colony showed that the truncated Cqm1 mutant protein does not seem to be expressed in larvae (Romão et al., 2006). These results suggest that if these proteins are expressed in larvae it is not certain that they could play their role in digestion. Biochemical and bioinformatic analyses (Gabrisko, 2013; Romão et al., 2006) have provided evidence that *Cx. quinquefasciatus* displays a set of α-glucosidases and the comparative pattern in susceptible and resistant larvae from the CqRL1/2362 colony are similar, except for the Cqm1 that is missing in resistant larvae (Fig. 3.9). Maybe the expression of other α-glucosidases in these larvae can compensate the lack of Cqm1 and this could be an explanation for the minor level of biological cost observed in these colonies, although, this hypothesis has not been proved. The specific role played by Cqm1 and the other α-glucosidases in digestion in this species is still unknown and this would be useful to understand the impact of those mutations more clearly in the biological fitness of these insects.

The CqRL1/2362 colony, for instance, has been maintained for more than 200 generations under laboratory conditions and the Cqm1 α-glucosidase, which is lacking, does not seem to be essential. A study performed with the progeny from the cross between resistant CqRL1/2362 and susceptible individuals in a 1:1 ratio, showed that the resistance alleles displayed a stable frequency of 0.5 during 10 generations while this colony was kept in the absence of *Ls* treatments (Amorim et al., 2010). The biological fitness of SPHAE and GEO field-derived colonies was not investigated, but these are also examples of a highly resistant colonies successfully established and maintained under laboratory conditions during several years (M.H. Silva-Filha, personal communication). This is consistent with the hypothesis that some resistance alleles would not be necessarily associated with crucial adverse effects on biological fitness and they could be maintained in the populations in the absence of selection pressure (Ffrench-Constant, 2007).

6. MANAGEMENT OF RESISTANCE

6.1. Integrated mosquito-control programmes

The major goal of integrated vector-control approach is to achieve the reduction of population density, based on the adoption of different methods, in a cost-effective manner (WHO, 1983, 2004). The utilisation of multiple strategies to impact the density of mosquitoes, contributes to the decrease of insecticide utilisation and, thus, for a lower selection pressure. The appearance of resistance has often been associated with the intensive use of larvicides and/or adulticides, as the major intervention adopted to control insects. This model has predominated since the introduction of synthetic insecticides and it has greatly contributed to the disseminated resistance reported worldwide which affects, in particularly, dipteran insects that have been targeted by control programmes (Brogdon and McAllister, 1998; Georghiou and Lagunes-Tejeda, 1991; Hemingway et al., 2004; WHO, 1986). In developing countries, the situation is more critical due to lack of resources and structure that are necessary to set up sustainable mosquito-control programmes. The choice of suitable insecticides and the monitoring of their efficacy are important aspects required for the success of interventions (Becker et al., 2003). After the large-scale utilisation of *Ls* for *Cx. quinquefasciatus* control, resistance was recorded in exposed populations in different countries (Section 5), highlighting the need to manage the resistance to this control agent.

6.2. Factors involved in the prevention of resistance

The mode of action of all *Ls* larvicides available, to date, relies on the Bin toxin that targets one specific class of midgut molecules, and this characteristic greatly favours the selection of resistance. Therefore, efforts to eliminate mosquito breeding sources and to reduce the utilisation of larvicides are important strategies to reduce the selection pressure. The adoption of control agents with different modes of action, in a system of rotation or mixtures, should be designed according to the local conditions. The Bin toxin is essentially distinct from other insecticidal compounds and it is safe to other organisms, thus, other control agents can be used in association with *Ls*. Monitoring the susceptibility of populations exposed to larvicides is crucial to detect early stages of resistance selection and to introduce management strategies. As described in Section 5, it is important to evaluate larval susceptibility to the products employed in a control programme through bioassays and also to adopt methods that determine the frequency of resistance alleles in populations.

Hence, when resistance to *Ls* is detected in an exposed population, its use should be stopped and alternative larvicides with different modes of action can be utilised to revert the resistance developed. Mosquitoes are *r*-strategist organisms and they are able to recover rapidly after suspension of control interventions. Therefore, the interruption of treatments *per se* allows for the immigration of susceptible individuals from the surrounding areas and also from untreated breeding sites, which leads to the dilution of the resistance alleles. The reversal of *Ls* resistance in a field population from China was recorded 6 months after stopping treatment, and the RR decreased from 22,672- to 5.78-fold (Yuan et al., 2000). The low-level of resistance (10-fold) observed in a treated population from Coque (Brazil) was replaced by full-susceptibility to *Ls*, 11 months after treatment suspension in that area (Silva-Filha and Regis, 1997). The reversal of *Ls* resistance is also facilitated by the recessive inheritance of the gene involved (Amorim et al., 2007, 2010; Chevillon et al., 2001; Nielsen-Leroux et al., 1995, 1997, 2002; Oliveira et al., 2004).

6.3. Candidates for managing Bin-toxin resistance

Ls strains such as 2362, 1593, C3-41, IAB881, IAB872 display a profile of cross-resistance, because they all produce the Bin toxins, even if their Bin toxins belong to different classes (Nielsen-LeRoux et al., 2001; Regis and Nielsen-LeRoux, 2000; Rodcharoen and Mulla, 1996; Silva-Filha et al., 2004). One exception is the LP1-G strain that produces the Bin4 toxin type that has low toxicity due to replacement of a leucine by a serine at position 93 of the BinA subunit (Humphreys and Berry, 1998; Priest et al., 1997). Alternatively, some strains can produce other classes of toxins, such as the Mtx toxins and Cry48Aa/Cry49Aa that can be used to overcome Bin resistance. These toxins do not confer a high level of activity of their respective native strains (see Section 2), but they have an important potential when they are expressed and delivered under optimal conditions. Their ability to overcome resistance to Bin toxin will be discussed further in this section.

Among the potential agents to be used in association with *Ls*, *Bti*-based biolarvicides are considered the most promising and in addition, they are commercially available. *Bti* is a proven alternative for the management of *Ls* resistance because its toxins and their modes of action are unrelated to the Bin toxin. The most common protoxins found in *Bti* crystals are Cry4Aa (125 kDa), Cry4Ba (135 kDa), Cry11Aa (68 kDa) and Cyt1Aa (28 kDa) (Berry et al., 2002). Some *Bti* strains can produce other dipteran-active toxins such as Cry10Aa (58 kDa) and Cyt2Ba (30 kDa) (Guerchicoff et al., 1997; Thorne et al., 1986). *Bti* does not display cross-resistance to Bin toxin and, in addition, its mode of action does not favour the selection of resistance. Some factors behind these features are the multiple composition of Cry and Cyt protoxins available in the *Bti* crystal; the synergistic role played by the Cyt1Aa toxin as a surrogate receptor for the Cry toxins (Pérez et al., 2005, 2007); the participation of other midgut molecules that are also receptors for the Cry toxins (Likitvivatanavong et al., 2011). These features endorse *Bti* as an excellent candidate to be used for the management of *Ls* resistance and, to date, there are no records of resistance in mosquito field populations exposed to this agent (Ferreira and Silva-Filha, 2013; Wirth, 2010). Cry toxins from other Diptera-active *Bt* strains also do not display cross-resistance to Bin toxin as will be described in the following topics.

6.3.1 Bacillus thuringiensis *mosquitocidal toxins*

Cx. pipiens larvae resistant to the Bin toxin are susceptible to *Bti*, as has been shown by studies based on resistant colonies and populations from United

States, Brazil, India, China and Thailand (Amorim et al., 2007; Nielsen-Leroux et al., 1995; Rao et al., 1995; Silva-Filha et al., 1995; Su and Mulla, 2004; Yuan et al., 2003). Individual Cry and Cyt toxins from *Bti* or other mosquitocidal *Bt* strains, such as serovar. *medellin* (*Btmed*) and serovar. *jegathesan* (*Btjeg*) are able to overcome Bin toxin resistance. The synergistic interactions among these toxins and Bin toxin have been demonstrated using mixtures of toxins produced in the respective native strains, or by their combined expression in recombinant *Bti* or *Ls* strains as described below.

The synergistic effect of Cyt toxins on Bin toxicity was first investigated using the Cyt1Ab1 from *Btmed*. In this case, a recombinant *Ls* 2297 strain that produced both Bin and Cyt1Ab1 exhibited a significant activity against the larvae from GEO- and SPHAE-resistant colonies (Thiery et al., 1998). Mixtures of Cyt1Aa or Cyt2Ba, from *Bti*, with the Bin toxin were shown to be effective against *Ae. aegypti* larvae that is a refractory species to Bin toxin (Wirth et al., 2000a) and to *Cx. pipiens* larvae that were resistant to the Bin toxin (Wirth et al., 2000c, 2001a, 2004). A laboratory selection trial of *Cx. quinquefasciatus* larvae using a mixture of Bin and Cyt1 toxins in a 3:1 ratio during 20 generations showed a RR of 1.4 while a RR > 1000 was observed for their counterparts selected with the Bin toxin only (Wirth et al., 2005), corroborating the role of the Cyt1Aa toxin to delay Bin toxin resistance.

The synergy of the whole *Bti* crystal, or its individual Cry toxins, with Bin toxin has also been demonstrated. The introduction of *Bti* toxins in *Ls* strains has been attempted, before the advent of Bin toxin resistance, aiming at the improving the spectrum of *Ls* toxicity since this is more restricted than *Bti* (Bar et al., 1991; Trisrisook et al., 1990). Later, working with Bin resistance, a study evaluated a set of mixtures of Bin with *Bti* and its toxins that showed an enhanced activity to Bin-resistant larvae and the potential of these mixtures to overcome the resistance to the Bin toxin (Wirth et al., 2004). Evaluation of the synergism of six different mixtures of wild *Ls* and *Bti* crystals to susceptible *Cx. quinquefasciatus* showed that the highest synergism factor found was observed using a 3:1 ratio of *Ls* to *Bti* (Sreshty et al., 2011). This combination also displayed faster cytopathological effects on the midgut epithelium and muscles.

The production of recombinant *Ls* strains containing *Bti* toxins and/or the Cry11Ba toxin from *Btjeg* showed an enhanced activity to larvae from Bin-resistant colonies and partially restored their susceptibility to the Bin toxin (Poncet et al., 1994, 1997; Servant et al., 1999). However, it is important to note that the introduction of *Bt* genes in *Ls* strains has been often marked by the low expression and/or instability of the respective foreign proteins (Federici et al., 2010; Gammon et al., 2006). Elucidation of the molecular factors involved in this issue could lead to a significant advancement for the construction of improved strains.

The integration of Bin toxin into *Bti* was also performed but there are fewer reports of this, perhaps due to the fact that *Bti* displays a worse field performance compared to *Ls*, especially in *Culex* and *Anopheles* spp breeding sites (Nicolas et al., 1987; Silapanuntakul et al., 1983). Nevertheless, *Bti* recombinant constructs were obtained and they successfully produced *Bti* and Bin toxins with improved toxicity. A recombinant *Bti*, producing an association of Cyt1Aa, Cry11Ba and the Bin toxin, showed improved toxicity to susceptible *Cx. quinquefasciatus* larvae (Park et al., 2003). Another construct expressing the complete set of *Bti* toxins and the Bin toxin showed improved toxicity to *Cx. quinquefasciatus*, compared to the toxicity of each of the strains used, and it also suppressed Bin resistance (Park et al., 2005). The integration of *Bt* toxins and Bin toxins is effective to delay or avoid the selection of Bin toxin resistance and recombinant bacteria have potential as tools to produce a set of toxins for mosquito control. The expression of *Bt* and *Ls* insecticidal toxins has also been attempted in other microorganisms found in the aquatic environment in order to achieve a combined expression of their toxins and a suitable delivery in these larval habitats (Tanapongpipat et al., 2003; Tandeau de Marsac et al., 1987; Thanabalu et al., 1992b; Yap et al., 1994). Products based in such recombinant bacteria have not been used in the field and it remains a promising perspective in view of the substantial advancements achieved, as previously reviewed (Federici et al., 2003, 2007, 2010; Park et al., 2005).

To date, the management of *Ls* resistance has been based on rotation or mixtures with *Bti*, since this agent is commercially available and its field effectiveness has been proved worldwide (Becker, 1997, 2000; Regis et al., 2001). The choice between the use of *Ls* and *Bti* in rotation or mixtures depends on operational issues of the control programme, and also if the strategy is applied for the reversal of the resistance, or for preventing its appearance. Both rotation and mixtures seem to be effective in the first scenario, but mixtures seem more efficient to delay the onset of resistance (Mulla et al., 2003). Laboratory experiments using *Cx. quinquefasciatus* colonies subjected to mixtures of *Ls* and *Bti* proved to be more efficient to delay resistance, compared to the scheme of rotation (Zahiri and Mulla, 2003; Zahiri et al., 2002). Association of *Ls* and *Bti* products has been used for mosquito control and, recently, commercial formulations containing the mixtures of crystals produced by each agent in a single product have been developed (Anderson et al., 2011; Kahindi et al., 2008; Mwangangi et al., 2011). These products aim to target a wider range of mosquito species and breeding sites. Trials have been carried out to control *Culex* and *Aedes* species that colonise breeding sites from urban areas (Anderson et al., 2011; Eritja, 2013; Guidi et al., 2013a), or other species that occur in wetlands from environmentally sensitive areas (Dritz et al., 2011). These multi-toxin products showed promising results and can be effectively used in mosquito-control programmes ensuring the environmentally safe action and a low potential for resistance selection.

6.3.2 *Other* L. sphaericus *mosquitocidal toxins*

Other insecticidal factors produced by *Ls* strains have proved to be suitable in principle for the management of Bin resistance due to their lack of cross-resistance with this toxin. The Mtx toxins do not provide an important contribution for the activity exhibited by their native strains, as previously described (Section 2). On the other hand when Mtx1 and Mtx2 are produced as recombinant proteins in *E. coli* they revealed a capacity to synergise the activity of Bin toxin towards *Cx. quinquefasciatus* (Wirth et al., 2007). Low amounts of Mtx1 and Mtx2 combined with the Bin toxin (1:1:8 ratio) provoked a significant reduction of the resistance levels to Bin toxin in a selected colony. Evaluation of Mtx1 showed a high activity against larvae from one susceptible and two highly Bin-resistant colonies selected in China, indicating its distinct mode of action and target sites (Wei et al., 2007).

The discovery of the Cry48Aa/Cry49Aa Bin toxin in strain IAB59 was supported by the earlier reports of the low cross-resistance displayed by this strain to the Bin toxin (Amorim et al., 2007; Nielsen-LeRoux et al., 2001; Pei et al., 2002; Yuan et al., 2003). Similarly to the Mtx toxins, Cry48Aa/Cry49Aa does not contribute to an enhanced activity of IAB59 final sporulated cultures due to the low expression of Cry48Aa, which is required in an equimolar ratio with Cry49Aa for optimal toxicity (Jones et al., 2007). The adequate production of these factors as recombinant proteins and the administration of their mixtures at an equimolar ratio show a remarkable toxicity to both susceptible and to Bin-resistant *Cx. quinquefasciatus* larvae (Jones et al., 2007, 2008). The Cry48Aa/Cry49Aa combination probably targets midgut molecules that are distinct from the receptors identified for the Bin toxin in view of its activity in Bin-resistant *Cx. quinquefasciatus* larvae lacking the Cqm1 receptor (de Melo et al., 2009). This toxin is an also an interesting candidate for the development of multi-toxin biolarvicides in view of its unique composition as described in Section 2. Studies of the morphological alterations provoked by this toxin on the midgut cells of susceptible and Bin-resistant larvae showed a complex combination of effects that are similar to those of a mixture of Bin and Cry11Aa toxins (de Melo et al., 2009).

6.3.3 Other control agents

In a broader view for the management of *Ls* resistance, the integration of control agents available, other than toxins from *Bt* and *Ls* presented above, should also be taken in account (Hurst et al., 2006; Keiser et al., 2005; Lacey and Lacey, 1990; Lingenfelser et al., 2010; Skovmand et al., 2011; Tietze et al., 1994). Biological control agents can be used with *Ls*, in view of its safety for other organisms that occur in the aquatic environment of mosquito breeding sites. Lacey reviewed this aspect and presented studies of the combined use of *Ls* with predators (fish, aquatic insects), entomopathogenic fungi and parasites (nematodes), as viable choices to be adopted (Lacey, 2007). In practise, the use of these agents has been much more limited by other factors related to their production, storage, transport and liberation but they can be successfully associated with *Ls*. Other control agents such as synthetic insecticides, insect growth regulators and spinosins can be considered to be use with *Ls* and *Bti* due to their unrelated modes of action

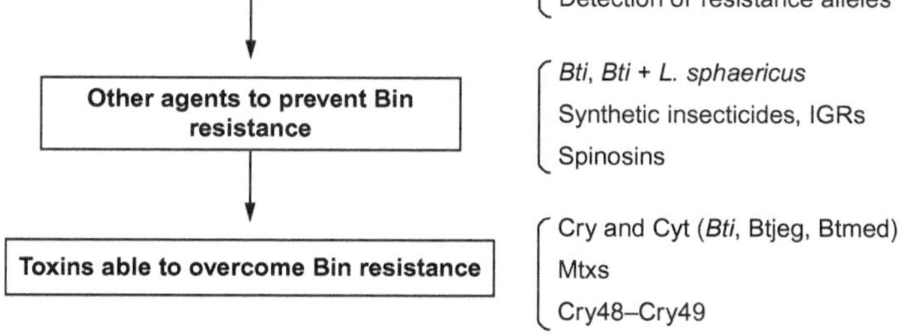

Figure 3.10 Strategies for preventing the resistance to the Bin toxin from *Lysinibacillus sphaericus* (*Ls*) and agents/toxins that can be used for its management. Insect growth regulator (IGR); *Bacillus thuringiensis* serovar. *israelensis* (*Bti*), serovar. *jegathesan* (*Btjeg*), serovar. *medellin* (*Btmed*); mosquitocidal toxins (Mtxs).

(Cetin et al., 2005; Giraldo-Calderon et al., 2008; Guidi et al., 2013b; Liu et al., 2004; Marcombe et al., 2011; Pridgeon et al., 2008; Tetreau et al., 2013). Spinosins, for instance, are also bacterial larvicides that have been recently registered in some countries for mosquito control (Hertlein et al., 2010). They show a relatively narrow toxicity spectrum, and have recently been introduced with promising results, alone or combination with other biolarvicides (Anderson et al., 2011; Cetin et al., 2005; Harbison et al., 2013; Jiang and Mulla, 2009; Kumar et al., 2011; Marcombe et al., 2011; Marina et al., 2012). The rational utilisation of *Ls* in the context of an integrated programme using suitable control strategies and including monitoring of the mosquito susceptibility can ensure its effectiveness for the control interventions and can prevent the onset of resistance. The major aspects concerning the management of resistance to *Ls* presented in this section are summarised in Fig. 3.10.

ACKNOWLEDGEMENTS

We thank all colleagues who provided suggestions for this work; publishers for the permissions for using figures presented in this chapter; Cláudia Maria Fontes de Oliveira, Constância Junqueira Ayres, Maria Alice Varjal de Melo-Santos and Rosângela Maria Barbosa from the Department of Entomology (CPqAM-FIOCRUZ) for the encouragement for writing this chapter.

REFERENCES

Abdullah, M.A., Valaitis, A.P., Dean, D.H., 2006. Identification of a *Bacillus thuringiensis* Cry11Ba toxin-binding aminopeptidase from the mosquito *Anopheles quadrimaculatus*. BMC Biochem. 7, 16.

Ahmed, H.K., Mitchell, W.J., Priest, F.G., 1995. Regulation of mosquitocidal toxin synthesis in *Bacillus sphaericus*. Appl. Microbiol. Biotechnol. 43, 310–314.

Ahmed, I., Yokota, A., Yamazoe, A., Fujiwara, T., 2007. Proposal of *Lysinibacillus boronitolerans* gen. nov. sp. nov., and transfer of *Bacillus fusiformis* to *Lysinibacillus fusiformis* comb. nov. and *Bacillus sphaericus* to *Lysinibacillus sphaericus* comb. nov. Int. J. Syst. Evol. Microbiol. 57, 1117–1125.

Alexander, B., Priest, F.G., 1990. Numerical classification and identification of *Bacillus sphaericus* including some strains pathogenic for mosquito larvae. J. Gen. Microbiol. 136, 367–376.

Almog, O., Gonzalez, A., Klein, D., Greenblatt, H.M., Braun, S., Shoham, G., 2003. The 0.93A crystal structure of sphericase: a calcium-loaded serine protease from *Bacillus sphaericus*. J. Mol. Biol. 332, 1071–1082.

Aly, C., Mulla, M.S., Federici, B.A., 1989. Ingestion, dissolution, and proteolysis of the *Bacillus sphaericus* toxin by mosquito larvae. J. Invertebr. Pathol. 53, 12–20.

Amorim, L.B., Oliveira, C.M.F., Rios, E.M., Regis, L., Silva-Filha, M.H.N.L., 2007. Development of *Culex quinquefasciatus* resistance to *Bacillus sphaericus* strain IAB59 needs long term selection pressure. Biol. Control 42, 155–160.

Amorim, L.B., De Barros, R.A., Chalegre, K.D., De Oliveira, C.M., Regis, L.N., Silva-Filha, M.H., 2010. Stability of *Culex quinquefasciatus* resistance to *Bacillus sphaericus* evaluated by molecular tools. Insect Biochem. Mol. Biol. 40, 311–316.
Anderson, J.F., Ferrandino, F.J., Dingman, D.W., Main, A.J., Andreadis, T.G., Becnel, J.J., 2011. Control of mosquitoes in catch basins in Connecticut with *Bacillus thuringiensis israelensis*, *Bacillus sphaericus*, [corrected] and spinosad. J. Am. Mosq. Control Assoc. 27, 45–55.
Andow, D.A., Olson, D.M., Hellmich, R.L., Alstad, D.N., Hutchison, W.D., 2000. Frequency of resistance to *Bacillus thuringiensis* toxin Cry1Ab in an Iowa population of European corn borer (Lepidoptera: Crambidae). J. Econ. Entomol. 93, 26–30.
Andrade, C.F.S., Campos, J.C., Cabrini, I., Marques Filho, C.A.M., Hibi, S., 2007. Suscetibilidade de populações de *Culex quinquefasciatus* Say (Diptera: Culicidae) sujeitas ao controle com *Bacillus sphaericus* Neide no rio Pinheiros, São Paulo. BioAssay 2, 1–4.
Anilkumar, K.J., Pusztai-Carey, M., Moar, W.J., 2008. Fitness costs associated with Cry1Ac-resistant *Helicoverpa zea* (Lepidoptera: Noctuidae): a factor countering selection for resistance to Bt cotton? J. Econ. Entomol. 101, 1421–1431.
Aquino De Muro, M., Mitchell, W.J., Priest, F.G., 1992. Differentiation of mosquito-pathogenic strains of *Bacillus sphaericus* from non-toxic varieties by ribosomal RNA gene restriction patterns. J. Gen. Microbiol. 138, 1159–1166.
Araújo, A.P., Melo-Santos, M.A.V., Carlos, S.O., Rios, E.M., Regis, L., 2007. Evaluation of an experimental product based on *Bacillus thuringiensis* sorovar. *israelensis* against *Aedes aegypti* larvae (Diptera: Culicidae). Biol. Control 41, 339–347.
Arenas, I., Bravo, A., Soberón, M., Gómez, I., 2010. Role of alkaline phosphatase from *Manduca sexta* in the mechanism of action of *Bacillus thuringiensis* Cry1Ab toxin. J. Biol. Chem. 285, 12497–12503.
Arredondo-Jimenez, J.I., Lopez, T., Rodriguez, M.H., Bown, D.N., 1990. Small scale field trials of *Bacillus sphaericus* (strain 2362) against anopheline and culicine mosquito larvae in southern Mexico. J. Am. Mosq. Control Assoc. 6, 300–305.
Atsumi, S., Miyamoto, K., Yamamoto, K., Narukawa, J., Kawai, S., Sezutsu, H., Kobayashi, I., Uchino, K., Tamura, T., Mita, K., Kadono-Okuda, K., Wada, S., Kanda, K., Goldsmith, M.R., Noda, H., 2012. Single amino acid mutation in an ATP-binding cassette transporter gene causes resistance to Bt toxin Cry1Ab in the silkworm *Bombyx mori*. Proc. Natl. Acad. Sci. U.S.A 109, E1591–E1598.
Bar, E., Lieman-Hurwitz, J., Rahamim, E., Keynan, A., Sandler, N., 1991. Cloning and expression of *Bacillus thuringiensis israelensis* delta-endotoxin DNA in *B. sphaericus*. J. Invertebr. Pathol. 57, 149–158.
Barbazan, P., Baldet, T., Darriet, F., Escaffre, H., Djoda, D.H., Hougard, J.M., 1997. Control of *Culex quinquefasciatus* (Diptera: Culicidae) with *Bacillus sphaericus* in Maroua, Cameroon. J. Am. Mosq. Control Assoc. 13, 263–269.
Barbazan, P., Baldet, T., Darriet, F., Escaffre, H., Djoda, D.H., Hougard, J.M., 1998. Impact of treatments with *Bacillus sphaericus* on *Anopheles populations* and the transmission of malaria in Maroua, a large city in a savannah region of Cameroon. J. Am. Mosq. Control Assoc. 14, 33–39.
Baumann, L., Baumann, P., 1991. Effects of components of the *Bacillus sphaericus* toxin on mosquito larvae and mosquito-derived tissue culture-grown cells. Curr. Microbiol. 23, 51–57.
Baumann, P., Unterman, B.M., Baumann, L., Broadwell, A.H., Abbene, S.J., Bowditch, R.D., 1985. Purification of the larvicidal toxin of *Bacillus sphaericus* and evidence for high-molecular-weight precursors. J. Bacteriol. 163, 738–747.
Baumann, P., Baumann, L., Bowditch, R.D., Broadwell, A.H., 1987. Cloning of the gene for the larvicidal toxin of *Bacillus sphaericus* 2362: evidence for a family of related sequences. J. Bacteriol. 169, 4061–4067.

Baumann, L., Broadwell, A.H., Baumann, P., 1988. Sequence analysis of the mosquitocidal toxin genes encoding 51.4- and 41.9-kilodalton proteins from *Bacillus sphaericus* 2362 and 2297. J. Bacteriol. 170, 2045–2050.
Baumann, P., Clark, M.A., Baumann, L., Broadwell, A.H., 1991. *Bacillus sphaericus* as a mosquito pathogen: properties of the organism and its toxins. Microbiol. Rev. 55, 425–436.
Becker, N., 1997. Microbial control of mosquitoes: management of the upper Rhine mosquito population as a model programme. Parasitol. Today 13, 485–487.
Becker, N., 1998. The use of *Bacillus thuringiensis* subs. *israelensis* (Bti) against mosquitoes, with special emphasis on the ecological impact. Isr. J. Entomol. 32, 63–69.
Becker, N., 2000. Bacterial control of vector mosquitoes and black flies. In: Charles, J.F., Delecluse, A., Nielsen-Leroux, C. (Eds.), Entomopathogenic Bacteria: From Laboratory to Filed Application. Kluwer Academic Publishers, Dordrecht, pp. 384–398.
Becker, N., 2003. Ice granules containing endotoxins of microbial agents for the control of mosquito larvae—a new application technique. J. Am. Mosq. Control Assoc. 19, 63–66.
Becker, N., Zgomba, M., Petric, D., Beck, M., Ludwig, M., 1995. Role of larval cadavers in recycling processes of *Bacillus sphaericus*. J. Am. Mosq. Control Assoc. 11, 329–334.
Becker, N., Petric, D., Dahl, C., Lane, J., Kaiser, A., 2003. Integrated pest management. In: Becker, N. (Ed.), Mosquitos and Their Control. Kluwer Academic/Plenum Publishers, New York, pp. 417–424.
Berrocal, E., Carey, C., Rodriguez, L., Calampa, C., Valdivia, L., 2000. A residual effect of *Bacillus sphaericus* for control of *Anopheles darlingi*, in Iquitos in Perú. J. Am. Mosq. Control Assoc. 16, 295–312.
Berry, C., 2012. The bacterium, *Lysinibacillus sphaericus*, as an insect pathogen. J. Invertebr. Pathol. 109, 1–10.
Berry, C., Hindley, J., 1987. *Bacillus sphaericus* strain 2362: identification and nucleotide sequence of the 41.9 kDa toxin gene. Nucleic Acids Res. 15, 5891.
Berry, W.J., Novak, M.G., Khounlo, S., Rowley, W.A., Melchior, G.L., 1987. Efficacy of *Bacillus sphaericus* and *Bacillus thuringiensis var. israelensis* for control of *Culex pipiens* and floodwater *Aedes* larvae in Iowa. J. Am. Mosq. Control Assoc. 3, 579–582.
Berry, C., Hindley, J., Ehrhardt, A.F., Grounds, T., De Souza, I., Davidson, E.W., 1993. Genetic determinants of host ranges of *Bacillus sphaericus* mosquito larvicidal toxins. J. Bacteriol. 175, 510–518.
Berry, C., O'neil, S., Ben-Dov, E., Jones, A.F., Murphy, L., Quail, M.A., Holden, M.T., Harris, D., Zaritsky, A., Parkhill, J., 2002. Complete sequence and organization of pBtoxis, the toxin-coding plasmid of *Bacillus thuringiensis* subsp. *israelensis*. App. Environ. Microbiol. 68, 5082–5095.
Berti, J., Gonzales, J., 2004. Evaluación de la efectividad y persistencia de la nueva formulación de *Bacillus sphaericus* contra larvas de *Anopheles aquasalis* Curry en los criaderos naturales del estado Sucre, Venezuela. Bol. Mal. Salud Amb. 44, 21–27.
Blanco-Castro, S.D., Martínez Arias, A., Cano Velasquez, O.R., Tello Granados, R., Mendonza, I., 2000. Introduction of *Bacillus sphaericus* strain-2362 (GRISELESF) for biological control of malaria vectors in Guatemala. Rev. Cubana Med. Trop. 52, 37–43.
Bone, L.W., Tinelli, R., 1987. *Trichostrongylus colubriformis*: larvicidal activity of toxic extracts from *Bacillus sphaericus* (strain 1593) spores. Exp. Parasitol. 64, 514–516.
Boonserm, P., Moonsom, S., Boonchoy, C., Promdonkoy, B., Parthasarathy, K., Torres, J., 2006. Association of the components of the binary toxin from *Bacillus sphaericus* in solution and with model lipid bilayers. Biochem. Biophys. Res. Commun. 342, 1273–1278.
Boonyos, P., Soonsanga, S., Boonserm, P., Promdonkoy, B., 2010. Role of cysteine at positions 67, 161 and 241 of a *Bacillus sphaericus* binary toxin BinB. BMB Rep. 43, 23–28.
Bourdeau, R.W., Malito, E., Chenal, A., Bishop, B.L., Musch, M.W., Villereal, M.L., Chang, E.B., Mosser, E.M., Rest, R.F., Tang, W.J., 2009. Cellular functions and

X-ray structure of anthrolysin O, a cholesterol-dependent cytolysin secreted by *Bacillus anthracis*. J. Biol. Chem. 284, 14645–14656.
Bowditch, R.D., Baumann, P., Yousten, A.A., 1989. Cloning and sequencing of the gene encoding a 125-kilodalton surface-layer protein from *Bacillus sphaericus* 2362 and of a related cryptic gene. J. Bacteriol. 171, 4178–4188.
Bowles, D.E., Meisch, M.V., Weathersbee, A.A., Jones, J.W., Efird, P., Bassi, D., 1990. Efficacy of *Bacillus sphaericus* formulations against *Psorophora columbiae* larvae in small rice plots. J. Am. Mosq. Control Assoc. 6, 631–634.
Bravo, A., Gómez, I., Conde, J., Muñoz-Garay, C., Sánchez, J., Miranda, R., Zhuang, M., Gill, S.S., Soberón, M., 2004. Oligomerization triggers binding of a *Bacillus thuringiensis* Cry1Ab pore-forming toxin to aminopeptidase N receptor leading to insertion into membrane microdomains. Biochim. Biophys. Acta 1667, 38–46.
Bravo, A., Gill, S.S., Soberón, M., 2007. Mode of action of *Bacillus thuringiensis* Cry and Cyt toxins and their potential for insect control. Toxicon 49, 423–435.
Broadwell, A.H., Baumann, P., 1986. Sporulation-associated activation of *Bacillus sphaericus* larvicide. Appl. Environ. Microbiol. 52, 758–764.
Broadwell, A.H., Baumann, P., 1987. Proteolysis in the gut of mosquito larvae results in further activation of the *Bacillus sphaericus* toxin. Appl. Environ. Microbiol. 53, 1333–1337.
Broadwell, A.H., Baumann, L., Baumann, P., 1990a. The 42- and 51-kilodalton mosquitocidal proteins of *Bacillus sphaericus* 2362: construction of recombinants with enhanced expression and in vivo studies of processing and toxicity. J. Bacteriol. 172, 2217–2223.
Broadwell, A.H., Baumann, L., Baumann, P., 1990b. Larvicidal properties of the 42 and 51 kilodalton *Bacillus sphaericus* proteins expressed in different bacterial hosts: evidence for a binary toxin. Curr. Microbiol. 21, 361–366.
Broadwell, A.H., Clark, M.A., Baumann, L., Baumann, P., 1990c. Construction by site-directed mutagenesis of a 39-kilodalton mosquitocidal protein similar to the larva-processed toxin of *Bacillus sphaericus* 2362. J. Bacteriol. 172, 4032–4036.
Brogdon, W.G., Mcallister, J.C., 1998. Insecticide resistance and vector control. Emerg. Infect. Dis. 4, 605–613.
Brown, I.D., Watson, T.M., Carter, J., Purdie, D.M., Kay, B.H., 2004. Toxicity of VectoLex (*Bacillus sphaericus*) products to selected Australian mosquito and nontarget species. J. Econ. Entomol. 97, 51–58.
Brownbridge, M., Margalit, J., 1987. Mosquito active strains of *Bacillus sphaericus* isolated from soil and mud samples collected in Israel. J. Invertebr. Pathol. 50, 106–112.
Burke, W.F., Mcdonald, K.O., Davidson, E.W., 1983. Effect of UV light on spore viability and mosquito larvicidal activity of *Bacillus sphaericus* 1593. Appl. Environ. Microbiol. 46, 954–956.
Cancino-Rodezno, A., Alexander, C., Villasenor, R., Pacheco, S., Porta, H., Pauchet, Y., Soberon, M., Gill, S.S., Bravo, A., 2010. The mitogen-activated protein kinase p38 is involved in insect defense against Cry toxins from *Bacillus thuringiensis*. Insect Biochem. Mol. Biol. 40, 58–63.
Carpusca, I., Schirmer, J., Aktories, K., 2004. Two-site autoinhibition of the ADP-ribosylating mosquitocidal toxin (MTX) from *Bacillus sphaericus* by its 70-kDa ricin-like binding domain. Biochemistry 43, 12009–12019.
Carpusca, I., Jank, T., Aktories, K., 2006. *Bacillus sphaericus* mosquitocidal toxin (MTX) and pierisin: the enigmatic offspring from the family of ADP-ribosyltransferases. Mol. Microbiol. 62, 621–630.
Cartaxo, M.F., Ayres, C.F., Weetman, D., 2011. Loss of genetic diversity in *Culex quinquefasciatus* targeted by a lymphatic filariasis vector control program in Recife. Brazil. Trans. R. Soc. Trop. Med. Hyg. 105, 491–499.

Cetin, H., Yanikoglu, A., Cilek, J.E., 2005. Evaluation of the naturally-derived insecticide spinosad against *Culex pipiens* L. (Diptera: Culicidae) larvae in septic tank water in Antalya, Turkey. J. Vector Ecol. 30, 151–154.

Cetin, H., Dechant, P., Yanikoglu, A., 2007. Field trials with tank mixtures of *Bacillus thuringiensis* subsp. *israelensis* and *Bacillus sphaericus* formulations against *Culex pipiens* larvae in septic tanks in Antalya, Turkey. J. Am. Mosq. Control Assoc. 23, 161–165.

Chalegre, K.D., Romão, T.P., Amorim, L.B., Anastacio, D.B., De Barros, R.A., De Oliveira, C.M., Regis, L., De-Melo-Neto, O.P., Silva-Filha, M.H., 2009. Detection of an allele conferring resistance to *Bacillus sphaericus* binary toxin in *Culex quinquefasciatus* populations by molecular screening. Appl. Environ. Microbiol. 75, 1044–1049.

Chalegre, K.D., Romão, T.P., Tavares, D.A., Santos, E.M., Ferreira, L.M., Oliveira, C.M. F., De-Melo-Neto, O.P., Silva-Filha, M.H.N.L., 2012. Novel mutations associated to *Bacillus sphaericus* resistance are identified in a polymorphic region of the *Culex quinquefasciatus cqm1* gene Appl. Environ. Microbiol. 78, 6321–6326.

Chan, S.W., Thanabalu, T., Wee, B.Y., Porter, A.G., 1996. Unusual amino acid determinants of host range in the Mtx2 family of mosquitocidal toxins. J. Biol. Chem. 271, 14183–14187.

Charles, J.F., 1987. Ultrastructural midgut events in Culicidae larvae fed with *Bacillus sphaericus* 2297 spore/crystal complex. Ann. Inst. Pasteur Microbiol. 138, 471–484.

Charles, J.F., Nicolas, L., 1986. Recycling of *Bacillus sphaericus* 2362 in mosquito larvae: a laboratory study. Ann. Inst. Pasteur Microbiol. 137B, 101–111.

Charles, J.F., Nielsen-Leroux, C., 2000. Mosquitocidal bacterial toxins: diversity, mode of action and resistance phenomena. Mem. Inst. Oswaldo Cruz 95 (Suppl. 1), 201–206.

Charles, J.F., Kalfon, A., Bourgouin, C., de Barjac, H., 1988. *Bacillus sphaericus* asporogenous mutants: morphology, protein pattern and larvicidal activity. Ann. Inst. Pasteur Microbiol. 139, 243–259.

Charles, J.F., Hamon, S., Baumann, P., 1993. Inclusion bodies and crystals of *Bacillus sphaericus* mosquitocidal proteins expressed in various bacterial hosts. Res. Microbiol. 144, 411–416.

Charles, J.F., Nielsen-Leroux, C., Delecluse, A., 1996. *Bacillus sphaericus* toxins: molecular biology and mode of action. Annu. Rev. Entomol. 41, 451–472.

Charles, J.F., Silva-Filha, M.H., Nielsen-Leroux, C., Humphreys, M.J., Berry, C., 1997. Binding of the 51- and 42-kDa individual components from the *Bacillus sphaericus* crystal toxin to mosquito larval midgut membranes from *Culex* and *Anopheles* sp. (Diptera: Culicidae). FEMS Microbiol. Lett. 156, 153–159.

Charles, J.F., Darboux, I., Pauron, D., Nielsen-Leroux, C., 2010. Mosquitocidal *Bacillus sphaericus*: toxins, genetics, mode of action, use, and resistance mechanisms. In: Gilbert, L., Gill, S.S. (Eds.), Insect Control: Biological and Synthetic Agents. Elsevier, London, pp. 283–307.

Chen, J., Aimanova, K.G., Fernandez, L.E., Bravo, A., Soberon, M., Gill, S.S., 2009a. *Aedes aegypti* cadherin serves as a putative receptor of the Cry11Aa toxin from *Bacillus thuringiensis* subsp. *israelensis*. Biochem. J. 424, 191–200.

Chen, J., Aimanova, K.G., Pan, S., Gill, S.S., 2009b. Identification and characterization of *Aedes aegypti* aminopeptidase N as a putative receptor of *Bacillus thuringiensis* Cry11A toxin. Insect Biochem. Mol. Biol. 39, 688–696.

Cheong, W.C., Yap, H.H., 1985. Bioassays of *Bacillus sphaericus* (strain 1593) against mosquitoes of public health importance in Malaysia. Southeast Asian J. Trop. Med. Public Health 16, 54–58.

Chevillon, C., Bernard, C., Marquine, M., Pasteur, N., 2001. Resistance to *Bacillus sphaericus* in *Culex pipiens* (Diptera: Culicidae): interaction between recessive mutants and evolution in southern France. J. Med. Entomol. 38, 657–664.

Chiou, C., Davidson, E.W., Thanabalu, T., Porter, A.G., Allen, J.P., 1999. Crystallization and preliminary X-ray diffraction studies of the 51 kDa protein of the mosquito-larvicidal binary toxin from *Bacillus sphaericus*. Acta Crystallogr. D Biol. Crystallogr. 55, 1083–1085.
Clark, M.A., Baumann, P., 1990. Deletion analysis of the 51-kilodalton protein of the *Bacillus sphaericus* 2362 binary mosquitocidal toxin: construction of derivatives equivalent to the larva-processed toxin. J. Bacteriol. 172, 6759–6763.
Clark, M.A., Baumann, P., 1991. Modification of the *Bacillus sphaericus* 51- and 42-kilodalton mosquitocidal proteins: effects of internal deletions, duplications, and formation of hybrid proteins. Appl. Environ. Microbiol. 57, 267–271.
Cokmus, C., Davidson, E.W., Cooper, K., 1997. Electrophysiological effects of *Bacillus sphaericus* binary toxin on cultured mosquito cells. J. Invertebr. Pathol. 69, 197–204.
Consoli, R.A., Santos Bde, S., Lamounier, M.A., Secundino, N.F., Rabinovitch, L., Silva, C.M., Alves, R.S., Carneiro, N.F., 1997. Efficacy of a new formulation of *Bacillus sphaericus* 2362 against *Culex quinquefasciatus* (Diptera: Culicidae) in Montes Claros, Minas Gerais, Brazil. Mem. Inst. Oswaldo Cruz 92, 571–573.
Correa, M., Yousten, A.A., 1995. *Bacillus sphaericus* spore germination and recycling in mosquito larval cadavers. J. Invertebr. Pathol. 66, 76–81.
Cuamba, N., Morgan, J.C., Irving, H., Steven, A., Wondji, C.S., 2010. High level of pyrethroid resistance in an *Anopheles funestus* population of the Chokwe district in Mozambique. PLoS One 5, e11010.
Darboux, I., Nielsen-Leroux, C., Charles, J.F., Pauron, D., 2001. The receptor of *Bacillus sphaericus* binary toxin in *Culex pipiens* (Diptera: Culicidae) midgut: molecular cloning and expression. Insect Biochem. Mol. Biol. 31, 981–990.
Darboux, I., Pauchet, Y., Castella, C., Silva-Filha, M.H., Nielsen-Leroux, C., Charles, J.F., Pauron, D., 2002. Loss of the membrane anchor of the target receptor is a mechanism of bioinsecticide resistance. Proc. Natl. Acad. Sci. U.S.A 99, 5830–5835.
Darboux, I., Charles, J.F., Pauchet, Y., Warot, S., Pauron, D., 2007. Transposon-mediated resistance to *Bacillus sphaericus* in a field-evolved population of *Culex pipiens* (Diptera: Culicidae). Cell. Microbiol. 9, 2022–2029.
Davidson, E.W., 1979. Ultrastructure of midgut events in the pathogenesis of *Bacillus sphaericus* strain SSII-1 infections of *Culex pipiens* quinquefasciatus larvae. Can. J. Microbiol. 25, 178–184.
Davidson, E.W., 1988. Binding of the *Bacillus sphaericus* (Eubacteriales: Bacillaceae) toxin to midgut cells of mosquito (Diptera: Culicidae) larvae: relationship to host range. J. Med. Entomol. 25, 151–157.
Davidson, E.W., 1989. Variation in binding of *Bacillus sphaericus* toxin and wheat germ agglutinin to larval midgut cells of six species of mosquitoes. J. Invertebr. Pathol. 53, 251–259.
Davidson, E.W., Myers, P., 1981. Parasporal inclusions in *Bacillus sphaericus*. FEMS Microbiol. Lett. 10, 261–265.
Davidson, E.W., Titus, M., 1987. Ultrastructural effects of the *Bacillus sphaericus* mosquito larvicidal toxin on cultured mosquito cells. J. Invertebr. Pathol. 50, 213–220.
Davidson, E.W., Morthon, H.L., Moffett, J.O., Singer, S., 1977. Effect of *Bacillus sphaericus* strain SSII-1 on honey bees, *Apis mellifera*. J. Invertebr. Pathol. 29, 344–346.
Davidson, E.W., Urbina, M., Payne, J., Mulla, M.S., Darwazeh, H., Dulmage, H.T., Correa, J.A., 1984. Fate of *Bacillus sphaericus* 1593 and 2362 spores used as larvicides in the aquatic environment. Appl. Environ. Microbiol. 47, 125–129.
Davidson, E.W., Shellabarger, C., Meyer, M., Bieber, A.L., 1987. Binding of the *Bacillus sphaericus* mosquito larvicidal toxin to cultured insect cells. Can. J. Microbiol. 33, 982–989.
Davidson, E.W., Oei, C., Meyer, M., Bieber, A.L., Hindley, J., Berry, C., 1990. Interaction of the *Bacillus sphaericus* mosquito larvicidal proteins. Can. J. Microbiol. 36, 870–878.
De Barjac, H., 1978. Toxicity of *Bacillus thuringiensis* var. *israelensis* for larvae of *Aedes aegypti* and *Anopheles stephensi*. C. R. Acad. Sci. Hebd. Seances Acad. Sci. D 286, 1175–1178.

De Barjac, H., 1990. Classification of *Bacillus sphaericus* trains and comparative toxicity to mosquito larvae. In: De Barjac, H., Sutherland, D. (Eds.), Bacterial Control of Mosquitoes and Black-Flies: Biochemistry, Genetics, and Applications of *Bacillus thuringiensis* and *Bacillus sphaericus*. Rutgers University Press, New Brunswick, pp. 228–236.

De Barjac, H., Charles, J.F., 1983. Une nouvelle toxine active sur les moustiques, présente dans des inclusions crystallines produites para *Bacillus sphaericus*. C. R. Acad. Sci. Paris Sér. III 296, 905–910.

De Barjac, H., Veron, M.C., Dumanoir, V., 1980. Biochemical and serological characterization of *Bacillus sphaericus* strains, pathogenic or non-pathogenic for mosquitoes (author's transl). Ann. Microbiol. 131B, 191–201.

De Barjac, H., Larget-Thiery, I., Cosmao Dumanoir, V., Ripouteau, H., 1985. Serological classification of *Bacillus sphaericus* strains in relation with toxicity to mosquito larvae. Appl. Microbiol. Biotechnol. 21, 85–90.

De La Torre, F., Bennardo, T., Sebo, P., Szulmajster, J., 1989. On the respective roles of the two proteins encoded by the *Bacillus sphaericus* 1593M toxin genes expressed in *Escherichia coli* and *Bacillus subtilis*. Biochem. Biophys. Res. Commun. 164, 1417–1422.

De Maagd, R.A., Bravo, A., Berry, C., Crickmore, N., Schnepf, H.E., 2003. Structure, diversity, and evolution of protein toxins from spore-forming entomopathogenic bacteria. Annu. Rev. Genet. 37, 409–433.

De Melo, J.V., Vasconcelos, R.H., Furtado, A.F., Peixoto, C.A., Silva-Filha, M.H., 2008. Ultrastructural analysis of midgut cells from *Culex quinquefasciatus* (Diptera: Culicidae) larvae resistant to *Bacillus sphaericus*. Micron 39, 1342–1350.

De Melo, J.V., Jones, G.W., Berry, C., Vasconcelos, R.H., De Oliveira, C.M., Furtado, A.F., Peixoto, C.A., Silva-Filha, M.H., 2009. Cytopathological effects of *Bacillus sphaericus* Cry48Aa/Cry49Aa toxin on binary toxin-susceptible and -resistant *Culex quinquefasciatus* larvae. Appl. Environ. Microbiol. 75, 4782–4789.

De Oliveira, C.M., Filho, F.C., Beltran, J.E., Silva-Filha, M.H., Regis, L., 2003. Biological fitness of a *Culex quinquefasciatus* population and its resistance to *Bacillus sphaericus*. J. Am. Mosq. Control Assoc. 19, 125–129.

Delecluse, A., Juarez-Perez, V., Berry, C., 2000. Vector-active toxins: structure and diversity. In: Charles, J.F., Delécluse, A., Nielsen-Leroux, C. (Eds.), Entomopathogenic Bacteria: From Laboratory to Field Application. Kluwer Academic Publishers, Dordrecht, pp. 101–125.

Dennett, J.A., Meisch, M.V., 2000. Effectiveness of aerial- and ground-applied *Bacillus* formulations against *Anopheles quadrimaculatus* larvae in Arkansas rice plots. J. Am. Mosq. Control Assoc. 16, 229–233.

Dennett, J.A., Meek, C.L., Meisch, M.V., 2001. Efficacy of VectoLex WDG against *Anopheles quadrimaculatus* and *Psorophora columbiae* larvae in Arkansas and Mississippi rice. J. Am. Mosq. Control Assoc. 17, 231–237.

Des Rochers, B., Garcia, R., 1984. Evidence for persistence and recycling of *Bacillus sphaericus*. Mosq. News 44, 160–165.

Dritz, D.A., Lawler, S.P., Evkhanian, C., Graham, P., Baracosa, V., Dula, G., 2011. Control of mosquito larvae in seasonal wetlands on a wildlife refuge using VectoMax CG. J. Am. Mosq. Control Assoc. 27, 398–403.

Du, Y., Nomura, Y., Satar, G., Hu, Z., Nauen, R., He, S.Y., Zhorov, B.S., Dong, K., 2013. Molecular evidence for dual pyrethroid-receptor sites on a mosquito sodium channel. Proc. Natl. Acad. Sci. U.S.A 110, 11785–11790.

Elangovan, G., Shanmugavelu, M., Rajamohan, F., Dean, D.H., Jayaraman, K., 2000. Identification of the functional site in the mosquito larvicidal binary toxin of *Bacillus sphaericus* 1593M by site-directed mutagenesis. Biochem. Biophys. Res. Commun. 276, 1048–1055.

El-Bendary, M., Priest, F.G., Charles, J.F., Mitchell, W.J., 2005. Crystal protein synthesis is dependent on early sporulation gene expression in *Bacillus sphaericus*. FEMS Microbiol. Lett. 252, 51–56.
Eritja, R., 2013. Laboratory tests on the efficacy of VBC60035, a combined larvicidal formulation of *Bacillus thuringiensis israelensis* (strain AM65-52) and *Bacillus sphaericus* (strain 2362) against *Aedes albopictus* in simulated catch basins. J. Am. Mosq. Control Assoc. 29, 280–283.
Federici, B.A., Park, H.W., Bideshi, D.K., Wirth, M.C., Johnson, J.J., 2003. Recombinant bacteria for mosquito control. J. Exp. Biol. 206, 3877–3885.
Federici, B.A., Park, H.W., Bideshi, D.K., Wirth, M.C., Johnson, J.J., Sakano, Y., Tang, M., 2007. Developing recombinant bacteria for control of mosquito larvae. J. Am. Mosq. Control Assoc. 23, 164–175.
Federici, B.A., Park, H.D., Bideshi, D.K., 2010. Overview of the basic biology of *Bacillus thuringiensis* with emphasis on genetic engineering of bacterial larvicides for mosquito control. Open J. Toxicol. 3, 83–100.
Fernández, L.E., Aimanova, K.G., Gill, S.S., Bravo, A., Soberón, M., 2006. A GPI-anchored alkaline phosphatase is a functional midgut receptor of Cry11Aa toxin in *Aedes aegypti* larvae. Biochem. J. 394, 77–84.
Fernandez-Luna, M.T., Lanz-Mendoza, H., Gill, S.S., Bravo, A., Soberón, M., Miranda-Rios, J., 2010. An alpha-amylase is a novel receptor for *Bacillus thuringiensis* ssp. *israelensis* Cry4Ba and Cry11Aa toxins in the malaria vector mosquito *Anopheles albimanus* (Diptera: Culicidae). Environ. Microbiol. 12, 746–757.
Ferreira, L.M., Silva-Filha, M.H.N.L., 2013. Bacterial larvicides for vector control: mode of action of toxins and implications for resistance. Biocontrol Sci. Technol. 23, 1137–1168.
Ferreira, L.M., Romão, T.P., De-Melo-Neto, O.P., Silva-Filha, M.H., 2010. The orthologue to the Cpm1/Cqm1 receptor in *Aedes aegypti* is expressed as a midgut GPI-anchored alpha-glucosidase, which does not bind to the insecticidal binary toxin. Insect Biochem. Mol. Biol. 40, 604–610.
Ferreira, L.M., Romão, T.P., Nascimento, N.A., Costa, M.D., Rezende, A.M., De-Melo-Neto, O.P., Silva-Filha, M.H., 2014. Non conserved residues between Cqm1 and Aam1 mosquito alpha-glucosidases are critical for the capacity of Cqm1 to bind the Binary (Bin) toxin from *Lysinibacillus sphaericus*. Insect Biochem. Mol. Biol. 50, 34–42.
Ffrench-Constant, R.H., 2007. Which came first: insecticides or resistance? Trends Genet. 23, 1–4.
Fillinger, U., Lindsay, S.W., 2006. Suppression of exposure to malaria vectors by an order of magnitude using microbial larvicides in rural Kenya. Trop. Med. Int. Health 11, 1629–1642.
Fillinger, U., Lindsay, S.W., 2011. Larval source management for malaria control in Africa: myths and reality. Malar. J. 10, 353.
Fillinger, U., Knols, B.G., Becker, N., 2003. Efficacy and efficiency of new *Bacillus thuringiensis* var *israelensis* and *Bacillus sphaericus* formulations against Afrotropical anophelines in Western Kenya. Trop. Med. Int. Health 8, 37–47.
Fillinger, U., Kannady, K., William, G., Vanek, M.J., Dongus, S., Nyika, D., Geissbuhler, Y., Chaki, P.P., Govella, N.J., Mathenge, E.M., Singer, B.H., Mshinda, H., Lindsay, S.W., Tanner, M., Mtasiwa, D., et al., 2008. A tool box for operational mosquito larval control: preliminary results and early lessons from the Urban Malaria Control Programme in Dar es Salaam, Tanzania. Malar. J. 7, 20.
Fillinger, U., Ndenga, B., Githeko, A., Lindsay, S.W., 2009. Integrated malaria vector control with microbial larvicides and insecticide-treated nets in western Kenya: a controlled trial. Bull. WHO 87, 655–665.
Floore, T., Wardz, R., 2009. Evaluation of *Bacillus sphaericus* against *Mansonia dyari* larvae in phosphate lakes in Polk County. Florida. J. Am. Mosq. Control Assoc. 25, 310–314.

Frachon, E., Hamon, S., Nicolas, L., de Barjac, H., 1991. Cellular fatty acid analysis as a potential tool for predicting mosquitocidal activity of *Bacillus sphaericus* strains. Appl. Environ. Microbiol. 57, 3394–3398.

From, C., Granum, P.E., Hardy, S.P., 2008. Demonstration of a cholesterol-dependent cytolysin in a noninsecticidal *Bacillus sphaericus* strain and evidence for widespread distribution of the toxin within the species. FEMS Microbiol. Lett. 286, 85–92.

Gabrisko, M., 2013. Evolutionary history of eukaryotic alpha-glucosidases from the alpha-amylase family. J. Mol. Evol. 76, 129–145.

Gahan, L.J., Pauchet, Y., Vogel, H., Heckel, D.G., 2010. An ABC transporter mutation is correlated with insect resistance to *Bacillus thuringiensis* Cry1Ac toxin. PLoS Genet. 6, e1001248.

Galardo, A.K., Zimmerman, R., Galardo, C.D., 2013. Larval control of *Anopheles* (Nyssorhinchus) *darlingi* using granular formulation of *Bacillus sphaericus* in abandoned gold-miners excavation pools in the Brazilian Amazon rainforest. Rev. Soc. Bras. Med. Trop. 46, 172–177.

Gammon, K., Jones, G.W., Hope, S.J., De Oliveira, C.M., Regis, L., Silva Filha, M.H., Dancer, B.N., Berry, C., 2006. Conjugal transfer of a toxin-coding megaplasmid from *Bacillus thuringiensis* subsp. *israelensis* to mosquitocidal strains of *Bacillus sphaericus*. Appl. Environ. Microbiol. 72, 1766–1770.

Geissbuhler, Y., Chaki, P., Emidi, B., Govella, N.J., Shirima, R., Mayagaya, V., Mtasiwa, D., Mshinda, H., Fillinger, U., Lindsay, S.W., Kannady, K., De Castro, M.C., Tanner, M., Killeen, G.F., 2007. Interdependence of domestic malaria prevention measures and mosquito-human interactions in urban Dar es Salaam, Tanzania. Malar. J. 6, 126.

Geissbuhler, Y., Kannady, K., Chaki, P.P., Emidi, B., Govella, N.J., Mayagaya, V., Kiama, M., Mtasiwa, D., Mshinda, H., Lindsay, S.W., Tanner, M., Fillinger, U., De Castro, M.C., Killeen, G.F., 2009. Microbial larvicide application by a large-scale, community-based program reduces malaria infection prevalence in urban Dar es Salaam, Tanzania. PLoS One 4, e5107.

Georghiou, G.P., Lagunes-Tejeda, A., 1991. The Occurrence of Resistance to Pesticides in Arthropods. Food and Agriculture Organization of the United Nations. Report FAO-AGP-MISC/91-1, 318 p.

Giraldo-Calderon, G.I., Perez, M., Morales, C.A., Ocampo, C.B., 2008. Evaluation of the triflumuron and the mixture of *Bacillus thuringiensis* plus *Bacillus sphaericus f*or control of the immature stages of *Aedes aegypti* and *Culex quinquefasciatus* (Diptera: Culicidae) in catch basins. Biomedica 28, 224–233.

Goldberg, L.H., Margalit, J., 1978. A bacterial spore demonstrating rapid larvicidal activity against *Anopheles segentii*, *Uranotaenia unguiculata*, *Culex univitatus*, *Aedes aegypti* and *Culex pipiens*. Mosq. News 37, 355–358.

Gonzalez, C.I., Bhattacharya, A., Wang, W., Peltz, S.W., 2001. Nonsense-mediated mRNA decay in *Saccharomyces cerevisiae*. Gene 274, 15–25.

Gould, F., Anderson, A., Jones, A., Sumerford, D., Heckel, D.G., Lopez, J., Micinski, S., Leonard, R., Laster, M., 1997. Initial frequency of alleles for resistance to *Bacillus thuringiensis* toxins in field populations of *Heliothis virescens*. Proc. Natl. Acad. Sci. U.S.A 94, 3519–3523.

Groves, R.L., Meisch, M.V., 1996. Laboratory and field plot bioassay of *Bacillus sphaericus* against Arkansas mosquito species. J. Am. Mosq. Control Assoc. 12, 220–224.

Guerchicoff, A., Ugalde, R.A., Rubinstein, C.P., 1997. Identification and characterization of a previously undescribed cyt gene in *Bacillus thuringiensis* subsp. *israelensis*. Appl. Environ. Microbiol. 63, 2716–2721.

Guidi, V., Patocchi, N., Luthy, P., Tonolla, M., 2011. Distribution of *Bacillus thuringiensis* subsp. *israelensis* in soil of a swiss wetland reserve after 22 years of mosquito control. Appl. Environ. Microbiol. 77, 3663–3668.

Guidi, V., Lehner, A., Luthy, P., Tonolla, M., 2013a. Dynamics of *Bacillus thuringiensis* var. *israelensis* and *Lysinibacillus sphaericus* spores in urban catch basins after simultaneous application against mosquito larvae. PLoS One 8, e55658.

Guidi, V., Luthy, P., Tonolla, M., 2013b. Comparison between diflubenzuron and a *Bacillus thuringiensis israelensis*- and *Lysinibacillus sphaericus*-based formulation for the control of mosquito larvae in urban catch basins in Switzerland. J. Am. Mosq. Control Assoc. 29, 138–145.

Guillet, P., Kurtak, D.C., Phillipon, B., Meyer, R., 1990. Use of *Bacillus thuringiensis* for onchorcercosis control in West Africa. In: de Barjac, H., Sutherland, D. (Eds.), Bacterial Control of Mosquitoes and Black-Flies, first ed. Rutgers University Press, New Brunswick, pp. 187–201.

Guo, Q.Y., Cai, Q.X., Yan, J.P., Hu, X.M., Zheng, D.S., Yuan, Z.M., 2013. Single nucleotide deletion of cqm1 gene results in the development of resistance to *Bacillus sphaericus* in *Culex quinquefasciatus*. J. Insect Physiol. 59, 967–973.

Harbison, J.E., Henry, M., Xamplas, C., Berry, R., 2013. Experimental use of Natular XRT tablets in a North Shore suburb of Chicago, IL. J. Am. Mosq. Control Assoc. 29, 237–242.

Hazes, B., Read, R.J., 1995. A mosquitocidal toxin with a ricin-like cell-binding domain. Nat. Struct. Biol. 2, 358–359.

Hemingway, J., Hawkes, N.J., Mccarroll, L., Ranson, H., 2004. The molecular basis of insecticide resistance in mosquitoes. Insect Biochem. Mol. Biol. 34, 653–665.

Hertlein, B.C., Levy, R., Miller, T.W.J., 1979. Recycling and selective retrieval of *Bacillus sphaericus* from soil in a mosquito habitat. J. Invertebr. Pathol. 15, 15–20.

Hertlein, M.B., Mavrotas, C., Jousseaume, C., Lysandrou, M., Thompson, G.D., Jany, W., Ritchie, S.A., 2010. A review of spinosad as a natural product for larval mosquito control. J. Am. Mosq. Control Assoc. 26, 67–87.

Hindley, J., Berry, C., 1987. Identification, cloning and sequence analysis of the *Bacillus sphaericus* 1593 41.9 kD larvicidal toxin gene. Mol. Microbiol. 1, 187–194.

Hire, R.S., Hadapad, A.B., Dongre, T.K., Kumar, V., 2009. Purification and characterization of mosquitocidal *Bacillus sphaericus* BinA protein. J. Invertebr. Pathol. 101, 106–111.

Hire, R.S., Hadapad, A.B., Vijayalakshmi, N., Dongre, T.K., 2010. Characterization of highly toxic indigenous strains of mosquitocidal organism *Bacillus sphaericus*. FEMS Microbiol. Lett. 305, 155–161.

Hougard, J.M., 1990. Formulations and persistence of *Bacillus sphaericus* in *Culex quinquefasciatus* larval sites in tropical Africa. In: De Barjac, H., Sutherland, D. (Eds.), Bacterial Control of Mosquitoes and Black-Flies: Biochemistry, Genetics and Applications of *Bacillus thuringiensis Israelensis* and *Bacillus sphaericus*. Rutgers University Press, New Brunswick, NJ, pp. 295–306.

Hougard, J.M., Back, C., 1992. Perspectives on the bacterial control of vectors in the tropics. Parasitol. Today 8, 364–366.

Hougard, J.M., Mbentengam, R., Lochouarn, L., Escaffre, H., Darriet, F., Barbazan, P., Quillevere, D., 1993. Campaign against *Culex quinquefasciatus* using *Bacillus sphaericus*: results of a pilot project in a large urban area of equatorial Africa. Bull. WHO 71, 367–375.

Hu, X., Li, J., Hansen, B.M., Yuan, Z., 2008. Phylogenetic analysis and heterologous expression of surface layer protein SlpC of *Bacillus sphaericus* C3-41. Biosci. Biotechnol. Biochem. 72, 1257–1263.

Hua, G., Zhang, R., Abdullah, M.A., Adang, M.J., 2008. *Anopheles gambiae* cadherin AgCad1 binds the Cry4Ba toxin of *Bacillus thuringiensis israelensis* and a fragment of AgCad1 synergizes toxicity. Biochemistry 47, 5101–5110.

Hua, G., Zhang, R., Bayyareddy, K., Adang, M.J., 2009. *Anopheles gambiae* alkaline phosphatase is a functional receptor of *Bacillus thuringiensis jegathesan* Cry11Ba toxin. Biochemistry 48, 9785–9793.

Humphreys, M.J., Berry, C., 1998. Variants of the *Bacillus sphaericus* binary toxins: implications for differential toxicity of strains. J. Invertebr. Pathol. 71, 184–185.

Hurst, T.P., Brown, M.D., Kay, B.H., Ryan, P.A., 2006. Evaluation of *Melanotaenia duboulayi* (Atheriniformes: Melanotaeniidae), *Hypseleotris galli* (Perciformes: Eleotridae), and larvicide VectoLex WG (*Bacillus sphaericus*) for integrated control of *Culex annulirostris*. J. Am. Mosq. Control Assoc. 22, 418–425.

Janecek, S., 1997. Alpha-amylase family: molecular biology and evolution. Prog. Biophys. Mol. Biol. 67, 67–97.

Jiang, Y., Mulla, M.S., 2009. Laboratory and field evaluation of spinosad, a biorational natural product, against larvae of *Culex* mosquitoes. J. Am. Mosq. Control Assoc. 25, 456–466.

Jones, J.W., Weathersbee 3rd., A.A., Efird, P., Meisch, M.V., 1990. Evaluation of *Bacillus sphaericus* 2362 against *Culex quinquefasciatus* in septic ditches. J. Am. Mosq. Control Assoc. 6, 496–499.

Jones, G.W., Nielsen-Leroux, C., Yang, Y., Yuan, Z., Dumas, V.F., Monnerat, R.G., Berry, C., 2007. A new Cry toxin with a unique two-component dependency from *Bacillus sphaericus*. FASEB J. 21, 4112–4120.

Jones, G.W., Wirth, M.C., Monnerat, R.G., Berry, C., 2008. The Cry48Aa-Cry49Aa binary toxin from *Bacillus sphaericus* exhibits highly restricted target specificity. Environ. Microbiol. 10, 2418–2424.

Jurat-Fuentes, J.L., Adang, M.J., 2004. Characterization of a Cry1Ac-receptor alkaline phosphatase in susceptible and resistant *Heliothis virescens* larvae. Eur. J. Biochem. 271, 3127–3135.

Kahindi, S.C., Midega, J.T., Mwangangi, J.M., Kibe, L.W., Nzovu, J., Luethy, P., Githure, J., Mbogo, C.M., 2008. Efficacy of vectobac DT and culinexcombi against mosquito larvae in unused swimming pools in Malindi, Kenya. J. Am. Mosq. Control Assoc. 24, 538–542.

Kale, A., Hire, R.S., Hadapad, A.B., D'souza, S.F., Kumar, V., 2013. Interaction between mosquito-larvicidal *Lysinibacillus sphaericus* binary toxin components: analysis of complex formation. Insect Biochem. Mol. Biol. 43, 1045–1054.

Kalfon, A., Charles, J.F., Bourgouin, C., de Barjac, H., 1984. Sporulation of *Bacillus sphaericus* 2297: an electron microscope study of crystal-like inclusion biogenesis and toxicity to mosquito larvae. J. Gen. Microbiol. 130, 893–900.

Karch, S., Charles, J.F., 1987. Toxicity, viability and ultrastructure of *Bacillus sphaericus* 2362 spore/crystal complex used in the field. Ann. Inst. Pasteur Microbiol. 138, 485–492.

Karch, S., Coz, J., 1986. Recycling of *Bacillus sphaericus* in dead larvae of *Culex pipiens* (Diptera: Culicidae). Cah. ORSTOM Entomol. Med. Parasitol. 24, 41–43.

Karch, S., Monteny, N., Jullien, J.L., Sinegre, G., Coz, J., 1990. Control of *Culex pipiens* by *Bacillus sphaericus* and role of nontarget arthropods in its recycling. J. Am. Mosq. Control Assoc. 6, 47–54.

Karch, S., Manzambi, Z.A., Salaun, J.J., 1991. Field trials with Vectolex (*Bacillus sphaericus*) and Vectobac (*Bacillus thuringiensis* (H-14)) against *Anopheles gambiae* and *Culex quinquefasciatus* breeding in Zaire. J. Am. Mosq. Control Assoc. 7, 176–179.

Karch, S., Asidi, N., Manzambi, Z.M., Salaun, J.J., 1992. Efficacy of *Bacillus sphaericus* against the malaria vector *Anopheles gambiae* and other mosquitoes in swamps and rice fields in Zaire. J. Am. Mosq. Control Assoc. 8, 376–380.

Keiser, J., Maltese, M.F., Erlanger, T.E., Bos, R., Tanner, M., Singer, B.H., Utzinger, J., 2005. Effect of irrigated rice agriculture on Japanese encephalitis, including challenges and opportunities for integrated vector management. Acta Trop. 95, 40–57.

Kellen, W.R., Clark, T.B., Lindegren, J.E., Ho, B.C., Rogoff, M.H., Singer, S., 1965. *Bacillus sphaericus* Neide as a pathogen of mosquitoes. J. Invertebr. Pathol. 7, 442–448.

Key, P.B., Scott, G.I., 1992. Acute toxicity of the mosquito larvicide, *Bacillus sphaericus*, to the grass shrimp, *Palaemonetes pugio*, and mummichog, *Fundulus heteroclitus*. Bull. Environ. Contam. Toxicol. 49, 425–430.

Killeen, G.F., Smith, T.A., Ferguson, H.M., Mshinda, H., Abdulla, S., Lengeler, C., Kachur, S.P., 2007. Preventing childhood malaria in Africa by protecting adults from mosquitoes with insecticide-treated nets. PLoS Med. 4, e229.

Knight, P.J., Crickmore, N., Ellar, D.J., 1994. The receptor for *Bacillus thuringiensis* CrylA(c) delta-endotoxin in the brush border membrane of the lepidopteran *Manduca sexta* is aminopeptidase N. Mol. Microbiol. 11, 429–436.

Kramer, V.L., 1990. Efficacy and persistence of *Bacillus sphaericus, Bacillus thuringiensis var. israelensis,* and *methoprene* against *Culiseta incidens* (Diptera: Culicidae) in tires. J. Econ. Entomol. 83, 1280–1285.

Kramer, L.D., Styer, L.M., Ebel, G.D., 2008. A global perspective on the epidemiology of West Nile virus. Annu. Rev. Entomol. 53, 61–81.

Krasikov, V.V., Karelov, D.V., Firsov, L.M., 2001. Alpha-glucosidases. Biochemistry (Mosc.) 66, 267–281.

Krych, V.K., Johnson, J.L., Yousten, A.A., 1980. Deoxyribonucleic acid homologies among strans of *Bacillus sphaericus*. Int. J. Syst. Bacteriol. 30, 253–256.

Kumar, A., Sharma, V.P., Sumodan, P.K., Thavaselvam, D., Kamat, R.H., 1994. Malaria control utilizing *Bacillus sphaericus* against *Anopheles stephensi* in Panaji, Goa. J. Am. Mosq. Control Assoc. 10, 534–539.

Kumar, A., Sharma, V.P., Thavaselvam, D., Sumodan, P.K., Kamat, R.H., Audi, S.S., Surve, B.N., 1996. Control of *Culex quinquefasciatus* with *Bacillus sphaericus* in Vasco City, Goa. J. Am. Mosq. Control Assoc. 12, 409–413.

Kumar, A.N., Murugan, K., Madhiyazhagan, P., Prabhu, K., 2011. Spinosad and neem seed kernel extract as bio-controlling agents for malarial vector, *Anopheles stephensi* and non-biting midge, *Chironomus circumdatus*. Asian Pac. J. Trop. Med. 4, 614–618.

Kunthic, T., Promdonkoy, B., Srikhirin, T., Boonserm, P., 2011. Essential role of tryptophan residues in toxicity of binary toxin from *Bacillus sphaericus*. BMB Rep. 44, 674–679.

Kuppusamy, M., Hoti, S.L., Balaraman, K., 1987. Residual activity of briquette & alginate formulations of *Bacillus sphaericus* against mosquito larvae. Indian J. Med. Res. 86, 591–596.

Lacey, L., 2007. *Bacillus thuringiensis* serovariety *israelensis* and *Bacillus sphaericus* for mosquito control. J. Am. Mosq. Control Assoc. 23, 133–163.

Lacey, L.A., Lacey, C.M., 1990. The medical importance of riceland mosquitoes and their control using alternatives to chemical insecticides. J. Am. Mosq. Control Assoc. Suppl. 2, 1–93.

Lacey, L., Mulla, M.S., 1990. Safety of *Bacillus thuringiensis* (H-14) and *Bacillus sphaericus* to non-target organisms in the aquatic environment. In: Laird, M., Lacey, L.A., Davidson, E.W. (Eds.), Safety of Microbial Insecticides. CRC Press, Boca Raton, FL, pp. 168–188.

Lacey, L.A., Siegel, J.P., 2000. Safety and ecotoxicology of entomopathogenic bacteria. In: Charles, J.F., Nielsen-Leroux, C., Delecluse, A. (Eds.), Entomopathogenic Bacteria: From Laboratory to Field Application. Kluwer Academic Publishers, Dordrecht, The Netherlands, pp. 253–273.

Lacey, L.A., Urbina, M., Heitzman, C., 1984. Sustained release formulations of *Bacillus sphaericus* and *Bacillus thuringiensis* (H-14) for the control of container breeding *Culex quinquefasciatus*. Mosq. News 44, 26–32.

Lacey, L.A., Heitzman, C.M., Meisch, M., Billodeaux, J., 1986. Beecomist-applied *Bacillus sphaericus* for the control of riceland mosquitoes. J. Am. Mosq. Control Assoc. 2, 548–551.

Lacey, L.A., Lacey, C.M., Peacock, B., Thiery, I., 1988a. Mosquito host range and field activity of *Bacillus sphaericus* isolate 2297 (serotype 25). J. Am. Mosq. Control Assoc. 4, 51–56.

Lacey, L.A., Ross, D.H., Lacey, C.M., 1988b. Experimental formulations of *Bacillus sphaericus* for the control of anopheline and culicine larvae. J. Indus. Microbiol. 3, 39–47.
Lecadet, M.M., 1996. La lutte bactériologique contre les insectes: une vielle histoire très actuelle. Ann. Inst. Pasteur/Actualit. 7, 207–216.
Li, T., Sun, F., Yuan, Z., Zhang, Y., Yu, J., Pang, Y., 2000. Coexpression of cyt1Aa of *Bacillus thuringiensis* subsp. *israelensis* with *Bacillus sphaericus* binary toxin gene in acrystalliferous strain of *B. thuringiensis*. Curr. Microbiol. 40, 322–326.
Li, J., Yang, L., Hu, X., Zheng, D., Yan, J., Yuan, Z., 2013. Nanoscale mono- and multilayer cylinder structures formed by recombinant S-layer proteins of mosquitocidal *Bacillus sphaericus* C3-41. Appl. Microbiol. Biotechnol. 97, 7275–7283.
Likitvivatanavong, S., Chen, J., Evans, A.M., Bravo, A., Soberón, M., Gill, S.S., 2011. Multiple receptors as targets of Cry toxins in mosquitoes. J. Agric. Food Chem. 59, 2829–2838.
Limpanawat, S., Promdonkoy, B., Boonserm, P., 2009. The C-terminal domain of BinA is responsible for *Bacillus sphaericus* binary toxin BinA-BinB interaction. Curr. Microbiol. 59, 509–513.
Lingenfelser, A., Rydzanicz, K., Kaiser, A., Becker, N., 2010. Mosquito fauna and perspectives for integrated control of urban vector-mosquito populations in Southern Benin (West Africa). Ann. Agric. Environ. Med. 17, 49–57.
Liu, J.W., Hindley, J., Porter, A.G., Priest, F.G., 1993. New high-toxicity mosquitocidal strains of *Bacillus sphaericus* lacking a 100-kilodalton-toxin gene. Appl. Environ. Microbiol. 59, 3470–3473.
Liu, J.W., Porter, A.G., Wee, B.Y., Thanabalu, T., 1996. New gene from nine *Bacillus sphaericus* strains encoding highly conserved 35.8-kilodalton mosquitocidal toxins. App. Environ. Microbiol. 62, 2174–2176.
Liu, H., Cupp, E.W., Micher, K.M., Guo, A., Liu, N., 2004. Insecticide resistance and cross-resistance in Alabama and Florida strains of *Culex quinquefasciatus*. J. Med. Entomol. 41, 408–413.
Lord, J.C., 1991. Sustained release pellets for control of *Culex larvae* with *Bacillus sphaericus*. J. Am. Mosq. Control Assoc. 7, 560–564.
Lozano, L.C., Ayala, J.A., Dussan, J., 2011. *Lysinibacillus sphaericus* S-layer protein toxicity against *Culex quinquefasciatus*. Biotechnol. Lett. 33, 2037–2041.
Luo, K., Sangadala, S., Masson, L., Mazza, A., Brousseau, R., Adang, M.J., 1997. The *Heliothis virescens* 170 kDa aminopeptidase functions as "receptor A" by mediating specific *Bacillus thuringiensis* Cry1A delta-endotoxin binding and pore formation. Insect Biochem. Mol. Biol. 27, 735–743.
Luo, W., Liu, C., Zhang, R., He, J., Han, B., 2014. Anticancer activity of binary toxins from *Lysinibacillus sphaericus* against human lung cancer cell line A549. Nat. Prod. Commun. 9, 107–110.
Maciel, A., Rocha, A., Marzochi, K.B., Medeiros, Z., Carvalho, A.B., Regis, L., Souza, W., Lapa, T., Furtado, A., 1996. Epidemiological study of bancroftian filariasis in Recife, northeastern Brazil. Mem. Inst. Oswaldo Cruz 91, 449–455.
Maheu-Giroux, M., Castro, M.C., 2013. Impact of community-based larviciding on the prevalence of malaria infection in Dar es Salaam, Tanzania. PLoS One 8, e71638.
Majambere, S., Lindsay, S.W., Green, C., Kandeh, B., Fillinger, U., 2007. Microbial larvicides for malaria control in the Gambia. Malar. J. 6, 76.
Marcombe, S., Darriet, F., Agnew, P., Etienne, M., Yp-Tcha, M.M., Yebakima, A., Corbel, V., 2011. Field efficacy of new larvicide products for control of multi-resistant *Aedes aegypti* populations in Martinique (French West Indies). Am. J. Trop. Med. Hyg. 84, 118–126.
Margalit, J., Dean, D., 1985. The story of *Bacillus thuringiensis* var. *israelensis* (B.t.i.). J. Am. Mosq. Control Assoc. 1, 1–7.

Marina, C.F., Bond, J.G., Munoz, J., Valle, J., Chirino, N., Williams, T., 2012. Spinosad: a biorational mosquito larvicide for use in car tires in southern Mexico. Parasit. Vectors 5, 95.
Marinotti, O., James, A.A., 1990. An α-glucosidase in the salivary glands of the vector mosquito, *Aedes aegypti*. Insect Biochem. Mol. Biol. 20, 619–623.
Marinotti, O., De Brito, M., Moreira, C.K., 1996. Apyrase and α-glucosidase in the salivary glands of *Aedes albopictus*. Comp. Biochem. Physiol. B Biochem. Mol. Biol. 113, 675–679.
Matanmi, B.A., Federici, B.A., Mulla, M.S., 1990. Fate and persistence of *Bacillus sphaericus* used as a mosquito larvicide in dairy wastewater lagoons. J. Am. Mosq. Control Assoc. 6, 384–389.
Maxwell, C.A., Mohammed, K., Kisumku, U., Curtis, C.F., 1999. Can vector control play a useful supplementary role against bancroftian filariasis? Bull. WHO 77, 138–143.
Melo-Santos, M.A., Araújo, A.P., Rios, E.M., Regis, L., 2009. Long lasting persistence of *Bacillus thuringiensis* serovar. *israelensis* larvicidal activity in *Aedes aegypti* (Diptera: Culicidae) breeding places is associated to bacteria recycling. Biol. Control 49, 186–191.
Menon, K.K.R., Rao, A.S., Amonkar, S.V., 1982. Toxic activity and histopathological effects of *Bacillus sphaericus* (ISPC-5) on mosquito larvae. Indian J. Exp. Biol. 20, 768–772.
Merritt, R.W., Lessard, J.L., Wessell, K.J., Hernandez, O., Berg, M.B., Wallace, J.R., Novak, J.A., Ryan, J., Merritt, B.W., 2005. Lack of effects of *Bacillus sphaericus* (Vectolex) on nontarget organisms in a mosquito-control program in southeastern Wisconsin: a 3-year study. J. Am. Mosq. Control Assoc. 21, 201–212.
Montero Lago, G., Diaz Perez, M., Marrero Figueroa, A., Castillo Gonzalez, F.A., 1991. The pilot project results of applications of the biolarvicide *Bacillus sphaericus* 2362 on mosquito breeding grounds of the town of Santa Cruz del Norte (La Habana Province). Rev. Cubana Med. Trop. 43, 39–44.
Moreno, J.E., Acevedo, P., Martínez, A., Sanchez, V., Petterson, L., 2010. Evaluación de la persistencia de una formulación comercial de *Bacillus sphaericus* en criaderos naturales de anofelinos vectores de malaria en estado Bolívar, Venezuela. Bol. Mal. Salud Amb. 50, 109–117.
Moser, J.B., Ramirez, X., González, J.E., Herrera, M., 2002. Evaluación de la efectividad de *Bacillus sphaericus* contra larvas de *Anopheles aquasalis* Curry (Diptera: Culicidae) en criaderos naturales del estado Sucre, Venezuela. Entomotropica 17, 1–5.
Moser, J.B., Herrera, M., Rivas, J.G., Puentes, N., Caraballo, R., Valero, J., 2012. Field trials on the efficacy and persistence of three formulations of *Bacillus sphaericus* against larvae of *Anopheles aquasalis* Curry in mangroves of Mariño municipality, Sucre state, Venezuela. Bol. Mal. Salud Amb. 52, 67–77.
Mulla, M.S., Darwazeh, H., Davidson, E.W., Dulmage, H.T., 1984a. Efficacy and persistence of the microbial agent *Bacillus sphaericus* against mosquito larvae in organically enriched habitats. Mosq. News 44, 166–173.
Mulla, M.S., Darwazeh, H., Davidson, E.W., Dulmage, H.T., 1984b. Larvicidal activity and field efficacy of *Bacillus sphaericus* strains against mosquito larvae and their safety for non target organisms. Mosq. News 44, 336–342.
Mulla, M.S., Darwazeh, H.A., Ede, L., Kennedy, B., Dulmage, H.T., 1985. Efficacy and field evaluation of *Bacillus thuringiensis* (H-14) and *B. sphaericus* against floodwater mosquitoes in California. J. Am. Mosq. Control Assoc. 1, 310–315.
Mulla, M.S., Axelrod, H., Darwazeh, H.A., Matanmi, B.A., 1988a. Efficacy and longevity of *Bacillus sphaericus* 2362 formulations for control of mosquito larvae in dairy wastewater lagoons. J. Am. Mosq. Control Assoc. 4, 448–452.
Mulla, M.S., Darwazeh, H.A., Tietze, N.S., 1988b. Efficacy of *Bacillus sphaericus* 2362 formulations against floodwater mosquitoes. J. Am. Mosq. Control Assoc. 4, 172–174.
Mulla, M.S., Rodcharoen, J., Ngamsuk, W., Tawatsin, A., Pan-Urai, P., Thavara, U., 1997. Field trials with *Bacillus sphaericus* formulations against polluted water mosquitoes in a suburban area of Bangkok, Thailand. J. Am. Mosq. Control Assoc. 13, 297–304.

Mulla, M.S., Su, T., Thavara, U., Tawatsin, A., Ngamsuk, W., Pan-Urai, P., 1999. Efficacy of new formulations of the microbial larvicide *Bacillus sphaericus* against polluted water mosquitoes in Thailand. J. Vector Ecol. 24, 99–110.

Mulla, M.S., Thavara, U., Tawatsin, A., Kong-Ngamsuk, W., Chompoosri, J., Su, T., 2001. Mosquito larval control with *Bacillus sphaericus*: reduction in adult populations in low-income communities in Nonthaburi Province, Thailand. J. Vector Ecol. 26, 221–231.

Mulla, M.S., Thavara, U., Tawatsin, A., Chomposri, J., Su, T., 2003. Emergence of resistance and resistance management in field populations of tropical *Culex quinquefasciatus* to the microbial control agent *Bacillus sphaericus*. J. Am. Mosq. Control Assoc. 19, 39–46.

Mulligan, S.F., Schaefer, C.H., Miura, T., 1978. Laboratory and field evaluation of *Bacillus sphaericus* as a mosquito (Diptera-Culicidae) control agent. J. Econ. Entomol. 71, 774–777.

Mulligan, S.F., Schaefer, C.H., Wilder, W.H., 1980. Efficacy and persistence of *Bacillus sphaericus* and *Bacillus thuringiensis* H.14 against mosquitos under laboratory and field conditions. J. Econ. Entomol. 73, 684–688.

Munhenga, G., Masendu, H.T., Brooke, B.D., Hunt, R.H., Koekemoer, L.K., 2008. Pyrethroid resistance in the major malaria vector *Anopheles arabiensis* from Gwave, a malaria-endemic area in Zimbabwe. Malar. J. 7, 247.

Mwangangi, J.M., Kahindi, S.C., Kibe, L.W., Nzovu, J.G., Luethy, P., Githure, J.I., Mbogo, C.M., 2011. Wide-scale application of Bti/Bs biolarvicide in different aquatic habitat types in urban and peri-urban Malindi, Kenya. Parasitol. Res. 108, 1355–1363.

Myers, P., Yousten, A.A., Davidson, E.W., 1979. Comparative studies of the mosquito-larval toxin of *Bacillus sphaericus* SSII-1 and 1593. Can. J. Microbiol. 25, 1227–1231.

Nakamura, L.K., 2000. Phylogeny of *Bacillus sphaericus*-like organisms. Int. J. Syst. Evol. Microbiol. 50 (Pt. 5), 1715–1722.

Narasu, M.L., Gopinathan, K.P., 1986. Purification of larvicidal protein from *Bacillus sphaericus* 1593. Biochem. Biophys. Res. Commun. 141, 756–761.

Narasu, M.L., Gopinathan, K.P., 1988. Effect of *Bacillus sphaericus* 1593 toxin on choline acetyl transferase and mitochondrial oxidative activities of the mosquito larvae. Indian J. Biochem. Biophys. 25, 253–256.

Neide, E., 1904. Botanische beschreibund einiger sporenbildenden bacterien. Zentralbl. Bakteriol. Parasitenkd. Infektionskr. Hyg. Abt. 12 (Pt. 1), 1–11.

Nene, V., Wortman, J.R., Lawson, D., Haas, B., Kodira, C., Tu, Z.J., Loftus, B., Xi, Z., Megy, K., Grabherr, M., Ren, Q., Zdobnov, E.M., Lobo, N.F., Campbell, K.S., Brown, S.E., et al., 2007. Genome sequence of *Aedes aegypti*, a major arbovirus vector. Science 316, 1718–1723.

Nicolas, L., Dossou-Yovo, J., 1987. Differential effects of *Bacillus sphaericus* strain 2362 on *Culex quinquefasciatus* and its competitor *Culex cinereus* in West Africa. Med. Vet. Entomol. 1, 23–27.

Nicolas, L., Darriet, F., Hougard, J.M., 1987. Efficacy of *Bacillus sphaericus 2362* against larvae of *Anopheles gambiae* under laboratory and field conditions in West Africa. Med. Vet. Entomol. 1, 157–162.

Nicolas, L., Lecroisey, A., Charles, J.F., 1990. Role of the gut proteinases from mosquito larvae in the mechanism of action and the specificity of the *Bacillus sphaericus* toxin. Can. J. Microbiol. 36, 804–807.

Nicolas, L., Nielsen-Leroux, C., Charles, J.F., Delécluse, A., 1993. Respective role of the 42- and 51-kDa components of the *Bacillus sphaericus* toxin overexpressed in *Bacillus thuringiensis*. FEMS Microbiol. Lett. 106, 275–280.

Nielsen-Leroux, C., Charles, J.F., 1992. Binding of *Bacillus sphaericus* binary toxin to a specific receptor on midgut brush-border membranes from mosquito larvae. Eur. J. Biochem. 210, 585–590.

Nielsen-Leroux, C., Charles, J.F., Thiery, I., Georghiou, G.P., 1995. Resistance in a laboratory population of *Culex quinquefasciatus* (Diptera: Culicidae) to *Bacillus sphaericus* binary toxin is due to a change in the receptor on midgut brush-border membranes. Eur. J. Biochem. 228, 206–210.

Nielsen-Leroux, C., Pasquier, F., Charles, J.F., Sinegre, G., Gaven, B., Pasteur, N., 1997. Resistance to *Bacillus sphaericus* involves different mechanisms in *Culex pipiens* (Diptera: Culicidae) larvae. J. Med. Entomol. 34, 321–327.

Nielsen-Leroux, C., Rao, D.R., Murphy, J.R., Carron, A., Mani, T.R., Hamon, S., Mulla, M.S., 2001. Various levels of cross-resistance to *Bacillus sphaericus* strains in *Culex pipiens* (Diptera: Culicidae) colonies resistant to *B. sphaericus* strain 2362. Appl. Environ. Microbiol. 67, 5049–5054.

Nielsen-Leroux, C., Pasteur, N., Pretre, J., Charles, J.F., Sheikh, H.B., Chevillon, C., 2002. High resistance to *Bacillus sphaericus* binary toxin in *Culex pipiens* (Diptera: Culicidae): the complex situation of West Mediterranean countries. J. Med. Entomol. 39, 729–735.

Nishiwaki, H., Nakashima, K., Ishida, C., Kawamura, T., Matsuda, K., 2007. Cloning, functional characterization, and mode of action of a novel insecticidal pore-forming toxin, sphaericolysin, produced by *Bacillus sphaericus*. Appl. Environ. Microbiol. 73, 3404–3411.

Oei, C., Hindley, J., Berry, C., 1990. An analysis of the genes encoding the 51.4- and 41.9-kDa toxins of *Bacillus sphaericus* 2297 by deletion mutagenesis: the construction of fusion proteins. FEMS Microbiol. Lett. 60, 265–273.

Oei, C., Hindley, J., Berry, C., 1992. Binding of purified *Bacillus sphaericus* binary toxin and its deletion derivatives to *Culex quinquefasciatus* gut: elucidation of functional binding domains. J. Gen. Microbiol. 138, 1515–1526.

Oliveira, C.M., Silva-Filha, M.H., Nielsen-Leroux, C., Pei, G., Yuan, Z., Regis, L., 2004. Inheritance and mechanism of resistance to *Bacillus sphaericus* in *Culex quinquefasciatus* (Diptera: Culicidae) from China and Brazil. J. Med. Entomol. 41, 58–64.

Opota, O., Charles, J.F., Warot, S., Pauron, D., Darboux, I., 2008. Identification and characterization of the receptor for the *Bacillus sphaericus* binary toxin in the malaria vector mosquito, *Anopheles gambiae*. Comp. Biochem. Physiol. B Biochem. Mol. Biol. 149, 419–427.

Opota, O., Gauthier, N.C., Doye, A., Berry, C., Gounon, P., Lemichez, E., Pauron, D., 2011. *Bacillus sphaericus* binary toxin elicits host cell autophagy as a response to intoxication. PLoS One 6, e14682.

Paing, M., Myat, M., Thu Sebastian, A.A., 1987. Laboratory evaluation of *Bacillus sphaericus* 1593–4 and preliminary field trials for control of *Culex quinquefasciatus* in septic tanks. J. Commun. Dis. 19, 164–167.

Park, H.W., Bideshi, D.K., Federici, B.A., 2003. Recombinant strain of *Bacillus thuringiensis* producing Cyt1A, Cry11B, and the *Bacillus sphaericus* binary toxin. Appl. Environ. Microbiol. 69, 1331–1334.

Park, H.W., Bideshi, D.K., Wirth, M.C., Johnson, J.J., Walton, W.E., Federici, B.A., 2005. Recombinant larvicidal bacteria with markedly improved efficacy against *Culex* vectors of West Nile virus. Am. J. Trop. Med. Hyg. 72, 732–738.

Park, H.W., Bideshi, D.K., Federici, B.A., 2007. The 20-kDa protein of *Bacillus thuringiensis* subsp. *israelensis* enhances *Bacillus sphaericus* 2362 bin toxin synthesis. Curr. Microbiol. 55, 119–124.

Park, H.W., Tang, M., Sakano, Y., Federici, B.A., 2009. A 1.1-kilobase region downstream of the bin operon in *Bacillus sphaericus* strain 2362 decreases bin yield and crystal size in strain 2297. Appl. Environ. Microbiol. 75, 878–881.

Partridge, M.R., Berry, C., 2002. Insecticidal activity of the *Bacillus sphaericus* Mtx1 toxin against *Chironomus riparus*. J. Invertebr. Pathol. 79, 135–136.

Pauchet, Y., Luton, F., Castella, C., Charles, J.F., Romey, G., Pauron, D., 2005. Effects of a mosquitocidal toxin on a mammalian epithelial cell line expressing its target receptor. Cell. Microbiol. 7, 1335–1344.
Payne, J.M., Davidson, E.W., 1984. Insecticidal activity of the crystalline parasporal inclusions and other components of the *Bacillus sphaericus* 1593 spore complex. J. Invertebr. Pathol. 43, 383–388.
Pei, G., Oliveira, C.M., Yuan, Z., Nielsen-Leroux, C., Silva-Filha, M.H., Yan, J., Regis, L., 2002. A strain of *Bacillus sphaericus* causes slower development of resistance in *Culex quinquefasciatus*. Appl. Environ. Microbiol. 68, 3003–3009.
Pena, G., Miranda-Rios, J., De La Riva, G., Pardo-Lopez, L., Soberon, M., Bravo, A., 2006. A *Bacillus thuringiensis* S-layer protein involved in toxicity against *Epilachna varivestis* (Coleoptera: Coccinellidae). Appl. Environ. Microbiol. 72, 353–360.
Penner, H., Wilamowski, A., 1996. Susceptibility of larvae of the sandfly *Phlebotomus papatasi* (Diptera: Psychodidae) to *Bacillus sphaericus*. Bull. Entomol. Res. 86, 173–175.
Pérez, C., Fernandez, L.E., Sun, J., Folch, J.L., Gill, S.S., Soberón, M., Bravo, A., 2005. *Bacillus thuringiensis* subsp. *israelensis* Cyt1Aa synergizes Cry11Aa toxin by functioning as a membrane-bound receptor. Proc. Natl. Acad. Sci. U.S.A 102, 18303–18308.
Pérez, C., Muñoz-Garay, C., Portugal, L.C., Sánchez, J., Gill, S.S., Soberón, M., Bravo, A., 2007. *Bacillus thuringiensis ssp. israelensis* Cyt1Aa enhances activity of Cry11Aa toxin by facilitating the formation of a pre-pore oligomeric structure. Cell. Microbiol. 9, 2931–2937.
Petersen, L.R., Fischer, M., 2012. Unpredictable and difficult to control—the adolescence of West Nile virus. N. Engl. J. Med. 367, 1281–1284.
Phannachet, K., Raksat, P., Limvuttegrijeerat, T., Promdonkoy, B., 2010. Production and characterization of N- and C-terminally truncated Mtx2: a mosquitocidal toxin from *Bacillus sphaericus*. Curr. Microbiol. 61, 549–553.
Poncet, S., Delecluse, A., Anello, D., Klier, A., Rapoport, G., 1994. Transfer and expression of the CryIVB and CryIVD genes of *Bacillus thuringiensis* subs. *israelensis* in *Bacillus sphaericus* 2297. FEMS Microbiol. Lett. 117, 91–96.
Poncet, S., Bernard, C., Dervyn, E., Cayley, J., Klier, A., Rapoport, G., 1997. Improvement of *Bacillus sphaericus* toxicity against dipteran larvae by integration, via homologous recombination, of the Cry11A toxin gene from *Bacillus thuringiensis* subsp. *israelensis*. Appl. Environ. Microbiol. 63, 4413–4420.
Poophati, S., Mani, C., Vignesh, V., Praba, V.L., Thirugnanasambantham, K., 2013. Genotypic diversity of mosquitocidal bacteria (*Bacillus sphaericus*, *B. thuringiensis*, and *B. cereus*) newly isolated from natural sources. Appl. Biochem. Biotechnol. 171, 2233–2246.
Porta, H., Cancino-Rodezno, A., Soberon, M., Bravo, A., 2011. Role of MAPK p38 in the cellular responses to pore-forming toxins. Peptides 32, 601–606.
Porter, A.G., Davidson, E.W., Liu, J.W., 1993. Mosquitocidal toxins of bacilli and their genetic manipulation for effective biological control of mosquitoes. Microbiol. Rev. 57, 838–861.
Prabhu, D.I., Sankar, S.G., Vasan, P.T., Piriya, P.S., Selvan, B.K., Vennison, S.J., 2013. Molecular characterization of mosquitocidal *Bacillus sphaericus* isolated from Tamil Nadu, India. Acta Trop. 127, 158–164.
Pradeepkumar, N., Sabesan, S., Kuppusamy, M., Balaraman, K., 1988. Effect of controlled release formulation of *Bacillus sphaericus* on *Mansonia* breeding. Indian J. Med. Res. 87, 15–18.
Pridgeon, J.W., Pereira, R.M., Becnel, J.J., Allan, S.A., Clark, G.G., Linthicum, K.J., 2008. Susceptibility of Aedes aegypti, *Culex quinquefasciatus* Say, and A*nopheles quadrimaculatus* Say to 19 pesticides with different modes of action. J. Med. Entomol. 45, 82–87.
Priest, F.G., Ebdrup, L., Zahner, V., Carter, P.E., 1997. Distribution and characterization of mosquitocidal toxin genes in some strains of *Bacillus sphaericus*. Appl. Environ. Microbiol. 63, 1195–1198.

Promdonkoy, B., Promdonkoy, P., Wongtawan, B., Wongtawan, B., Boonserm, P., Panyim, S., 2008. Cys31, Cys47, and Cys195 in BinA are essential for toxicity of a binary toxin from *Bacillus sphaericus*. Curr. Microbiol. 56, 334–338.

Ragoonanansingh, R.N., Njunwa, K.J., Curtis, C.F., Becker, N., 1992. A field study of *Bacillus sphaericus* for the control of culicine and anopheline mosquito larvae in Tanzania. Bull. Soc. Vector Ecol. 17, 45–50.

Ramoska, W.A., Singer, S., Levy, R., 1977. Bioassay of three strains of *Bacillus sphaericus* on field-collected mosquito larvae. J. Invertebr. Pathol. 30, 151–154.

Ranson, H., N'guessan, R., Lines, J., Moiroux, N., Nkuni, Z., Corbel, V., 2011. Pyrethroid resistance in African anopheline mosquitoes: what are the implications for malaria control? Trends Parasitol. 27, 91–98.

Rao, D.R., Mani, T.R., Rajendran, R., Joseph, A.S., Gajanana, A., Reuben, R., 1995. Development of a high level of resistance to *Bacillus sphaericus* in a field population of *Culex quinquefasciatus* from Kochi, India. J. Am. Mosq. Control Assoc. 11, 1–5.

Regis, L.N., Nielsen-Leroux, C., 2000. Management of resistance to bacterial vector control. In: Charles, J.F., Nielsen-Leroux, C., Delecluse, A. (Eds.), Entomopathogenic Bacteria: From Laboratory to Field Application. Kluwer Academic Publisher, Dordrecht, pp. 419–439.

Regis, L., Silva-Filha, M.H., De Oliveira, C.M., Rios, E.M., Da Silva, S.B., Furtado, A.F., 1995. Integrated control measures against *Culex quinquefasciatus*, the vector of filariasis in Recife. Mem. Inst. Oswaldo Cruz 90, 115–119.

Regis, L., Furtado, A.F., Oliveira, C.M., Bezerra, C.B., Silva, L.R., Araujo, J., Maciel, A., Silva-Filha, M.H., Silva, S.B., 1996. Integrated control of the filariasis vector with community participation in an urban area of Recife, Pernambuco, Brazil. Cad. Saude Publica 12, 473–482.

Regis, L., Da Silva, S.B., Melo-Santos, M.A., 2000a. The use of bacterial larvicides in mosquito and black fly control programmes in Brazil. Mem. Inst. Oswaldo Cruz 95 (Suppl. 1), 207–210.

Regis, L., Oliveira, C.M., Silva-Filha, M.H., Silva, S.B., Maciel, A., Furtado, A.F., 2000b. Efficacy of *Bacillus sphaericus* in control of the filariasis vector *Culex quinquefasciatus* in an urban area of Olinda, Brazil. Trans. R. Soc. Trop. Med. Hyg. 94, 488–492.

Regis, L., Silva-Filha, M.H., Nielsen-Leroux, C., Charles, J.F., 2001. Bacteriological larvicides of dipteran disease vectors. Trends Parasitol. 17, 377–380.

Reinert, D.J., Carpusca, I., Aktories, K., Schulz, G.E., 2006. Structure of the mosquitocidal toxin from *Bacillus sphaericus*. J. Mol. Biol. 357, 1226–1236.

Rinkevich, F.D., Du, Y., Dong, K., 2013. Diversity and convergence of sodium channel mutations involved in resistance to pyrethroids. Pestic. Biochem. Physiol. 106, 93–100.

Rivera, P.L.E., Lopez, M., Valle, S., Delgado, M., Lopez, D., Larios, F., 1997. Evaluatión de la efetividad biolarvicida y residualidad de *Bacillus sphaericus* (cepa 2362) para el control de *Anopheles albimanus* en la costa del Lago Xolotlan, Managua, Nicaragua. Rev. Nica. Entomol. 42, 7–14.

Robert, L.L., Perich, M.J., Schlein, Y., Jacobson, R.L., Wirtz, R.A., Lawyer, P.G., Githure, J.I., 1997. Phlebotomine sand fly control using bait-fed adults to carry the larvicide *Bacillus sphaericus* to the larval habitat. J. Am. Mosq. Control Assoc. 13, 140–144.

Robert, L.L., Perich, M.J., Schlein, Y., Jacobson, J.L., 1998. *Bacillus sphaericus* inhibits hatching of phlebotomine sand fly eggs. J. Am. Mosq. Control Assoc. 14, 351–352.

Robertson, J.L., Preisler, H.K., Ng, S.S., Hinkle, L.A., Gelernter, W.D., 1995. Natural variations: a complicating factor in bioassays with chemical and microbial pesticides. J. Econ. Entomol. 88, 1–10.

Rodcharoen, J., Mulla, M.S., 1994. Resistance development in *Culex quinquefasciatus* to the microbial agent *Bacillus sphaericus*. J. Econ. Entomol. 87, 1133–1140.

Rodcharoen, J., Mulla, M.S., 1996. Cross-resistance to *Bacillus sphaericus* strains in *Culex quinquefasciatus*. J. Am. Mosq. Control Assoc. 12, 247–250.

Rodcharoen, J., Mulla, M.S., 1997. Biological fitness of *Culex quinquefasciatus* (Diptera: Culicidae) susceptible and resistant to *Bacillus sphaericus*. J. Med. Entomol. 34, 5–10.

Rodcharoen, J., Mulla, M.S., Chaney, J.D., 1991. Microbial larvicides for the control of nuisance aquatic midges (Diptera: Chironomidae) inhabiting mesocosms and man-made lakes in California. J. Am. Mosq. Control Assoc. 7, 56–62.

Rodrigues, I.B., Tadei, W.P., Dias, J.M., 1998. Studies on the *Bacillus sphaericus* larvicidal activity against malarial vector species in Amazonia. Mem. Inst. Oswaldo Cruz 93, 441–444.

Rodrigues, I.B., Tadei, W.P., Dias, J.M., 1999. Larvicidal activity of *Bacillus sphaericus* 2362 against *Anopheles nuneztovari*, *Anopheles darlingi* and *Anopheles braziliensis* (Diptera, Culicidae). Rev. Inst. Med. Trop. Sao Paulo 41, 101–105.

Rodrigues, I.B., Tadei, W.P., Santos, R.L.C., Santos, S., Baggio, J.B., 2008. Controle da malária: eficácia de formulados de *Bacillus sphaericus* 2362 contra espécies de *Anopheles* em criadouros artificiais-tanque de piscicultura e criadouros de olaria. Rev. Pat. Trop. 37, 161–176.

Rojas, J.E., Mazzarri, M., Sojo, M., Garcia, A.G., 2001. Effectiveness of *Bacillus sphaericus* strain 2362 on larvae of *Anopheles nuneztovari*. Invest. Clin. 42, 131–146.

Romão, T.P., De Melo Chalegre, K.D., Key, S., Ayres, C.F., Fontes De Oliveira, C.M., De-Melo-Neto, O.P., Silva-Filha, M.H., 2006. A second independent resistance mechanism to *Bacillus sphaericus* binary toxin targets its alpha-glucosidase receptor in *Culex quinquefasciatus*. FEBS J. 273, 1556–1568.

Romão, T.P., De-Melo-Neto, O.P., Silva-Filha, M.H., 2011. The N-terminal third of the BinB subunit from the *Bacillus sphaericus* binary toxin is sufficient for its interaction with midgut receptors in *Culex quinquefasciatus*. FEMS Microbiol. Lett. 321, 167–174.

Romi, R., Toma, L., Severini, F. Di, Luca, M., 2003. Susceptibility of Italian populations of *Aedes albopictus* to temephos and to other insecticides. J. Am. Mosq. Control Assoc. 19, 419–423.

Rungrod, A., Tjahaja, N.K., Soonsanga, S., Audtho, M., Promdonkoy, B., 2009. *Bacillus sphaericus* Mtx1 and Mtx2 toxins co-expressed in *Escherichia coli* are synergistic against *Aedes aegypti* larvae. Biotechnol. Lett. 31, 551–555.

Rydzanicz, K., Lonc, E., Becker, N., 2009. Current procedures of the integrated urban vector-mosquito control as an example in Cotonou (Benin, West Africa) and Wroclaw area (Poland). Wiad. Parazytol. 55, 335–340.

Saavedra-Rodriguez, K., Urdaneta-Marquez, L., Rajatileka, S., Moulton, M., Flores, A.E., Fernandez-Salas, I., Bisset, J., Rodriguez, M., Mccall, P.J., Donnelly, M.J., Ranson, H., Hemingway, J., Black, W.C.T., 2007. A mutation in the voltage-gated sodium channel gene associated with pyrethroid resistance in Latin American *Aedes aegypti*. Insect Mol. Biol. 16, 785–798.

Sanitt, P., Promdonkoy, B., Boonserm, P., 2008. Targeted mutagenesis at charged residues in *Bacillus sphaericus* BinA toxin affects mosquito-larvicidal activity. Curr. Microbiol. 57, 230–234.

Sarkar, M., Borkotoki, A., Baruah, I., Bhattacharyya, I.K., Srivastava, R.B., 2009. Molecular analysis of knock down resistance (kdr) mutation and distribution of kdr genotypes in a wild population of *Culex quinquefasciatus* from India. Trop. Med. Int. Health 14, 1097–1104.

Schirmer, J., Just, I., Aktories, K., 2002a. The ADP-ribosylating mosquitocidal toxin from *Bacillus sphaericus*: proteolytic activation, enzyme activity, and cytotoxic effects. J. Biol. Chem. 277, 11941–11948.

Schirmer, J., Wieden, H.J., Rodnina, M.V., Aktories, K., 2002b. Inactivation of the elongation factor Tu by mosquitocidal toxin-catalyzed mono-ADP-ribosylation. Appl. Environ. Microbiol. 68, 4894–4899.

Schwartz, J.L., Potvin, L., Coux, F., Charles, J.F., Berry, C., Humphreys, M.J., Jones, A.F., Bernhart, I., Dalla Serra, M., Menestrina, G., 2001. Permeabilization of model lipid membranes by *Bacillus sphaericus* mosquitocidal binary toxin and its individual components. J. Membr. Biol. 184, 171–183.
Schwede, T., Kopp, J., Guex, N., Peitsch, M.C., 2003. SWISS-MODEL: an automated protein homology-modeling server. Nucleic Acids Res. 31, 3381–3385.
Sebo, P., Bennardo, T., De La Torre, F., Szulmajster, J., 1990. Delineation of the minimal portion of the *Bacillus sphaericus* 1593M toxin required for the expression of larvicidal activity. Eur. J. Biochem. 194, 161–165.
Servant, P., Rosso, M.L., Hamon, S., Poncet, S., Del Cluse, A., Rapoport, G., 1999. Production of Cry11A and Cry11Ba toxins in *Bacillus sphaericus* confers toxicity towards *Aedes aegypti* and *resistant Culex* populations. Appl. Environ. Microbiol. 65, 3021–3026.
Seyoum, A., Abate, D., 1997. Larvicidal efficacy of *Bacillus thuringiensis* var. *israelensis* and *Bacillus sphaericus* on *Anopheles arabiensis* in Ethiopia. World J. Microbiol. Biotechnol. 13, 21–24.
Shadduck, J.A., Singer, S., Lause, S., 1980. Lack of mammalian pathogenicity of entomocidal isolates of *Bacillus sphaericus*. Environ. Entomol. 9, 403–407.
Shanmugavelu, M., Rajamohan, F., Kathirvel, M., Elangovan, G., Dean, D.H., Jayaraman, K., 1998. Functional complementation of nontoxic mutant binary toxins of *Bacillus sphaericus* 1593M generated by site-directed mutagenesis. Appl. Environ. Microbiol. 64, 756–759.
Shi, Y., Yuan, Z., Cai, Q., Yu, J., Yan, J., Pang, Y., 2001. Cloning and expression of the binary toxin gene from *Bacillus sphaericus* IAB872 in a crystal-minus *Bacillus thuringiensis* subsp. *israelensis*. Curr. Microbiol. 43, 21–25.
Shililu, J.I., Tewolde, G.M., Brantly, E., Githure, J.I., Mbogo, C.M., Beier, J.C., Fusco, R., Novak, R.J., 2003. Efficacy of *Bacillus thuringiensis israelensis, Bacillus sphaericus* and temephos for managing *Anopheles* larvae in Eritrea. J. Am. Mosq. Control Assoc. 19, 251–258.
Siegel, J.P., Novak, R.J., 1997. Field trials of VectoLex CG, a *Bacillus sphaericus* larvicide, in Illinois waste tires and catch basins. J. Am. Mosq. Control Assoc. 13, 305–310.
Siegel, J.P., Novak, R.J., 1999. Duration of activity of the microbial larvicide VectoLex CG (*Bacillus sphaericus*) in Illinois catch basins and waste tires. J. Am. Mosq. Control Assoc. 15, 366–370.
Siegel, J.P., Shadduck, J.A., 1990. Mammalian safety of *Bacillus sphaericus*. In: De Barjac, H., Sutherland, D. (Eds.), Bacterial Control of Mosquitoes and Black Flies: Biochemistry, Genetics, and Applications of *Bacillus thuringiensis israelensis* and *Bacillus sphaericus*. Rutgers University Press, New Brunswick, NJ, pp. 202–217.
Siegel, J.P., Smith, A.R., Novak, R.J., 2001. Recovery of commercially produced *Bacillus thuringiensis* var. *israelensis* and *Bacillus sphaericus* from tires and prevalence of bacilli in artificial and natural containers. J. Am. Mosq. Control Assoc. 17, 33–41.
Silapanuntakul, S., Pantuwatana, S., Bhumiratana, A., Charoensiri, K., 1983. The comparative persistence of toxicity of *Bacillus sphaericus* strain 1593 and *Bacillus thuringiensis* serotype H-14 against mosquito larvae in different kinds of environments. J. Invertebr. Pathol. 42, 387–392.
Silva Filha, M.H.N.L., Peixoto, C.A., 2003. Immunocytochemical localization of the *Bacillus sphaericus* toxin components in *Culex quinquefasciatus* (Diptera: Culicidae) larvae midgut. Pestic. Biochem. Physiol. 77, 138–146.
Silva-Filha, M.H., Regis, L., 1997. Reversal of low-level resistance to *Bacillus sphaericus* in a field population of the southern house mosquito (Diptera:Culicidae) from an urban area of Recife. Brazil. J. Econ. Entomol. 90, 299–303.
Silva-Filha, M.H.N.L., Regis, L., Nielsen-Leroux, C., Charles, J.-F., 1995. Low level resistance to *Bacillus sphaericus* in a field-treated population of *Culex quinquefasciatus* (Diptera: Culicidae). J. Econ. Entomol. 88, 525–530.

Silva-Filha, M.H., Nielsen-Leroux, C., Charles, J.F., 1997. Binding kinetics of *Bacillus sphaericus* binary toxin to midgut brush-border membranes of *Anopheles* and *Culex* sp. mosquito larvae. Eur. J. Biochem. 247, 754–761.

Silva-Filha, M.H., Nielsen-Leroux, C., Charles, J.F., 1999. Identification of the receptor for *Bacillus sphaericus* crystal toxin in the brush border membrane of the mosquito *Culex pipiens* (Diptera: Culicidae). Insect Biochem. Mol. Biol. 29, 711–721.

Silva-Filha, M.H., Regis, L., Oliveira, C.M., Furtado, A.E., 2001. Impact of a 26-month *Bacillus sphaericus* trial on the preimaginal density of *Culex quinquefasciatus* in an urban area of Recife, Brazil. J. Am. Mosq. Control Assoc. 17, 45–50.

Silva-Filha, M.H., Oliveira, C.M., Regis, L., Yuan, Z., Rico, C.M., Nielsen-Leroux, C., 2004. Two *Bacillus sphaericus* binary toxins share the midgut receptor binding site: implications for resistance of *Culex pipiens* complex (Diptera: Culicidae) larvae. FEMS Microbiol. Lett. 241, 185–191.

Silva-Filha, M.H.N.L., Chalegre, K.D., Anastacio, D.B., Oliveira, C.M.F., Silva, S.B., Acioli, R.V., Hibi, S., Oliveira, D.C., Parodi, E.S.M., Marques Filho, C.A.M., Furtado, A.F., Regis, L., 2008. *Culex quinquefasciatus* field populations subjected to treatment with *Bacillus sphaericus* did not display high resistance levels. Biol. Control 44, 227–234.

Sinègre, G., Babinot, M., Quermel, J.M., Gavon, B., 1994. First field occurrence of *Culex pipiens* resistance to *Bacillus sphaericus* in Southern France. In: Abstracts of the VIIIth European Meeting of Vector Ecology. Barcelone, September 5-8 1994, p. 17.

Singer, S., 1973. Insecticidal activity of recent bacterial isolates and their toxins against mosquito larvae. Nature 244, 110–111.

Singer, S., 1977. Isolation and development of bacterial pathogens in vectors. In: Biological Regulation of Vectors. DHEW Publication No. (NIH) 77–1180, Bethesda, pp. 3–18.

Singer, S., 1980. *Bacillus sphaericus* for the control of mosquitoes. Biotechnol. Bioeng. 22, 1335–1355.

Singh, G.J., Gill, S.S., 1988. An electron microscope study of the toxic action of *Bacillus sphaericus* in *Culex quinquefasciatus* larvae. J. Invertebr. Pathol. 52, 237–247.

Singh, O.P., Bali, P., Hemingway, J., Subbarao, S.K., Dash, A.P., Adak, T., 2009. PCR-based methods for the detection of L1014 kdr mutation in *Anopheles culicifacies* sensu lato. Malar. J. 8, 154.

Singkhamanan, K., Promdonkoy, B., Chaisri, U., Boonserm, P., 2010. Identification of amino acids required for receptor binding and toxicity of the *Bacillus sphaericus* binary toxin. FEMS Microbiol. Lett. 303, 84–91.

Singkhamanan, K., Promdonkoy, B., Srikhirin, T., Boonserm, P., 2013. Amino acid residues in the N-terminal region of the BinB subunit of *Lysinibacillus sphaericus* binary toxin play a critical role during receptor binding and membrane insertion. J. Invertebr. Pathol. 114, 65–70.

Skovmand, O., Bauduin, S., 1997. Efficacy of a granular formulation of *Bacillus sphaericus* against *Culex quinquefasciatus* and *Anopheles gambiae* in West African countries. J. Vector Ecol. 22, 43–51.

Skovmand, O., Guillet, P., 2000. Sedimentation of *Bacillus sphaericus* in tap water and sewage water. J. Invertebr. Pathol. 75, 243–250.

Skovmand, O., Sanogo, E., 1999. Experimental formulations of *Bacillus sphaericus* and *B. thuringiensis israelensis* against *Culex quinquefasciatus* and *Anopheles gambiae* (Diptera: Culicidae) in Burkina Faso. J. Med. Entomol. 36, 62–67.

Skovmand, O., Ouedraogo, T.D., Sanogo, E., Samuelsen, H., Toe, L.P., Baldet, T., 2009. Impact of slow-release *Bacillus sphaericus* granules on mosquito populations followed in a tropical urban environment. J. Med. Entomol. 46, 67–76.

Skovmand, O., Ouedraogo, T.D., Sanogo, E., Samuelsen, H., Toe, L.P., Bosselmann, R., Czajkowski, T., Baldet, T., 2011. Cost of integrated vector control with improved sanitation and road infrastructure coupled with the use of slow-release *Bacillus sphaericus* granules in a tropical urban setting. J. Med. Entomol. 48, 813–821.

Smith, A.W., Camara-Artigas, A., Allen, J.P., 2004. Crystallization of the mosquito-larvicidal binary toxin produced by *Bacillus sphaericus*. Acta Crystallogr. D Biol. Crystallogr. 60, 952–953.
Smith, A.W., Camara-Artigas, A., Brune, D.C., Allen, J.P., 2005. Implications of high-molecular-weight oligomers of the binary toxin from *Bacillus sphaericus*. J. Invertebr. Pathol. 88, 27–33.
Souza-Neto, J.A., Machado, F.P., Lima, J.B., Valle, D., Ribolla, P.E., 2007. Sugar digestion in mosquitoes: identification and characterization of three midgut alpha-glucosidases of the neo-tropical malaria vector *Anopheles aquasalis* (Diptera: Culicidae). Comp. Biochem. Physiol. A Mol. Integr. Physiol. 147, 993–1000.
Sreshty, M.A., Kumar, K.P., Murty, U.S., 2011. Synergism between wild-type *Bacillus thuringiensis* subsp. *israelensis* and *B. sphaericus* strains: a study based on isobolographic analysis and histopathology. Acta Trop. 118, 14–20.
Srisucharitpanit, K., Inchana, P., Rungrod, A., Promdonkoy, B., Boonserm, P., 2012. Expression and purification of the active soluble form of *Bacillus sphaericus* binary toxin for structural analysis. Protein Expr. Purif. 82, 368–372.
Srisucharitpanit, K., Yao, M., Chimnaronk, S., Promdonkoy, B., Tanaka, I., Boonserm, P., 2013. Crystallization and preliminary X-ray crystallographic analysis of the functional form of BinB binary toxin from *Bacillus sphaericus*. Acta Crystallogr. F Struct. Biol. Crystal. Commun. 69, 170–173.
Srisucharitpanit, K., Yao, M., Promdonkoy, B., Chimnaronk, S., Tanaka, I., Boonserm, P., 2014. Crystal structure of BinB: a receptor binding component of the binary toxin from *Lysinibacillus sphaericus*. Proteins. http://dx.doi.org/10.1002/prot.24636. [Epub ahead of print].
Sternberg, M., Grue, C., Conquest, L., Grassley, J., King, K., 2012. Efficacy, fate, and potential effects on salmonids of mosquito larvicides in catch basins in Seattle, Washington. J. Am. Mosq. Control Assoc. 28, 206–218.
Su, T.S., 2008. Evaluation of water-soluble pouches of *Bacillus sphaericus* applied as prehatch treatment against *Culex* mosquitoes in simulated catch basins. J. Am. Mosq. Control Assoc. 24, 54–60.
Su, T., Mulla, M.S., 1999. Field evaluation of new water-dispersible granular formulations of *Bacillus thuringiensis* ssp. *israelensis* and *Bacillus sphaericus* against *Culex* mosquitoes in microcosms. J. Am. Mosq. Control Assoc. 15, 356–365.
Su, T., Mulla, M.S., 2004. Documentation of high-level *Bacillus sphaericus* 2362 resistance in field populations of *Culex quinquefasciatus* breeding in polluted water in Thailand. J. Am. Mosq. Control Assoc. 20, 405–411.
Suarez, M.F., Morales, C.A., 1999. Impact of *Bacillus sphaericus* (Vectolex(r) CG) in the control of *Anopheles albimanus* and *Culex* spp. in Buenaventura, Colombia. J. Am. Mosq. Control Assoc. 15, 407–415.
Sun, M., Luo, X., Dai, J., Qu, K., Liu, Z., Yu, L., Chen, Y., Yu, Z., 1996. Evaluation of *Bacillus thuringiensis* and *Bacillus sphaericus* strains from chinese soils toxic to mosquito larvae. J. Invertebr. Pathol. 68, 74–77.
Sundararaj, R., Reuben, R., 1991. Evaluation of a microgel droplet formulation of *Bacillus sphaericus* 1593 M (Biocide-S) for control of mosquito larvae in rice fields in southern India. J. Am. Mosq. Control Assoc. 7, 556–559.
Tabashnik, B.E., Finson, N., Johnson, M.W., Heckel, D.G., 1994. Cross-Resistance to *Bacillus thuringiensis* Toxin CryIF in the Diamondback Moth (*Plutella xylostella*). Appl. Environ. Microbiol. 60, 4627–4629.
Tabashnik, B.E., Carrière, Y., Dennehy, T.J., Morin, S., Sisterson, M.S., Roush, R.T., Shelton, A.M., Zhao, J.Z., 2003. Insect resistance to transgenic Bt crops: lessons from the laboratory and field. J. Econ. Entomol. 96, 1031–1038.
Tabashnik, B.E., Fabrick, J.A., Henderson, S., Biggs, R.W., Yafuso, C.M., Nyboer, M.E., Manhardt, N.M., Coughlin, L.A., Sollome, J., Carriere, Y., Dennehy, T.J., Morin, S.,

2006. DNA screening reveals pink bollworm resistance to Bt cotton remains rare after a decade of exposure. J. Econ. Entomol. 99, 1525–1530.
Tanapongpipat, S., Luxananil, P., Promdonkoy, B., Chewawiwat, N., Audtho, M., Panyim, S., 2003. A plasmid encoding a combination of mosquito-larvicidal genes from *Bacillus thuringiensis* subsp. *israelensis* and *Bacillus sphaericus* confers toxicity against a broad range of mosquito larvae when expressed in Gram-negative bacteria. FEMS Microbiol. Lett. 228, 259–263.
Tandeau De Marsac, N., De La Torre, F., Szulmajster, J., 1987. Expression of the larvicidal gene of *Bacillus sphaericus* 1593M in the cyanobacterium *Anàcystis nidulans* R2. Mol. Gen. Genet. 209, 396–398.
Tangsongcharoen, C., Boonserm, P., Promdonkoy, B., 2011. Functional characterization of truncated fragments of *Bacillus sphaericus* binary toxin BinB. J. Invertebr. Pathol. 106, 230–235.
Tchicaya, E.S., Koudou, B.G., Keiser, J., Adja, A.M., Cisse, G., Tanner, M., Tano, Y., Utzinger, J., 2009. Effect of repeated application of microbial larvicides on malaria transmission in central Cote d'Ivoire. J. Am. Mosq. Control Assoc. 25, 382–385.
Tetreau, G., Patil, C.D., Chandor-Proust, A., Salunke, B.K., Patil, S.V., Despres, L., 2013. Production of the bioinsecticide *Bacillus thuringiensis* subsp. *israelensis* with deltamethrin increases toxicity towards mosquito larvae. Lett. Appl. Microbiol. 57, 151–156.
Thanabalu, T., 1992. Cloning and characterization of a gene encoding a 100 kDa toxin from *Bacillus sphaericus* SSII-1 and expression of insecticidal toxins in *Caulobacter crescentus*. PhD, National University of Singapore.
Thanabalu, T., Porter, A.G., 1995. Efficient expression of a 100-kilodalton mosquitocidal toxin in protease-deficient recombinant *Bacillus sphaericus*. Appl. Environ. Microbiol. 61, 4031–4036.
Thanabalu, T., Porter, A.G., 1996. A *Bacillus sphaericus* gene encoding a novel type of mosquitocidal toxin of 31.8 kDa. Gene 170, 85–89.
Thanabalu, T., Hindley, J., Jackson-Yap, J., Berry, C., 1991. Cloning, sequencing, and expression of a gene encoding a 100-kilodalton mosquitocidal toxin from *Bacillus sphaericus* SSII-1. J. Bacteriol. 173, 2776–2785.
Thanabalu, T., Hindley, J., Berry, C., 1992a. Proteolytic processing of the mosquitocidal toxin from *Bacillus sphaericus* SSII-1. J. Bacteriol. 174, 5051–5056.
Thanabalu, T., Hindley, J., Brenner, S., Oei, C., Berry, C., 1992b. Expression of the mosquitocidal toxins of *Bacillus sphaericus* and *Bacillus thuringiensis subsp. israelensis* by recombinant *Caulobacter crescentus*, a vehicle for biological control of aquatic insect larvae. Appl. Environ. Microbiol. 58, 905–910.
Thanabalu, T., Berry, C., Hindley, J., 1993. Cytotoxicity and ADP-ribosylating activity of the mosquitocidal toxin from *Bacillus sphaericus* SSII-1: possible roles of the 27- and 70-kilodalton peptides. J. Bacteriol. 175, 2314–2320.
Thiery, I., de Barjac, H., 1989. Selection of the most potent *Bacillus sphaericus* strains based on activity ratios determined on 3 mosquito species. Appl. Microbiol. Biotechnol. 31, 577–581.
Thiery, I., Back, C., Barbazan, P., Sinègre, G., 1996. Applications de *Bacillus thuringiensis* et de *B. sphaericus* dans la démoustication et la lutte contre les vecteurs de maladies tropicales. Ann. Institut Pasteur/Actualit. 7, 247–260.
Thiery, I., Hamon, S., Delecluse, A., Orduz, S., 1998. The introduction into *Bacillus sphaericus* of the *Bacillus thuringiensis* subsp. *medellin* Cyt1Ab1 gene results in higher susceptibility of resistant mosquito larva populations to *B. sphaericus*. Appl. Environ. Microbiol. 64, 3910–3916.
Thompson, J.D., Higgins, D.G., Gibson, T.J., 1994. CLUSTAL W: improving the sensitivity of progressive multiple sequence alignment through sequence weighting, position-specific gap penalties and weight matrix choice. Nucleic Acids Res. 22, 4673–4680.

Thorne, L., Garduno, F., Thompson, T., Decker, D., Zounes, M., Wild, M., Walfield, A.M., Pollock, T.J., 1986. Structural similarity between the lepidoptera- and diptera-specific insecticidal endotoxin genes of *Bacillus thuringiensis* subsp. "*kurstaki*" and "*israelensis*". J. Bacteriol. 166, 801–811.

Tietze, N.S., Hester, P.G., Shaffer, K.R., Prescott, S.J., Schreiber, E.T., 1994. Integrated management of waste tire mosquitoes utilizing *Mesocyclops longisetus* (Copepoda: Cyclopidae), *Bacillus thuringiensis* var. *israelensis*, *Bacillus sphaericus*, and methoprene. J. Am. Mosq. Control Assoc. 10, 363–373.

Townson, H., Nathan, M.B., Zaim, M., Guillet, P., Manga, L., Bos, R., Kindhauser, M., 2005. Exploiting the potential of vector control for disease prevention. Bull. WHO 83, 942–947.

Treiber, N., Reinert, D.J., Carpusca, I., Aktories, K., Schulz, G.E., 2008. Structure and mode of action of a mosquitocidal holotoxin. J. Mol. Biol. 381, 150–159.

Tripet, F., Wright, J., Lanzaro, G., 2006. A new high-performance PCR diagnostic for the detection of pyrethroid knockdown resistance kdr in *Anopheles gambiae*. Am. J. Trop. Med. Hyg. 74, 658–662.

Trisrisook, M., Pantuwatana, S., Bhumiratana, A., Panbangred, W., 1990. Molecular cloning of the 130-kilodalton mosquitocidal delta-endotoxin gene of *Bacillus thuringiensis* subsp. *israelensis* in *Bacillus sphaericus*. Appl. Environ. Microbiol. 56, 1710–1716.

Vadlamudi, R.K., Weber, E., Ji, I., Ji, T.H., Bulla Jr., L.A., 1995. Cloning and expression of a receptor for an insecticidal toxin of *Bacillus thuringiensis*. J. Biol. Chem. 270, 5490–5494.

Valaitis, A.P., Jenkins, J.L., Lee, M.K., Dean, D.H., Garner, K.J., 2001. Isolation and partial characterization of gypsy moth BTR-270, an anionic brush border membrane glycoconjugate that binds *Bacillus thuringiensis* Cry1A toxins with high affinity. Arch. Biochem. Biophys. 46, 186–200.

Vasquez, M.I., Violaris, M., Hadjivassilis, A., Wirth, M.C., 2009. Susceptibility of *Culex pipiens* (Diptera: Culicidae) field populations in Cyprus to conventional organic insecticides, *Bacillus thuringiensis* subsp. *israelensis*, and methoprene. J. Med. Entomol. 46, 881–887.

Wagner, E., Lykke-Andersen, J., 2002. mRNA surveillance: the perfect persist. J. Cell Sci. 115, 3033–3038.

Wahba, M.M., 2000. The influence of *Bacillus sphaericus* on the biology and histology of *Phlebotomus papatasi*. J. Egypt. Soc. Parasitol. 30, 315–323.

Wati, M.R., Thanabalu, T., Porter, A.G., 1997. Gene from tropical *Bacillus sphaericus* encoding a protease closely related to subtilisins from *Antarctic bacilli*. Biochim. Biophys. Acta 1352, 56–62.

Wei, S., Cai, Q., Cai, Y., Yuan, Z., 2007. Lack of cross-resistance to Mtx1 from *Bacillus sphaericus* in *B. sphaericus*-resistant *Culex quinquefasciatus* (Diptera: Culicidae). Pest Manag. Sci. 63, 190–193.

Weiser, J., 1984. A mosquito-virulent *Bacillus sphaericus* in adult *Simulium damnosum* from northern Nigeria. Zentralbl. Mikrobiol. 139, 57–60.

Wermelinger, E.D., Zanuncio, J.C., Rangel, E.F., Cecon, P.R., Rabinovitch, L., 2000. Toxicity of *Bacillus* species to larvae of *Lutzomyia longipalpis* (L. & N.) (Diptera: Psychodidae: Phlebotominae). An. Soc. Entomol. Bras. 26, 609–614.

White, J.P., Lotay, H.K., 1980. Minimal nutritional requirements of *Bacillus sphaericus* NCTC9602 and 26 others strains of this species: the majority grow and sporulate with acetate as sole major source of carbon. J. Gen. Microbiol. 118, 13–19.

WHO, 1983. Integrated Vector Control: 7th Rep. WHO Exp. Comm.VCB, Technical Reportys Series, No. 668.

WHO, 1985. Informal Consultation on the Development of *Bacillus sphaericus* as Microbial Larvicide. WHO, Geneva, pp. 1–24, TDR/BCV/SPHAERICUS/85.3.

WHO, 1986. Resistance of Vectors and Reservoirs of Disease to Pesticides: World Health Organization technical report series 737. pp. 1–87.

WHO, 1993. Report on the Workshop on the Large-Scale Use of *Bacillus sphaericus* to Control *Culex quinquefasciatus* in Urban Environments. Centre Pasteur Cameroun, Maroua, Cameroun.

WHO, 2004. Global Strategic Framework for Integrated Vector Management. WHO, p. 12, Document WHO/CDS/CPE/PVC/2004.10.

Wickremesingue, R.S.B., Mendis, C.L., 1980. *Bacillus sphaericus* spores from Sri-Lanka demostrating rapid larvicidal activity on *Culex quinquefasciatus*. Mosq. News 40, 387–389.

Wilkinson, M.F., Shyui, A.B., 2002. RNA surveillance by nuclear scanning? Nat. Cell Biol. 4, E144–E147.

Wirth, M.C., 2010. Mosquito resistance to bacterial larvicidal proteins. Open J. Toxicol. 3, 101–115.

Wirth, M.C., Federici, B.A., Walton, W.E., 2000a. Cyt1A from *Bacillus thuringiensis* synergizes activity of *Bacillus sphaericus* against *Aedes aegypti* (Diptera: Culicidae). App. Environ. Microbiol. 66, 1093–1097.

Wirth, M.C., Georghiou, G.P., Malik, J.I., Abro, G.H., 2000b. Laboratory selection for resistance to *Bacillus sphaericus* in *Culex quinquefasciatus* (Diptera: Culicidae) from California, USA. J. Med. Entomol. 37, 534–540.

Wirth, M.C., Walton, W.E., Federici, B.A., 2000c. Cyt1A from *Bacillus thuringiensis* restores toxicity of *Bacillus sphaericus* against resistant *Culex quinquefasciatus* (Diptera: Culicidae). J. Med. Entomol. 37, 401–407.

Wirth, M.C., Delecluse, A., Walton, W.E., 2001a. Cyt1Ab1 and Cyt2Ba1 from *Bacillus thuringiensis* subsp. *medellin* and *B. thuringiensis* subsp. *israelensis* Synergize *Bacillus sphaericus* against *Aedes aegypti* and resistant *Culex quinquefasciatus* (Diptera: Culicidae). Appl. Environ. Microbiol. 67, 3280–3284.

Wirth, M.C., Ferrari, J.A., Georghiou, G.P., 2001b. Baseline susceptibility to bacterial insecticides in populations of *Culex pipiens* complex (Diptera: Culicidae) from California and from the Mediterranean Island of Cyprus. J. Econ. Entomol. 94, 920–928.

Wirth, M.C., Jiannino, J.A., Federici, B.A., Walton, W.E., 2004. Synergy between toxins of *Bacillus thuringiensis* subsp. *israelensis* and *Bacillus sphaericus*. J. Med. Entomol. 41, 935–941.

Wirth, M.C., Jiannino, J.A., Federici, B.A., Walton, W.E., 2005. Evolution of resistance toward *Bacillus sphaericus* or a mixture of *B. sphaericus*+Cyt1A from *Bacillus thuringiensis*, in the mosquito, *Culex quinquefasciatus* (Diptera: Culicidae). J. Invertebr. Pathol. 88, 154–162.

Wirth, M.C., Yang, Y., Walton, W.E., Federici, B.A., Berry, C., 2007. Mtx toxins synergize *Bacillus sphaericus* and Cry11Aa against susceptible and insecticide-resistant *Culex quinquefasciatus* larvae. Appl. Environ. Microbiol. 73, 6066–6071.

Wirth, M.C., Berry, C., Walton, W.E., Federici, B.A., 2014. Mtx toxins from *Lysinibacillus sphaericus* enhance mosquitocidal cry-toxin activity and suppress cry-resistance in *Culex quinquefasciatus*. J. Invertebr. Pathol. 115, 62–67.

Woodburn, M.A., Yousten, A.A., Hilu, K.H., 1995. Random amplified polymorphic DNA fingerprinting of mosquito-pathogenic and nonpathogenic strains of *Bacillus sphaericus*. Int. J. Syst. Bacteriol. 45, 212–217.

Wraight, S.P., Molloy, D.P., Singer, S., 1987. Studies on the culicine mosquito host range of *Bacillus sphaericus* and *Bacillus thuringiensis var. israelensis* with notes on the effects of temperature and instar on bacterial efficacy. J. Invertebr. Pathol. 49, 291–302.

Xu, Z., Liu, F., Chen, J., Huang, F., Andow, D.A., Wang, Y., Zhu, Y.C., Shen, J., 2009. Using an F(2) screen to monitor frequency of resistance alleles to Bt cotton in field populations of *Helicoverpa armigera* (Hubner) (Lepidoptera: Noctuidae). Pest Manag. Sci. 65, 391–397.

Yadav, R.S., Sharma, V.P., Upadhyay, A.K., 1997. Field trial of *Bacillus sphaericus* strain B-101 (serotype H5a, 5b) against filariasis and Japanese encephalitis vectors in India. J. Am. Mosq. Control Assoc. 13, 158–163.

Yang, Y., Chen, H., Wu, S., Xu, X., Wu, Y., 2006. Identification and molecular detection of a deletion mutation responsible for a truncated cadherin of *Helicoverpa armigera*. Insect Biochem. Mol. Biol. 36, 735–740.
Yang, Y., Chen, H., Wu, Y., Wu, S., 2007a. Mutated cadherin alleles from a field population of *Helicoverpa armigera* confer resistance to *Bacillus thuringiensis* toxin Cry1Ac. Appl. Environ. Microbiol. 73, 6939–6944.
Yang, Y., Wang, L., Gaviria, A., Yuan, Z., Berry, C., 2007b. Proteolytic stability of insecticidal toxins expressed in recombinant bacilli. Appl. Environ. Microbiol. 73, 218–225.
Yap, H.H., Tan, H.T., Yahaya, A.M., Baba, R., Chong, N.L., 1991. Small-scale field trials of *Bacillus sphaericus* (strain 2362) formulations against *Mansonia mosquitoes in Malaysia*. J. Am. Mosq. Control Assoc. 7, 24–29.
Yap, W.H., Thanabalu, T., Porter, A.G., 1994. Expression of mosquitocidal toxin genes in a gas-vacuolated strain of *Ancylobacter aquaticus*. Appl. Environ. Microbiol. 60, 4199–4202.
Yoshida, K., Hidaka, H., Miyado, S., Shibata, U., Saito, K., Yamada, Y., 1977. Purification and some properties of *Bacillus sphaericus* protease. Agric. Biol. Chem. 41, 745–754.
Young, M.D., Undeen, A.H., Dame, D.A., Wing, S.R., 1990. The effect of *Bacillus sphaericus* upon the susceptibility of *Anopheles quadrimaculatus* to *Plasmodium berghei*. J. Am. Mosq. Control Assoc. 6, 139–140.
Yousten, A.A., 1984a. *Bacillus sphaericus*: microbiological factors related to its potential as a mosquito larvicide. Adv. Biotechnol. Processes 3, 315–343.
Yousten, A.A., 1984b. Bacteriophage typing of mosquito pathogenic strains of *Bacillus sphaericus*. J. Invertebr. Pathol. 43, 124–125.
Yousten, A.A., Davidson, E.W., 1982. Ultrastructural analysis of spores and parasporal crystals formed by *Bacillus sphaericus* 2297. Appl. Environ. Microbiol. 44, 1449–1455.
Yousten, A.A., De Barjac, H., Hedrick, J., Cosmao Dumanoir, V., Myers, P., 1980. Comparison between bacteriophage typing and serotyping for the differentiation of *Bacillus sphaericus* strains. Ann. Microbiol. (Paris) 131B, 297–308.
Yousten, A.A., Benfield, E.F., Campbell, R.P., Foss, S.S., Genthner, F.J., 1991. Fate of *Bacillus sphaericus* 2362 spores following ingestion by nontarget invertebrates. J. Invertebr. Pathol. 58, 427–435.
Yousten, A.A., Genthner, F.J., Benfield, E.F., 1992. Fate of *Bacillus sphaericus* and *Bacillus thuringiensis* serovar *israelensis* in the aquatic environment. J. Am. Mosq. Control Assoc. 8, 143–148.
Yousten, A.A., Benfield, E.F., Genthner, F.J., 1995. *Bacillus sphaericus* mosquito pathogens in the aquatic environment. Mem. Inst. Oswaldo Cruz 90, 125–129.
Yuan, Z., Neilsen-Leroux, C., Pasteur, N., Delecluse, A., Charles, J.F., Frutos, R., 1999. Cloning and expression of the binary toxin genes of *Bacillus sphaericus* C3-41 in a crystal minus *B. thuringiensis subsp. israelensis*. Wei Sheng Wu Xue Bao 39, 29–35.
Yuan, Z.M., Zhang, Y.M., Liu, E.Y., 2000. High-level field resistance to *Bacillus sphaericus* C3-41 in *Culex quinquefasciatus* from Southern China. Biocontrol Sci. Technol. 10, 43–51.
Yuan, Z., Rang, C., Maroun, R.C., Juarez-Perez, V., Frutos, R., Pasteur, N., Vendrely, C., Charles, J.F., Nielsen-Leroux, C., 2001. Identification and molecular structural prediction analysis of a toxicity determinant in the *Bacillus sphaericus* crystal larvicidal toxin. Eur. J. Biochem. 268, 2751–2760.
Yuan, Z.M., Pei, G.F., Regis, L., Nielsen-Leroux, C., Cai, Q.X., 2003. Cross-resistance between strains of *Bacillus sphaericus* but not *B. thuringiensis israelensis* in colonies of the mosquito *Culex quinquefasciatus*. Med. Vet. Entomol. 17, 251–256.
Zahiri, N.S., Mulla, M.S., 2003. Susceptibility profile of *Culex quinquefasciatus* (Diptera: Culicidae) to *Bacillus sphaericus* on selection with rotation and mixture of B. sphaericus and *B. thuringiensis israelensis*. J. Med. Entomol. 40, 672–677.

Zahiri, N.S., Su, T., Mulla, M.S., 2002. Strategies for the management of resistance in mosquitoes to the microbial control agent *Bacillus sphaericus*. J. Med. Entomol. 39, 513–520.
Zhao, J., Jin, L., Yang, Y., Wu, Y., 2010. Diverse cadherin mutations conferring resistance to *Bacillus thuringiensis* toxin Cry1Ac in *Helicoverpa armigera*. Insect Biochem. Mol. Biol. 40, 113–118.
Zheng, L., Whang, L.H., Kumar, V., Kafatos, F.C., 1995. Two genes encoding midgut-specific maltase-like polypeptides from *Anopheles gambiae*. Exp. Parasitol. 81, 272–283.

CHAPTER FOUR

Discovery and Development of Insect-Resistant Crops Using Genes from *Bacillus thuringiensis*

Kenneth E. Narva, Nicholas P. Storer, Thomas Meade
Dow AgroSciences, LLC., Indianapolis, Indiana, USA

Contents

Advances in Insect Physiology, Volume 47
ISSN 0065-2806
http://dx.doi.org/10.1016/B978-0-12-800197-4.00004-X

Abstract

Bacillus thuringiensis (*Bt*) is a ubiquitous, spore-forming soil bacterium that is well known for production of insecticidal proteins that are active on a wide range of pest insects. The potential of *Bt* to be used as an insecticide was recognized in the early twentieth century and since that time many *Bt*-based biopesticides have been commercialized. The advent of modern molecular biology tools made it possible to engineer plants to express the genes coding for *Bt* insecticidal proteins as a safe, convenient and highly effective means to protect plants from insect damage. The first *Bt* crop was commercialized in 1995, and today *Bt* corn, cotton and soybean are cultivated on ca. 76 million hectares in 27 countries. First generation products containing single *Bt* genes were followed by broader spectrum products containing multiple *Bt* genes with the most recent generation of products contain multiple *Bt* genes encoding proteins that target the same pest(s) but with differences in their mechanism of action (i.e. gene pyramids) as a means of increasing product durability. Developing *Bt* crops is a long and expensive process that by recent estimates averages 13 years at a cost of $136 million. The process of obtaining approvals by government regulatory agencies is among the most critical in the later stages of the development process and represents ca. 25% of the total cost in bringing a *Bt* crop to the market. Multiple factors drive the search for novel insect resistance (IR) traits and *Bt* remains a significant focus of new IR trait discovery.

1. INTRODUCTION

Bacillus thuringiensis (*Bt*) is a ubiquitous, spore-forming soil bacterium that is well known for production of parasporal crystalline inclusions during the stationary phase of cell growth. These parasporal inclusions are comprised of insecticidal proteins known as δ-endotoxins, including those classified as Cry (crystalline) or Cyt (cytolytic) proteins. The parasporal crystalline inclusions produced by *Bt* are composed of a diversity of proteins across distinct phylogenetic groups of sequences (Crickmore et al., 1998) and (http://www.lifesci.sussex.ac.uk/home/Neil_Crickmore/Bt/). Collectively, Cry proteins are active on a wide range of insects including those among the orders of Lepidoptera, Diptera and Coleoptera (van Frankenhuyzen, 2009). *Bt* also produces soluble insecticidal proteins during the cell vegetative growth phase before the onset of sporulation that are named Vips (vegetative insecticidal proteins) (Estruch et al., 1996; Warren, 1997; http://www.lifesci.sussex.ac.uk/home/Neil_Crickmore/Bt/).

Various subspecies of *Bt* have historically been developed for use as foliar applied biopesticides (Sanahuja et al., 2011) and have a long history of safe use (Siegel, 2001). With the advent of modern molecular biology tools, it has become possible to engineer plants to express the genes coding for *Bt* insecticidal proteins as a safe, convenient and highly effective means to protect plants from insect damage. The development of insect resistant crops has rapidly progressed since the commercial introduction of *Bt* potato in 1995 and Bt corn and cotton in 1996 (http://www.epa.gov/oppbppd1/biopesticides/pips/pip_list.htm). Today insect resistance (IR) traits based on *Bt* proteins have achieved a high rate of world-wide adoption (James, 2013). A current challenge for *Bt* trait seed producers is protecting the long-term durability of *Bt* trait technology. Innovative insect resistance management (IRM) strategies include the use of genetically engineered (GE) crops containing combinations of *Bt* genes encoding novel insecticidal proteins (i.e. pyramids). This chapter provides an overview of the history of *Bt* biopesticides leading to *Bt* crop development, the success of *Bt*-based IR traits and future prospects for *Bt* as a source of IR trait technology.

2. *Bt*-BASED BIOPESTICIDES

2.1. History of use of *Bt* for insect control

Bt has a long history of safe use as a biopesticide for insect control (Siegel, 2001). For an elegant review of the early historical events in the discovery and development of insecticidal bacteria with significant attention directed at *Bt* see Federici (2005). The bacterium that became known as *Bt* was first reported in Japanese literature by Ishiwata (1901) during study of bacterial disease of silkworms. Later, Berliner (1915) described a similar *Bacillus* bacterium that killed flour moths and named the organism *B. thuringiensis* for the Thuringia region in Germany where the bacterial disease was discovered. Research into the utility of *Bt* as an insecticide followed (Mattes, 1927) and activity in field trials against the European corn borer, *Ostrinia nubilalis* (Hübner), was reported in 1930 (Husz, 1930). This work led to the development of a *Bt* product known as "Sporeine" that was commercialized in the late 1930s (Federici, 2005). The potential for *Bt* to be used as an insecticide became more widely appreciated years after these early studies. Publications by Hannay (1953) on the *Bt* parasporal crystal bodies and demonstration that the parasporal crystals were capable of killing silkworms (Angus, 1954)set the stage for an increase in research focused on developing *Bt* as an insect control agent.

2.2. Biopesticides based on *Bt*

Advances in the applied science of *Bt* were aided by systematic characterization of the insecticidal properties of *Bt* strains. A system for naming strains based on flagellar serotype (de Barjac and Bonnefoi, 1962, 1968) and establishment of standardized bioassay techniques based on *B. thuringiensis* HD-1 (Dulmage, 1981) provided the basis for characterizing strains and comparing insecticidal properties among *Bt* isolates. This led to the development of successful commercial *Bt* products in the 1960s, most notably Dipel™ (Abbot Laboratories) and Thuricide™ (Sandoz Corporation), both of which were based on the HD-1 isolate of *Bt* subspecies kurstaki (serotype H 3a3b) (Federici, 2005). These products controlled lepidopertan pests important in agriculture and forestry such as the cabbage looper, *Trichoplusia ni* (Hübner), corn earworm or bollworm, *Helicoverpa zea* (Boddie), the tobacco budworm, *Heliothis virescens* (F.), the diamondback moth, *Plutella xylostella* (L.), the gypsy moth, *Lymantria dispar* (L.) and the spruce budworm, *Choristoneura fumiferana* (Clemens).

The success ofDipel and Thuricide led to the development in the United States of 177 registered products containing viable *Bt* between the years 1961 and 1995. *Bt*-based biopesticide products have an excellent mammalian safety record based on laboratory studies and extensive field experience (Siegel, 2001). Examples of *Bt*-based biopesticide products are shown in Table 4.1. For a listing of currently registered *Bt* biopestides, refer to the United States Environmental Agency website (http://www.epa.gov/pesticides/biopesticides/).

Efforts to increase *Bt* strain productivity through optimized fermentation and formulation processes drove the development of improved products that replaced earlier product offerings (Kaur, 2000). Further, the discovery of *Bt* strains with activity on different orders of insects provided the opportunity to expand the range of pests controlled by *Bt* biopesticides. While many of the most successful products for control of lepidopteran pests were based on *Bt kurstaki* strains, novel *Bt* subspecies were discovered with activity against other insect orders. Importantly, *Bt* subspecies *israelinsis* (H 14) (Goldberg

Table 4.1 Commercialized *Bt* biopesticides

Trade name	Bt subsp. strain	Producer	Specificity
Bactospeine	kurstaki HD-1	Abbott	Lepidoptera
Biobit	kurstaki HD-1	Abbott	Lepidoptera
Dipel	kurstaki HD-1	Abbott	Lepidoptera
Florbac	aizawai	Abbott	Lepidoptera
Costar	kurstaki SA-12	Thermo trilogy	Lepidoptera
Del®n	kurstaki SA-11	Thermo trilogy	Lepidoptera
Javelin	kurstaki SA-11	Thermo trilogy	Lepidoptera
Thuricide	kurstaki HD-1	Thermo trilogy	Lepidoptera
Tekar	israelensis	Thermo trilogy	Diptera
Bactimos	israelensis	Abbott	Diptera
Vectolex GC	B. sphaericus	Abbott	Diptera
Acrobe	israelensis	American cyanamide	Diptera
Novodor	tenebrionis	Abbott	Coleoptera
Trident	tenebrionis	Thermo trilogy	Coleoptera

From Kaur (2000).

and Margalit, 1977); recently reviewed by Ben-Dov (2014) was found to be active on mosquitoes and black flies, while *Bt* subspecies *morrisoni* (H 8a8b, variety *tenebrioinis*) was active on the larvae of coleopteran species (Krieg et al., 1983). The expanded range of pests controlled by various *Bt* subspecies suggested that additional new strains could be found with unique pesticidal properties. This prompted a vigorous world-wide effort to discover novel strains with new insecticidal activity profiles (see for example Feitelson et al., 1992; Jung et al., 1998; Wasano and Ohba, 1998). Efforts to discover *Bt* isolates with novel biological activity and characterization of the insecticidal proteins that are responsible for strain activity continue today (Arrieta and Espinoza, 2006; Bravo et al., 1998; Noguera and Ibarra, 2010; Vidal-Quist et al., 2009).

2.3. Molecular era—First cloned *Bt* insecticidal protein genes

Plasmid-based DNA cloning became a routine laboratory procedure in the late 1970s (Bolivar et al., 1977), making it possible to isolate and study recombinant genes and proteins. The fact that parasporal protein inclusions were known to be responsible for the insecticidal activity of *Bt* led researchers to use molecular biology techniques to search for the genes encoding these proteins. Schnepf and Whiteley (1981) cloned the first *Bt* gene encoding an insecticidal protein from the *Bt* strain in Dipel, *Bt kurstaki* HD-1. Under the revised nomenclature system for *Bt* insecticidal proteins (Crickmore et al., 1998) this gene later became known as *cry1Aa1*. Further molecular biology work demonstrated that genes coding for different insecticidal proteins were located on distinct restriction endonuclease fragments of DNA from *Bt* HD-1 (Kronstad et al., 1983). These results were important in establishing that *Bt* strains most often contain multiple genes coding for insecticidal proteins. The characterization of the genes encoding Cry proteins in *Bt kurstaki* HD-1 (now designated Cry1Aa, Cry1Ab, Cry1Ac and Cry2Aa) was followed quickly by the isolation of genes coding for many additional Cry proteins including *cry3Aa* (see for example Herrnstadt et al., 1987; Hofte and Whiteley, 1989; Schnepf et al., 1998; Sekar et al., 1987). Cry3Aa is notable as the first example of a *Bt* protein with activity on a coleopteran pest, the Colorado potato beetle, *Leptinotarsa decemlineata* (Say). Characterization of recombinant Cry proteins in *E. coli* or acrystalliferous *Bt* strains using shuttle plasmids (Arantes and Lereclus, 1991; Lecadet et al., 1992) provided a means to investigate the genetic basis for different strain-level insecticidal activity. This set the stage to develop new pest control technology based on recombinant *Bt* insecticidal proteins.

2.4. Transconjugation, recombinant strains and alternative delivery systems for *Bt*-based biopesticides

Several approaches have been used to develop *Bt* biopesticides improved for properties such as increased toxicity, expanded range of target pests, or for delaying the development of resistant insect populations by combining insecticidal proteins that target the same pests but differ in their mechanism of action, such as by acting at different binding sites. Ecogen Corporation developed methods for conjugal transfer of Cry protein encoding native *Bt* plasmids, e.g. *Bt* strain 2424 that expresses both *cry*1A and *cry*3A genes for control of lepidopteran and coleopteran pests (Carlton and Gawron-Burke, 1993). However, strain construction by plasmid conjugation is limited by factors including plasmid incompatibility, location of *cry* genes on large, non-transmissible plasmids and segregational loss of plasmids in transconjugant strains. Ecogen addressed these challenges by developing recombinant DNA technology and site-specific recombination systems to introduce *cry* genes into *Bt* recipient host strains and subsequently eliminate the antibiotic selectable marker resistant genes that might cause environmental safety concerns (Baum et al., 1998). Several examples of recombinant *Bt* strains are listed in Table 4.2.

Mycogen Corporation used a different approach to produce novel biopesticides based on over-expression of *cry* genes in recombinant *Pseudomonas fluorescens* (Gaertner et al., 1993). This gene expression system used recombinant DNA technology to express Cry proteins at high levels under high-density cell culture fermentation conditions. The recombinant bacteria were fixed in a proprietary treatment that rendered cells non-viable without impacting the activity of the insecticidal proteins. The fixed,

Table 4.2 Bt biopesticides based on novel/recombinant strains

Product	Bt subsp. strain or genes	Producer	Specificity
Transconjugant strains	**Bt subsp. strain**		
Agree	aizawai	Thermo trilogy	Lepidoptera
Condor	kurstaki	Ecogen	Lepidoptera
Cutlass	kurstaki	Ecogen	Lepidoptera
Design	aizawai	Ecogen	Lepidoptera
Foil	kurstaki	Ecogen	Lepidoptera/ Coleoptera
Recombinant strains	**genes**		
Raven	*cry1Ac* (x2), *cry3A* + *cry3Bb* (recombinant)	Ecogen	Lepidoptera/ Coleoptera
CRYMAX	*cry1Ac* (x3), *cry2A* + *cry1C* (recombinant)	Ecogen	Lepidoptera
Lepinox	*cry1Aa*, *cry1Ac* (x2), *cry2A* + *cry1F-1Ac* (recombinant)	Ecogen	Lepidoptera
Maatch	kurstaki *cry1A* and aizawai *cry1C*	Mycogen	Lepidoptera
M/C	aizawai *cry1C*	Mycogen	Lepidoptera
M-Peril	kurstaki *cry1Ac*	Mycogen	Lepidoptera
MVP	kurstaki *cry1Ac*	Mycogen	Lepidoptera
MTRAK	*cry3Aa*	Mycogen	Coleoptera

bioencapsulated proteins were more persistent under environmental field conditions. These were the first recombinant biopesticides approved for field tests and commercialization by the United States Environmental Protection Agency in 1991 (http://www.epa.gov/pesticides/biopesticides/). Mycogen Corporation marketed products based on Cry1Ac for control of Lepidoptera (MVP™) and Cry3Aa for control of Colorado potato beetle (MTRAK™), along with combinations of Cry1Ac and Cry1C (Maatch™) for broad spectrum control of Lepidoptera.

Crop Genetics International used yet another approach for delivering *Bt* biopesticides in recombinant endophytic bacteria (Dimock et al., 1993). The probability of endophytes surviving outside the plant host are low, thereby providing a level of biological containment. Lampel et al. (1994) engineered *Clavibacter xyli* subspecies *cynodontis* (CXC), a bacterial endophyte that inhabits the xylem of Bermuda grass, to express a chromosomally integrated *cry*1Ac gene. *C. xyli* can colonize other grasses including maize. Colonized maize expressing Cry1Ac showed reduced feeding damage by *O. nubilalis* though the level of protection to insect feeding damage did not translate to increased grain yield (Tomasino et al., 1995).

3. DISCOVERY, CHARACTERIZATION AND DEVELOPMENT OF INSECTICIDAL PROTEIN GENES AS CROP TRAITS

3.1. Diversity of *Bt* insecticidal proteins

Bt produces a variety of crystalline and soluble insecticidal proteins that comprise various primary sequence homology groups (Schnepf et al., 1998). To date, over 750 unique *Bt* proteins ranging in size from ca. 14 kDa to over 140 kDa have been described that are classified into at least 73 distinct homology groups. Most *Bt* insecticidal proteins fall within three main phylogenetic groups: Cry, Cyt or Vip (Crickmore et al., 1998); (http://www.lifesci.sussex.ac.uk/home/Neil_Crickmore/*Bt*/intro.html).

The Cry class of pesticidal proteins contains the largest number of sequences, many of which share conserved amino acid sequence and structural similarity. The Cry family also includes binary, two component insecticidal proteins, some of which share similarity to the *Lysinibacillus sphaericus* Bin proteins, as well as proteins related to the Mtx families of toxins (Berry, 2012) and parasporins with cytotoxicity to human cancer cells (Ohba et al., 2009).

The Cyt family comprises a group of generally cytolytic proteins with no sequence homology to the Cry proteins (http://www.lifesci.sussex.ac.uk/

home/Neil_Crickmore/*Bt*/intro.html). Cyt proteins can synergize Cry proteins (Ben-Dov, 2014; Chang et al., 1993; Wu et al., 1994) in a manner that depends on the binding interaction of Cyt and Cry proteins (Perez et al., 2005, 2007). Cyt proteins share varying levels of sequence homology with proteins originating from a variety of microbial pathogens (Soberon et al., 2013).

Vips are soluble proteins produced during the logarithmic phase of *Bt* growth. To date, four main groups, Vip1, Vip 2, Vip3 and Vip4, have been described. The soluble Vips, Vip1Aa1 and Vip2Aa1, are approximately 100 and 52 kDa molecular weight, respectively, and act together as a binary toxin (Warren, 1997). Vip1Aa is homologous to the CdtB toxin component of *Clostridium difficile*, the Ib component of *Clostridium perfringens* iota toxin and the protective antigen of *B. anthracis*. Vip2Aa is an ADP-ribosylase with a high degree of sequence and structural similarity to the enzymatic domains of CdtA of *C. difficile* and iota toxin of *C. perfringens* (de Maagd et al., 2003; Han et al., 1999). Vip3 proteins are approximately 80 kDa proteins that are active on lepidopteran pests. The biological activity of Vip4 has not been published.

The number and diversity of genes encoding *Bt* insecticidal proteins continues to rapidly expand as researchers world-wide search for new *Bt* isolates with novel biological activity (Fig. 4.1).

3.2. Biological activity of *Bt* insecticidal proteins

A highly valued benefit of *Bt* insecticidal proteins is the relatively narrow spectrum of activity against susceptible insects. *Bt* insecticidal proteins are highly active on insect larvae but have little or no activity on adult insects (Betz et al., 2000).

Insecticidal activity of *Bt* Cry proteins across insect orders was recently reviewed by van Frankenhuyzen (2009, 2013). These reviews are based on over 25 years of published data on biological specificity of Cry and Cyt proteins. Much of these data are incorporated into the *Bt* Toxin Specificity Database (http://www.glfc.cfs.nrcan.gc.ca/bacillus). Information contained in the *Bt* Toxin Specificity Database is focused on spore free preparations of crystals or insecticidal proteins that were obtained through expression of cloned genes or purified from strains expressing a single insecticidal protein. As the number of *Bt* insecticidal protein sequences has grown, many different pests were found to be susceptible to *Bt* proteins including orders not previously tested such as Hymenoptera, Hemiptera and Rhadbditida

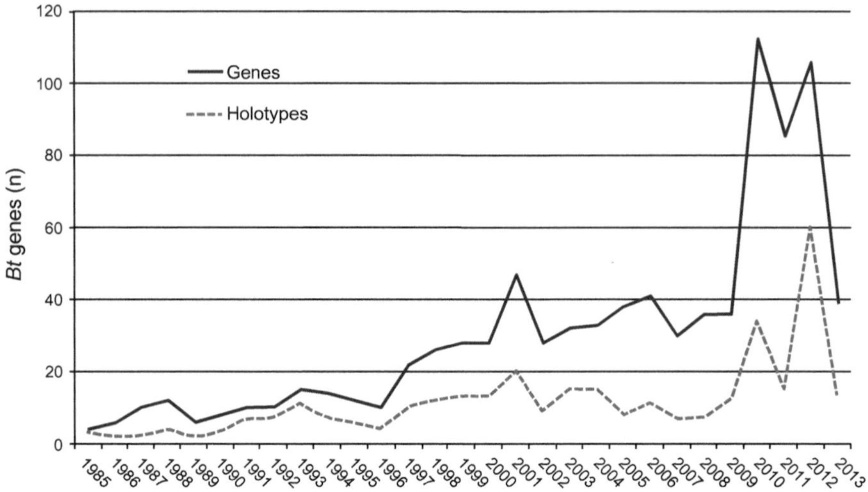

Figure 4.1 Discovery of *Bt* genes recorded on the *Bt* Toxin Nomenclature Website maintained by the *Bt* toxin nomenclature committee (Crickmore et al., 2014). The total number of new Cry, Cyt and Vip genes recognized by the committee in a given year is shown as *Genes*. The total number of new gene classes (as defined by the committee) recognized in a given year is shown as *Holotypes*. (See the colour plate.)

(nematodes). Data in the *Bt* Toxin Specificity Database also reveal cross-order activity in 13 primary rank families across three classes of insecticidal proteins (Cry, Cyt and Vip) (van Frankenhuyzen, 2013). Cross-order activity is an important consideration in selecting *Bt* proteins for commercialization because it necessitates the appropriate design of studies to characterize risk associated with activity outside the primary insect specificity range. The data also reflect that variation in factors such as assay conditions, methods of protein preparation and quantitation, pre-ingestion protein activation and insect population differences or life stage, to name a few, make comparison of protein insecticidal potency difficult. This highlights the need for standardized assays for estimating insecticidal protein expression levels and potency, factors important to IRM.

The class of *Bt* Vips have been more recently discovered (Estruch et al., 1996; Warren, 1997). Vip1Aa1 and Vip2Aa1 act together as a binary toxin that is highly potent against the western corn rootworm (WCR), *Diabrotica virgifera virgifera* LeConte, coleopteran pest that feeds on corn roots. Members of the Vip3 group of proteins have received more attention owing to excellent activity on economically important lepidopteran pests such as the black cutworm, *Agrotis ipsilon* (Hufnagel), *H. zea*, *H. virescens*, the fall armyworm, *Spodoptera frugiperda* (J. E. Smith) and the beet armyworm, *Spodoptera exigua* (Hübner) (Estruch et al., 1996; Fang et al., 2007; Hernandez-Martinez et al., 2013; Lee et al., 2003). The Vip3 proteins are very different in primary sequence compared to the lepidopteran-active Cry protein group, and binding studies suggest a different mechanism of action compared to three domain Cry proteins (Bergamasco et al., 2013; Lee et al., 2003, 2006; Sena et al., 2009). The novel mechanism of action for Vip3 proteins makes this group attractive for commercial applications when combined with Cry proteins as gene pyramids for IRM.

Cyt toxins (reviewed by Butko, 2003; Soberon et al., 2013; Chapter 3) are a subclass of *Bt* insecticidal crystal proteins that are named for their general cytolytic activity. Cyt proteins show selective toxicity against mosquitoes and blackflies. However, examples of coleopteran-active Cyt proteins are Cyt1Aa activity against *Chrysomela scripta* F. (Federici and Bauer, 1998) and the ability of Cyt1Ba (Payne et al., 1995) and Cyt2Ca1 (Rupar et al., 2000) to kill WCR larvae.

3.3. *Bt* insecticidal protein structure and function: Cry proteins

In terms of structure–function relationships, the most well studied *Bt* proteins are members of the three domain Cry δ-endotoxins. These proteins range in size from approximately 70–130 kDa. Many Cry proteins are produced as protoxins requiring activation by proteolytic removal of the C-terminal crystallization domain to produce the core insecticidal protein (Schnepf et al., 1998). Primary protein sequence analysis reveals five conserved sequence blocks and a high degree of sequence variability between conserved blocks three and five (Hofte et al., 1988; Schnepf et al., 1998). In contrast, the C-terminal crystallization domain sequences tend to be highly conserved among subclasses. The correlation of bioactivity spectrum with sequence variability among the activated forms of different *Bt* δ-endotoxins led to early hypotheses that the "hypervariable" regions between conserved blocks three and five are responsible for differences in insect specificity.

The first three-dimensional *Bt* crystal structures determined were of Cry3Aa1 (Li et al., 1991) and Cry1Aa1 (Grochulski et al., 1995; Fig. 4.2). The Cry1 and Cry3 structures are remarkably similar and are comprised of three distinct domains with the following features (for reviews see de Maagd et al., 2003; Pigott and Ellar, 2007). Domain 1 is a bundle of seven alpha helices where helix five is surrounded by six amphipathic helices. This domain has been implicated in pore formation and shares homology with other pore forming proteins including hemolysins and colicins. Domain 2

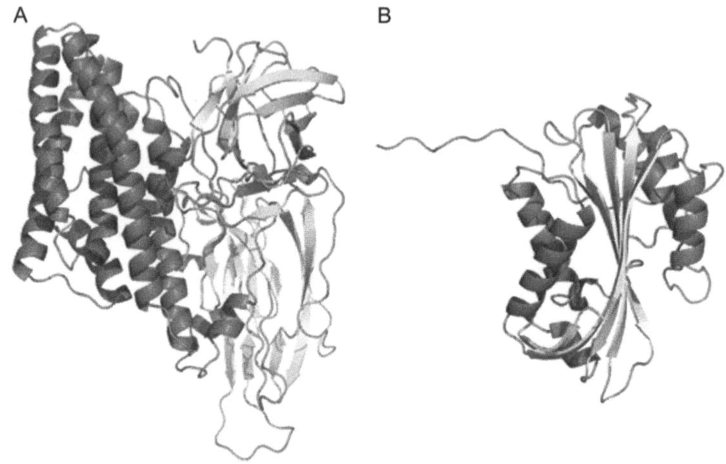

Figure 4.2 Protein crystal structures of representative *Bt* insecticidal proteins. (A) Three dimensional structure of Cry1Aa1 (PDB code: 1CIY), a three domain Cry protein. (B) Three-dimensional structure of the cytolytic crystal protein Cyt2Aa (PDB code: 1CBY). (See the colour plate.)

is comprised of three anti-parallel beta sheets. This domain shares homology with certain carbohydrate-binding proteins including vitelline and jacaline. The loops of this domain play important roles in binding insect midgut receptors. Domain 3 is a beta sandwich of two anti-parallel beta sheets. Structurally this domain is related to carbohydrate-binding domains of proteins such as glucanases, galactose oxidase, sialidase and others. This domain binds certain classes of receptor proteins and perhaps participates in pore formation. Conserved *Bt* sequence blocks two and three map near the N-terminus and C-terminus of domain 2, respectively. Hence, these conserved sequence blocks 2 and 3 are approximate boundary regions between the three functional domains. For greater detail of the structure and function relationships of these toxins, the reader is referred to Chapter 3.

Several other Cry protein structures have been determined (Table 4.3), including diverse structures for Cyt1Aa (Cohen et al., 2011), Cyt2Aa1 (Li et al., 1996; Fig. 4.2) and binary (Cry34Ab1/Cry35Ab1) proteins (see Table 4.3 for PDB accession numbers).

3.4. Cry protein mechanism of action

Cry proteins intoxicate insects by disrupting midgut epithelial tissues following oral ingestion (see Chapter 3 for greater detail). The mode of action of Cry involves pore formation. The mechanism of action, i.e. the molecular

Table 4.3 *Bt* insecticidal protein crystal structures available in the Protein Data Bank (PDB) (Website: http://www.rcsb.org/pdb/home/home.do)

Protein	PDB accession	Structure	Citation	Year
Cyt2A1	1CBY	Non-three domain	Li et al. (1996)	1996
Cry1Aa1	1CIY	Three domain	Grochulski et al. (1995)	1995
Cry3Aa1	1DLC	Three domain	Li et al. (1991)	1991
Cry2Aa	1I5P	Three domain	Morse et al. (2001)	
Cry3Bb1	1JI6	Three domain	Galitsky et al. (2001)	2001
Cry4Ba	1W99	Three domain	Boonserm et al. (2005)	2005
Cry4Aa	2C9K	Three domain	Boonserm et al. (2006)	2006
Cyt2Ba	2RCI	Non-Three domain	Cohen et al. (2008)	2008
Cry8Ea1	3EB7	Three domain	Guo et al. (2009)	2009
Cyt1Aa	3RON	Non-Three domain	Cohen et al. (2011)	2011
Cry5Ba1	4D8M	Three domain	Hui et al. (2012)	2012
Cry34Ab1	4JOX	Non-Three domain, binary with Cry35Ab1	unpublished	2014
Cry35Ab1	4JP0	Non-Three domain, binary with Cry34Ab1	unpublished	2014

events that lead to pore formation, can be summarized as follows. Cry proteins are often produced as protoxins that are first solubilized in the insect midgut and then proteolytically processed to yield smaller, activated polypeptides. The activated Cry proteins then bind to specific receptors on the surface of insect midgut epithelial cells. Receptor binding is followed by assembly of activated Cry proteins into pores that result in colloid osmotic lysis of midgut cells due to an influx of solutes from the midgut lumen. Cell lysis leads to disruption of the midgut epithelium and, ultimately, death of the insect larva. This is often considered the "classical" model for *Bt* mechanism of action (Fig. 4.3). However, many details of this model remain unresolved. Two models have been researched in recent years that propose more detailed mechanistic steps leading to insect death. These models are the sequential binding model leading to pore formation (reviewed in Soberon et al., 2009; Soberon et al., 2010) that builds on the classical pore formation model and the signalling pathway model wherein *Bt* protein

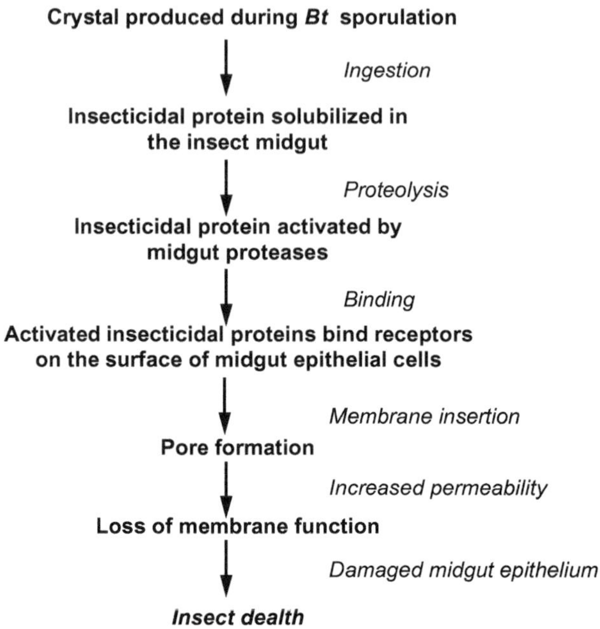

Figure 4.3 Schematic representation of the steps leading to pore formation and insect death according the "classical" model of Bt mechanism of action (Vachon et al., 2012).

binding to receptors is proposed to activate signalling pathways that lead to necrosis and cell death (Zhang et al., 2005, 2006). Critical review of *Bt* protein mechanism of action data continues to support the classical pore formation model as a sufficient description of how Cry proteins function, as the molecular events following receptor binding that lead to pore formation in insect midgut cell membranes remain poorly understood (Vachon et al., 2012).

3.5. *Bt* insecticidal protein structure and function: Cyt proteins

(The reader is referred to Chapter 3 for more detailed review of Cyt protein structure–function.) Cyt2Aa1 (Fig. 4.2) exemplifies the general fold of the Cyt group of proteins with known structures (Cohen et al., 2008, 2011; Li et al., 1996). Cyt proteins have a structure wherein two outer layers of alpha helix hairpins surround a beta sheet. Cyt proteins function through interactions with non-saturated membrane lipids including phosphatidylcholine, phospahtidylehtanolamine and sphingomylin (Ben-Dov, 2014; Thomas and Ellar, 1983). Cyt proteins are proposed to exert their insecticidal effect by formation of multimeric pores or by a less-specific detergent mechanism (Butko, 2003).

3.6. *Bt* insecticidal protein receptors

Despite the lack of a full understanding of *Bt* insecticidal protein mechanism of action, considerable information is available on the role of insect midgut receptors that bind Cry proteins (Gomez et al., 2007; Heckel et al., 2007; Likitvivatanavong et al., 2011a,b; Pigott and Ellar, 2007; Chapter 3). The demonstration of high-affinity binding sites on midgut membranes has led to the characterization of a number of functional Cry protein receptors. In Lepidoptera these membrane receptors include cadherin-like proteins, aminopeptidases (APNs), alkaline phosphatases (ALPs),and ABC transporters.

Several coleopteran midgut proteins other than cadherins have been demonstrated to function as *Bt* Cry protein receptors. These include a sodium solute symporter for Cry3Aa in *Tribolium castaneum* (Herbst) that contains cadherin repeats (Contreras et al., 2013). ADAM (A Disintegrin And Metalloprotease) was demonstrated to be a functional receptor for Cry3Aa in *L. decemlineata* (Ochoa-Campuzano et al., 2007). ADAMs belong to the metzincin subgroup of the zinc protease superfamily. ADAMs are modular transmembrane proteases implicated in the control of membrane adhesion. Cry3Aa domain 2 loop 1 was shown to be involved in ADAM recognition by competition with a synthetic peptide.

Lastly, in the nematode *Caenorhabditis elegans* (Maupas), glycolipids were identified as receptors for *Bt* Cry5Ba (Griffitts et al., 2005). *C. elegans* mutants resistant to Cry5Ba were determined to have lost glycolipid carbohydrates. It was further shown that Cry5Ba binds glycolipids and that binding is dependent on carbohydrates for toxicity *in vivo*.

3.7. Mechanisms of resistance to *Bt* insecticidal proteins

(The reader is referred to Chapter 8 for a detailed review of resistance to Bt proteins.) Field selection for insect populations resistant to *Bt* insecticidal proteins is a concern for the long-term durability of commercialized *Bt* products. As a result, significant research has been directed at characterizing *Bt*-resistant insect colonies selected in laboratory experiments to understand the genetic and molecular basis of *Bt* resistance. Resistance to *Bt* insecticidal proteins could possibly occur at any step in the mechanism of action outlined in Fig. 4.3. Among the different *Bt*-resistant insects several different mechanisms of resistance have been characterized (reviewed in Heckel et al.,

2007; Pardo-Lopez et al., 2013) including altered activation of Cry proteins by midgut proteases (Keller et al., 1996; Li et al., 2004; Oppert et al., 1997), protein sequestration by glycolipids (Ma et al., 2012) or esterases (Gunning et al., 2005), elevated immune response (Hernandez-Martinez et al., 2010; Rahman et al., 2004) or by reduced *Bt* insecticidal protein binding to insect midgut membranes.

The most common type of resistance to *Bt* insecticidal proteins, referred to as "Mode 1" resistance (Tabashnik et al., 1998), is characterized by a high level of resistance (>500-fold) to a Cry toxin, recessive inheritance and reduced Cry protein binding to insect midgut brush border membranes. Among the insect colonies resistant to *Bt* insecticidal proteins are multiple examples of reduced binding resulting from mutations in receptor molecules or reduced transcription of receptor genes (Heckel et al., 2007; Pardo-Lopez et al., 2013). Different resistant insect species are known to have receptor mutations in cadherin, APN or the ABCC2 transporter (Baxter et al., 2011; Gahan et al., 2001, 2010; Herrero et al., 2005; Jurat-Fuentes et al., 2004).

The first report of genetic linkage to cadherin-mediated resistance in Lepidoptera was in the Cry1A-resistant *H. virescens* YHD2 strain. Cadherin in this strain is interrupted by a retrotransposon resulting in high levels of resistance to Cry1Ac (Gahan et al., 2001; Jurat-Fuentes et al., 2004). The second example of cadherin-mediated resistance was in a Cry1Ac-resistant strain of the pink bollworm, *Pectinophora gossypiella* (Saunders), a pest of cotton (Morin et al., 2003). This strain harboured three mutant alleles of a cadherin encoding gene linked with resistance to *Bt* toxin Cry1Ac. The mutations all disrupted cadherin gene alleles upstream of the Cry protein binding region. In *H. armigera*, strain GYBT a deletion in a gene coding for cadherin resulted in high levels of resistance to activated Cry1Ac (Xu et al., 2005). Last, Cry1Ab-resistant sugarcane borer, *Diatraea saccharalis* (F.), with high levels of resistance to Cry1Ab, exhibited reduced levels of cadherin. RNAi was used to validate the role of cadherin in reduced susceptibility to Cry1Ab in *D. saccharalis* (Yang et al., 2011).

The first report implicating GPI-anchored APN in Cry protein resistance was in *S. frugiperda* where resistance to Cry1C correlated with a lack of APN expression (Herrero et al., 2005). These results are consistent with RNAi down regulation of *Spodoptera litura* (F.) APN, resulting in tolerance to Cry1C (Rajagopal et al., 2002). It was later demonstrated in *H. armigera* that a deletion in APN1 conferred resistance to Cry1Ac (Zhang et al., 2009). In resistant strains of the *O. nubilalis*, two mutations in the APN-P gene were identified by expressed sequence tag analysis (Khajuria et al., 2011). Lastly, Cry1Ac resistance in *T. ni* was found not to result from mutations in APN, but rather that downregulation of APN at the transcriptional level by a trans-regulatory mechanism resulted in Cry protein resistance. Down-regulation of APN was genetically linked to the Cry-resistance phenotype but was not caused by mutations in APN1.

The discovery of ABCC2 as a resistance determinant for *Bt* insecticidal proteins is more recent. In laboratory selected *H. virescens*, Cry protein resistance was genetically linked to mutant alleles of ABCC2 with a 22-base pair deletion (Gahan et al., 2010). In *P. xylostella* and *T. ni*, resistance to Cry1Ac mapped to a single homologous locus for ABCC2 (Baxter et al., 2011). Together these results suggest parallel evolutionary responses that raise questions on how ABCC2 interacts with other mechanisms of resistance to *Bt*.

4. DISCOVERY AND DEVELOPMENT OF *Bt* CROPS

4.1. The discovery and development process

The discovery and development process employed by the major developers of *Bt* crops has been the subject of recent reviews (Mumm, 2013; Privalle et al., 2012). Company websites are also a good source of information on current products and the innovation in their respective discovery and development pipelines.

Details of how each company manages its pipeline vary but all use a staging system that is similar to that illustrated in Fig. 4.4. The genetic basis for the desired trait is identified in the *Discovery* stage. In the *Proof of Concept* stage, genes are tested in plants to assess their potential to deliver the desired trait phenotype. Successful candidates are advanced to the *Early Development* stage which marks the start of the effort to produce a specific GE event for commercialization. This is also the stage in which studies are initiated that will be included in regulatory submissions to government agencies. Testing of events under more diverse environmental conditions and in more genetic backgrounds occurs in the *Advanced Development* stage with the goal of identifying a single event for commercialization. Regulatory studies are completed in this stage and regulatory packages are submitted to government agencies. In the *Pre-Launch* stage, plans are made for commercial introduction of the final product pending authorization by the relevant government regulatory agencies.

The best estimates of the time and cost of discovering and developing a *Bt* crop comes from a 2011 study conducted by the consulting firm Phillips

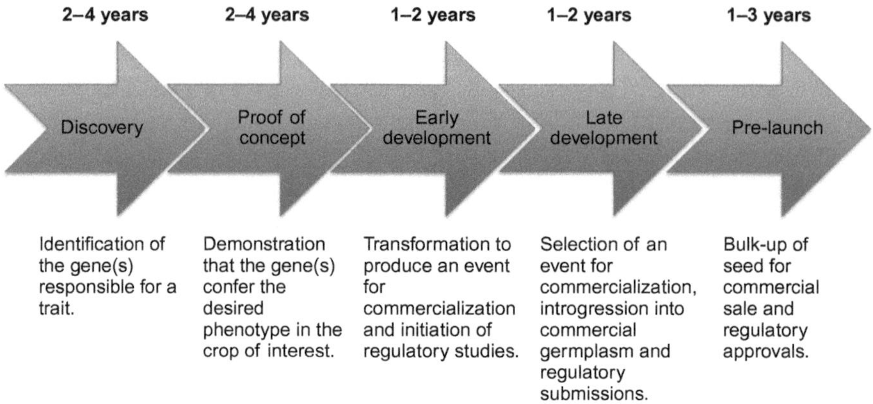

Figure 4.4 Generalized discovery and development staging system for a *Bt* crop. The websites of *Bt* crop developers are typically a good source of information on their specific discovery and development staging systems as well as the innovation that is in their pipelines. (See the colour plate.)

McDougall for the industry association CropLife International based on data provided by major developers of biotech crops (i.e. BASF Corporation, Bayer CropScience, Dow AgroSciences, DuPont/Pioneer Hi-Bred, Monsanto Company and Syngenta AG) (McDougall, 2011). For new biotech crops introduced between 2008 and 2012, the average time required to move through a pipeline from discovery to commercialization was 13.1 years at an average cost of $136 million. Discovery followed by Proof of Concept were the most expensive stages ($31 and $28 million, respectively) but the collective costs of meeting regulatory requirements was $35.1 million representing 25.8% of the total cost of bringing a biotech product to market.

4.2. Gene discovery

In the years following the isolation of the first gene coding for *Bt* Cry1Aa1 in 1981 (Schnepf and Whiteley, 1981), significant effort has been aimed at the discovery of new *Bt* strains and new genes coding for *Bt* insecticidal proteins. Today this research is driven by the wide range of potential applications of *Bt* biopesticides and *Bt* trait technology along with the rapidly increasing adoption of *Bt* crops (James, 2013).

The industrial process for *Bt* insecticidal protein gene discovery begins with the conception of a commercially important product idea to improve upon existing technology or address an unmet need for pest control. Product attributes considered important for insect control are pest spectrum, insecticidal protein potency and the low likelihood of cross resistance to other *Bt* products on the market. Lack of cross resistance is important for sustaining the utility of *Bt* products. The *Bt* discovery process often follows an insecticidal activity-driven approach beginning with *Bt* strain characterization. *Bt* strains are cultured under varying conditions and characterized for insecticidal activity on economically important insects. *Bt* strains with novel or superior insecticidal properties are chosen as a source of genes encoding insecticidal proteins.

Early *Bt* gene discovery efforts used standard biochemical fractionation and recombinant DNA technology to identify, characterize and clone the genes encoding insecticidal proteins. These techniques were based on common molecular biology methods such as DNA restriction fragment length polymorphism (RFLP) to identify novel genes in the genomes of highly active *Bt* strains (see for example Kronstad and Whiteley, 1986). Albeit a relatively time consuming and low throughput process, early *Bt* gene discovery work focused on *Bt* strains that had been well characterized for biological activity, and as a result, genes encoding commercially important proteins such as Cry1Ab, Cry1Ac, Cry1Fa, Cry2Ab and Cry3Aa were discovered. These proteins are still being used in commercial GE, *Bt* products today, although the rising pressure of field-evolved resistance to *Bt* crops expressing these and other Cry proteins is threatening the long-term utility of some *Bt* proteins in certain agricultural systems (Tabashnik et al., 2013).

Advances in molecular biology techniques and tools accelerated the rate of *Bt* insecticidal gene discovery during the 1990s. Polymerase chain reaction (PCR) represented a major advance in the ability to characterize DNA (Saiki et al., 1988). Methods to apply PCR for *Bt cry* gene identification were first reported by Carozzi et al. (1991). This led to the further development of methods for the rapid genotyping of *Bt* strains for *cry* gene content (reviewed in Porcar and Juarez-Perez, 2003). PCR-based methods can partially predict *Bt* strain insecticidal activity based on known Cry proteins, and can also detect novel cry sequence variants when coupled with multiplexed primer reactions, restriction fragment length polymorphisms or DNA sequencing of PCR amplicons. The use of *cry*-specific primers on DNA microarrays has also been used to rapidly characterize *cry* genes in native *Bt* isolates (Letowski et al., 2005).

Next-generation DNA sequencing technology represents the most recent advance in the ability to rapidly discover new genes coding for *Bt* insecticidal proteins. The affordability and high-throughput data generation of next-generation sequencing platforms promise to enable sequencing of many more *Bt* genomes in the very near future. As of March 7, 2014, there are 12 completed genomes publicly available on NCBI (http://www.ncbi.nlm.nih.gov/genome/genomes/486). The challenges of data analysis and identification of genes encoding new *Bt* insecticidal proteins are being addressed (Ye et al., 2012). Recently, a pangenomic study of *Bt* was reported (Fang et al., 2011) in which chromosomes and plasmids encoding Cry proteins were sequenced to a high degree of coverage for seven *Bt* strains. The pangenomic approach, which does not intend to assemble all genomes to completion, but rather to interrogate sequence space across multiple *Bt* strains, coupled with advances in the scale and throughput of insect bioassays, represents a powerful approach to identify genes encoding novel *Bt* insecticidal proteins. One can surmise that the growing number of new protein sequences in the *Bt* nomenclature database are at least in part due to the impact of next-generation sequencing (Fig. 4.1).

4.3. First demonstrated success of *Bt* Cry GE plants

While *Bt* biopesticides are environmentally safe, disadvantages of the technology include the relatively short timeframe of effectiveness under environmental conditions, the need for repeated application over time, and the inability to impact insects with specialized feeding behaviour such as those that feed on plant sap or below ground on plant roots. The development of biolistic and *Agrobacterium*-mediated plant transformation technology created the possibility to deliver and express *Bt* genes encoding insect control proteins within the plant for the duration of the plant growth cycle. Early attempts to express *Bt cry* genes in plants resulted in poor expression of Cry1A proteins and yet plant tolerance to insect feeding was achieved. Fischhoff et al. (1987) transformed tomato with truncated Cry1Ac resulting GE plants resistant to *H. zea* feeding damage. Vaeck et al. (1987) transformed tobacco with Cry1A resulting in GE plants resistant to *Manduca sexta* (L.). Because Cry proteins were expressed to very high levels in *Bt* strains, attention was given to the differences in gene structure and codon usage between *Bt* and the target host plants as a cause for low Cry protein expression observed in GE plants. Perlak et al. (1991) were successful in increasing Cry protein levels up to 100-fold by creating synthetic Cry-encoding transgenes with a codon usage biased toward that favoured by plants and lacking mRNA destabilizing sequences such as polyadenylation signal sequences, ATTTA sequences and A+T rich regions. Partially modified or fully modified transgenes encoding Cry1Ab or Cry1Ac resulted in higher expression and a higher proportion of GE tobacco or tomato plants tolerant to *M. sexta* damage. Similar success was achieved with a modified Cry3Aa gene expressed in GE potato that was resistant to *L. decemlineata* (Adang et al., 1993). Further gene expression improvements have been realized by increasing promoter strength, better polyadenylation termination and enhanced expression by including introns in the 5′ untranslated region of the mRNA (Koziel et al., 1993). These early successes in generating GE plants expressing *Bt* proteins set the stage for an industry-wide trend among seed producers to produce GE *Bt* crops.

4.4. Transformation technologies

The ability to successfully transform a plant depends on several factors including the availability of target tissues that are competent for propagation or regeneration, an efficient method for delivery of DNA, the ability to select for transformed cells and the ability to recover fertile, GE plants (Hansen and Wright, 1999). Many different plant tissue types are amenable to transformation including immature embryos, embryogenic suspension cultures, embryogenic shoot tips, immature cotyledonary-nodes, hypocotyls and leaf tissue (Lee et al., 2013). The selection of a tissue type for use in a transformation system depends on many factors including simplicity and accessibility (e.g. free from patent restrictions), but in the end it is critical that fertile, GE plants are produced.

Agrobacterium-mediated transformation and particle bombardment are the two most commonly used methods of DNA delivery. *Agrobacterium*-mediated transformation uses the gene-transfer machinery of the bacterium to introduce a specific piece of DNA (i.e. T-DNA) into the host cell which ultimately integrates into the genome. *Agrobacterium*-mediated transformation can be used to deliver DNA to both dicots and monocots, can deliver relatively large pieces of DNA, and typically a small number of T-DNA copies are integrated into the host genome at a single location in the genome (Hansen and Wright, 1999; Smith and Hood, 1995). (Note: the unique integration of DNA into the host genome is called an *event*.)

Particle bombardment and other physical delivery approaches do not rely on a biological mechanism for the delivery of DNA. Instead, particles of various materials are coated with DNA and physically introduced into target cells. The particles used for DNA introduction are typically gold or tungsten but silicon fibre "whiskers" have also been used (Hansen and Wright, 1999; Petolino and Arnold, 2009). Unlike *Agrobacterium*-mediated transformation, particle bombardment often creates complex events containing multiple copies and/or fragments of DNA and insertion of the DNA into multiple genomic regions (Finer and Dhillon, 2007).

Selectable marker genes (genes allowing transformed cells, tissues or plants to be differentiated from non-transformed ones) are an important component of plant transformation systems. Positive selectable marker genes are most commonly used in the production of GE crops and include antibiotic resistance (e.g. the *nptII* gene which confers resistance to the antibiotics kanamycin and neomycin), herbicide tolerance (e.g. the *pat* gene which confers tolerance to the herbicide glufosinate) or other genes (e.g. the *pmi* gene which enables plants to use mannose as a carbon source in tissue culture systems) which enhance survival of plant cells containing and expressing them (Rosellini, 2012). In some cases, plants are transformed with the selectable marker on a separate piece of DNA so that the plants produced in these systems have the genes of interest and the selectable marker genes integrated as two-independent events. In these cases, the selectable marker can be removed from the commercial product through traditional breeding processes. However, in most cases, the selectable marker gene is integrated with the genes of interest and is therefore contained in the commercialized event. In cases where the selectable marker gene confers herbicide tolerance, its presence in the commercial product is desirable.

4.5. Introgression and testing

Germplasm that is amenable to the transformation and tissue culture regeneration process is typically not the high-performing germplasm used in today's intensive production agriculture. It is necessary to introduce a GE event into elite germplasm via process of breeding and selection. The use of molecular markers can dramatically enhance the speed and effectiveness of this introduction by minimizing the transfer of alleles from the GE donor line and maximizing the recovery of alleles from the elite germplasm (Mumm, 2013). Throughout the process of introgression, a *Bt* event is evaluated in increasingly diverse germplasm and environments for performance of the IR trait and the germplasm.

5. REGULATION

GE crops undergo comprehensive regulatory reviews for human health and environmental safety by agencies throughout the world. Indeed, GE crops and food receive far greater regulatory and scientific scrutiny than any conventional counterpart (Fedoroff, 2011). While most regulatory authorities profess to regulate the products of genetic engineering rather than the process itself, oversight is usually triggered by the employment of recombinant DNA techniques to introduce new traits into a crop. Commercial launch of a GE crop requires authorization for commercial cultivation in the country (or countries) of production, and import, food and feed approvals in that country's trading partners.

When evaluating the safety of GE crops, regulatory systems cover two broad areas of consideration. First, regulators examine the potential for harm arising from the intended direct effect(s) of the genetic modification, in the present discussion being the expression of the *Bt* proteins that provide protection from targeted insect pest feeding damage. This assessment generally considers environmental effects such as toxicity to beneficial organisms (e.g. predators, parasitoids and pollinators) that feed in or on the *Bt* crop and soil fauna and flora. The assessment of direct effects also includes assessment of the safety of the *Bt* protein in food and feed, including toxicity and potential to be an allergen. Second, regulators examine the crops and food for any potential unintended effects on human health or the environment arising from the genetic transformation itself or unintended indirect effects of the added gene(s) and trait(s). The potential for unintended effects is considered to arise from effects of the transformation process itself on the crop genome, such as gene disruption in the region where the transgene is inserted, and from pleiotropic effects of the transgenic protein(s) on plant metabolic processes. Either of these could lead theoretically to altered crop composition or agronomic properties.

When developing the environmental, food and feed safety profile of a GE crop, developers must demonstrate to the regulatory agencies that the GE crop does not have any new or altered risks relative to its non-GE counterpart in respects other than those that derive directly from the action of the inserted gene(s) and trait(s) (Codex 2008). In the case of an insect-protection event, the developers must show that the crop is compositionally and agronomically similar to its non-GE counterparts in the absence of the target pest(s) and that the only differences observed are related to the action of the trait to reduce pest injury. With the demonstration that the GE and non-GE counterparts are compositionally and agronomically equivalent with no harmful unintentional changes, the risk assessment conducted by the regulator can focus on the specific trait(s) added.

Table 4.4 lists the types of information that are generally provided to regulators in support of the safety assessment of a GE insect-protected crop.

Table 4.4 List of studies typically conducted in support of safety assessments for human health and the environment of GE crops

Test material	Study type	Food/feed/ import	Environmental/ cultivation
Crop	Molecular characterization	✓	✓
Crop	Inheritance	✓	✓
Crop/protein	Detection methods (ELISA, PCR)	✓	✓
Crop	Protein expression	✓ (Grain, seed after harvest)	✓ (Leaf, pollen, stalk, root at various growth stages)
Crop	Field efficacy		✓
Crop	Compositional analysis	✓	
Crop	Agronomic properties		✓
Protein	Acute oral toxicity (mouse gavage)	✓	
Protein	Homology to known toxins	✓	
Protein	Protein biochemistry (digestive stability, thermolability as indicators of potential allergenicity)	✓	
Protein	Homology to known allergens	✓	
Crop	Animal feeding with grain	✓	✓
Protein	Soil degradation		✓
Protein	Spectrum of Cry protein activity		✓
Protein	Non-target organism hazard testing		✓
Crop	Field non-target organism surveys		✓ (If exposure and hazard data suggest potential effects under field conditions)
Protein	Endangered and threatened species assessment		✓
Crop	Weediness potential		✓
Protein	Potential effects of gene flow (if wild relatives present in area of proposed release)		✓

5.1. Product identification and characterization

Studies are conducted to characterize the transformation event being developed for commercialization. These studies include an analysis of the genetic insert to ensure that the intended genetic elements are present and intact and that unintended elements (like the backbone DNA sequence of a plasmid for *Agrobacterium*-mediated transformation) are absent. Production of mRNA and the gene product are also characterized. Studies of the inheritance of the transgene across generations ensure that it is inherited in the expected manner (typically Mendelian segregation). Both DNA and protein detection methods are developed to enable identification of plants and plant tissues containing the transgene(s) (*Bt* gene, selectable marker gene) and accompanying regulatory elements. Finally, the efficacy of the product under field conditions is characterized to ensure that the intended phenotype is present.

5.2. Human health assessment

The human health assessment of insect-protected GE crops includes characterization of both the introduced protein and the food/feed derived from the crop, including where appropriate processed products. The protein safety assessment includes information on the source of the protein and history of safe exposure of the protein in its natural state (including toxicity and allergenicity) and its insecticidal mode of action in the target pest. In the case of insecticidal proteins derived from *Bt* strains, there is a considerable body of evidence of safe history, dating back to the organisms discovery a hundred years ago and its development as a biological insecticide over 60 years ago (Sanahuja et al., 2011). *Bt* is a very common soil and phylloplane microorganism to which humans and animals have always been exposed with no known adverse effects. Furthermore, the mechanism of action of the insecticidal proteins has been characterized to involve binding to specific receptors in the midgut of sensitive insects, receptors that are not present in mammalian digestive tracts. Indeed, *Bt* proteins are generally very selective in their toxicity to specific orders of insects or insects within a specific order even though other insects within or outside that order may also have related receptor proteins. Additionally, *Bt* proteins are rapidly degraded by digestive enzymes and the acidic condition of human stomachs (Mendelsohn et al., 2003).

Bioinformatic approaches are used to investigate any amino acid sequence homology to known toxins or allergens. The assessment of the allergenic potential of the protein considers not only sequence homology to known allergens, but also biochemical properties, such as digestibility and heat lability that may be characteristics of some allergens, to understand if the introduced protein may be a novel allergen. If there is significant homology to a known allergen, additional testing can be conducted to understand whether the introduced protein may elicit the allergenic response in sensitive individuals (Ladics et al., 2011). *Bt* is not known to be a source of allergic responses despite its ubiquity, and therefore these proteins have very low potential to be allergens.

Finally, acute oral toxicity of the novel protein is assessed through gavage with a large quantity of the protein in a model organism, usually a mouse. Through all these tests, only proteins with no evidence of toxicity or allergenicity are developed for use in GE crops.

To complete the dataset required for the human health risk assessment for the GE protein, data are provided on the expression levels of the proteins in the harvested grain. For crops where the consumption is of processed products, further analysis of protein levels after processing can be provided (Hammond and Jez, 2011). Since the expressed proteins are not associated with any hazard to human health, the expression data provide additional assurance that there will be no harmful effects when the food is consumed.

In addition to information on the human health and food safety of the insecticidal proteins, regulators also review information on unintended effects to the crop of the transformation itself. Extensive data are provided on the nutritional profile of the GE crop, its nearest non-GE isoline and a broader set of varieties of the crop grown under diverse agricultural conditions. This compositional analysis includes quantification of lipids (including the fatty-acid profile), proteins (including the amino acid profile), carbohydrates, vitamins, minerals and anti-nutrients. Feeding studies using the grain in rapidly growing animals such as broiler chickens and, in some cases, rats provide further information on the food safety and nutritional value of the crop.

Most of the required regulatory data generated for the human health and food safety assessments address extremely remote risks that are no greater than for any variety of a crop developed through conventional breeding and crop improvement techniques (Herman and Price, 2013). For example, the probability of introducing to the human diet a novel allergen is extremely low given the very small proportion of proteins that are allergens and that these relatively few proteins are clustered in a small number of protein families (Radauer et al., 2008). Similarly, in more than 140 studies of the composition of GE crops, not a single crop has shown evidence of a harmful change in nutritional value or anti-nuturients (Herman and Price, 2013). The variation in composition of crops developed through conventional crop improvement techniques is many times greater, due to the introduction of multiple new alleles and genes, most of which are uncharacterized, using conventional techniques, compared with the one well-characterized-specific intended change introduced through genetic engineering (DiLeo et al., 2014; Herman and Price, 2013; Herman et al., 2009; Ricroch, 2013; Ricroch et al., 2011).

5.3. Environmental effects

While environmental regulations and frameworks differ among countries according to their local laws and environmental protection goals, regulatory requirements relating to environmental release of GE crops tend to be similar across countries that permit commercial cultivation of these crops. Regulators seek to ensure that the environmental effects of a GE cropping system are not more harmful to the environment than the conventional cropping system that they would supplant. Under some regulatory regimes, agencies also consider the economic, human health and environmental benefits of the technology.

As with the human health and food safety studies, environmental safety studies cover both the direct effects of the GE protein itself and the effects of any unintended changes to the crop. For insect-protected GE crops, the data generated include sensitivity of representative non-target organisms that may occur in or around agricultural production fields, focusing on beneficial species such as predators, parasitoids and pollinators. Such studies may also include charismatic species, such as monarch butterflies. Hazard testing for non-target organisms resulting from exposure to a new transgene product is often accomplished following the tiered-testing paradigm (Romeis et al., 2011). Under this approach, the non-target organism of concern (or a surrogate that is functionally or phylogenetically similar to the organism of concern), is tested in a bioassay with the purified transgene product at a concentration many fold higher than the highest estimated exposure in the field (Tier 1). If the test population is not affected at this high concentration, or if the effects are moderate (for example, less than 50% mortality) then there is a high likelihood that exposure to the transgene product will not have significant effects under field conditions. If, however, effects are seen at this high concentration, further bioassays are conducted using more realistic exposure levels, perhaps using the tissues from the GE plant rather than purified transgene product (Tier 2). Again, if the test population is not affected at realistic exposure levels, of if effects observed would be acceptable (for example less than would occur with alternative pest control tools), additional testing is not warranted. If significant effects are seen in Tier 2, additional testing can be conducted with whole plants in a green house or field cages (Tier 3). Such tests allow more realistic spatial processes to function that may more accurately reflect actual exposure under field conditions. Finally, if these lower tier studies indicate potential for unacceptable harm, a field study may be warranted whereby natural populations are monitored under the same conditions as the proposed environmental release (Tier 4). Progressing through the tiers increases the ecological relevance of the study to the actual proposed release but decreases the ability to detect effects due to greater variability in the test system. Methods or guidance for testing many non-target organisms at several of the tiers are available in published literature (e.g. Romeis et al., 2011) or from regulatory agencies (e.g. U.S. Environmental Protection Agency, 2007).

Complementing such hazard testing, exposure analysis is accomplished by measuring the expression of the GE protein in representative tissues of the crop that are fed upon by herbivores. This can include leaf, stalk, pollen, flowers and fruits, depending on the tissues that are consumed. Expression is measured at several time points in the life cycle of the plant to provide a comprehensive assessment of the potential exposure of non-target organisms. Data are also generated on the environmental fate of the GE protein, typically examining the rate of degradation of the protein in agricultural soils (Shan, 2011).

It is reasonable to expect that some non-target species may be sensitive to the GE protein, especially those that are phylogenetically related to the target pest species. For example, larvae of the monarch butterfly and some other Lepidoptera are known to be sensitive to *Bt* proteins in the Cry1 class, which are targeted at lepidopteran pests. Similarly, larvae of certain Chrysomelidae are known to be sensitive to *Bt* proteins in the Cry3 class. The risk to such organisms is characterized by integrating their estimated sensitivity to high end estimate of exposure levels. Usually, conservative assumptions are made that over-estimate the sensitivity and over-estimate the exposure. If this characterization indicates that there is not a very low likelihood of a harmful effect to the population of the non-target organism, field studies may be warranted to investigate whether the estimated effects actually occur under field conditions.

When conducting non-target organism studies with GE crops, it is important to use relevant comparators to understand the significance of any effect. Typically the comparators are the nearest non-GE isoline that is managed in accordance with conventional pest management practices. Additional comparators may include the isoline that is not treated with insecticides (providing a worst-case evaluation of the effects of the GE crop) and additional varieties of the crop that are typically grown. These provide estimates of the typical differences among varieties of a crop and therefore the context to assess the biological significance of any effects measured with the GE crop.

In addition to non-target organisms, environmental assessments of GE crops include assessment of its agronomic properties when grown according to normal agricultural practices. Such properties may include the growth habit of the crop (e.g. time to flower, crop height and yield), observations of susceptibility to pests (other than the targeted pests), diseases and other environmental stressors. Such data are interpreted for any indications that the transformation may have increased the potential of the crop to become a weed for example through increased persistence, ecological competitiveness or ability to spread outside of agricultural areas (Raybould et al., 2012).

For crop species that are grown in the same area as sexually compatible wild relatives, regulators will typically consider whether gene flow from the crop to the wild relatives may occur, and what the consequences of such gene flow could be on the population of the wild relative. In the case of insect protection traits, the assessment might include an assessment of whether addition of the trait may reduce feeding on the wild relatives by insects such that plant becomes more invasive. Generally, however, insect feeding is not an important limiter of wild plant populations, and the addition of an insecticidal trait would not have a biologically significant effect. Furthermore, hybrids of wild plant populations and crops are generally less fit than native wild plants due to the agronomic properties of the crop that have been bred for generations to make them suitable for cultivation and harvest.

In some regulatory systems, it is necessary to perform environmental risk assessments for GE crops when the requested approval is not for commercial environmental release, but instead for importation of grain that is for food and feed use. In these situations, the assessment specifically considers the potential for inadvertent environmental release of the GE crop. Because in these situations there is very low exposure potential, a conclusion about acceptability of risk may usually be reached with very limited environmental data.

5.4. Considerations for stacks

Traditional breeding, which combines characterized and uncharacterized traits, has generated products with a long history of safe use. Human and animal diets have always included multiple food combinations, with no documented adverse health effects from interactions. Different crops are grown in adjacent spaces, and crop varieties have been crossed to generate new genomic combinations, with the recognized principle that combination is not inherently unsafe.

There is no reason to expect GE traits or genes to interact in a different manner compared with native traits or genes. Combined event products ("breeding stacks") contain two or more biotechnology-derived events combined through conventional breeding. Where the individual events have been determined to be as safe as the conventional counterparts and no trait interaction is expected, the combined event can be considered equally safe as food and feed (although many regulatory frameworks require confirmatory data for example on efficacy or crop composition). When seeking cultivation approval for a combination of two or more *Bt* proteins, additional information on impacts to target and non-target species may be required by the cultivating country. Where no trait interaction is anticipated, analysis of existing data from individual events can be used to assess the effects of the combined event on target and non-target species. If laboratory tests indicate trait interaction (e.g. synergism or antagonism), or an interaction is expected, additional testing of the protein combination may be warranted, similar to the non-target organism data generated for single events discussed above. Combined insecticidal events may also be subject to product-specific oversight relating to IRM.

5.5. Continued regulatory oversight of commercialized GE events

Upon completion of regulatory review in the country or countries where a GE crop is to be cultivated, regulatory agencies will issue a decision on its permissibility for unconfined environmental released. With similar approvals from any countries that typically import the crop, GE crops may be commercially released. Depending on the regulatory framework and agencies involved, the decision to permit commercialization can take different forms. For example, when the USDA "deregulates" a product, the regulators have no further oversight of the product. On the other hand, EPA, which registers the *Bt* proteins expressed by GE crops, continues regulatory oversight. The EPA may require on-going studies on the environmental effects of a *Bt* crop when grown on a commercial scale. Such studies generally are confirmatory in nature, providing additional data on exposure and effects of the *Bt* proteins. The European Food Safety Authority requires technology providers to conduct on-going general surveillance for changes in the agricultural ecosystem that may be attributable to the release of GE crop.

However, such post-market monitoring is rarely scientifically justified. The regulatory risk assessment prior to launch is in most cases sufficiently thorough that unanticipated effects are known not to occur. General surveillance is not hypothesis-driven, and collection of environmental data provides no information as to the cause of any changes, and whether such changes are harmful or undesirable. Without a testable hypothesis, general surveillance has little utility and is unlikely to identify environmental effects resulting from the GE crop. Post market monitoring (PMM) is only warranted when pre-market risk assessment identifies potentially unacceptable risks, and these risks can only be tested using large scale studies. In these rare instances, post market monitoring can help determine actual levels of harm and the efficacy of mitigation measures under the field conditions reflective of commercialization. For additional information on PMM and policy considerations, see FAO Expert Consultation on Genetically Modified Organisms in Crop Production and Their Effects on the Environment (2005).

Several regulatory agencies around the world require the technology provider to implement resistance management programs that are designed to slow the adaptation of target pest populations to GE *Bt* crops thereby extending their utility and their benefits to the environment. Even where these programs are not required, technology providers nevertheless will implement measures to protect the durability of the products (Head and Greenplate, 2012; MacIntosh, 2010).

6. INSECT RESISTANCE MANAGEMENT

The potential for targeted insect populations to evolve resistance to *Bt* crops was recognized prior to when the first commercial crops were released (Alstad and Andow, 1995; Gould, 1998; Roush and Shelton, 1997; Tabashnik et al., 1990). These concerns have led to the development of proactive resistance management programs that are designed to delay the onset of resistance and slow its spread. Today, such programs are in place for all *Bt* crops in all geographies where they are grown. Resistance management programs are focused on the primary pest species that are of greatest importance to the continued value of the *Bt* crop.

The primary tactic to delay resistance is the use of refuges, or host plants that do not contain *Bt* genes and allow the persistence of susceptible pests. Susceptible individuals are thereby able to mate with any resistant individuals that emerge from the *Bt* crops and maintain susceptible alleles in the population.

To be fully effective in delaying resistance development in a field population, refuges should produce sufficient insects to overwhelm any resistant insects—a ratio of 500 susceptible to 1 resistant has been used as a rule of thumb (U.S. Environmental Protection Agency, 2001b). The refuge should be in sufficiently close proximity to *Bt* fields that normal insect dispersal will promote mating between refuge-produced and *Bt*-produced insects. Adult insect emergence from the refuge should occur at the same time as emergence of resistant insects from *Bt* crops.

Different forms of refuge are used in resistance management programs. Natural refuges can be composed of crop or non-crop host plants, often of different species from the *Bt* crop, but nevertheless of sufficient abundance, proximity and temporal overlap to promote mating of susceptible insects with resistant insects from the *Bt* crop. Natural refuges consisting of crop and non-crop hosts of *H. virescens* and *H. zea* provide the refuge for *Bt* cotton in the south and southeastern United States (Gould et al., 2002; Gustafson et al., 2006; Jackson et al., 2004) and for *H. armigera* in China (Qiao et al., 2010). Structured refuges are specifically grown in association with *Bt* crops, and consist of non-*Bt* varieties, usually of the same species as the *Bt* crop. The recommended amount and layout of the refuge vary by pest species and crop. For example, for *O. nubilalis* in the U.S. Corn Belt and single-gene *Bt* maize hybrids, non-*Bt* maize must be on an area that is at least 20% of the area of the *Bt* crop and the refuge must be planted within ½ mile (~800 m) of each *Bt* field (U.S. Environmental Protection Agency, 2001a). For *D. saccharalis* in the Argentina corn belt and single-gene *Bt* maize, the recommendation is for 10% refuge within 800 m of the *Bt* maize field. For WCR in the U.S. Corn Belt and pyramided *Bt* maize, 5% refuge is required which must be planted within or adjacent to the *Bt* maize field. Recently, refuge provided as seed blends with *Bt* seeds that produce two or more *Bt* proteins against each key target pest have been released to simplify the refuge planting and management by growers and to ensure that the required refuge is present (Onstad et al., 2011).

Population genetics theory and simulation models indicate that refuges are extremely effective for *Bt* crops that provide a "high dose" against the key target pest(s) and when resistance alleles are initially rare (Alstad and Andow, 1995; Gould, 1998). In these cases, the *Bt* crop kills nearly all susceptible larvae and 95% or more of larvae that are heterozygous for resistance alleles. When resistance alleles are rare, most of the resistance alleles are carried by heterozygotes and so removed by the high-dose *Bt* crop, greatly delaying resistance.

An alternative (or additional) strategy to the high dose to remove heterozygous insects is to pyramid more than one *Bt* protein active against the same target pest. If each protein differs in their target insect midgut receptors, one protein can kill insects that are heterozygous or homozygous for resistance alleles to the other protein. This provides dramatic delays in resistance development (Storer et al., 2012c) provided that resistance is not already developing to one of the component proteins (Tabashnik and Gould, 2012). Understanding the receptors involved in the mode of action of each *Bt* protein, or at least understanding differences in binding sites as well as other direct or indirect indicators of cross-resistance potential, therefore can be important in designing appropriate resistance management programs. This also applies in the situation where crops containing different *Bt* proteins active against the same pest are deployed in the same agricultural environment, a situation which did not apply when the first *Bt* crops were developed but is now the norm.

The design of refuge-based resistance management must balance the biological risks of resistance (which depend on the properties of the *Bt* crop, the adoption of the product by farmers, the genetics of resistance and the ecological interactions between the target pests and their host crops) with economic and practical realities of crop production. Resistance management programs are intended to delay but not prevent resistance, and the length of the delay sought must also reflect the continued development of new pest management tools including GE crops producing novel insecticidal mechanisms. Refuges, to be effective, must allow survival and development of pest insects. These insects cause yield loss and economic costs. For example, Marra et al. (2012) estimated that for every 1% decline in expected maize yield in the United States, maize prices are expected to increase 4.2%. Practical considerations also have to be taken into account (MacIntosh, 2010). The larger the refuge required, the smaller the benefit to growers using the technology. It should be expected that a larger refuge would lower grower acceptance of the product, and for those growers who do plant it, compliance with the refuge is likely to be lower. Grower compliance with requirements for 50% refuge for single-gene *Bt* maize in the southern United States (cotton-growing region) (U.S. Environmental Protection Agency, 2011) is much lower than for a 20% refuge in the northern United States. There the intended durability benefits of larger refuges may not be fully realized. Most models indicate that blended refuge with pyramided trait products, while less durable than a separate refuge with which there is 100% compliance, provides superior durability compared with a larger structured refuge with single trait products (Carroll et al., 2012; Ives et al., 2011).

7. *Bt* CROPS—A SNAPSHOT OF TODAY

7.1. Commercialized *Bt* proteins

GE corn and cotton comprise the majority of *Bt* crops cultivated in the Americas. The *Bt* proteins in current GE products target lepidopteran pests of corn, cotton, or soy and coleopteran pests of corn. GE potato expressing Cry3Aa was developed for control of *L. decemlineata* but is not marketed today.

A relatively small number of thc known *Bt* insecticidal proteins have been developed for insect resistant GE crops. The Cry proteins developed for control of foliage-feeding Lepidoptera include several well characterized or modified three domain Cry proteins. These are Cry1Ab, Cry1Ac, Cry1Fa2, Cry2Ab, Cry2Ae and Cry1A.105. Vip3Aa has also been developed for control of lepidopteran pests in both corn and cotton.

7.1.1 Cry1Ab

Cry1Ab is one of the most well studied 3-domain Cry proteins. The *cry1Ab1* gene and encoded protein were first cloned and characterized by Wabiko et al. (1986) from *Bt* subspecies *Berliner*. Cry1Ab is produced as a protoxin that is activated by proteases in the insect midgut and functions by the classical pore formation mechanism of action described in Section 3.4. Cry1Ab is noted for broad spectrum lepidopteran activity that includes economically important stalk boring pests such as *Ostrinia* spp. and *Diatraea* spp. (van Frankenhuyzen, 2009). Much is known regarding functional aspects and mechanisms of IR to Cry1Ab and cross-resistance with other Cry proteins (Heckel et al., 2007; Schnepf et al., 1998).

7.1.2 Cry1Ac

Cry1Ac from *Bt kurstaki* strain HD73 was first described by Adang et al. (1985). Cry1Ac is another example of a classical three domain insecticidal protein (Derbyshire et al. unpublished; PDB 4ARX, 4ARY) with a long history of study (Schnepf et al., 1998). Cry1Ac is one of the most potent Cry proteins against a range of Lepidoptera: Noctuidae, resulting in widespread deployment in GE crops, especially cotton. Cry1Ac was recently approved for commercialization in soybean in Latin America for control of lepidopteran pests including velvetbean caterpillar, *Anticarsia gemmatalis* Hübner and soybean looper, *Chrysodeixis includens* (Walker) (http://www.ctnbio.gov.br/index.php/content/view/15558.html).

7.1.3 Cry1Fa2

Cry1Fa2 was discovered by Payne and Sick (1993). Cry1Fa2 displays sequence characteristics of classical three domain Cry proteins. The lepidopteran pest spectrum of Cry1Fa2 is quite broad and includes high potency against *Spodoptera* spp. (van Frankenhuyzen, 2009), an attribute that makes Cry1Fa2 valuable in combination with Cry1A proteins with low potency on *Spodoptera* spp.

7.1.4 Cry1A.105

Developed by Monsanto Co., Cry1A.105 is a chimeric protein comprising parts of four domains from other Cry proteins previously used in GE plants. The amino acid sequences of domains 1 and 2 are identical with the respective domains from Cry1Ab and Cry1Ac proteins, domain 3 is almost identical to the Cry1F protein, and the C-terminal domain is identical to the Cry1Ac protein (http://cera-gmc.org/index.php?action=gm_crop_database&mode=ShowProd&data=MON89034). As a result, the Cry1A.105 chimeric protein combines most of the insecticidal properties displayed by the Cry1A and Cry1F proteins.

7.1.5 Cry2Ab

Cry2Ab is a three domain protein that diverges in primary sequence and predicted structure from Cry1A proteins. Cry2Ab was the first dual spectrum Cry protein described with activity against both lepidopteran and dipteran insects (Widner and Whiteley, 1989). Cry2Ab is valued for broad spectrum lepidopteran activity and the ability to control resistant insect populations (summarized in Schnepf et al., 1998). Based on lack of shared midgut binding sites (Hernandez-Rodriguez et al., 2008), Cry2Ab proteins are candidates for pyramiding with Cry1 proteins in insect resistant crops. Cry2Ab is often targeted to chloroplasts in order to increase expression levels and reduce negative plant phenotypes in GE plants (Corbin and Romano, 2006).

7.1.6 Cry2Ae

Cry2Ae is related in sequence to Cry2Ab. Cry2Ae represents a different mechanism of action from Cry1 proteins based on lack of shared binding sites (Caccia et al., 2010; Gouffon et al., 2011). Cry2Ae, like Cry2Ab, is a candidate for combining with Cry1 proteins in pyramids for resistance management.

7.1.7 Vip3Aa

Vip3Aa is a soluble insecticidal protein expressed during the vegetative growth phase of *Bt* growth, prior to sporulation (Estruch et al., 1996). Vip3Aa is broadly active on lepidopteran pests of corn and cotton. Vip3Aa is different in sequence and mechanism of action compared with Cry1 or Cry2A proteins (Lee et al., 2006). As a result, Vip3Aa is useful in combination with Cry1 or Cry2 proteins to slow the development of resistant insect populations.

Proteins developed in GE corn for control of WCR are mCry3Aa, eCry3A.1ab, Cry3Bb and Cry34Ab1/35Ab1. These proteins and other GE approaches for the control of WCR were recently reviewed by Narva et al. (2013).

7.1.8 mCry3Aa (modified Cry3Aa)

Cry3Aa, the first reported coleopteran-active *Bt* insecticidal protein (Herrnstadt et al., 1987; Sekar et al., 1987) is ineffective against *Diabrotica* spp. (Herrnstadt et al., 1987; MacIntosh et al., 1990; Slaney et al., 1992). To improve activity on WCR, Walters et al. (2008) engineered Cry3Aa to contain a chymotrypsin/cathespsin G protease site at a location in domain 1 known to enhance Cry3Aa proteolytic activation (Carroll et al., 1989, 1997). This modification enhanced Cry3Aa cleavage in WCR and improved binding to WCR midgut membranes. The mCry3Aa insecticidal activity against WCR larvae was superior to native Cry3Aa.

7.1.9 eCry3.1Ab

The eCry3.1Ab protein is a hybrid resulting from exchange of the domain 3 variable region from a lepidopteran-active toxin, Cry1Ab with domain 3 from Cry3Aa (Walters et al., 2010). The resulting protein, eCry3.1Ab, has higher activity against WCR than Cry3Aa. Another interesting feature of Cry3A.1Ab is that it binds WCR midgut brush border membrane vesicles

(BBMVs) at sites independent of mCry3Aa binding. The lack of competitive binding between mCry3A and eCry3.1Ab suggests that these proteins might act by independent mechanisms on WCR midguts.

7.1.10 Cry3Bb1

Cry3Bb1 is active on coleopteran pests such as *L. decemlineata* (Donovan et al., 1992). Cry3Bb1 is a three domain *Bt* protein with structural similarity to many other Cry proteins (Galitsky et al., 2001) and functions by forming ion channels in membranes (Von Tersch et al., 1994). The version of Cry3Bb1 expressed in events MON863 and MON88017 from Monsanto is a modified protein with six amino acid residue changes compared with the native sequence (Vaughn et al., 2005).

7.1.11 Cry34Ab1/Cry35Ab1

The second major class of *Bt* toxins developed for protection of WCR injury in maize are the binary *Bt* crystal proteins Cry34Ab1 and Cry35Ab1 (Cry34Ab1/Cry35Ab1) that function together as oral intoxicants of WCR larvae (Ellis et al., 2002; Herman et al., 2002). Cry34Ab1/Cry35Ab1 are structurally different from the Cry3-type proteins described above. Cry34Ab1 is one example of a family of 14 kDa proteins that have no protein sequence homology beyond the *Bt* Cry34 group (Schnepf et al., 2005), whereas Cry35Ab1 is a member of a family of 44 kDa *Bt* proteins that share low sequence homology to *Bt* Cry36Aa1, *Bacillus sphaericus* mosquitocidal binary proteins BinA and BinB and *B. sphaericus* mosquitocidal binary protein, Cry49Aa1 (Jones et al., 2007). Crystal structures for Cry34Ab1 and Cry35Ab1 were recently solved (Cry34Ab1 PDB accession 4JOX; Cry35Ab1 PDB accession 4JP0; Kelker et al., 2014). Cry34Ab1/Cry35Ab1 appears to function by disrupting the WCR midgut epithelium. Consistent with this observation, Cry34Ab1/Cry35Ab1 formed ion channels in artificial lipid membranes (Masson et al., 2004). Li et al. (2013) recently demonstrated Cry34Ab1/Cry35Ab1-specific binding to WCR midgut BBMV and a lack of competitive binding between Cry34Ab1/Cry35Ab1 and the coleopteran-active proteins Cry3Aa, Cry6Aa and Cry8Ba. Lastly, Cry34Ab1/Cry35Ab1 are not cross resistant with Cry3Bb1 insect resistant traits; Gassmann et al. (2011) recently demonstrated that field-derived WCR populations with reduced susceptibility to Cry3Bb1 corn are susceptible to Cry34Ab1/Cry35Ab1 maize.

7.2. Global adoption of *Bt* crops

Bt potato was the first *Bt* crop commercialized in the United States (1995) followed by *Bt* corn and cotton in 1996 (Betz et al., 2000). *Bt* crops have been rapidly adopted by growers around the world and in 2013 were grown on nearly 76 million hectares in 27 countries by 18 million farmers (James, 2013). While our focus on describing *Bt* crops will be on traits commercialized in the United States, in recent years developing countries have planted more biotech crops than industrialized countries (James, 2013). In fact, recently commercialized products like PowerCore® maize and Intacta® soybean were developed primarily to meet the needs of growers in Latin America.

Economists Graham Brookes and Peter Barfoot have, since 2005, provided the most comprehensive assessments of the global economic and environmental impacts of biotech crops including *Bt* crops (Barfoot and Brookes, 2014; Brookes and Barfoot, 2014). They calculate the aggregate impact on global farm income in 2012 for Bt corn and cotton at $6.7 billion and $5.3 billion respectively (Brookes and Barfoot, 2014). Economic gains are tied to increased yields due to effective pest control and decreased costs of production related to expenses associated with insecticide use. Reductions in insecticide use are most often associated with *Bt* cotton which in 2012 was estimated at 16.8 million kg (a 40% reduction of insecticides targeted at lepidopteran pests) but the reductions in insecticide use associated with *Bt* corn are also impressive (2012: 7.6 million kg representing an 86.5 % reduction of insecticides targeted at lepidopteran and coleopteran pests) (Barfoot and Brookes, 2014). They also calculate the reduced greenhouse gas emissions associated with reduced insecticide applications in *Bt* crops (and the increase in reduced or no-till production systems associated herbicide tolerant crops) to be the equivalent of taking 940,000 cars off the road (Barfoot and Brookes, 2014).

A benefit associated specifically with *Bt* corn is the reduction in mycotoxin contamination in corn kernels attributed to reduced insect damage (Wu, 2006). Insect feeding on the ear provides a point of entry for fungi (primarily *Fusarium* and *Aspergillus* spp.) known to produce these potent toxins (and carcinogens) and the insects themselves may be important vectors of these fungi (Abbas et al., 2013). In a recent review by Abbas et al. (2013), a strong association between *Bt* corn and reductions in fumonisins (mycotoxins produced by *Fusarium* spp.) was apparent while the association with reductions in aflatoxin (produced by *Aspergillus* spp.) was less clear.

7.3. Commercialized products

Information on the status of deregulated biotech traits is maintained by several organizations including the Biotechnology Industry Organization (BIO) (http://www.biotradestatus.com/), the Center for Environmental Risk Assessment (CERA, 2012) and the International Service for the Acquisition of Agri-Biotech Applications (ISAAA, http://www.isaaa.org/gmapprovaldatabase/). In all cases, the information contained in these online resources is a summary of publicly available information provided by various governmental regulatory authorities. Tables 4.5, 4.7 and 4.9 are summaries of information obtained from these sites for *Bt* events that have been approved for cultivation in the United States and which, to the best of our knowledge, have been commercialized.

7.3.1 Bt *corn*

The first *Bt* corn events were approved for cultivation in the United States in 1995 and commercialized in 1996 targeting control of *O. nubilalis* and other stalk boring Lepidoptera such as southwestern corn borer, *Diatraea grandiosella* Dyer and *D. sacharralis* (Archer et al., 2001; Buntin et al., 2004). *O. nubilalis* had not been a target for insecticide applications so the rapid adoption of Bt corn was based on the yield improvements created by controlling pest populations that were below the economic threshold justifying insecticide applications (Catangui, 2003; Shelton et al., 2002).

Events expressing Cry1Ab (176, MON810 and Bt11), Cry1Ac (DBT418), Cry9C (CBH351), Cry1Fa (TC1507) and Vip3Aa (MIR162) have all been commercialized (Table 4.5) and 4 (MON810, Bt11, TC1507 and MIR162) remain in the market today (Table 4.6). Events 176 (Cry1Ab) and DBT418 (Cry1Ac) were discontinued based on the superior performance of MON810 and Bt11 (Buntin et al., 2004; Walker et al., 2000). The removal of event 176 from commercial production was hastened when it became the focus of concerns over the potential for impact on susceptible, non-target Lepidoptera consuming plant tissue onto which pollen from *Bt* corn had fallen (Hellmich et al., 2001) (Sears et al., 2001). StarLink® (event CBH-351 expressing Cry9C) was approved for cultivation and for use in animal feed and industrial non-food uses. It was removed from commercial use when traces of the product were found in the food supply (Fox, 2001).

MON810 and Bt11, the single *Bt* gene events expressing Cry1Ab in the market today, provide a high level of control of *O. nubilalis* and other stalk

Table 4.5 Corn events approved for cultivation and commercialized

Developer	Event name	OECD unique identifier	*Bt* protein(s)	Pest spectrum	Year approved (cultivation—USA)	Non-IR genes[a]
Syngenta	176	SYN-EV176-9	Cry1Ab	Lepidoptera	1995	*pat*
Monsanto	MON810	MON-00810-6	Cry1Ab	Lepidoptera	1996	*nptII*
Syngenta	Bt11	SYN-BT011-1	Cry1Ab	Lepidoptera	1996	*pat*
Dekalb Genetics Corporation	DBT418	DKB-89614-9	Cry1Ac	Lepidoptera	1997	*bar*
Aventis CropScience	CBH-351[b]	ACS-ZM004-3	Cry9C	Lepidoptera	1998	*pat*
Dow AgroSciences DuPont Pioneer	TC1507	DAS-01507-1	Cry1Fa	Lepidoptera	2001	*pat*
Monsanto	MON863	MON-00863-5	Cry3Bb1	Coleoptera	2003	*nptII*
Dow AgroSciences DuPont Pioneer	DAS-59122-7	DAS-59122-7	Cry34Ab1 Cry35Ab1	Coleoptera	2005	*pat*
Monsanto	MON88017	MON-88017-3	Cry3Bb1	Coleoptera	2005	*cp4 epsps*
Syngenta	MIR604	SYN-IR604-5	mCry3A	Coleoptera	2007	*pmi*
Monsanto	MON89034	MON-89034-3	Cry1A.105 Cry2Ab	Lepidoptera	2008	
Syngenta	MIR162	SYN-IR162-4	Vip3Aa20	Lepidoptera	2010	*pmi*
Syngenta	5307	SYN-05307-1	eCry3.1Ab	Coleoptera	2012	*pmi*

[a]Non-IR genes *pat*: a selectable marker which confers tolerance to the herbicide glufosinate ammonium in plant tissue. *nptII*: a selectable marker which confers the ability to metabolize the antibiotics neomycin and kanamycin in plant tissue. *bar*: a selectable marker which confers tolerance to the herbicide glufosinate ammonium in plant tissue. *cp4 epsps*: a selectable marker confers tolerance to the herbicide glyphosate in plant tissue. *pmi*: a selectable marker which confers the ability to utilize mannose as a carbon source in plant tissue.

[b]Approved for environmental release and use as animal feed only.

boring Lepidoptera and moderate levels of control of *H.* zea (Buntin et al., 2004; Farinós et al., 2004; He et al., 2003). TC1507, a single-gene event expressing Cry1Fa, also provides a high level of control of *O. nubilalis* and other stalk boring Lepidoptera (Farinos et al., 2012) but differs from the Cry1Ab events in that it also provides protection against *Spodoptera* species including *S. frugiperda* (Hardke et al., 2011), black cutworm (Kullik et al., 2011) and western bean cutworm (*Striacosta albicosta* (Smith)) (Eichenseer et al., 2008; Fig. 4.5A). While TC1507 does provide suppression of *H. zea*, it does not provide the same level of protection as MON810 (Buntin, 2008). MIR162, a single-gene event expressing Vip3Aa, does not provide protection against *O. nubilalis* but provides a high level of control against other stalk boring Lepidoptera, fall armyworm, black cutworm, western bean cutworm (Syngenta Seeds) and the highest level of control of single-gene events on *H. zea* (Burkness et al., 2010).

Cry1A.105 and Cry2Ab are lepidopteran-active *Bt* proteins expressed in event MON89034 (Table 4.5). This combination creates a trait with a broad spectrum and strong performance in controlling many important lepidopteran pests (Reay-Jones et al., 2009; Rule et al., 2014; Shelton et al., 2013; Siebert et al., 2012). It also provides a pyramid of two modes-of-action on certain pests thereby decreasing the risk of resistance (Ghimire et al., 2011; Storer et al., 2012a). This reduced risk for resistance development makes practical the use of a blended refuge (i.e. refuge-in-a-bag or RIB); where a bag of *Bt*-traited seed contains a small percentage of seed without the *Bt* trait (Storer et al., 2012b). The "built-in" nature of a blended refuge insures that a refuge is deployed thereby increasing the durability of the *Bt* trait. Enhanced spectrum and the ability to deploy blended refuges by virtue of mechanism of action pyramids are factors that drive the combination of *Bt* traits by breeding (Table 4.6) (Rule et al., 2014; Siebert et al., 2012; Storer et al., 2012b).

The first *Bt* event targeting corn rootworm (*Diabrotica* spp.) was commercialized in 2003 (MON863 expressing Cry3Bb) and today there are five events in the market targeting control of this pest complex (Tables 4.5 and 4.6; Fig. 4.5). Products containing these events are sold primarily in North America where their superior performance compared with insecticide treatments led to rapid and widespread adoption (Petzold-Maxwell et al., 2013). Increased protection from damage under high-rootworm pressure and the increased durability associated with pyramiding mechanisms-of-action has led to the development of products like SmartStax®, which combines events expressing Cry3Bb (MON88017) and the binary insecticidal protein

Table 4.6 *Bt* corn products currently sold in the United States

		Bt event(s)									Non-*Bt* events[a]			
		Dow AgroSciences / DuPont Pioneer		Monsanto			Syngenta				Monsanto		Syngenta	
Company	**Product(s)**	**TC1507**	**DAS-59122-7**	**MON89034**	**MON88017**	**MON810**	**Bt11**	**MIR604**	**MIR162**	**5307**	**NK603**	**MON87460**	**GA21**	**3272**
Dow AgroSciences	Herculex® I	■												
	Herculex® RW		■											
	Herculex® XTRA	■	■											
	Herculex® XTRA Roundup Ready® Corn 2	■	■								■			
	Herculex® I Roundup Ready® Corn 2	■									■			
	SmartStax® Refuge Advanced®	■	■	■	■									
	PowerCore®	■		■							■			

DuPont Pioneer	Herculex® I													
	Herculex® RW													
	Herculex® XTRA													
	Herculex® I Roundup Ready Corn 2													
	Herculex® RW Roundup Ready® Corn 2													
	Herculex® XTRA Roundup Ready® Corn 2													
	Optimum® AcreMax®													
	Optimum® AcreMax® RW													
	Optimum® AcreMax® 1													

Continued

Table 4.6 *Bt* corn products currently sold in the United States—cont'd

Company	Product(s)	*Bt* event(s)									Non-*Bt* events			
		Dow AgroSciences		Monsanto			Syngenta				Monsanto		Syngenta	
		DuPont Pioneer												
		TC1507	DAS-59122-7	MON89034	MON88017	MON810	Bt11	MIR604	MIR162	5307	NK603	MON87460	GA21	3272
	Optimum® AcreMax® XTRA (GT)	■	■			■					■			
	Optimum® AcreMax® Xtreme	■	■			■		■			■			
	Optimum® Leptra™	■				■			■		■			
	Optimum® Intrasect® Xtreme	■	■			■		■			■			
	Optimum® Intrasect®	■				■					■			
	Optimum® Intrasect® XTRA	■	■			■					■			
	Optimum® TRIsect™	■						■			■			

Monsanto	YieldGard® Corn Borer Corn													
	YieldGard® Corn Borer with Roundup Ready® Corn 2													
	Genuity® SmartStax® Genuity® SmartStax® RIB Complete™													
	YieldGard® VT Rootworm													
	Genuity® VT Double PRO® Genuity® VT Double PRO® RIB Complete™													
	Genuity® VT Triple PRO® Genuity® VT													

Continued

Table 4.6 *Bt* corn products currently sold in the United States—cont'd

Company	Product(s)	*Bt* event(s)									Non-*Bt* events			
		Dow AgroSciences		Monsanto			Syngenta				Monsanto		Syngenta	
		DuPont Pioneer												
		TC1507	DAS-59122-7	MON89034	MON88017	MON810	Bt11	MIR604	MIR162	5307	NK603	MON87460	GA21	3272
	Triple PRO® RIB Complete™													
	Genuity® DroughtGard™ with VT Triple PRO®													
Syngenta	Agrisure® CB/LL													
	Agrisure® GT/CB/LL													
	Agrisure® RW													
	Agrisure® CB/LL/RW													
	Agrisure® 3000GT													

	Enogen/ Agrisure® 3000GT						■	■					■	■
	Agrisure® Viptera®								■					
	Agrisure® Viptera® 3110						■		■				■	
	Agrisure® Viptera® 3111						■	■	■				■	
	Agrisure® 3122	■	■				■	■					■	
	Agrisure® Viptera® 3220 Agrisure® Viptera® 3220 E-Z Refuge™	■					■		■				■	
	Agrisure® Duracade™ 5122 E-Z Refuge®	■					■	■		■			■	
	Agrisure® Duracade™ 5222 E-Z Refuge®	■					■	■	■	■			■	
	Attribute®						■							
	Attribute® II						■		■				■	

[a]Non-IR events NK603: glyphosate tolerance trait. MON87460: DroughtGard® drought tolerance trait. GA21: glyphosate tolerance trait. 3272: Enogen® ethanol production trait. Shaded boxes indicate that the event is present in the product. See Table 4.5 for *Bt* genes and pest spectrum associated with *Bt* events.

Figure 4.5 Insect-protected *Bt* corn. (A) Contrast of corn root systems after heavy pressure from larval western corn rootworm. Top row is non-*Bt* plants showing severe root damage. Bottom row is plants from the same genetic background as the plants in the top row but containing event DAS-59122-7. (B) Contrast of ears after heavy pressure from larval western bean cutworm. Top row is ears from non-*Bt* plants showing severe damage. Bottom row is ears from the same genetic background as the plants in the top row but containing event TC1507. *Photo credit: Bader Rutter.* (See the colour plate.)

Cry34Ab/Cry35Ab (DAS-59122-7) (Head et al., 2014). Recent reports of the decrease in effectiveness of events expressing Cry3Bb (MON863 and MON88017) and mCry3A (MIR604) contrasted with the continued effectiveness of DAS-59122-7 are additional evidence that Cry3s and Cry34/Cry35 represent distinct modes-of-action and validate the value of the SmartStax pyramid (Gassmann, 2012; Gassmann et al., 2014). Another pyramid for rootworm control that is entering the market is Duracade™; the combination of mCry3A (MIR604) and eCry3.1Ab (5307) (Table 4.6). *The Handy Bt Trait Table* (DiFonzo and Cullen, 2013) is a useful and frequently updated guide to the *Bt* corn products available in the United States.

7.3.2 Bt *cotton*

Bt cotton was commercially introduced in the United States in 1996 and was rapidly adopted due in large part to its effectiveness in controlling *H. virescens*, a pest that had developed resistance to virtually all insecticides available at the time (Blanco, 2012). *Bt* cotton was also introduced in Australia in 1996 where the primary target was *H. armigera* (Downes and

Mahon, 2012). In 2013, *Bt* cotton was grown in 15 countries and high-adoption rates in the major cotton producing countries speak to the value of the technology (James, 2013; Naranjo, 2010; Fig. 4.6).

Since the introduction of BollGard® cotton (MON531 expressing Cry1Ac), events expressing Cry1Ab, Cry1Fa, Cry2Ab, Cry2Ae and Vip3Aa have been commercialized and, with the exception of Cry1Ac, have been introduced as pyramids (Tables 4.7 and 4.8). Control of *H. virescens* with Cry1Ac-expressing events has been consistently strong with no evidence of resistance development in the United States (Blanco, 2012). *Bt* cotton expressing Cry1Ac is also highly effective at controlling pink bollworm and in the United States; it has been a pillar of the pink bollworm eradication program (Tabashnik et al., 2012). Increased durability provided by different mechanisms-of action on key pests as well as increased consistency in the control of bollworm was seen with the introduction of pyramided varieties like BollGard®II and WideStrike® (2003 and 2005, respectively), but under high-bollworm pressure the yield of *Bt* cotton varieties is enhanced by insecticide applications (Luttrell and Jackson, 2012). The recent addition of Vip3Aa to the pyramid of Cry1Ac and Cry1Fa (Table 4.8) should bring additional levels of performance against bollworm and *Spodoptera* species (Bommireddy et al., 2011).

7.3.3 Bt *soybean*

MON87701 is the only *Bt* event that has been commercialized in soybean (Table 4.9). Monsanto commercialized this event in Brazil under the trade name Intacta RR2 PRO™ which is a breeding stack with MON89788, an event conferring the Roundup Ready 2 Yield™ trait. Published studies show that events expressing Cry1Ac in soybean are efficacious against important North and South American soybean pests including *C. includens*, *A. gemmatalis*, *Crocidosema aporema* (Walsingham), *Rachiplusia nu* (Guenée) and the yellow woolybear, *Spilosoma virginica* (F.) (Macrae et al., 2005; McPherson and MacRae, 2009). A more recent study by Bernardi et al. (2014) using soybeans containing the events in Intacta RR2 PRO showed that the relatively low intrinsic potency of Cry1Ac on *Spodoptera* translated to poor control of *S. frugiperda*, *Spodoptera cosmioides* Walker and *Spodoptera eridania* (Stoll); economically important pests of soybeans in Brazil.

Foliage feeding Lepidoptera frequently reach economically damaging levels in South American soybean production, so *Bt* soybean potentially delivers a clear benefit to growers (de Freitas Bueno et al., 2011). The value

Figure 4.6 Insect-protected Bt cotton. Contrast of yield after heavy lepidopteran pest pressure. Rows on the left are plants containing events 281-24-236 and 3006-210-23 (i.e. WideStrike®) showing a large number of open bolls and harvestable lint. Rows on the right are plants of the same genetic background as plants on the left but without the Bt events. *Photo credit: Eileen Crosby and Jim Steadman, Bader Rutter.* (See the colour plate.)

of controlling these pests in North American soybean production is less clear as the levels and timing of defoliation typically created by lepidopteran pests do not negatively impact yield (Hammond et al., 2000). In the case of *Bt* soybean, McPherson and MacRae (2009) conducted 3 years of replicated field trials comparing several Cry1Ac-expressing soybean events with non-GE checks. While they observed significant reductions in larval counts and foliar damage on *Bt* compared with non-*Bt* soybean lines, they never translated to increased yields.

7.3.4 Bt *potato*

Bt potato expressing Cry3Aa was developed by Monsanto and commercialized by Monsanto subsidiary NatureMark in 1995. Multiple events were deregulated and commercialized as NewLeaf® potatoes (expressing Cry3Aa), NewLeaf Plus® potatoes (expressing Cry3Aa and the full length replicase gene of potato leafroll virus) conferring high levels of control of the Colorado potato beetle and potato leafroll virus (Lawson et al., 2001) and NewLeaf® Y potatoes (expressing Cry3Aa and potato potyvirus Y coat protein) (Table 4.9).

Cry3Aa-expressing potatoes provided control of *L. decemlineata* that was superior to the commercial insecticides available during the time this

Table 4.7 Cotton events approved for cultivation and commercialized

Developer	Event name	OECD unique identifier	*Bt* protein(s)	Pest spectrum	Year approved (cultivation—USA)	Non-IR genes[a]
Monsanto	MON531	MON-000531-6	Cry1Ac	Lepidoptera	1995	*nptII* *aad*
Calgene Inc.	31807 31808		Cry1Ac	Lepidoptera	1998	*bxn* *nptII*
Dow AgroSciences	281-24-236[b]	DAS-24236-5	Cry1Fa	Lepidoptera	2004	*pat*
Dow AgroSciences	3006-210-23[b]	DAS-21023-5	Cry1Ac	Lepidoptera	2004	*pat*
Monsanto	MON15985[c]	MON-15985-7	Cry1Ac Cry2Ab	Lepidoptera	2005	*nptII* *aad* *uidA*
Bayer CropSciences	GHB119[b]	BCS-GH005-8	Cry2Ae	Lepidoptera	2011	*bar*
Bayer CropSciences	T304-40[b]	BCS-GH004-7	Cry1Ab	Lepidoptera	2011	*bar*
Syngenta	COT 102[b]	SYN-IR102-7	Vip3Aa	Lepidoptera	2011	*aph4*

[a]Non-IR genes *nptII*: a selectable marker which confers the ability to metabolize the antibiotics neomycin and kanamycin in plant tissue. *aad*: a selectable marker used in the creation of the gene construct and is not expressed in plants. *bxn*: confers tolerance to the herbicide bromoxinil in plants. *pat*: a selectable marker which confers tolerance to the herbicide glufosinate ammonium in plant tissue. *uidA*(*GUS*): a scorable marker enabling visual identification of transformed plants. *bar*: a selectable marker which confers tolerance to the herbicide glufosinate ammonium in plant tissue. *aph4*: a selectable marker which confers the ability to metabolize the antibiotic hygromycin in plant tissue.
[b]Commercialized only as a breeding stack.
[c]Created via biolistic transformation of germplasm containing Event MON531 (Cry1Ac) with Cry2Ab2.

Table 4.8 *Bt* cotton products currently sold in the United States

		Bt event(s)						Herbicide tolerance events[a]		
		Monsanto	Dow AgroSciences		Syngenta	Bayer CropSciences		Monsanto	Bayer CropSciences	
Company	**Product(s)**	**MON15985**	**3006-210-23**	**281-240-236**	**COT102**	**T304-40**	**GBH119**	**MON88913**	**LLCotton25**	**GBH614**
Bayer CropScience	LibertyLink® × BollGard® II									
Bayer CropScience	LibertyLink® × BollGard® II × GlyTol®									
Bayer CropScience	GlyTol® × TwinLink®									
Dow AgroSciences	WideStrike®									
Dow AgroSciences	WideStrike® Roundup Ready® FLEX									
Dow AgroSciences	WideStrike® 3									
Monsanto	Genuity® BollGard II®									
Monsanto	Genuity® BollGard II® with Roundup Ready® FLEX									

[a]Herbicide tolerance events MON88913: glyphosate tolerance trait. LLCotton25: glufosinate ammonium tolerance trait. GBH614: glyphosate tolerance trait.
Shaded boxes indicate that the event is present in the product. See Table 4.7 for *Bt* genes and pest spectrum associated with *Bt* events.

Table 4.9 Soybean and potato events approved for cultivation and commercialized

Crop	Developer	Event name	OECD unique identifier	*Bt* protein(s)	Pest spectrum	Year approved (cultivation—USA)	Non-IR genes[a]
Potato	Monsanto	BT6 BT10 BT12 BT16 BT17 BT18 BT23	NMK-89812-3 NMK-89175-5 NMK-89601-8 NMK-89167-6 NMK-89593-9 NMK-89906-7 NMK-89675-1	Cry3A	Coleoptera	1995	*nptII*
Potato	Monsanto	ATBT04-6 ATBT04-30 ATBT04-36 SPBT02-5	NMK-89761-6 NMK-89613-2 NMK-89279-1 NMK-89576-1	Cry3A	Coleoptera	1996	*nptII*
Potato	Monsanto	RBMT21-129 RBMT21-350 RBMT22-082	NMK-89684-1 NMK-89185-6 NMK-89896-6	Cry3A	Coleoptera	1998	*nptII* *aad* *plrv orf1* *plrv orf2*
Potato	Monsanto	RBMT15-101 SEMT15-02 SEMT15-15	NMK-89653-6 NMK-89935-9 NMK-89930-4	Cry3A	Coleoptera	1999	*nptII* *aad* *pvy CP*
Soybean	Monsanto	MON87701[b]	MON-87701-2	Cry1Ac	Lepidoptera	2011	

[a]Non-IR genes: *nptII*: a selectable marker which confers the abilty to metabolize the antibiotics neomycin and kanamycin in plant tissue. *aad*: a selectable marker used in the creation of the gene construct and is not expressed in plants. *plrv orf1*: a viral protein which confers resistance to the potato leafroll virus. *plrv orf2*: a viral protein which confers resistance to the potato leafroll virus. *pvy CP*: a viral coat protein which confers resistance to potato potyvirus Y.
[b]Commercialized only as a breeding stack.

product was being developed (Reed et al., 2001). Imidaclprid, a neonicitinoid insecticide, was introduced about the same time as NewLeaf® potatoes and provided growers with an effective alternative for control of both *L. decemlineata* and aphid vectors of plant viruses (Boiteau et al., 1997). The availability of this alternative coupled with the decision of large, institutional potato consumers to not accept NewLeaf® potatoes led NatureMark to discontinue the product line in 2001.

8. *Bt* CROPS—PROSPECTS FOR THE FUTURE

8.1. Novel *Bt* proteins

The large number of known *Bt* insecticidal proteins will continue to be evaluated for agricultural applications as new information is generated on the spectrum, potency and potential utility in pyramids to control resistant insects. Looking forward, it is highly probable that next-generation sequencing will accelerate the discovery of many more insecticidal proteins from *Bt* and its subspecies, as well as related bacilli (e.g. *B. popilliae* Zhang et al., 1997) and other bacterial genera, e.g. *Clostridium bifermentans* (Barloy et al., 1996), where new examples of Cry-type proteins are likely to be present.

In addition to identification of new, naturally occurring insecticidal proteins, there are several approaches to improve insecticidal properties of existing proteins through knowledge-driven protein engineering (reviewed in part in Pardo-Lopez et al., 2009). Some of these approaches have delivered novel Cry proteins that are used in GE crops. A brief review of approaches to engineer *Bt* proteins for improved insecticidal attributes follows.

8.1.1 Protease activation

Bt Cry proteins are converted to an active form by insect midgut proteases. Proteolytic removal of the C-terminal crystallization domain is necessary for activation of many Cry proteins (Schnepf et al., 1998). Continuing to understand the ability of different insects to activate *Bt* proteins will be important to engineer new Cry protein variants to control pests of economic concern. As mentioned previously, mCry3Aa for control of WCR is an engineered version of Cry3Aa with a chymotrypsin/cathepsin G protease recognition site inserted in domain 1. The mCry3Aa protein has consistently improved activity against WCR compared with the parent Cry3Aa protein (Walters et al., 2008). For Cry1A proteins, removal of the amino-terminal α-helix in Cry1a proteins can sometimes overcome IR (Soberon et al.,

2007; Tabashnik et al., 2011), though this technology has yet to be developed for commercial application.

8.1.2 Site directed mutagenesis

Site directed mutagenesis has been used as a method to probe structure–function relationships in Cry proteins (see reviews by Dean et al., 1996; Schnepf et al., 1998). Most mutations result in deleterious effects on insecticidal activity. Notable exceptions are point mutations in Cry3Aa domain 2 that improved activity against *T. molitor* (Wu et al., 2000) and rational design of Cry1Aa loop modifications to generate a protein more similar to Cry4Ba and thereby convert Cry1Aa from a lepidopteran-active protein to a mosquito-active protein (Liu and Dean, 2006).

Protein structure information, when available, greatly aids the rational design of improved pesticidal proteins by site-specific mutagenesis. One commercially successful example of structure-based protein engineering is Cry3Bb1. Cry3Bb1 contains six amino acid residue changes compared with the native Cry3Bb1 sequence (Vaughn et al., 2005). The engineered version of Cry3Bb1 is more active on WCR than the parent Cry3Bb1 protein. A more recent example of structure-based protein engineering was the insertion of a 12 amino acid pea aphid gut-binding peptide in loop regions of the *Bt* cytolytic toxin, Cyt2Aa, that resulted in enhanced binding and toxicity against both the pea aphid, *Acyrthosiphon pisum* (Harris) and the green peach aphid, *Myzus persicae* (Sulzer) (Chougule et al., 2013).

8.1.3 Gene shuffling

DNA shuffling (Stemmer, 1994) is an *in vitro* approach to protein evolution with many applications. When applied to *Bt* insecticidal protein improvement, this approach depends on innovative high-throughput screens to be successful. Two recent studies used similar approaches of DNA shuffling combined with phage display screening by membrane binding to enhance the insecticidal activity of Cry8Ka1 against cotton boll weevil *Anthonomus grandis grandis* Boheman (Oliveira et al., 2011) and Cry1Ia against sugarcane giant borer *Telchin licus licus* (Craveiro et al., 2010).

8.1.4 Domain 3 exchange

Three domain Cry proteins appear to have co-evolved domains 1 and 2 together, and in some cases domain 3 as well. However, phylogenetic analysis suggests evidence for evolutionary exchange of domain 3 by homologous recombination between some Cry proteins (reviewed in de Maagd et al.,

2001). Domain 3 exchange is a proven route to generate Cry proteins with broader spectrum of biological activity or higher insecticidal potency. The first published example of domain 3 exchange by homologous recombination generated a Cry1Aa/Cry1Ac hybrid protein with improved activity on *H. virescens* (Caramori et al., 1991). Later, de Maagd et al. (1996b) created Cry1 hybrid proteins containing Cry1C domain 3 to improve activity on *S. exigua*. A Cry1Ab/Cry1C hybrid improved *S. exigua* activity 10-fold over the parent Cry1C protein (de Maagd et al., 1996a). As mentioned previously, Cry1A.105 is a commercialized chimera of Cry1Ab domains 1 and 2 fused to Cry1Fa domain 3. Domain 3 swapping was also used to improve coleopteran activity for Cry1Ba/Cry1Ia hybrid proteins (Naimov et al., 2001). Last, Cry3A.1Ab combines Cry3Aa domains 1 and 2 with domain 3 from Cry1Ab for improved activity on WCR (Walters et al., 2010).

9. CONCLUSIONS

It is safe to say that the introduction of *Bt* crops revolutionized agricultural pest control. Success in the discovery of *Bt* genes for use in crops leveraged a long history of development of *Bt* as a biopesticide. Application of all of the tools of modern biotechnology was required to genetically engineer crops capable of producing these proteins, and new approaches in many fields, including plant breeding and regulatory science, were developed to bring products to the market. The result is crops with a previously unknown capacity to resist pest damage leading to significant economic, environmental and societal benefits (James, 2013).

Nearly 20 years after the introduction of the first *Bt* crops, the future of this technology remains bright. Insuring that farmers and society can continue to reap its benefits remains a priority for technology providers as is evidenced by the development of pyramided products designed to slow the inevitable evolutionary response of pest populations to the selection pressure imposed by this highly effective control. Evidence of the on-going commitment to commercializing mechanism of action pyramids is seen in recently announced products in the late stages of development including: DuPont Pioneer's corn event DP4114, a molecular stack of the *Bt* proteins contained in Herculex® XTRA, which will only be commercialized as a pyramid with other modes of action; Monsanto's corn event MON87411, a pyramid of Cry3Bb and a novel RNAi mechanism of action, that will be commercialized as a pyramid with Dow AgroSciences'

DAS-59122-7; and Dow AgroSciences' soybean event DAS-81419-2, a pyramid of Cry1Fa and Cry1Ac for control of lepidopteran pests of soybean.

Publicly disclosed pipelines of the major technology providers list IR corn, soybean and cotton efforts in virtually all stages reflecting their long-term investments in bringing new IR traits technologies to the market. Given the success of *Bt* crops to date, and the untapped potential for discovering *Bt* proteins with novel modes of action and pest spectra, we expect to see new *Bt*-based IR traits well into the future.

ACKNOWLEDGEMENTS

We would like to thank E. Bouquet, M. Garcia, T. Hey, S. Jayne, R. Maciak and A. Mel for their review and critique of earlier versions of this chapter. We would also like to thank the editors of this volume for their constructive comments and patience.

REFERENCES

Abbas, H.K., Zablotowicz, R.M., Weaver, M.A., Shier, W.T., Bruns, H.A., Bellaloui, N., Accinelli, C., Abel, C.A., 2013. Implications of Bt traits on mycotoxin contamination in maize: overview and recent experimental results in southern United States. J. Agric. Food Chem. 61, 11759–11770.

Adang, M.J., Staver, M.J., Rocheleau, T.A., Leighton, J., Barker, R.F., Thompson, D.V., 1985. Characterized full-length and truncated plasmid clones of the crystal protein of *Bacillus thuringiensis* subsp. *kurstaki* HD-73 and their toxicity to *Manduca sexta*. Gene 36, 289–300.

Adang, M., Brody, M., Cardineau, G., Eagan, N., Roush, R., Shewmaker, C., Jones, A., Oakes, J., McBride, K., 1993. The reconstruction and expression of a *Bacillus thuringiensis* cryIIIA gene in protoplasts and potato plants. Plant Mol. Biol. 21, 1131–1145.

Alstad, D.N., Andow, D.A., 1995. Managing the evolution of insect resistance to transgenic plants. Science 268, 1894–1896.

Angus, T.A., 1954. A bacterial toxin paralysing silkworm larvae. Nature 173, 545–546.

Arantes, O., Lereclus, D., 1991. Construction of cloning vectors for *Bacillus thuringiensis*. Gene 108, 115–119.

Archer, T.L., Patrick, C., Schuster, G., Cronholm, G., Bynum Jr., E.D., Morrison, W.P., 2001. Ear and shank damage by corn borers and corn earworms to four events of Bacillus thuriengiensis transgenic maize. Crop Prot. 20, 139–144.

Arrieta, G., Espinoza, A.M., 2006. Characterization of a *Bacillus thuringiensis* strain collection isolated from diverse Costa Rican natural ecosystems. Rev. Biol. Trop. 54, 13–27.

Barfoot, P., Brookes, G., 2014. Key global environmental impacts of genetically modified (GM) crop use 1996–2012. GM Crops Food Biotechnol. Agric. Food Chain 5, 10–21.

Barloy, F., Delecluse, A., Nicolas, L., Lecadet, M.M., 1996. Cloning and expression of the first anaerobic toxin gene from *Clostridium bifermentans* subsp. *malaysia*, encoding a new mosquitocidal protein with homologies to *Bacillus thuringiensis* delta-endotoxins. J. Bacteriol. 178, 3099–3105.

Baum, J.A., Johnson, T.B., Carlton, B.C., 1998. *Bacillus thuringiensis*: natural and recombinant bioinsecticide products. In: Hall, F.R., Menn, J.J. (Eds.), Biopesticides: Use and Delivery, pp. 189–209.

Baxter, S.W., Badenes-Perez, F.R., Morrison, A., Vogel, H., Crickmore, N., Kain, W., Wang, P., Heckel, D.G., Jiggins, C.D., 2011. Parallel evolution of *Bacillus thuringiensis* toxin resistance in lepidoptera. Genetics 189, 675–679.

Ben-Dov, E., 2014. *Bacillus thuringiensis* subsp. *israelensis* and its Dipteran-specific toxins. Toxins 6, 1222–1243.

Bergamasco, V.B., Mendes, D.R., Fernandes, O.A., Desiderio, J.A., Lemos, M.V., 2013. *Bacillus thuringiensis* Cry1Ia10 and Vip3Aa protein interactions and their toxicity in *Spodoptera* spp. (Lepidoptera). J. Invertebr. Pathol. 112, 152–158.

Berliner, E., 1915. Ueber die schlaVsucht der *Ephestia kuhniella* und *Bac. thuringiensis* n. sp. Z. Angew. Entomol. 2, 21–56.

Bernardi, O., Sorgatto, R.J., Barbosa, A.D., Domingues, F.A., Dourado, P.M., Carvalho, R.A., Martinelli, S., Head, G.P., Omoto, C., 2014. Low susceptibility of Spodoptera cosmioides, Spodoptera eridania and Spodoptera frugiperda (Lepidoptera: Noctuidae) to genetically-modified soybean expressing Cry1Ac protein. Crop Prot. 58, 33–40.

Berry, C., 2012. The bacterium, *Lysinibacillus sphaericus*, as an insect pathogen. J. Invertebr. Pathol. 109, 1–10.

Betz, F.S., Hammond, B.G., Fuchs, R.L., 2000. Safety and advantages ofBacillus thuringiensis-protected plants to control insect pests. Regul. Toxicol. Pharmacol. 32, 156–173.

Blanco, C.A., 2012. *Heliothis virescens* and Bt cotton in the United States. GM Crops Food Biotechnol. Agric. Food Chain 3, 201–212.

Boiteau, G., Osborn, W.P.L., Drew, M.E., 1997. Residual activity of lmidacloprid controlling colorado potato beetle (Coleoptera: Chrysomelidae) and three species of potato colonizing aphids (Homoptera: Aphidae). J. Econ. Entomol. 90, 309–319.

Bolivar, F., Rodriguez, R.L., Betlach, M.C., Boyer, H.W., 1977. Construction and characterization of new cloning vehicles I. Ampicillin-resistant derivatives of the plasmid pMB9. Gene 2, 75–93.

Bommireddy, P.L., Leonard, B.R., Temple, J., Price, P., Emfinger, K., Cook, D., Hardke, J.T., 2011. Field performance and seasonal efficacy profiles of transgenic cotton lines expressing Vip3A and VipCot against Helicoverpa zea (Boddie) and heliothis virescens (F.). J. Cotton Sci. 15, 251–259.

Boonserm, P., Davis, P., Ellar, D.J., Li, J., 2005. Crystal structure of the mosquito-larvicidal toxin Cry4Ba and its biological implications. J. Mol. Biol. 348, 363–382.

Boonserm, P., Mo, M., Angsuthanasombat, C., Lescar, J., 2006. Structure of the functional form of the mosquito larvicidal Cry4Aa toxin from Bacillus thuringiensis at a 2.8-angstrom resolution. J. Bacteriol. 188, 3391–3401.

Bravo, A., Sarabia, S., Lopez, L., Ontiveros, H., Abarca, C., Ortiz, A., Ortiz, M., Lina, L., Villalobos, F.J., Pena, G., Nunez-Valdez, M.E., Soberon, M., Quintero, R., 1998. Characterization of cry genes in a Mexican *Bacillus thuringiensis* strain collection. Appl. Environ. Microbiol. 64, 4965–4972.

Brookes, G., Barfoot, P., 2014. Economic impact of GM crops: the global income and production effects 1996–2012. GM Crops Food Biotechnol. Agric. Food Chain 5, 37–47.

Buntin, G.D., 2008. Corn expressing Cry1Ab or Cry1F endotoxin for fall armyworm and corn earworm (Lepidoptera: Noctuidae) management in field corn for grain production. Fla. Entomol. 91, 523–530.

Buntin, G.D., All, J.N., Lee, R.D., Wilson, D.M., 2004. Plant-incorporated *Bacillus thuringiensis* resistance for control of fall armyworm and corn earworm (Lepidoptera: Noctuidae) in corn. J. Econ. Entomol. 97, 1603–1611.

Burkness, E.C., Dively, G., Patton, T., Morey, A.C., Hutchison, W.D., 2010. Novel Vip3A *Bacillus thuringiensis* (Bt) maize approaches high-dose efficacy against *Helicoverpa zea* (Lepidoptera: Noctuidae) under field conditions: implications for resistance management. GM Crops 1, 337–343.

Butko, P., 2003. Cytolytic toxin Cyt1A and its mechanism of membrane damage: data and hypotheses. Appl. Environ. Microbiol. 69, 2415–2422.
Caccia, S., Hernandez-Rodriguez, C.S., Mahon, R.J., Downes, S., James, W., Bautsoens, N., Van Rie, J., Ferre, J., 2010. Binding site alteration is responsible for field-isolated resistance to *Bacillus thuringiensis* Cry2A insecticidal proteins in two *Helicoverpa* species. PLoS One 5, e9975.
Caramori, T., Albertini, A.M., Galizzi, A., 1991. In vivo generation of hybrids between two *Bacillus thuringiensis* insect-toxin-encoding genes. Gene 98, 37–44.
Carlton, B.C., Gawron-Burke, C., 1993. Genetic Improvement of *Bacillus thuringiensis* for Bioinsecticide Development. Marcel Dekker Inc., New York, NY.
Carozzi, N.B., Kramer, V.C., Warren, G.W., Evola, S., Koziel, M.G., 1991. Prediction of insecticidal activity of *Bacillus thuringiensis* strains by polymerase chain reaction product profiles. Appl. Environ. Microbiol. 57, 3057–3061.
Carroll, J., Li, J., Ellar, D.J., 1989. Proteolytic processing of a coleopteran-specific delta-endotoxin produced by *Bacillus thuringiensis* var. *tenebrionis*. Biochem. J. 261, 99–105.
Carroll, J., Convents, D., Van Damme, J., Boets, A., Van Rie, J., Ellar, D.J., 1997. Intramolecular proteolytic cleavage of *Bacillus thuringiensis* Cry3A delta-endotoxin may facilitate its coleopteran toxicity. J. Invertebr. Pathol. 70, 41–49.
Carroll, M.W., Head, G., Caprio, M.A., 2012. When and where a seed mix refuge makes sense for managing insect resistance to Bt plants. Crop Prot. 38, 74–79.
Catangui, M.A., 2003. Transgenic *Bacillus thuringiensis* corn hybrid performance against univoltine ecotype European corn borer (Lepidoptera: Crambidae) in South Dakota. J. Econ. Entomol. 96, 957–968.
CERA, 2012. GM Crop Database. Center for Environmental Risk Assessment (CERA), ILSI Research Foundation, Washington D.C.http://cera-gmc.org/index.php?action=gm_crop_database.
Chang, C., Yu, Y.M., Dai, S.M., Law, S.K., Gill, S.S., 1993. High-level cryIVD and cytA gene expression in *Bacillus thuringiensis* does not require the 20-kilodalton protein, and the coexpressed gene products are synergistic in their toxicity to mosquitoes. Appl. Environ. Microbiol. 59, 815–821.
Chougule, N.P., Li, H., Liu, S., Linz, L.B., Narva, K.E., Meade, T., Bonning, B.C., 2013. Retargeting of the *Bacillus thuringiensis* toxin Cyt2Aa against hemipteran insect pests. Proc. Natl. Acad. Sci. U. S. A. 110, 8465–8470.
Cohen, S., Dym, O., Albeck, S., Ben-Dov, E., Cahan, R., Firer, M., Zaritsky, A., 2008. High-resolution crystal structure of activated Cyt2Ba monomer from Bacillus thuringiensis subsp. israelensis. J. Mol. Biol. 380, 820–827.
Cohen, S., Albeck, S., Ben-Dov, E., Cahan, R., Firer, M., Zaritsky, A., Dym, O., 2011. Cyt1Aa toxin: crystal structure reveals implications for its membrane-perforating function. J. Mol. Biol. 413, 804–814.
Contreras, E., Schoppmeier, M., Real, M.D., Rausell, C., 2013. Sodium solute symporter and cadherin proteins act as *Bacillus thuringiensis* Cry3Ba toxin functional receptors in *Tribolium castaneum*. J. Biol. Chem. 288, 18013–18021.
Corbin, D.R., Romano, C.P., 2006. Plants transformed to express Cry2A δ-endotoxins. U.S. patent number 7,064,249.
Craveiro, K.I., Gomes Junior, J.E., Silva, M.C., Macedo, L.L., Lucena, W.A., Silva, M.S., de Souza Junior, J.D., Oliveira, G.R., de Magalhaes, M.T., Santiago, A.D., Grossi-de-Sa, M.F., 2010. Variant Cry1Ia toxins generated by DNA shuffling are active against sugarcane giant borer. J. Biotechnol. 145, 215–221.
Crickmore, N., Zeigler, D.R., Feitelson, J., Schnepf, E., Van Rie, J., Lereclus, D., Baum, J., Dean, D.H., 1998. Revision of the nomenclature for the *Bacillus thuringiensis* pesticidal crystal proteins. Microbiol. Mol. Biol. Rev. 62, 807–813.

Crickmore, N., Baum, J., Bravo, A., Lereclus, D., Narva, K., Sampson, K., Schnepf, E., Sun, M., Zeigler, D.R., 2014. Bacillus thuringiensis toxin nomenclature. http://www.btnomenclature.info/.
Dean, D.H., Rajamohan, F., Lee, M.K., Wu, S.J., Chen, X.J., Alcantara, E., Hussain, S.R., 1996. Probing the mechanism of action of Bacillus thuringiensis insecticidal proteins by site-directed mutagenesis—a minireview. Gene 179, 111–117.
de Barjac, H., Bonnefoi, A., 1962. Essai de classification biochemique et serologique de 24 sourches de *Bacillus* du type *B. thuringiensis*. Entomophaga 7, 5–31.
de Barjac, H., Bonnefoi, A., 1968. A classification of strains of *Bacillus thuringiensis* Berliner with a key to their differentiation. J. Invertebr. Pathol. 11, 335–347.
de Freitas Bueno, R.C.O., de Freitas Bueno, A., Moscardi, F., Postali Parra, J.R., Hoffmann-Campo, C.B., 2011. Lepidopteran larva consumption of soybean foliage: basis for developing multiple-species economic thresholds for pest management decisions. Pest Manag. Sci. 67, 170–174.
de Maagd, R.A., Kwa, M.S., van der Klei, H., Yamamoto, T., Schipper, B., Vlak, J.M., Stiekema, W.J., Bosch, D., 1996a. Domain III substitution in *Bacillus thuringiensis* delta-endotoxin CryIA(b) results in superior toxicity for *Spodoptera exigua* and altered membrane protein recognition. Appl. Environ. Microbiol. 62, 1537–1543.
de Maagd, R.A., van der Klei, H., Bakker, P.L., Stiekema, W.J., Bosch, D., 1996b. Different domains of *Bacillus thuringiensis* delta-endotoxins can bind to insect midgut membrane proteins on ligand blots. Appl. Environ. Microbiol. 62, 2753–2757.
de Maagd, R.A., Bravo, A., Crickmore, N., 2001. How *Bacillus thuringiensis* has evolved specific toxins to colonize the insect world. Trends Genet. 17, 193–199.
de Maagd, R.A., Bravo, A., Berry, C., Crickmore, N., Schnepf, H.E., 2003. Structure, diversity, and evolution of protein toxins from spore-forming entomopathogenic bacteria. Annu. Rev. Genet. 37, 409–433.
DiFonzo, C., Cullen, E.M., 2013. Handy Bt Trait Table. University of Wisconsin. http://labs.russell.wisc.edu/cullenlab/extension/extension-publications/.
DiLeo, M.V., Bakker, M.d., Chu, E.Y., Hoekenga, O.A., 2014. An assessment of the relative influences of genetic background, functional diversity at major regulatory genes, and transgenic constructs on the tomato fruit metabolome. Plant Genome 7. http://dx.doi.org/10.3835/plantgenome2013.06.0021.
Dimock, M., Turner, J., Lampel, J., 1993. Endophytic microorganisms for delivery of genetically engineered microbial pesticides in plants. In: Kim, L. (Ed.), Advanced Engineered Biopesticides. Marcel Dekker Inc., New York, NY
Donovan, W.P., Rupar, M.J., Slaney, A.C., Malvar, T., Gawron-Burke, M.C., Johnson, T.B., 1992. Characterization of two genes encoding *Bacillus thuringiensis* insecticidal crystal proteins toxic to Coleoptera species. Appl. Environ. Microbiol. 58, 3921–3927.
Downes, S., Mahon, R., 2012. Successes and challenges of managing resistance in *Helicoverpa armigera* to Bt cotton in Australia. GM Crops Food 3, 228–234.
Dulmage, H.D., 1981. Insecticidal activity of isolates of *Bacillus thuringiensis* and their potential for pest control. In: Burges, H.D. (Ed.), Microbial Control of Pests and Diseases, 1970-1980. Academic Press, London, pp. 193–222.
Eichenseer, H., Strohbehn, R., Burks, J., 2008. Frequency and severity of western bean cutworm (Lepidoptera: Noctuidae) ear damage in transgenic corn hybrids expressing different *Bacillus thuringiensis* cry toxins. J. Econ. Entomol. 101, 555–563.
Ellis, R.T., Stockhoff, B.A., Stamp, L., Schnepf, H.E., Schwab, G.E., Knuth, M., Russell, J., Cardineau, G.A., Narva, K.E., 2002. Novel *Bacillus thuringiensis* binary insecticidal crystal proteins active on western corn rootworm, *Diabrotica virgifera virgifera* LeConte. Appl. Environ. Microbiol. 68, 1137–1145.

Estruch, J.J., Warren, G.W., Mullins, M.A., Nye, G.J., Craig, J.A., Koziel, M.G., 1996. Vip3A, a novel *Bacillus thuringiensis* vegetative insecticidal protein with a wide spectrum of activities against lepidopteran insects. Proc. Natl. Acad. Sci. U. S. A. 93, 5389–5394.

Fang, J., Xu, X., Wang, P., Zhao, J.Z., Shelton, A.M., Cheng, J., Feng, M.G., Shen, Z., 2007. Characterization of chimeric *Bacillus thuringiensis* Vip3 toxins. Appl. Environ. Microbiol. 73, 956–961.

Fang, Y., Li, Z., Liu, J., Shu, C., Wang, X., Zhang, X., Yu, X., Zhao, D., Liu, G., Hu, S., Zhang, J., Al-Mssallem, I., Yu, J., 2011. A pangenomic study of *Bacillus thuringiensis*. J. Genet. Genomics 38, 567–576.

FAO Expert Consultation on Genetically modified organisms in crop production and their effects on the environment: methodologies for monitoring and the way ahead. 2005. Food and Agriculture Organization of the United Nations. Rome, Italy. "http://www.fao.org/docrep/008/ae738e/ae738e00.htm" "TopOfPage"http://www.fao.org/docrep/008/ae738e/ae738e00.htm#TopOfPage.

Farinós, G.P., De La Poza, M., Hernández-Crespo, P., Ortego, F., Castañera, P., 2004. Resistance monitoring of field populations of the corn borers Sesamia nonagrioides and Ostrinia nubilalis after 5 years of Bt maize cultivation in Spain. Entomol. Exp. Appl. 110, 23–30.

Farinos, G.P., De la Poza, M., Ortego, F., Castanera, P., 2012. Susceptibility to the Cry1F toxin of field populations of *Sesamia nonagrioides* (Lepidoptera: Noctuidae) in Mediterranean maize cultivation regions. J. Econ. Entomol. 105, 214–221.

Federici, B.A., 2005. Insecticidal bacteria: an overwhelming success for invertebrate pathology. J. Invertebr. Pathol. 89, 30–38.

Federici, B.A., Bauer, L.S., 1998. Cyt1Aa protein of *Bacillus thuringiensis* is toxic to the cottonwood leaf beetle, chrysomela scripta, and suppresses high levels of resistance to Cry3Aa. Appl. Environ. Microbiol. 64, 4368–4371.

Fedoroff, N.V., 2011. Burdensome and unnecessary regulation. GM Crops 2, 87–88.

Feitelson, J.S., Payne, J., Kim, L., 1992. *Bacillus thuringiensis*: insects and beyond. Nat. Biotechnol. 10, 271–275.

Finer, J., Dhillon, T., 2007. Transgenic plant production. In: Stewart, C.N. (Ed.), Plant Biotechnology and Genetics: Principles, Techniques, and Applications. John Wiley & Sons, Inc., Hoboken, NJ, pp. 245–274

Fischhoff, D.A., Bowdish, K.S., Perlak, F.J., Marrone, P.G., McCormick, S.M., Niedermeyer, J.G., Dean, D.A., Kusano-Kretzmer, K., Mayer, E.J., Rochester, D.E., Rogers, S.G., Fraley, R.T., 1987. Insect tolerant transgenic tomato plants. Nat. Biotechnol. 5, 807–813.

Fox, J.L., 2001. EPA re-evaluates StarLink license. Nat. Biotechnol. 19, 11.

Gaertner, F.H., Quick, T.C., Thompson, M.A., 1993. Cell cap: an encapsulation system for insecticidal biotoxin proteins. In: Kim, L. (Ed.), Advanced Engineered Biopesticides. Marcel Dekker Inc., New York, NY, pp. 73–84.

Gahan, L.J., Gould, F., Heckel, D.G., 2001. Identification of a gene associated with Bt resistance in *Heliothis virescens*. Science 293, 857–860.

Gahan, L.J., Pauchet, Y., Vogel, H., Heckel, D.G., 2010. An ABC transporter mutation is correlated with insect resistance to *Bacillus thuringiensis* Cry1Ac toxin. PLoS Genet. 6, e1001248.

Galitsky, N., Cody, V., Wojtczak, A., Ghosh, D., Luft, J.R., Pangborn, W., English, L., 2001. Structure of the insecticidal bacterial delta-endotoxin Cry3Bb1 of *Bacillus thuringiensis*. Acta Crystallogr. D Biol. Crystallogr. 57, 1101–1109.

Gassmann, A.J., 2012. Field-evolved resistance to Bt maize by western corn rootworm: predictions from the laboratory and effects in the field. J. Invertebr. Pathol. 110, 287–293.

Gassmann, A.J., Petzold-Maxwell, J.L., Keweshan, R.S., Dunbar, M.W., 2011. Field-evolved resistance to bt maize by Western corn rootworm. PLoS One 6, e22629.

Gassmann, A.J., Petzold-Maxwell, J.L., Clifton, E.H., Dunbar, M.W., Hoffmann, A.M., Ingber, D.A., Keweshan, R.S., 2014. Field-evolved resistance by western corn rootworm to multiple Bacillus thuringiensis toxins in transgenic maize. Proc. Natl. Acad. Sci. 111, 5141–5146.

Ghimire, M.N., Huang, F., Leonard, R., Head, G.P., Yang, Y., 2011. Susceptibility of Cry1Ab-susceptible and -resistant sugarcane borer to transgenic corn plants containing single or pyramided Bacillus thuringiensis genes. Crop Prot. 30, 74–81.

Goldberg, L.J., Margalit, J., 1977. A bacterial spore demonstrating rapid larvicidal activity against *Anopheles sergentii*, Uranotaeniaunguiculata, *Culex univittatus*, *Aedes aegypti*, and *Culex pipiens*. Mosq. News 37, 317–324.

Gomez, I., Pardo-Lopez, L., Munoz-Garay, C., Fernandez, L.E., Perez, C., Sanchez, J., Soberon, M., Bravo, A., 2007. Role of receptor interaction in the mode of action of insecticidal Cry and Cyt toxins produced by *Bacillus thuringiensis*. Peptides 28, 169–173.

Gouffon, C., Van Vliet, A., Van Rie, J., Jansens, S., Jurat-Fuentes, J.L., 2011. Binding sites for Bacillus thuringiensis Cry2Ae toxin on heliothine brush border membrane vesicles are not shared with Cry1A, Cry1F, or Vip3A toxin. Appl. Environ. Microbiol. 77, 3182–3188.

Gould, F., 1998. Sustainability of transgenic insecticidal cultivars: integrating pest genetics and ecology. Annu. Rev. Entomol. 43, 701–726.

Gould, F., Blair, N., Reid, M., Rennie, T.L., Lopez, J., Micinski, S., 2002. *Bacillus thuringiensis*-toxin resistance management: stable isotope assessment of alternate host use by *Helicoverpa zea*. Proc. Natl. Acad. Sci. U. S. A. 99, 16581–16586.

Griffitts, J.S., Haslam, S.M., Yang, T., Garczynski, S.F., Mulloy, B., Morris, H., Cremer, P.S., Dell, A., Adang, M.J., Aroian, R.V., 2005. Glycolipids as receptors for *Bacillus thuringiensis* crystal toxin. Science 307, 922–925.

Grochulski, P., Masson, L., Borisova, S., Pusztai-Carey, M., Schwartz, J.L., Brousseau, R., Cygler, M., 1995. *Bacillus thuringiensis* CryIA(a) insecticidal toxin: crystal structure and channel formation. J. Mol. Biol. 254, 447–464.

Gunning, R.V., Dang, H.T., Kemp, F.C., Nicholson, I.C., Moores, G.D., 2005. New resistance mechanism in *Helicoverpa armigera* threatens transgenic crops expressing *Bacillus thuringiensis* Cry1Ac toxin. Appl. Environ. Microbiol. 71, 2558–2563.

Guo, S., Ye, S., Liu, Y., Wei, L., Xue, J., Wu, H., Song, F., Zhang, J., Wu, X., Huang, D., Rao, Z., 2009. Crystal structure of Bacillus thuringiensis Cry8Ea1: an insecticidal toxin toxic to underground pests, the larvae of Holotrichia parallela. J. Struct. Biol. 168, 259–266.

Gustafson, D.I., Head, G.P., Caprio, M.A., 2006. Modeling the impact of alternative hosts on *Helicoverpa zea* adaptation to bollgard cotton. J. Econ. Entomol. 99, 2116–2124.

Hammond, B.G., Jez, J.M., 2011. Impact of food processing on the dietary risk assessment for proteins introduced into biotechnology-derived soybean and corn crops. Food Chem. Toxicol. 49, 711–721.

Hammond, R.B., Higley, L.G., Pedigo, L.P., Bledsoe, L., Spomer, S.M., DeGooyer, T.A., 2000. Simulated Insect Defoliation on Soybean: Influence of Row Width. J. Econ. Entomol. 93, 1429–1436.

Han, S., Craig, J.A., Putnam, C.D., Carozzi, N.B., Tainer, J.A., 1999. Evolution and mechanism from structures of an ADP-ribosylating toxin and NAD complex. Nat. Struct. Mol. Biol. 6, 932–936.

Hannay, C.L., 1953. Crystalline inclusions in aerobic sporeforming bacteria. Nature 172, 1004.

Hansen, G., Wright, M.S., 1999. Recent advances in the transformation of plants. Trends Plant Sci. 4, 226–231.

Hardke, J.T., Leonard, B.R., Huang, F., Jackson, R.E., 2011. Damage and survivorship of fall armyworm (Lepidoptera: Noctuidae) on transgenic field corn expressing Bacillus thuringiensis Cry proteins. Crop Prot. 30, 168–172.

He, K., Wang, Z., Zhou, D., Wen, L., Song, Y., Yao, Z., 2003. Evaluation of transgenic Bt corn for resistance to the Asian corn borer (Lepidoptera: Pyralidae). J. Econ. Entomol. 96, 935–940.

Head, G.P., Greenplate, J., 2012. The design and implementation of insect resistance management programs for Bt crops. GM Crops Food 3, 144–153.

Head, G., Carroll, M., Clark, T., Galvan, T., Huckaba, R.M., Price, P., Samuel, L., Storer, N.P., 2014. Efficacy of SmartStax(r) insect-protected corn hybrids against corn rootworm: the value of pyramiding the Cry3Bb1 and Cry34/35Ab1 proteins. Crop Prot. 57, 38–47.

Heckel, D.G., Gahan, L.J., Baxter, S.W., Zhao, J.Z., Shelton, A.M., Gould, F., Tabashnik, B.E., 2007. The diversity of Bt resistance genes in species of Lepidoptera. J. Invertebr. Pathol. 95, 192–197.

Hellmich, R.L., Siegfried, B.D., Sears, M.K., Stanley-Horn, D.E., Daniels, M.J., Mattila, H.R., Spencer, T., Bidne, K.G., Lewis, L.C., 2001. Monarch larvae sensitivity to Bacillus thuringiensis- purified proteins and pollen. Proc. Natl. Acad. Sci. U. S. A. 98, 11925–11930.

Herman, R.A., Price, W.D., 2013. Unintended compositional changes in genetically modified (gm) crops: 20 years of research. J. Agric. Food Chem. 61, 11695–11701.

Herman, R.A., Scherer, P.N., Young, D.L., Mihaliak, C.A., Meade, T., Woodsworth, A.T., Stockhoff, B.A., Narva, K.E., 2002. Binary insecticidal crystal protein from *Bacillus thuringiensis*, strain PS149B1: effects of individual protein components and mixtures in laboratory bioassays. J. Econ. Entomol. 95, 635–639.

Herman, R.A., Chassy, B.M., Parrott, W., 2009. Compositional assessment of transgenic crops: an idea whose time has passed. Trends Biotechnol. 27, 555–557.

Hernandez-Martinez, P., Naseri, B., Navarro-Cerrillo, G., Escriche, B., Ferre, J., Herrero, S., 2010. Increase in midgut microbiota load induces an apparent immune priming and increases tolerance to *Bacillus thuringiensis*. Environ. Microbiol. 12, 2730–2737.

Hernandez-Martinez, P., Hernandez-Rodriguez, C.S., Rie, J.V., Escriche, B., Ferre, J., 2013. Insecticidal activity of Vip3Aa, Vip3Ad, Vip3Ae, and Vip3Af from *Bacillus thuringiensis* against lepidopteran corn pests. J. Invertebr. Pathol. 113, 78–81.

Hernandez-Rodriguez, C.S., Van Vliet, A., Bautsoens, N., Van Rie, J., Ferre, J., 2008. Specific binding of *Bacillus thuringiensis* Cry2A insecticidal proteins to a common site in the midgut of *Helicoverpa* species. Appl. Environ. Microbiol. 74, 7654–7659.

Herrero, S., Gechev, T., Bakker, P.L., Moar, W.J., de Maagd, R.A., 2005. Bacillus thuringiensis Cry1Ca-resistant Spodoptera exigua lacks expression of one of four aminopeptidase N genes. BMC Genomics 6, 96.

Herrnstadt, C., Gilroy, T.E., Sobieski, D.A., Bennett, B.D., Gaertner, F.H., 1987. Nucleotide sequence and deduced amino acid sequence of a coleopteran-active delta-endotoxin gene from *Bacillus thuringiensis* subsp. *san diego*. Gene 57, 37–46.

Hofte, H., Whiteley, H.R., 1989. Insecticidal crystal proteins of *Bacillus thuringiensis*. Microbiol. Rev. 53, 242–255.

Hofte, H., Van Rie, J., Jansens, S., Van Houtven, A., Vanderbruggen, H., Vaeck, M., 1988. Monoclonal antibody analysis and insecticidal spectrum of three types of lepidopteran-specific insecticidal crystal proteins of *Bacillus thuringiensis*. Appl. Environ. Microbiol. 54, 2010–2017.

Hui, F., Scheib, U., Hu, Y., Sommer, R.J., Aroian, R.V., Ghosh, P., 2012. Structure and glycolipid binding properties of the nematicidal protein Cry5B. Biochemistry 51, 9911–9921.

Husz, B., 1930. Field experiments on the application of *Bacillus thuringiensis* against the corn borer. Int. Corn Borer Invest. Sci. Rep. 3, 91–98.

Ishiwata, S., 1901. On a type of severe Xacherie (sotto disease). Dainihon Sanshi Kaiho 114, 1–5.

Ives, A.R., Glaum, P.R., Ziebarth, N.L., Andow, D.A., 2011. The evolution of resistance to two-toxin pyramid transgenic crops. Ecol. Appl. 21, 503–515.

Jackson, R.E., Bradley Jr., J.R., Van Duyn, J.W., Gould, F., 2004. Comparative production of *Helicoverpa zea* (Lepidoptera: Noctuidae) from transgenic cotton expressing either one or two *Bacillus thuringiensis* proteins with and without insecticide oversprays. J. Econ. Entomol. 97, 1719–1725.

James, C., 2013. Global Status of Commercialize/Biotech Crops: 2013. ISAAA, Ithaca, NY, ISAAA Brief No. 46.

Jones, G.W., Nielsen-Leroux, C., Yang, Y., Yuan, Z., Dumas, V.F., Monnerat, R.G., Berry, C., 2007. A new Cry toxin with a unique two-component dependency from *Bacillus sphaericus*. FASEB J. 21, 4112–4120.

Jung, Y.C., Kim, S.U., Cote, J.C., Lecadet, M.M., Chung, Y.S., Bok, S.H., 1998. Characterization of a new *Bacillus thuringiensis* Subsp. *higo* strain isolated from rice bran in Korea. J. Invertebr. Pathol. 71, 95–96.

Jurat-Fuentes, J.L., Gahan, L.J., Gould, F.L., Heckel, D.G., Adang, M.J., 2004. The HevCaLP protein mediates binding specificity of the Cry1A class of *Bacillus thuringiensis* toxins in *Heliothis virescens*. Biochemistry 43, 14299–14305.

Kaur, S., 2000. Molecular approaches towards development of novel *Bacillus thuringiensis* biopesticides. World J. Microbiol. Biotechnol. 16, 781–793.

Keller, M., Sneh, B., Strizhov, N., Prudovsky, E., Regev, A., Koncz, C., Schell, J., Zilberstein, A., 1996. Digestion of delta-endotoxin by gut proteases may explain reduced sensitivity of advanced instar larvae of *Spodoptera littoralis* to CryIC. Insect Biochem. Mol. Biol. 26, 365–373.

Khajuria, C., Buschman, L.L., Chen, M.S., Siegfried, B.D., Zhu, K.Y., 2011. Identification of a novel aminopeptidase P-like gene (OnAPP) possibly involved in Bt toxicity and resistance in a major corn pest (*Ostrinia nubilalis*). PLoS One 6, e23983.

Koziel, M.G., Beland, G.L., Bowman, C., Carozzi, N.B., Crenshaw, R., Crossland, L., Dawson, J., Desai, N., Hill, M., Kadwell, S., Launis, K., Lewis, K., Maddox, D., McPherson, K., Meghji, M.R., Merlin, E., Rhodes, R., Warren, G.W., Wright, M., Evola, S.V., 1993. Field performance of elite transgenic maize plants expressing an insecticidal protein derived from Bacillus thuringiensis. Nat. Biotechnol. 11, 194–200.

Krieg, A., Huger, A.M., Langenbruch, G.A., Schnetter, W., 1983. Bacillus thuringiensis var. tenebrionis: a new pathotype effective against larvae of Coleoptera. Z. Angew. Entomol. 96, 500–508.

Kronstad, J.W., Whiteley, H.R., 1986. Three classes of homologous *Bacillus thuringiensis* crystal-protein genes. Gene 43, 29–40.

Kronstad, J.W., Schnepf, H.E., Whiteley, H.R., 1983. Diversity of locations for *Bacillus thuringiensis* crystal protein genes. J. Bacteriol. 154, 419–428.

Kullik, S.A., Sears, M.K., Schaafsma, A.W., 2011. Sublethal effects of Cry 1F Bt corn and clothianidin on black cutworm (Lepidoptera: Noctuidae) larval development. J. Econ. Entomol. 104, 484–493.

Ladics, G.S., Cressman, R.F., Herouet-Guicheney, C., Herman, R.A., Privalle, L., Song, P., Ward, J.M., McClain, S., 2011. Bioinformatics and the allergy assessment of agricultural biotechnology products: industry practices and recommendations. Regul. Toxicol. Pharmacol. 60, 46–53.

Lampel, J.S., Canter, G.L., Dimock, M.B., Kelly, J.L., Anderson, J.J., Uratani, B.B., Foulke, J.S., Turner, J.T., 1994. Integrative cloning, expression, and stability of the cryIA(c) gene from *Bacillus thuringiensis* subsp. *kurstaki* in a recombinant strain of *Clavibacter xyli* subsp. *cynodontis*. Appl. Environ. Microbiol. 60, 501–508.

Lawson, E.C., Weiss, J.D., Thomas, P.E., Kaniewski, W.K., 2001. NewLeaf Plus(r) Russet Burbank potatoes: replicase-mediated resistance to potato leafroll virus. Mol. Breed. 7, 1–12.

Lecadet, M.M., Chaufaux, J., Ribier, J., Lereclus, D., 1992. Construction of novel *Bacillus thuringiensis* strains with different insecticidal activities by transduction and transformation. Appl. Environ. Microbiol. 58, 840–849.

Lee, M.K., Walters, F.S., Hart, H., Palekar, N., Chen, J.S., 2003. The mode of action of the *Bacillus thuringiensis* vegetative insecticidal protein Vip3A differs from that of Cry1Ab delta-endotoxin. Appl. Environ. Microbiol. 69, 4648–4657.

Lee, M.K., Miles, P., Chen, J.S., 2006. Brush border membrane binding properties of *Bacillus thuringiensis* Vip3A toxin to *Heliothis virescens* and *Helicoverpa zea* midguts. Biochem. Biophys. Res. Commun. 339, 1043–1047.

Lee, H., Park, S.Y., Zhang, Z.J., 2013. An overview of genetic transformation of soybean. In: Board, J. (Ed.), A Comprehensive Survey of International Soybean Research - Genetics. Physiology, Agronomy and Nitrogen Relationships. InTech, pp. 439–506. http://www.intechopen.com/books/a-comprehensive-survey-of-international-soybean-research-genetics-physiology-agronomy-and-nitrogen-relationships/an-overview-of-genetic-transformation-of-soybean.

Letowski, J., Bravo, A., Brousseau, R., Masson, L., 2005. Assessment of cry1 gene contents of *Bacillus thuringiensis* strains by use of DNA microarrays. Appl. Environ. Microbiol. 71, 5391–5398.

Li, J.D., Carroll, J., Ellar, D.J., 1991. Crystal structure of insecticidal delta-endotoxin from *Bacillus thuringiensis* at 2.5 A resolution. Nature 353, 815–821.

Li, J., Koni, P.A., Ellar, D.J., 1996. Structure of the mosquitocidal delta-endotoxin CytB from *Bacillus thuringiensis* sp. *kyushuensis* and implications for membrane pore formation. J. Mol. Biol. 257, 129–152.

Li, H., Oppert, B., Higgins, R.A., Huang, F., Zhu, K.Y., Buschman, L.L., 2004. Comparative analysis of proteinase activities of *Bacillus thuringiensis*-resistant and -susceptible *Ostrinia nubilalis* (Lepidoptera: Crambidae). Insect Biochem. Mol. Biol. 34, 753–762.

Li, H., Olson, M., Lin, G., Hey, T., Tan, S.Y., Narva, K.E., 2013. Bacillus thuringiensis Cry34Ab1/Cry35Ab1 interactions with western corn rootworm midgut membrane binding sites. PLoS One 8, e53079.

Likitvivatanavong, S., Chen, J., Bravo, A., Soberon, M., Gill, S.S., 2011a. Cadherin, alkaline phosphatase, and aminopeptidase N as receptors of Cry11Ba toxin from *Bacillus thuringiensis* subsp. *jegathesan* in *Aedes aegypti*. Appl. Environ. Microbiol. 77, 24–31.

Likitvivatanavong, S., Chen, J., Evans, A.M., Bravo, A., Soberon, M., Gill, S.S., 2011b. Multiple receptors as targets of Cry toxins in mosquitoes. J. Agric. Food Chem. 59, 2829–2838.

Liu, X.S., Dean, D.H., 2006. Redesigning *Bacillus thuringiensis* Cry1Aa toxin into a mosquito toxin. Protein Eng. Des. Sel. 19, 107–111.

Luttrell, R.G., Jackson, R.E., 2012. *Helicoverpa zea* and Bt cotton in the United States. GM Crops Food Biotechnol. Agric. Food Chain 3, 213–227.

Ma, G., Schmidt, O., Keller, M., 2012. Pre-feeding of a glycolipid binding protein LEC-8 from *Caenorhabditis elegans* revealed enhanced tolerance to Cry1Ac toxin in *Helicoverpa armigera*. Results Immunol. 2, 97–103.

MacIntosh, S.C., 2010. Managing the risk of insect resistance to transgenic insect control traits: practical approaches in local environments. Pest Manag. Sci. 66, 100–106.

MacIntosh, S.C., McPherson, S.L., Perlak, F.J., Marrone, P.G., Fuchs, R.L., 1990. Purification and characterization of *Bacillus thuringiensis* var. *tenebrionis* insecticidal proteins produced in *E. coli*. Biochem. Biophys. Res. Commun. 170, 665–672.

Macrae, T.C., Baur, M.E., Boethel, D.J., Fitzpatrick, B.J., Gao, A.G., Gamundi, J.C., Harrison, L.A., Kabuye, V.T., Mcpherson, R.M., Miklos, J.A., Paradise, M.S., Toedebusch, A.S., Viegas, A., 2005. Laboratory and field evaluations of transgenic soybean exhibiting high-dose expression of a synthetic Bacillus thuringiensis cry1A gene for control of Lepidoptera. J. Econ. Entomol. 98, 577–587.

Marra, M.C., Piggott, N.E., Goodwin, B.K., 2012. The impact of corn rootworm protected biotechnology traits in the United States. AgBioForum 15, 217–230.

Masson, L., Schwab, G., Mazza, A., Brousseau, R., Potvin, L., Schwartz, J.L., 2004. A novel *Bacillus thuringiensis* (PS149B1) containing a Cry34Ab1/Cry35Ab1 binary toxin specific for the western corn rootworm *Diabrotica virgifera virgifera* LeConte forms ion channels in lipid membranes. Biochemistry 43, 12349–12357.

Mattes, O., 1927. Parasitare krankheiten der mehimottenlarven und versuche uber ihre verwendbarkeit als biologisches bekamfungsmittel. Ges Naturw Marburg Schrift. 381.

McDougall, P., 2011. The Cost and Time Involved in the Discovery, Development and Authorisation of a New Plant Biotechnology Derived Trait. A Consultancy Study for CropLife International. http://www.croplife.org/PhillipsMcDougallStudy.

McPherson, R.M., MacRae, T.C., 2009. Evaluation of transgenic soybean exhibiting high expression of a synthetic Bacillus thuringiensis cry1A transgene for suppressing lepidopteran population densities and crop injury. J. Econ. Entomol. 102, 1640–1648.

Mendelsohn, M., Kough, J., Vaituzis, Z., Matthews, K., 2003. Are Bt crops safe? Nat. Biotechnol. 21, 1003–1009.

Morin, S., Biggs, R.W., Sisterson, M.S., Shriver, L., Ellers-Kirk, C., Higginson, D., Holley, D., Gahan, L.J., Heckel, D.G., Carriere, Y., Dennehy, T.J., Brown, J.K., Tabashnik, B.E., 2003. Three cadherin alleles associated with resistance to *Bacillus thuringiensis* in pink bollworm. Proc. Natl. Acad. Sci. U. S. A. 100, 5004–5009.

Morse, R.J., Yamamoto, T., Stroud, R.M., 2001. Structure of Cry2Aa suggests an unexpected receptor binding epitope. Structure 9, 409–417.

Mumm, R.H., 2013. A look at product development with genetically modified crops: examples from maize. J. Agric. Food Chem. 61, 8254–8259.

Naimov, S., Weemen-Hendriks, M., Dukiandjiev, S., de Maagd, R.A., 2001. *Bacillus thuringiensis* delta-endotoxin Cry1 hybrid proteins with increased activity against the Colorado potato beetle. Appl. Environ. Microbiol. 67, 5328–5330.

Naranjo, S.E., 2010. Impacts of Bt transgenic cotton on integrated pest management. J. Agric. Food Chem. 59, 5842–5851.

Narva, K., Siegfried, B., Storer, N., 2013. Transgenic approaches to western corn rootworm control. In: Vilcinskas, A. (Ed.), Yellow Biotechnology II. Springer, Berlin, pp. 135–162.

Noguera, P.A., Ibarra, J.E., 2010. Detection of new cry genes of *Bacillus thuringiensis* by use of a novel PCR primer system. Appl. Environ. Microbiol. 76, 6150–6155.

Ochoa-Campuzano, C., Real, M.D., Martinez-Ramirez, A.C., Bravo, A., Rausell, C., 2007. An ADAM metalloprotease is a Cry3Aa *Bacillus thuringiensis* toxin receptor. Biochem. Biophys. Res. Commun. 362, 437–442.

Ohba, M., Mizuki, E., Uemori, A., 2009. Parasporin, a new anticancer protein group from *Bacillus thuringiensis*. Anticancer Res. 29, 427–433.

Oliveira, G.R., Silva, M.C., Lucena, W.A., Nakasu, E.Y., Firmino, A.A., Beneventi, M.A., Souza, D.S., Gomes Jr., J.E., de Souza Jr., J.D., Rigden, D.J., Ramos, H.B., Soccol, C.R., Grossi-de-Sa, M.F., 2011. Improving Cry8Ka toxin activity towards the cotton boll weevil (*Anthonomus grandis*). BMC Biotechnol. 11, 85.

Onstad, D.W., Mitchell, P.D., Hurley, T.M., Lundgren, J.G., Porter, R.P., Krupke, C.H., Spencer, J.L., DiFonzo, C.D., Baute, T.S., Hellmich, R.L., Buschman, L.L., Hutchison, W.D., Tooker, J.F., 2011. Seeds of change: corn seed mixtures for resistance management and integrated pest management. J. Econ. Entomol. 104, 343–352.

Oppert, B., Kramer, K.J., Beeman, R.W., Johnson, D., McGaughey, W.H., 1997. Proteinase-mediated insect resistance to *Bacillus thuringiensis* toxins. J. Biol. Chem. 272, 23473–23476.

Pardo-Lopez, L., Munoz-Garay, C., Porta, H., Rodriguez-Almazan, C., Soberon, M., Bravo, A., 2009. Strategies to improve the insecticidal activity of Cry toxins from *Bacillus thuringiensis*. Peptides 30, 589–595.

Pardo-Lopez, L., Soberon, M., Bravo, A., 2013. *Bacillus thuringiensis* insecticidal three-domain Cry toxins: mode of action, insect resistance and consequences for crop protection. FEMS Microbiol. Rev. 37, 3–22.

Payne, J., Sick, A.J., 1993. *Bacillus thuringiensis* isolate active against lepidopteran pests, and genes encoding novel lepidopteran-active toxins. U.S. patent number 5,188,960.

Payne, J.M., Narva, K.E., Uyeda, K.A., Stalder, C.J., Michaels, T.M., 1995. *Bacillus thuringiensis* isolate PS201T6 toxin. U.S. patent number 5,436,002.

Perez, C., Fernandez, L.E., Sun, J., Folch, J.L., Gill, S.S., Soberon, M., Bravo, A., 2005. *Bacillus thuringiensis* subsp. *israelensis* Cyt1Aa synergizes Cry11Aa toxin by functioning as a membrane-bound receptor. Proc. Natl. Acad. Sci. U. S. A. 102, 18303–18308.

Perez, C., Munoz-Garay, C., Portugal, L.C., Sanchez, J., Gill, S.S., Soberon, M., Bravo, A., 2007. *Bacillus thuringiensis* ssp. *israelensis* Cyt1Aa enhances activity of Cry11Aa toxin by facilitating the formation of a pre-pore oligomeric structure. Cell. Microbiol. 9, 2931–2937.

Perlak, F.J., Fuchs, R.L., Dean, D.A., McPherson, S.L., Fischhoff, D.A., 1991. Modification of the coding sequence enhances plant expression of insect control protein genes. Proc. Natl. Acad. Sci. U. S. A. 88, 3324–3328.

Petolino, J., Arnold, N., 2009. Whiskers-mediated maize transformation. In: Scott, M.P. (Ed.), Transgenic Maize. Humana Press, Clifton, NJ, pp. 59–67.

Petzold-Maxwell, J.L., Meinke, L.J., Gray, M.E., Estes, R.E., Gassmann, A.J., 2013. Effect of Bt maize and soil insecticides on yield, injury, and rootworm survival: implications for resistance management. J. Econ. Entomol. 106, 1941–1951.

Pigott, C.R., Ellar, D.J., 2007. Role of receptors in *Bacillus thuringiensis* crystal toxin activity. Microbiol. Mol. Biol. Rev. 71, 255–281.

Porcar, M., Juarez-Perez, V., 2003. PCR-based identification of *Bacillus thuringiensis* pesticidal crystal genes. FEMS Microbiol. Rev. 26, 419–432.

Privalle, L.S., Chen, J., Clapper, G., Hunst, P., Spiegelhalter, F., Zhong, C.X., 2012. Development of an agricultural biotechnology crop product: testing from discovery to commercialization. J. Agric. Food Chem. 60, 10179–10187.

Qiao, F., Huang, J., Rozelle, S., Wilen, J., 2010. Natural refuge crops, buildup of resistance, and zero-refuge strategy for Bt cotton in China. Sci. China Life Sci. 53, 1227–1238.

Radauer, C., Bublin, M., Wagner, S., Mari, A., Breiteneder, H., 2008. Allergens are distributed into few protein families and possess a restricted number of biochemical functions. J. Allergy Clin. Immunol. 121, 847–852, e847.

Rahman, M.M., Roberts, H.L., Schmidt, O., 2004. The development of the endoparasitoid *Venturia canescens* in Bt-tolerant, immune induced larvae of the flour moth *Ephestia kuehniella*. J. Invertebr. Pathol. 87, 129–131.

Rajagopal, R., Sivakumar, S., Agrawal, N., Malhotra, P., Bhatnagar, R.K., 2002. Silencing of midgut aminopeptidase N of *Spodoptera litura* by double-stranded RNA establishes its role as *Bacillus thuringiensis* toxin receptor. J. Biol. Chem. 277, 46849–46851.

Raybould, A., Higgins, L., Horak, M., Layton, R., Storer, N., De La Fuente, J., Herman, R., 2012. Assessing the ecological risks from the persistence and spread of feral populations of insect-resistant transgenic maize. Transgenic Res. 21, 655–664.

Reay-Jones, F.P.F., Wiatrak, P., Greene, J.K., 2009. Evaluating the performance of transgenic corn producing Bacillus thuringiensis toxins in South Carolina. J. Agric. Urban Entomol. 26, 77–86.

Reed, G.L., Jensen, A.S., Riebe, J., Head, G., Duan, J.J., 2001. Transgenic Bt potato and conventional insecticides for Colorado potato beetle management: comparative efficacy and non-target impacts. Entomol. Exp. Appl. 100, 89–100.

Ricroch, A.E., 2013. Assessment of GE food safety using '-omics' techniques and long-term animal feeding studies. Nat. Biotechnol. 30, 349–354.

Ricroch, A.E., Berge, J.B., Kuntz, M., 2011. Evaluation of genetically engineered crops using transcriptomic, proteomic and metabolomic profiling techniques. Plant Physiol. 155, 1752–1761.

Romeis, J., Hellmich, R., Candolfi, M., Carstens, K., De Schrijver, A., Gatehouse, A., Herman, R., Huesing, J., McLean, M., Raybould, A., Shelton, A., Waggoner, A., 2011. Recommendations for the design of laboratory studies on non-target arthropods for risk assessment of genetically engineered plants. Transgenic Res. 20, 1–22.

Rosellini, D., 2012. Selectable markers and reporter genes: a well furnished toolbox for plant science and genetic engineering. Crit. Rev. Plant Sci. 31, 401–453.

Roush, R.T., Shelton, A.M., 1997. Assessing the odds: the emergence of resistance to Bt transgenic plants. Nat. Biotechnol. 15, 816–817.

Rule, D.M., Nolting, S.P., Prasifka, P.L., Storer, N.P., Hopkins, B.W., Scherder, E.F., Siebert, M.W., Hendrix, W.H., 2014. Efficacy of pyramided Bt proteins Cry1F, Cry1A.105, and Cry2Ab2 expressed in SmartStax corn hybrids against lepidopteran insect pests in the northern United States. J. Econ. Entomol. 107, 403–409.

Rupar, M.J., Donovan, W.P., Tan, Y., Annette, S., 2000. Bacillus thuringiensis CryET29 compositions toxic to coleopteran insects and Ctenocephalides spp. U.S. patent number 6,093,69.

Saiki, R.K., Gelfand, D.H., Stoffel, S., Scharf, S.J., Higuchi, R., Horn, G.T., Mullis, K.B., Erlich, H.A., 1988. Primer-directed enzymatic amplification of DNA with a thermostable DNA polymerase. Science 239, 487–491.

Sanahuja, G., Banakar, R., Twyman, R.M., Capell, T., Christou, P., 2011. *Bacillus thuringiensis*: a century of research, development and commercial applications. Plant Biotechnol. J. 9, 283–300.

Schnepf, H.E., Whiteley, H.R., 1981. Cloning and expression of the *Bacillus thuringiensis* crystal protein gene in *Escherichia coli*. Proc. Natl. Acad. Sci. U. S. A. 78, 2893–2897.

Schnepf, E., Crickmore, N., Van Rie, J., Lereclus, D., Baum, J., Feitelson, J., Zeigler, D.R., Dean, D.H., 1998. *Bacillus thuringiensis* and its pesticidal crystal proteins. Microbiol. Mol. Biol. Rev. 62, 775–806.

Schnepf, H.E., Lee, S., Dojillo, J., Burmeister, P., Fencil, K., Morera, L., Nygaard, L., Narva, K.E., Wolt, J.D., 2005. Characterization of Cry34/Cry35 binary insecticidal proteins from diverse *Bacillus thuringiensis* strain collections. Appl. Environ. Microbiol. 71, 1765–1774.

Sears, M.K., Hellmich, R.L., Stanley-Horn, D.E., Oberhauser, K.S., Pleasants, J.M., Mattila, H.R., Siegfried, B.D., Dively, G.P., 2001. Impact of Bt corn pollen on monarch butterfly populations: a risk assessment. Proc. Natl. Acad. Sci. U. S. A. 98, 11937–11942.

Sekar, V., Thompson, D.V., Maroney, M.J., Bookland, R.G., Adang, M.J., 1987. Molecular cloning and characterization of the insecticidal crystal protein gene of *Bacillus thuringiensis* var. *tenebrionis*. Proc. Natl. Acad. Sci. U. S. A. 84, 7036–7040.

Sena, J.A., Hernandez-Rodriguez, C.S., Ferre, J., 2009. Interaction of *Bacillus thuringiensis* Cry1 and Vip3A proteins with *Spodoptera frugiperda* midgut binding sites. Appl. Environ. Microbiol. 75, 2236–2237.

Shan, G., 2011. Immunoassays in Agricultural Biotechnology. John Wiley & Sons, Hoboken, NJ.

Shelton, A.M., Zhao, J.-Z., Roush, R.T., 2002. Economic, ecological, food safety, and social consequences of the deployment of BT transgenic plants. Annu. Rev. Entomol. 47, 845–881.

Shelton, A.M., Olmstead, D.L., Burkness, E.C., Hutchison, W.D., Dively, G., Welty, C., Sparks, A.N., 2013. Multi-state trials of Bt sweet corn varieties for control of the corn earworm (Lepidoptera: Noctuidae). J. Econ. Entomol. 106, 2151–2159.

Siebert, M.W., Nolting, S.P., Hendrix, W., Dhavala, S., Craig, C., Leonard, B.R., Stewart, S.D., All, J., Musser, F.R., Buntin, G.D., Samuel, L., 2012. Evaluation of corn hybrids expressing Cry1F, Cry1A.105, Cry2Ab2, Cry34Ab1/Cry35Ab1, and Cry3Bb1 against southern United States insect pests. J. Econ. Entomol. 105, 1825–1834.
Siegel, J.P., 2001. The mammalian safety of *Bacillus thuringiensis*-based insecticides. J. Invertebr. Pathol. 77, 13–21.
Slaney, A.C., Robbins, H.L., English, L., 1992. Mode of action of *Bacillus thuringiensis* toxin CryIIIA: an analysis of toxicity in *Leptinotarsa decemlineata* (Say) and *Diabrotica undecimpunctata howardi* Barber. Insect Biochem. Mol. Biol. 22, 9–18.
Smith, R.H., Hood, E.E., 1995. Agrobacterium-tumefaciens transformation of monocotyledons. Crop Sci. 35, 301–309.
Soberon, M., Pardo-Lopez, L., Lopez, I., Gomez, I., Tabashnik, B.E., Bravo, A., 2007. Engineering modified Bt toxins to counter insect resistance. Science 318, 1640–1642.
Soberon, M., Gill, S.S., Bravo, A., 2009. Signaling versus punching hole: how do *Bacillus thuringiensis* toxins kill insect midgut cells? Cell. Mol. Life Sci. 66, 1337–1349.
Soberon, M., Pardo, L., Munoz-Garay, C., Sanchez, J., Gomez, I., Porta, H., Bravo, A., 2010. Pore formation by Cry toxins. Adv. Exp. Med. Biol. 677, 127–142.
Soberon, M., Lopez-Diaz, J.A., Bravo, A., 2013. Cyt toxins produced by *Bacillus thuringiensis*: a protein fold conserved in several pathogenic microorganisms. Peptides 41, 87–93.
Stemmer, W.P., 1994. DNA shuffling by random fragmentation and reassembly: in vitro recombination for molecular evolution. Proc. Natl. Acad. Sci. U. S. A. 91, 10747–10751.
Storer, N.P., Kubiszak, M.E., Ed King, J., Thompson, G.D., Santos, A.C., 2012a. Status of resistance to Bt maize in Spodoptera frugiperda: lessons from Puerto Rico. J. Invertebr. Pathol. 110, 294–300.
Storer, N.P., Thompson, G.D., Head, G.P., 2012b. Application of pyramided traits against Lepidoptera in insect resistance management for *Bt* crops. GM Crops Food Biotechnol. Agric. Food Chain 3, 154–162.
Storer, N.P., Thompson, G.D., Head, G.P., 2012c. Application of pyramided traits against Lepidoptera in insect resistance management for Bt crops. GM Crops Food 3, 154–162.
Tabashnik, B.E., Gould, F., 2012. Delaying corn rootworm resistance to Bt corn. J. Econ. Entomol. 105, 767–776.
Tabashnik, B.E., Cushing, N.L., Finson, N., Johnson, M.W., 1990. Field development of resistance to *Bacillus thuringiensis* in diamondback moth (Lepidoptera: Plutellidae). J. Econ. Entomol. 83, 1671–1676.
Tabashnik, B.E., Liu, Y., Malvar, T., Heckel, D.G., Masson, L., Ferré, J., 1998. Insect resistance to *Bacillus thuringiensis*: uniform or diverse? Philos. Trans. R. Soc. Lond. B Biol. Sci. 353, 1751–1756.
Tabashnik, B.E., Huang, F., Ghimire, M.N., Leonard, B.R., Siegfried, B.D., Rangasamy, M., Yang, Y., Wu, Y., Gahan, L.J., Heckel, D.G., Bravo, A., Soberon, M., 2011. Efficacy of genetically modified Bt toxins against insects with different genetic mechanisms of resistance. Nat. Biotechnol. 29, 1128–1131.
Tabashnik, B.E., Morin, S., Unnithan, G.C., Yelich, A.J., Ellers-Kirk, C., Harpold, V.S., Sisterson, M.S., Ellsworth, P.C., Dennehy, T.J., Antilla, L., Liesner, L., Whitlow, M., Staten, R.T., Fabrick, J.A., Li, X., Carriere, Y., 2012. Sustained susceptibility of pink bollworm to Bt cotton in the United States. GM Crops Food 3, 194–200.
Tabashnik, B.E., Brevault, T., Carriere, Y., 2013. Insect resistance to Bt crops: lessons from the first billion acres. Nat. Biotechnol. 31, 510–521.
Thomas, W.E., Ellar, D.J., 1983. *Bacillus thuringiensis* var *israelensis* crystal delta-endotoxin: effects on insect and mammalian cells in vitro and in vivo. J. Cell Sci. 60, 181–197.

Tomasino, S.F., Leister, R.T., Dimock, M.B., Beach, R.M., Kelly, J.L., 1995. Field performance of *Clavibacter xyli* subsp. *cynodontis* expressing the insecticidal protein gene cryIA(c) of *Bacillus thuringiensis* against European corn borer in field corn. Biol. Control 5, 442–448.

U.S. Environmental Protection Agency, 2001a. Biopesticides Registration Action Document—*Bacillus thuringiensis* (*Bt*) Plant-Incorporated Protectants. Office of Pesticide Programs, Biopesticides and Pollution Prevention Division, Washington D.C.

U.S. Environmental Protection Agency, 2001b. A set of scientific issues being considered by the environmental protection agency regarding: Bt plant-pesticides risk and benefit assessments: insect resistance management: FIFRA Scientific Advisory Panel Report No. 2000-07a.

U.S. Environmental Protection Agency, 2007. White Paper on Tier-Based Testing for the Effects of Proteinaceous Insecticidal Plant-Incorporated Protectants on Non-Target Arthropods for Regulatory Risk Assessments. Washington D.C. http://www.epa.gov/opp00001/biopesticides/pips/non-target-arthropods.pdf.

U.S. Environmental Protection Agency, 2011. Updated EPA Review of ABSTC's 2010 Corn Insect Resistance Management Compliance Assurance Program. Memo from J. Martinez to M. Mendelsohn, Office of Chemical Safety and Pollution Prevention, Biopesticides and Pollution Prevention Division, Washington D.C.

Vachon, V., Laprade, R., Schwartz, J.L., 2012. Current models of the mode of action of *Bacillus thuringiensis* insecticidal crystal proteins: a critical review. J. Invertebr. Pathol. 111, 1–12.

Vaeck, M., Reynaerts, A., Hofte, H., Jansens, S., De Beuckeleer, M., Dean, C., Zabeau, M., Montagu, M.V., Leemans, J., 1987. Transgenic plants protected from insect attack. Nature 328, 33–37.

van Frankenhuyzen, K., 2009. Insecticidal activity of *Bacillus thuringiensis* crystal proteins. J. Invertebr. Pathol. 101, 1–16.

van Frankenhuyzen, K., 2013. Cross-order and cross-phylum activity of *Bacillus thuringiensis* pesticidal proteins. J. Invertebr. Pathol. 114, 76–85.

Vaughn, T., Cavato, T., Brar, G., Coombe, T., DeGooyer, T., Ford, S., Groth, M., Howe, A., Johnson, S., Kolacz, K., Pilcher, C., Purcell, J., Romano, C., English, L., Pershing, J., 2005. A method of controlling corn rootworm feeding using a *Bacillus thuringiensis* protein expressed in transgenic maize. Crop Sci. 45, 931–938.

Vidal-Quist, J.C., Castanera, P., Gonzalez-Cabrera, J., 2009. Diversity of *Bacillus thuringiensis* strains isolated from citrus orchards in spain and evaluation of their insecticidal activity against *Ceratitis capitata*. J. Microbiol. Biotechnol. 19, 749–759.

Von Tersch, M.A., Slatin, S.L., Kulesza, C.A., English, L.H., 1994. Membrane-permeabilizing activities of *Bacillus thuringiensis* coleopteran-active toxin CryIIIB2 and CryIIIB2 domain I peptide. Appl. Environ. Microbiol. 60, 3711–3717.

Wabiko, H., Raymond, K.C., Bulla Jr., L.A., 1986. *Bacillus thuringiensis* entomocidal protoxin gene sequence and gene product analysis. DNA 5, 305–314.

Walker, K.A., Hellmich, R.L., Lewis, L.C., 2000. Late-Instar European Corn Borer (Lepidoptera: Crambidae) tunneling and survival in transgenic corn hybrids. J. Econ. Entomol. 93, 1276–1285.

Walters, F.S., Stacy, C.M., Lee, M.K., Palekar, N., Chen, J.S., 2008. An engineered chymotrypsin/cathepsin G site in domain I renders *Bacillus thuringiensis* Cry3A active against Western corn rootworm larvae. Appl. Environ. Microbiol. 74, 367–374.

Walters, F.S., deFontes, C.M., Hart, H., Warren, G.W., Chen, J.S., 2010. Lepidopteran-active variable-region sequence imparts coleopteran activity in eCry3.1Ab, an engineered *Bacillus thuringiensis* hybrid insecticidal protein. Appl. Environ. Microbiol. 76, 3082–3088.

Warren, G.W., 1997. Vegetative insecticidal proteins: novel proteins for control of corn pests. In: Carozzi, N.B., Koziel, M. (Eds.), Advances in Insect Control, the Role of Transgenic Plants. Taylors & Francis Ltd., London, pp. 109–121.

Wasano, N., Ohba, M., 1998. Assignment of delta-endotoxin genes of the four lepidoptera-specific *Bacillus thuringiensis* strains that produce spherical parasporal inclusions. Curr. Microbiol. 37, 408–411.

Widner, W.R., Whiteley, H.R., 1989. Two highly related insecticidal crystal proteins of *Bacillus thuringiensis* subsp. *kurstaki* possess different host range specificities. J. Bacteriol. 171, 965–974.

Wu, F., 2006. Mycotoxin reduction in Bt corn: potential economic, health, and regulatory impacts. Transgenic Res. 15, 277–289.

Wu, D., Johnson, J.J., Federici, B.A., 1994. Synergism of mosquitocidal toxicity between CytA and CryIVD proteins using inclusions produced from cloned genes of *Bacillus thuringiensis*. Mol. Microbiol. 13, 965–972.

Wu, S.J., Koller, C.N., Miller, D.L., Bauer, L.S., Dean, D.H., 2000. Enhanced toxicity of *Bacillus thuringiensis* Cry3A delta-endotoxin in coleopterans by mutagenesis in a receptor binding loop. FEBS Lett. 473, 227–232.

Xu, X., Yu, L., Wu, Y., 2005. Disruption of a cadherin gene associated with resistance to Cry1Ac {delta}-endotoxin of *Bacillus thuringiensis* in *Helicoverpa armigera*. Appl. Environ. Microbiol. 71, 948–954.

Yang, Y., Zhu, Y.C., Ottea, J., Husseneder, C., Leonard, B.R., Abel, C., Luttrell, R., Huang, F., 2011. Down regulation of a gene for cadherin, but not alkaline phosphatase, associated with Cry1Ab resistance in the sugarcane borer *Diatraea saccharalis*. PLoS One 6, e25783.

Ye, W., Zhu, L., Liu, Y., Crickmore, N., Peng, D., Ruan, L., Sun, M., 2012. Mining new crystal protein genes from *Bacillus thuringiensis* on the basis of mixed plasmid-enriched genome sequencing and a computational pipeline. Appl. Environ. Microbiol. 78, 4795–4801.

Zhang, J., Hodgman, T.C., Krieger, L., Schnetter, W., Schairer, H.U., 1997. Cloning and analysis of the first cry gene from *Bacillus popilliae*. J. Bacteriol. 179, 4336–4341.

Zhang, X., Candas, M., Griko, N.B., Rose-Young, L., Bulla Jr., L.A., 2005. Cytotoxicity of *Bacillus thuringiensis* Cry1Ab toxin depends on specific binding of the toxin to the cadherin receptor BT-R1 expressed in insect cells. Cell Death Differ. 12, 1407–1416.

Zhang, X., Candas, M., Griko, N.B., Taussig, R., Bulla Jr., L.A., 2006. A mechanism of cell death involving an adenylyl cyclase/PKA signaling pathway is induced by the Cry1Ab toxin of *Bacillus thuringiensis*. Proc. Natl. Acad. Sci. U. S. A. 103, 9897–9902.

Zhang, S., Cheng, H., Gao, Y., Wang, G., Liang, G., Wu, K., 2009. Mutation of an aminopeptidase N gene is associated with *Helicoverpa armigera* resistance to *Bacillus thuringiensis* Cry1Ac toxin. Insect Biochem. Mol. Biol. 39, 421–429.

CHAPTER FIVE

Progress Towards RNAi-Mediated Insect Pest Management

James A. Baum, James K. Roberts
Monsanto Company, Chesterfield, Missouri, USA

Contents

Abstract

Gene suppression via RNA interference (RNAi) provides an alternative strategy for insect pest management. The ingestion by insects of double-stranded RNAs targeting essential insect genes can trigger RNAi and lead to growth inhibition, developmental aberrations, reduced fecundity, and mortality. This RNAi response is particularly acute in certain coleopteran species, most notably the western corn rootworm, a devastating pest impacting corn production in the United States. The development of next-generation rootworm-protected corn hybrids includes an RNAi-based trait that provides a mode of action distinct from those of *Bacillus thuringiensis* insecticidal protein-based traits currently used for rootworm pest management. Unfortunately, many insect species including important lepidoptera and hemiptera pests appear largely recalcitrant in their response to environmental RNA, suggesting biological barriers that thus far limit the utility of RNAi for agricultural pest management. This review will highlight recent

Advances in Insect Physiology, Volume 47
ISSN 0065-2806
http://dx.doi.org/10.1016/B978-0-12-800197-4.00005-1

efforts to understand the barriers to RNA delivery in recalcitrant insect species, describe recent advances in the commercial development of insect-protected crops and biological insecticides utilizing RNAi, and discuss this strategy in the context of an integrated pest management approach.

1. INTRODUCTION

Eukaryotic cells possess a conserved pathway by which exogenously applied and endogenously expressed double-stranded (ds) RNAs direct the degradation of complementary endogenous messenger RNA (mRNA) transcripts within a cell resulting in sequence-specific gene suppression. This phenomenon is referred to as RNA interference (RNAi) (Fire et al., 1998; Hannon, 2002). In plants and animals, RNAi provides one line of defence against RNA viruses and foreign dsRNA molecules. Small endogenous RNAs known as micro RNAs are also processed by a related pathway to regulate tissue-specific patterns of gene expression primarily via translational regulation (Bartel, 2009). Long non-coding RNAs also play a prominent role in the epigenetic regulation of gene expression (Lee, 2012). It is now clear that far more of the genome is transcribed than previously thought (Djebali et al., 2012) and that RNA, in addition to being the obligate messenger and facilitator of protein synthesis in the cell, is also a central player in the regulation of eukaryotic gene expression.

The general mechanism of dsRNA-mediated degradation of mRNA transcripts is understood (Tomari and Zamore, 2005). Long dsRNAs are a substrate for RNAse III-like proteins referred to as Dicer or Dicer-like proteins. Dicer appears to preferentially initiate dsRNA cleavage at the ends of the dsRNA, making successive cleavages to generate 21- to 24-bp silencing (si) RNA duplexes (Elbashir et al., 2001). The resulting siRNA duplexes are loaded into a multiprotein complex called the RNA-induced silencing complex (RISC) where the passenger (sense) strand is removed and the guide (antisense) strand remains to target mRNA for silencing. The guide strand in the RISC enables Watson–Crick base pairing of the complex to complementary mRNA transcripts and enzymatic cleavage of the target mRNA by a class of proteins referred to as Argonaute proteins, thereby preventing mRNA translation. Accordingly, this mechanism of gene suppression is highly sequence specific. The ability to selectively down-regulate genes via RNAi has proved to be valuable, particularly in insects for which genetic tools are not readily available to study gene function (Belles, 2010).

The surprising observation that ingested dsRNAs can trigger gene suppression in the nematode, *Caenorhabditis elegans*, (Timmons and Fire, 1998; Timmons et al., 2001) offered hope that the oral delivery of dsRNA could modulate gene expression in other invertebrates, including insects, for the purpose of pest management.

The ability of herbivorous insects to adapt to insecticide use in agricultural systems presents an ongoing challenge for pest management. Over 20 years ago, more than 500 species of arthropods were documented with resistance to one or more pesticides (Georghiou et al., 1991); an updated dataset can be found at http://www.pesticideresistance.com/index.php. The development of insecticides with new modes of action (MOAs) is a priority, but so is the implementation of resistance management strategies to prolong the use of existing insecticides for use in agriculture and public health (http://www.irac-online.org/). Insect-protected crops, expressing insecticidal proteins derived from the bacterium, *Bacillus thuringiensis* (Bt), represent a significant fraction of the >170 M ha of transgenic crops cultivated worldwide (James, 2013) and provide excellent control of many economically important insect pest species. Consequentially, such crops also impose strong selective pressure on insects to adapt. As is the case for synthetic- and biological insecticides, alternative MOAs for insect-protected crops are needed, either because some insect species are not sensitive to Bt insecticidal proteins or because some have evolved field resistance to efficacious Bt proteins (Storer et al., 2010). To that end, RNAi-mediated insect control represents a significant opportunity.

In 2007, papers from Baum et al. and Mao et al. demonstrated that transgenic plants expressing insect-derived dsRNAs could impact the growth and development of insect herbivores. Corn plants expressing a dsRNA hairpin that targets the vacuolar ATPase A subunit gene in western corn rootworm (WCR), *Diabrotica virgifera* virgifera, caused severe rootworm stunting and exhibited significant protection from rootworm feeding damage, consistent with artificial diet feeding assays demonstrating the insecticidal activity of such dsRNA species (Baum et al., 2007). *Arabidopsis* plants expressing a dsRNA hairpin that targets a cytochrome P450 monooxygenase gene in the cotton bollworm, *Helicoverpa armigera*, led to decreased bollworm tolerance to the cotton sesquiterpene aldehyde, gossypol (Mao et al., 2007), consistent with the proposed role of this enzyme in gossypol detoxification. The former example has advanced towards commercial development as a component of next-generation rootworm-protected corn hybrids (Kupferschmidt, 2013) and will be discussed further in this review.

Numerous studies have since explored the sensitivity of diverse insect species to ingested dsRNAs. In addition, several useful reviews have been published on the subject of RNAi and insects to which we refer the reader (e.g. Aronstein et al., 2011; Belles, 2010; Burand and Hunter, 2013; Gu and Knipple, 2013; Huvenne and Smagghe, 2010; Li et al., 2013b; Terenius et al., 2011; Yu et al., 2013; Zhang et al., 2012a, 2013a). Rather than recapitulate these publications, this review will focus on the development and use of RNAi strategies for insect control in agriculture, highlight efforts to understand the barriers to RNA delivery in recalcitrant insect species, describe recent advances in the commercial development of insect-protected crops and biological insecticides utilizing RNAi, and discuss this strategy in the context of an integrated pest management (IPM) approach.

2. ENVIRONMENTAL RNAi

Definitions have been proposed to discuss the various aspects of RNAi in plants and animals (Huvenne and Smagghe, 2010; Whangbo and Hunter, 2008). Cell-autonomous RNAi refers to the RNAi response that individual cells carry out when encountering dsRNA, a response that is executed by a broadly conserved or core RNAi machinery found in eukaryotic cells. Non-cell autonomous RNAi includes the phenomenon of systemic RNAi—the movement of a silencing signal, presumably siRNA and/or dsRNA, from cell to cell and from one part of an organism to another. Non-cell autonomous RNAi also includes the phenomenon of environmental RNAi which, as its name suggests, refers to the triggering of RNAi by environmental exposure to dsRNA by soaking or feeding. Environmental RNAi may or may not be followed by systemic movement of the silencing signal, perhaps a key step in determining the biological activity of a dsRNA.

Components of the core RNAi machinery are readily identified in insect species whose genomes have been sequenced (e.g. Honeybee Genome Sequencing Consortium, 2006; International Silkworm Genome Consortium, 2008; The International Aphid Genomics Consortium, 2010; Tomoyasu et al., 2008; Tribolium Genome Sequence Consortium, 2008) and evidence for functional RNAi reported in a wide range of insect species encompassing the orders Coleoptera, Diptera, Dictyoptera, Hemiptera, Hymenoptera, Isoptera, Lepidoptera, Neuroptera, and Orthoptera. In order to be useful as an insect control agent in agriculture, environmental RNAi must first be operational: dsRNA must be delivered to the insect either by ingestion or by penetration of the insect cuticle in order to trigger an RNAi response in recipient cells. For transgenic plant applications and for most topical spray applications, delivery via ingestion is likely the dominant route of entry. An overall picture of the responsiveness of different insect species and orders to ingested dsRNAs is emerging (for reviews, see Huvenne and Smagghe, 2010; Li et al., 2013b; Terenius et al., 2011). In feeding studies, dsRNAs are supplied by droplet feeding, incorporated into an artificial diet, or applied to leaf/plant tissue suitable for insect feeding. While the dose of dsRNA consumed is not known in many instances, this variable is less important than the concentration of dietary dsRNA presented to the insect: the effective concentration of a dsRNA active ultimately determines its utility for insect control. In reviewing these studies, we determined or estimated, whenever possible, the dietary concentration of dsRNA tested in parts per million (ppm) or parts per billion (ppb) in order to normalize the results and compare among studies. A summary of published studies evaluating the effects of ingested dsRNAs or siRNAs on insect species is presented in Table 5.1.

3. INSECT SENSITIVITY TO ENVIRONMENTAL RNAi

3.1. Coleoptera

A number of coleopteran species, including the WCR (*D. virgifera* virgifera), southern corn rootworm, *Diabrotica undecimpunctata* howardi, Colorado potato beetle (CPB), *Leptinotarsa decemlineata*, and canola flea beetle, *Phyllotreta striolata*, are remarkably sensitive to ingested dsRNAs with LC_{50} values in the range of 1–10 ppb (Table 5.1, Baum et al., 2007; Bolognesi et al., 2012; Zhao et al., 2008). This sensitivity to environmental RNAi extends to both the larval and adult stages (Rangasamy and Siegfried, 2012; Zhao et al., 2008). To our knowledge, no insect species outside the Order Coleoptera approaches this level of sensitivity to ingested dsRNA, with most studies of successful oral delivery in other insect orders reporting LC_{50} values >10 ppm. This potent environmental RNAi response is not necessarily shared among all coleopteran species, however, as studies with the red flour beetle, *Tribolium castaneum*, and cotton boll weevil, *Anthonomus grandis*, suggest (Baum et al., 2007; Whyard et al., 2009).

Some aspects of the RNAi response in corn rootworm are instructive. Of the 290 gene targets screened by Baum et al. (2007), approximately 2/5 caused rootworm mortality or stunting in surface overlay diet bioassays when applied at the relatively low concentration of ~50 ng/cm^2, or roughly equivalent to 0.1 ppm in diet (see Baum et al., 2011 for a listing of efficacious

Table 5.1 Sensitivity of insect species to ingested dsRNAs

Organism	Target gene product	Stage	Assay	Mortality or stunting	LC_{50} or (concentration tested)	mRNA silencing	Reference
Coleoptera							
Diabrotica virgifera virgifera	Multiple targets	Neonates	Artificial diet	Yes	1–10 ppb[a]	Yes	Baum et al. (2007)
	V-ATPase A	Neonates	Transgenic plant	Yes	ND	Yes	Baum et al. (2007)
	Snf7	Neonates	Artificial diet	Yes	4.3 ppb	Yes	Bolognesi et al. (2012)
	Snf7	Neonates	Transgenic plant	Yes	ND	Yes	Bolognesi et al. (2012)
Diabrotica undecimpuctata howardii	Snf7	Neonates	Artificial diet	Yes	1.2 ppb	Yes	Bolognesi et al. (2012)
	V- ATPase A and E	Neonates	Artificial diet	Yes	(~0.1 ppm)[a]	Yes	Baum et al. (2007)
	α-Tubulin	Neonates	Artificial diet	Yes	(~0.1 ppm)[a]	Yes	Baum et al. (2007)
Leptinotarsa decemlineata	V-ATPase A and E	Neonates	Artificial diet	Yes	~10 ppb[a]	–	Baum et al. (2007)
	Multiple targets	Neonates	Leaf tissue	Yes	ND	Yes	Zhu et al. (2011)
Phyllotreta striolata	Arginine kinase	Adults	Leaf tissue	Yes	0.8 ppb	Yes	Zhao et al. (2008)
Tribolium castaneum	V-ATPase E	Neonates	Flour	Yes	2.5 ppm	Yes	Whyard et al. (2009)

Diptera							
Aedes aegypti	V-ATPase A	Adults	Artificial diet	–	(1000 ppm)	Yes	Coy et al. (2012)
	Multiple targets	First instars	Water	Yes	(200, 500 ppm)	Yes	Singh et al. (2013)
	ATP-dependent efflux pump	Second instars	Water	–	30 ppm	Yes	Figueira-Mansur et al. (2013)
Anopholes gambiae	Chitin synthase 2	Third instars	Artificial diet	Yes	–	Yes	Zhang et al. (2010)
Anopheles stephensi	3-HKT	First instars	*Chlamydomonas*	Yes	ND	Yes	Kumar et al. (2013)
Bactrocera dorsalis	Multiple targets	Adults	Artificial diet	Yes/No	(2000 ppm)	Yes/No	Li et al. (2011b)
Glossina morsitans morsitans	Tsetse EP	Male adults	Blood meal	No	(>400 ppm)	Yes	Walshe et al. (2009)
	Transferrin	Male adults	Blood meal	No	(>400 ppm)	No	Walshe et al. (2009)
Hemiptera							
Acyrthosiphon pisum	Aquaporin	Six-day-old nymphs	Artificial diet	No	(1000–5000 ppm)	Yes	Shakesby et al. (2009)
	V-ATPase E	First instars	Artificial diet	Yes	3.4 ppm	Yes	Whyard et al. (2009)

Continued

Table 5.1 Sensitivity of insect species to ingested dsRNAs—cont'd

Organism	Target gene product	Stage	Assay	Mortality or stunting	LC_{50} or (concentration tested)	mRNA silencing	Reference
	V-ATPase E	Neonates	Artificial diet	No	ND	No	Christiaens et al. (2014)
	Hunchback (hb)	Neonates	Artificial diet	Yes	(750 ppm)	Yes	Mao and Zeng (2012)
Bactericerca cockerelli	Multiple targets	Adults	Artificial diet	Yes	(500–1000 ppm)	Yes	Wuriyanghan et al. (2011)
Bemisia tabaci	V-ATPase subunit A, rpL19	Adults	Artificial diet	Yes	3,11 ppm	Yes	Upadhyay et al. (2011)
Lygus lineolaris	Inhibitor of apotosis	Neonates	Artificial diet	No	(1000 ppm)	No	Allen and Walker (2012)
Myzus persicae	RACK-1, COO2	Nymphs	Transgenic plant	Yes	ND	Yes	Pitino et al. (2011)
	Hunchback (hb)	Neonates	Transgenic plant	Yes	ND	Yes	Mao and Zeng (2014)
Nilaparvata lugens	Trehalose PO4 synthase	Third instars	Artificial diet	?	(500 ppm)	Yes	Chen et al. (2010)
	V-ATPase E	2nd instars	Artificial diet	No	(50 ppm)	Yes	Li et al. (2011a)
	Multiple targets	Neonates	Transgenic plant	No	ND	Yes	Zha et al. (2011)

Peregrinus maidis	V-ATPase B and D	Third instars	Artificial diet	Yes	(500 ppm)	Yes	Yao et al. (2013)
Rhodnius prolixus	Nitroporin 2	Second instars	Artificial diet	NA	(1000 ppm)	Yes	Araujo et al. (2006)
Sitobion avenae	Multiple targets	Third instars	Artificial diet	Yes	(3 – 7.5 ppm)	Yes	Zhang et al. (2013b)
Hymenoptera							
Apis mellifera	Vitellogenin	Second instars	Natural diet	Nonspecific	(500–3000 ppm)	Yes	Nunes and Simões (2009)
Solenopsis invicta	PBAN/pyrokinin	Fourth instars	Artificial diet	Yes	(1000 ppm)	ND	Vander Meer and Choi (2013)
	GNBP	Workers	Artificial diet	Yes	(200 ppm)	ND	Zhao and Chen (2013)
Isoptera							
Reticulitermes flavipes	Cellulase	Workers	Paper discs	Yes	(5.1 μg/cm^2)	Yes	Zhou et al. (2008)
	Hexamerin	Workers	Paper discs	Yes	(2.2 μg/cm^2)	Yes	Zhou et al. (2008)
Lepidoptera							
Chilo infuscatellus	CiHR3 moulting factor	Third instars	Corn kernels	Yes	(250 ppm)	Yes	Zhang et al. (2012c)
Epiphyas postvittana	Carboxylesterase	Third instars	Droplet	No	(4000 ppm)	Yes	Turner et al. (2006)
	Pheromone bp	Third instars	Droplet	No	(4000 ppm)	Yes	Turner et al. (2006)

Continued

Table 5.1 Sensitivity of insect species to ingested dsRNAs—cont'd

Organism	Target gene product	Stage	Assay	Mortality or stunting	LC_{50} or (concentration tested)	mRNA silencing	Reference
Helicoverpa armigera	CYP6AE14	Third instars	Transgenic plant	Yes	ND	Yes	Mao et al. (2007)
	GST	Third instars	Transgenic plant	No	ND	Yes	Mao et al. (2007)
	CYP6AE14	Third instars	Transgenic plant	Yes	ND	Yes	Mao et al. (2011, 2013)
	AchE receptor	Neonates	Artificial diet	Yes	(~0.35 ppm)	Yes	Kumar et al. (2009)
	AchE receptor	Neonates	Leaf tissue	Yes	(~0.35 ppm)	–	Kumar et al. (2009)
	Ecdysone receptor EcR	Second instars	Transgenic plant	Yes	ND	Yes	Zhu et al. (2012)
	Ecdysone receptor EcR	Third instars	Artificial diet (Ec)[b]	Yes	ND	Yes	Zhu et al. (2012)
	HaHR3 moulting factor	Third instars	Transgenic plant	Yes	ND	Yes	Xiong et al. (2013)
	HaHR3 moulting factor	Third instars	Artificial diet (Ec)	Yes	ND	Yes	Xiong et al. (2013)

	CYP6B6	Third instars	Artificial diet (Ec)	Yes	ND	Yes	Zhang et al. (2013c)
	Ultraspiracle protein, EcR	Third instars	Artificial diet	Yes	(1000 ppm)	Yes	Yang and Han (2014)
Manduca sexta	V-ATPase E	Neonates	Artificial diet	Yes	11 ppm	Yes	Whyard et al. (2009)
Ostrinia nubilalis	Chitinase	Neonates	Artificial diet	Yes	(2500 ppm)	Yes	Khajuria et al. (2010)
Plutella xylostella	CYP6BG1	Fourth instars	Droplet		(~800 ppm)	Yes	Bautista et al. (2009)
	Rieske protein	Second instars	Leaf tissue	Yes	(6 μg/cm^2)	Yes	Gong et al. (2011)
	AchE receptor	Second instars	Leaf tissue	Yes	53.7 ppm	-	Gong et al. (2013)
Sesamia nonagriodes	JH esterase JHER	First to sixth instars	Artificial diet (Ec)	No	ND	Yes	Kontogiannatos et al. (2013)
Spodoptera exigua	Chitin synthase A	Neonates	Artificial diet (Ec)	Yes	ND	Yes	Tian et al. (2009)
	β1 integrin subunit	Fourth instars	Leaf tissue	Yes	100–200 ppm	–	Surakasi et al. (2011)
	Ecdysone receptor EcR	Second instars	Transgenic plant	Yes	ND	Yes	Zhu et al. (2012)

Continued

Table 5.1 Sensitivity of insect species to ingested dsRNAs—cont'd

Organism	Target gene product	Stage	Assay	Mortality or stunting	LC_{50} or (concentration tested)	mRNA silencing	Reference
Spodoptera litura	Aminopeptidase N	Neonates	Artificial diet	No	ND	No	Rajagopal et al. (2002)
Spodoptera frugiperda	Allatostatin C	Fifth instars	Droplet	NA	(600 ppm)	Yes	Griebler et al. (2008)
	Allototropin 2	Fifth instars	Droplet	NA	(600 ppm)	Yes	Griebler et al. (2008)
	SfT6 serine protease	Fourth instars	Droplet	NA	(600 ppm)	Yes	Rodríguez-Cabrera et al. (2010)
Orthoptera							
Gryllus bimaculatus	Sulfakinins	Adults	Droplet	NA	(100–600 ppm)	–	Meyering-Vos and Müller (2007)
Locusta migratoria	Multiple targets	Fourth instars	Artificial diet	No	(~240 ppm)	No	Luo et al. (2013)
Schistocerca gregaria	Tubulin, GAPDH	Adults	Artificial diet	No	ND	No	Wynant et al. (2012)

[a]Estimated from sample overlay assays in which 20 μL samples are infused into 200 μL artificial diet.
[b]Ec = dsRNA expressed in *E. coli*.
ND, not determined or reported; NA, not applicable.

targets). While the suppression of certain gene transcripts did not result in a phenotypic response (Baum et al., 2007, supplemental figure 3), it is clear that the number of specific gene targets available for successful environmental RNAi is large. Sectioning of the V-ATPase A coding region into six ~300 bp dsRNAs did not reveal dramatic differences in efficacy, suggesting that a single dsRNA of this size provides a reasonable sampling of target knockdown efficacy and phenotype. One of the targets identified by Baum et al. (2007), a *Snf7* ortholog, was selected for a more detailed study of the RNAi response (Bolognesi et al., 2012). *Snf7* dsRNAs shorter than 50 bp exhibited dramatically reduced activity in the corn rootworm feeding assay. This study employed a single *Snf7* 27-mer sequence embedded in neutral sequences of varying length, ruling out the possibility that differences in siRNA composition accounted for differences in activity. An *in situ* study of RNA uptake further demonstrated that a Cy3-labelled 240 bp *Snf7* dsRNA was taken up by rootworm midgut epithelial cells while a Cy3-labelled 21 bp *Snf7* siRNA was not, corroborating the size dependency of the RNAi response observed in feeding assays (Bolognesi et al., 2012). Likewise, injection studies with corn rootworm larvae have demonstrated the inability of siRNAs to produce an RNAi response leading to rootworm mortality (Khajuria et al., 2013). Finally, the RNAi response in corn rootworm appears to be systemic as judged by qPCR analysis of *Snf7* mRNA transcripts in isolated midgut and cadaver tissues (Bolognesi et al., 2012).

Elements of the rootworm RNAi response can be found in another coleopteran species, the red flour beetle *T. castaneum*, which has become a model system for studying systemic RNAi in insects (Miller et al., 2012). Injection studies with a transgenic *gfp-Tribolium* line permitted visualization of silencing as suppression of green fluorescent protein (gfp) fluorescence. The RNAi response was observed to be dose-dependent, systemic, and likewise dependent on the size of the dsRNA. Injection of ~60 ng of a 520-bp *gfp* dsRNA/last instar larva resulted in detectable silencing of *gfp*, demonstrating that this coleopteran species is very sensitive to systemic RNAi. The efficiency of systemic RNAi appears to drop off with shorter dsRNA molecules although the precise breakpoint was not determined. Adopting a strategy similar to that used for characterization of the corn rootworm *Snf7* dsRNA (Bolognesi et al., 2012), a 30-bp *gfp* dsRNA fused to a 30-bp neutral *Ultrabithorax* (Ubx) dsRNA was shown to be more effective than the 30-bp *gfp* dsRNA alone in triggering a systemic RNAi response, presumably due to less efficient cellular uptake of the smaller dsRNA. Comparing a series of siRNAs and larger dsRNAs targeting the *Tc-achaete-scute-homolog* (*Tc-ASH*) and *Ultrabithorax* (*UBX*) genes in *Tribolium*, Wang et al. (2013) reported that dsRNAs provided greater silencing for longer periods of time resulting in developmental phenotypes. Interestingly, the silencing observed with injected siRNAs (0.4–0.5 μg) was short-lived, lasting from days 2 to 4, and did not result in a phenotypic response. The dose-dependent systemic RNAi response observed in *Tribolium* is consistent with the concentration-dependent environmental RNAi response observed in *Diabrotica* that permitted calculation of LC_{50} values (Baum et al., 2007). The steepness of those concentration–response curves may reflect a threshold effect, as has been suggested for *Tribolium* (Miller et al., 2012). Like other insect species, neither *Tribolium* nor *Diabrotica* contains a recognizable RNA-dependent RNA polymerase that could amplify the production of dsRNAs for a systemic response (Baum et al., 2007; Tomoyasu et al., 2008). Despite the apparent absence of this amplification mechanism, corn rootworm larvae are remarkably sensitive to ingested dsRNAs and require no more than a 3-h exposure to dietary dsRNA to observe a lethal phenotype (Bolognesi et al., 2012). Comparative genomic analysis of the *Tribolium* and *Diabrotica* genomes may shed light on the apparent differences in sensitivity to ingested dsRNAs observed between these two species.

3.2. Diptera

Belles (2010) provides some historical context for the pioneering RNAi studies conducted with the fruit fly, *Drosophila melanogaster*, a model system for understanding both the mechanism of RNAi and its role in mediating antiviral immunity in invertebrates (Nayak et al., 2013). Reverse genetic studies of gene function have been enabled through the use of cultured cells, employing both genome-wide (Boutros and Ahringer, 2008; Boutros et al., 2004) and pathway-specific screens (e.g. Clemens et al., 2000). Functional studies of genes involved in *Drosophila* development have relied on micro-injection of dsRNA into embryonic tissue (e.g. Kennerdell and Carthew, 1998; Koizumi et al., 2007; Pilot et al., 2006). Libraries of transgenic *Drosophila* lines containing conditionally expressed dsRNA hairpins have been generated for use in whole-animal RNAi screens (e.g. Dietzl et al., 2007). Abundant tools for both *in vivo* and *in vitro* RNAi studies in *Drosophila*, as well as relevant literature, can be found at http://www.flyrnai.org/.

Drosophila appears deficient in systemic RNAi when confronted with endogenously expressed dsRNA hairpins (Roignant et al, 2003), but is capable of mounting a systemic RNAi response to viral infection (Saleh et al., 2009). The systemic RNAi response to injected dsRNAs is also limited. Using transgenic *Drosophila* lines expressing Gal4-regulated enhanced green fluorescent protein (EGFP) or GFP, Miller et al. (2008) evaluated the systemic RNAi response in last instar larvae and reported that only hemocytes were responsive to injected (hemolymph) dsRNA, whereas in *Tribolium* larvae, virtually all cell types were responsive. Co-expression of an EGFP dsRNA hairpin resulted in down-regulation of EGFP in all tissues examined, demonstrating that the limited systemic RNAi response observed in *Drosophila* larvae is not due to the absence of cell-autonomous RNAi machinery. *Drosophila* is likewise recalcitrant to environmental RNA, but larvae fed with dsRNAs formulated with transfection agents exhibited both specific target gene knockdown and significant mortality (Whyard et al., 2009), suggesting that the presence of a robust systemic RNAi response may not be a prerequisite for RNAi-mediated insect control.

Extensive RNAi studies have likewise been conducted with dipteran vectors of human diseases to study gene function related to vector biology and vector–pathogen interactions (reviewed in Barnard et al., 2012; Belles, 2010; Manzano-Roman et al., 2012). Although the majority of studies with mosquito species have relied on microinjection of dsRNAs into the hemolymph, other methods such as topical delivery of dsRNA to adults (Pridgeon et al., 2008) and deployment of a recombinant densovirus-mediated RNAi system (Gu et al., 2011) have been described. Several studies have reported success in the oral delivery of dsRNAs to dipteran species, including the mosquitoes, *Aedes aegypti* and *Anopheles gambiae*, and the tsetse fly, *Glossina morsitans morsitans* (Table 5.1). Ingestion of a 10% sucrose solution containing dsRNA (1000 ppm) targeting the vacuolar ATPase subunit A gene resulted in significant knockdown of the target transcript in *A. aegypti* adults as early as 12 h after the onset of feeding, but no mortality data were reported (Coy et al., 2012). Silencing of a P-glycoprotein (P-gp; ATP-dependent efflux pump) gene in *A. aegypti* larvae following ingestion of dsRNA resulted in enhanced sensitivity to the insecticide temephos (Figueira-Mansur et al., 2013), consistent with evidence that P-gp proteins are involved in resistance to organophosphate insecticides. Singh et al. (2013) evaluated several dsRNAs in feeding assays with *A. aegypti* and observed both silencing and mortality/stunting of larvae treated with the *β-tubulin* and *chitin synthase-1* dsRNAs at dietary concentrations of 200–500 ppm. DsRNAs targeting *chitin synthase-2* induced silencing but had no apparent effect on larval survival while silencing of *heat shock protein*

83 manifested itself by increased larval mortality under heat stress. In these studies with mosquito larvae, it is not clear whether the route of RNA delivery is oral or via penetration of the larval cuticle. DsRNA complexed with chitosan and packaged in a gel-based diet was used to silence chitin synthase genes in *A. gambiae* larvae resulting in increased larval sensitivity to agents that either inhibit chitin biosynthesis in insects (diflubenzuron) or act to disrupt the organization or integrity of the peritrophic matrix (calcofluor white, dithiothreitol) (Zhang et al., 2010). Expression of dsRNAs in transgenic *Chlamydomonas* has been proposed as an alternative delivery vehicle for RNAi-mediated control of mosquito larvae (Kumar et al., 2013).

3.3. Lepidoptera

Terenius et al. (2011) provided a useful overview of the status of RNAi studies in lepidopteran species and noted that, with respect to oral delivery, gene suppression only appears to be successful when high concentrations of dsRNA are provided in the diet. In the first reported example, Turner et al. (2006) achieved a significant suppression of several target genes in the brown apple moth, *Epiphyas postvittana* (Walker), by droplet feeding a 4000 ppm dsRNA solution. Subsequently, silencing via oral delivery of dsRNA was reported in a wide range of lepidopteran species including the tobacco hornworm, *Manduca sexta*, diamondback moth (DBM), *Plutella xylostella*, beet armyworm, *Spodoptera exigua*, fall armyworm, *Spodoptera frugiperda*, European corn borer, *Ostrinia nubilalis*, sugarcane stem borer, *Chilo infuscatellus*, and the cotton bollworm, *H. armigera* (Bautista et al., 2009; Gong et al., 2011; Khajuria et al., 2010; Mao et al., 2007; Rodríguez-Cabrera et al., 2010; Surakasi et al., 2011; Tian et al., 2009; Whyard et al., 2009; Xiong et al., 2013; Zhang et al., 2012c; Zhu et al., 2012).

For many early RNAi studies, the object was not to kill an insect but to selectively down-regulate a gene to study its function in a metabolic or developmental process (Belles, 2010). For example, environmental RNAi was used to examine the role of two chitinase genes in regulating chitin content in the peritrophic matrix of the European corn borer (Khajuria et al., 2010), to demonstrate that the cytochrome P450 gene, *CYPBG1*, in DBM is involved in larval resistance to the insecticide permethrin (Bautista et al., 2009), to study the role of the beet armyworm β1 subunit integrin (β*Se1*) in development and cellular immunity (Surakasi et al., 2011), and to demonstrate that a serine protease gene in the fall armyworm plays an important role in the processing of the *B. thuringiensis* Cry1Ca1 insecticidal protein and the insect's pathogen response to Bt toxins (Rodríguez-Cabrera et al., 2010). In these studies, high concentrations of dsRNA (50–2500 ppm) were delivered by droplet feeding, incorporation into artificial diet, or by soaking leaf tissue prior to feeding. Surakasi et al. (2011) reported concentration-dependent mortality upon silencing of the βSe1 subunit gene in the beet armyworm from which one can interpolate an LC_{50} of 100–200 ppm for this topical application. Silencing of the vacuolar ATPase E subunit gene in *M. sexta* resulted in concentration-dependent mortality and an LC_{50} of only 11 ppm (Whyard et al., 2009). Even so, this level of activity is still three orders of magnitude lower than that observed among the sensitive coleopteran species.

Transgenic plants expressing insect-specific dsRNAs have been reported to impact the growth and survival of certain lepidopteran species, most notably the cotton bollworm, *H. armigera*. Larvae of *H. armigera* fed on tobacco plants expressing a dsRNA targeting the *H. armigera* ecdysone receptor (EcR) gene showed elevated mortality (~40% compared to 10% in the *gfp* control group), growth reduction, and significant suppression of the *EcR* transcript (Zhu et al., 2012). Larvae of another lepidopteran species, the beet armyworm *S. exigua*, also showed elevated mortality and development aberrations when fed on the same transgenic plant material, presumably because the *EcR* target sequences in these two species share regions of >21 bp sequence identity (Zhu et al., 2012). Likewise, tobacco plants expressing a dsRNA targeting a moult-regulating transcription factor, HaHR3, were shown to suppress the *HaHR3* transcript in fed *H. armigera* larvae. In this study, larval mortality of 22–30% and >50% mass reduction was observed after 5 days of feeding on transgenic leaf discs (Xiong et al., 2013). Mao et al. (2007) was the first to use RNAi as a means to alter an insect's ability to cope with xenobiotic compounds, in this case gossypol. Transgenic cotton plants expressing a dsRNA hairpin derived from the *H. armigera* gossypol-inducible cytochrome P450 gene *CYP6AE14* showed increased tolerance to the cotton bollworm, *H. armigera* (Mao et al., 2011) but were not lethal to the larvae. This response can be enhanced by co-delivery of a cysteine proteinase to damage the larval peritrophic matrix, leading to higher gossypol accumulation in the midgut and a modest increase in larval stunting when fed on cotton (Mao et al., 2013). This strategy of targeting detoxification mechanisms in the insect midgut to mitigate plant feeding damage seems promising, particularly since it may not require a systemic RNAi response in the insect. In principle, this strategy is no different than the use of RNAi to restore sensitivity to insecticides among resistant insect species (Bautista et al., 2009; Figueira-Mansur et al., 2013; Tang et al., 2012).

Difficulty in delivering sufficient dsRNA to lepidopteran gut epithelial cells may be inferred from the paper of Gong et al. (2011) which describes the use of chemically modified siRNAs to target the Rieske iron-sulphur protein gene in the DBM. Both 2′-O-methoxy nucleotides and deoxythymidine were incorporated into the sense and antisense strands to stabilize the siRNAs. When provided to the larvae on cabbage leaves, again at relatively high concentrations (6 $\mu g/cm^2$), several siRNAs caused suppression of the target gene transcript and mortality. This approach was extended to include 5′ PEG modification of siRNAs designed to target the acetyl cholinesterase (AchE) receptor in the DBM (Gong et al., 2013). One such siRNA, Si-ace2_001, exhibited an LC_{50} of 53.7 ppm when sprayed on DBM-infested cabbage leaf discs. Kumar et al. (2009) also employed modified siRNAs targeting the AchE receptor of *H. armigera*, in this case substituting a pair of deoxythymidine residues at the 3′ end of both the sense- and antisense strands. Larvae showed reduced growth (stunting) and increased pupal malformation when fed diet containing siRNA at ~0.35 ppm. These results suggest that there might not be a strict dsRNA size dependency to the environmental RNAi response in lepidopterans, at least with chemically modified siRNAs. Larger dsRNAs expressed and encapsulated in *Escherichia coli* have also been reported to impact the growth and survival of *S. exigua* and *H. armigera* larvae (Tian et al., 2009; Xiong et al., 2013; Zhang et al., 2013c; Zhu et al., 2012) but neither the concentration of dsRNA in diet nor the effect of dsRNA alone is reported in these studies. Yang and Han (2014) reported that *E. coli*-encapsulated dsRNAs appear to be more effective than naked dsRNA (1000 ppm in diet) in blocking pupation and causing mortality of *H. armigera* larvae. Suppression of a juvenile hormone esterase-related gene in the corn stalk borer, *Sesamia nonagriodes*, via bacterial delivery of dsRNA did not result in a phenotype (Kontogiannatos et al., 2013). In the case of the sugarcane stem borer, *C. infuscatellus*, naked dsRNA (at 250 ppm) and bacterial-expressed dsRNA applied to corn kernels as a diet appeared equally effective in promoting silencing and stunting larval growth (Zhang et al., 2012c).

In some cases, the insect gut may be bypassed by the use of topical sprays that rely on penetration or adsorption through the insect cuticle (Wang et al., 2011). Topical application of dsRNAs (at 50 ppm) targeting larval stage-specific transcripts in the Asian corn borer, *Ostrinia furnacalis*, led to significant larval mortality and gene silencing at 5 days post-sprays. The route of dsRNA penetration into the larvae is not known but could involve transit to the hemolymph via the tracheoles. Many of the treatments in this study caused significant mortality in the absence of significant gene silencing at day 3 (Wang et al., 2011), suggesting that either a non-RNAi mechanism is at work or the method used to measure transcript knockdown (qRT-PCR on whole insects) could not detect localized tissue-specific silencing leading to mortality.

3.4. Hemiptera

RNAi techniques have been used successfully in a wide variety of hemipteran species encompassing phloem feeders such as aphids and piercing–sucking insects such as plant bugs (Li et al., 2013b). Since many hemipteran species are relatively small and fragile as nymphs, oral delivery of dsRNAs for gene silencing has been an attractive alternative to microinjection. As was the case with lepidopterans, early studies were not necessarily focused on killing insects but on studying gene function (e.g. Belles, 2010; Paim et al., 2013). Across all studies with hemipterans, the dietary concentrations of dsRNA required for silencing and/or lethal phenotypes vary widely, even between studies with the same organism, but tend to be at least three orders of magnitude higher than effective concentrations used with sensitive coleopteran species. Examples of gene silencing following ingestion of dsRNA include the *nitrophorin 2* (*Np2*) gene in the triatomine bug, *Rhodnius prolixus* (Araujo et al., 2006), the *aquaporin 1* (*ApAQP1*) gene in the pea aphid, *Acyrthosiphon pisum* (Shakesby et al., 2009), the vacuolar ATPase subunit E gene in *A. pisum* (Whyard et al., 2009), the trehalose phosphate synthase (*tps*) gene in brown planthopper, *Nilaparvata lugens* (Chen et al., 2010), the gap gene *hunchback* in *A. pisum* (Mao and Zeng, 2012), and the vacuolar ATPase subunit E gene in *N. lugens* (Li et al., 2011a). Drawing from a list of efficacious gene targets identified for the WCR, Upadhyay et al. (2011) reported silencing of the vacuolar ATPase subunit A—and ribosomal protein L9 genes in the whitefly, *Bemisia tabaci*, as well as mortality with LC_{50} values in the 3–11 ppm range. Focusing on gene targets that are highly or specifically expressed in the midgut, Wuriyanghan et al. (2011) demonstrated target gene suppression and lethality in the potato–tomato psyllid, *Bactericerca cockerelli*, when dsRNAs were presented at high concentrations (500–1000 ppm) in a 15% sucrose diet. Both studies also reported lethality upon ingestion of siRNAs, an effect that has not been reported in coleopteran species. Ingestion of dsRNA (750 ppm) targeting the *hunchback* gene led to target gene suppression and increased mortality of *A. pisum* nymphs fed on an artificial diet (Mao and Zeng, 2012). Tobacco plants expressing a dsRNA targeting the orthologous *hunchback* gene of the peach aphid, *Myzus persicae*, caused ~30% target mRNA suppression and ~13% inhibition of aphid reproduction (Mao and Zeng, 2014). Feeding large dsRNAs for multiple gene targets at dietary concentrations of 7.5 ppm resulted in target gene suppression and increased mortality in the grain aphid, *Sitobion avenae* F. (Zhang et al., 2013b). In this study, feeding 3–7.5 ppm dsRNA of *C002*, a gene encoding an unknown protein required for normal plant foraging behaviour in *A. pisum* (Mutti et al., 2008), was also reported to cause increased mortality, although it is not clear from the literature whether silencing of this gene should even impact feeding on artificial diet (Mutti et al., 2008). In contrast, *Inhibitor of Apoptosis* (*IAP*) dsRNA presented in diet at 1000 ppm had no effect on the survival of tarnished plant bug, *Lygus lineolaris*, nymphs (Allen and Walker, 2012) despite evidence that silencing of *IAP* via dsRNA injection results in increased *Lygus* mortality (Walker and Allen, 2011). Consistent with many studies, dsRNA injection was far more effective than dsRNA feeding in silencing target genes and causing a lethal phenotype in the corn planthopper, *Peregrinus maidis* (Yao et al., 2013). In this study, it is worth noting that early (day 2) and prolonged (day 6) suppression of the vacuolar ATPase subunit B transcript by diet feeding (500 ppm) did not result in the expected lethal phenotype observed via injection while late silencing (day 6) of the vacuolar ATPase subunit D transcript resulted in a detectable lethal phenotype, but at a time point when the majority (~60%) of the control insects had already died. This illustrates the difficulties encountered with artificial diet feeding assays conducted with many hemipteran species: phenotypes caused by environmental RNAi can be slow to emerge and keeping insects alive on an artificial diet for >7 days can be challenging.

A number of research groups apparently eschewed artificial diet feeding assays altogether and evaluated insect feeding on transgenic plants expressing dsRNAs designed to target essential insect genes. Pitino et al. (2011) selected two target genes in the aphid, *M. persicae*, for knockdown using transgenic *Arabidopsis thaliana*; receptor of activated kinase C (*Rack-1*) and *C002*. When fed on transgenic lines expressing the target dsRNAs, *M. persicae* nymphs exhibited a >50% knockdown in target gene expression after 16 days, but only a small reduction in progeny when compared to nymphs fed on the dsGFP control line or on *Rack-1* lines that did not suppress the *Rack-1* gene. Injection of *C002* dsRNA into *A. pisum* nymphs results in both rapid gene silencing and 100% mortality after 7 days of feeding on plants (Mutti et al., 2006), suggesting that the delivery of dsRNA via plant expression was simply inadequate for control of this species. A similar outcome was reported using transgenic rice plants expressing dsRNAs targeting three midgut-expressed genes in *N. lugens* (Zha et al., 2011). In this study, significant reductions in target transcript levels (as high as 73% for the *Nltry* gene) were detected 2–4 days after feeding but no lethal phenotype was observed.

Comparisons across related studies reveal marked variation in the hemipteran response to environmental RNAi. For example, Chen et al. (2010) reported significant silencing of the trehalose phosphate synthase (*tps*) gene in *N. lugens* at dsRNA diet concentrations of 500 ppm but no silencing at concentrations of <100 ppm while Li et al. (2011a) reported significant silencing of the vacuolar ATPase subunit E gene in *N. lugens* at dsRNA diet concentrations of only 50 ppm. Silencing of the same essential gene in two distantly related hemipteran species can apparently yield different phenotypes: a dsRNA silencing the vacuolar ATPase subunit E gene in *A. pisum* yielded a reported LC_{50} of 3.4 ppm (Whyard et al., 2009) while a dsRNA silencing the *N. lugens* ortholog caused no phenotype (Li et al., 2011a). Moreover, different groups working with the same insect species and gene target have reported conflicting outcomes. Whereas Whyard et al. (2009) reported oral efficacy with a dsRNA targeting the vacuolar ATPase subunit E gene in *A. pisum* and Mutti et al. (2006) reported lethality following injection of *C002* dsRNA in *A. pisum*, Christiaens et al. (2014) failed to observe a phenotypic response with either dsRNA and delivery method in this species. In this case, the decision to deploy a long *COO2* dsRNA rather than pre-diced siRNAs could have influenced the outcome of the injection experiment. Other factors that could account for these disparate results include insect colony source and health, viral load, life stage, dsRNA stability in different diets, and target gene propensity for silencing.

3.5. Other agricultural pests

Environmental RNAi has been studied in a number of other arthropod pests for which systemic RNAi has already been demonstrated via injections studies. Feeding studies with the invasive fire ant, *Solenopsis invicta*, have demonstrated that dsRNAs ingested by worker ants can be brought back to the colony and shared with the brood (trophallaxis), leading to significant mortality in the brood (larval) population. Worker ant feeding on a 10% sucrose diet containing 1000 ppm dsRNA targeting the PBAN/pyrokinin gene resulted in increased mortality among the nursing fourth instar larvae

(Vander Meer and Choi, 2013). In contrast, a dsRNA targeting a guanine nucleotide-binding protein (GNBP) was reported to be directly toxic to *S. invicta* worker ants when presented in diet at 200 ppm (Zhao and Chen, 2013). Although not an insect, spider mites can be a significant agricultural pest in both vegetable- and row crops. Using a dsRNA permeated leaf disc assay, Kwon et al. (2013) demonstrated gene silencing and significant mortality in the two-spotted spider mite, *Tetranychus urticae*, by targeting several genes at applied concentrations of 160 ppm dsRNA. Most of the efficacious targets surveyed, including genes encoding the COPI coatomer β′ subunit, V-ATPAse subunits A and B, and ribosomal protein S4, had been previously identified as efficacious targets in the coleopteran species *D. virgifera* virgifera (Baum et al., 2007; Khajuria et al., 2013) and *L. decemlineata* (Baum et al., 2007; Zhu et al., 2011). The mite, *Varroa destructor*, an ectoparasite of the honey bee, *Apis mellifera*, and a major suspect contributing to the phenomenon known as bee colony collapse disorder, is also sensitive to environmental RNAi but the route of delivery is unusual in this case. Immersion of *Varroa* mites in a saline solution containing dsRNA can lead to gene silencing (Campbell et al., 2010), but so can feeding dsRNA to the honey bees upon which the mites subsequently feed (Garbian et al., 2012). This unique interspecies transfer of dsRNA from bee to mite presents a strategy for control of this debilitating agricultural pest. Finally, it is worth noting here that locust species (order Orthoptera) display a highly sensitive systemic RNAi response but are refractory to environmental RNAi. As little as 15 pg of dsRNA per mg body mass (~10 ng/insect) is required for significant knockdown of target gene expression in the desert locust, *Schistocerca gregaria* (Wynant et al., 2012). As is the case for *T. castaneum*, the systemic response is dose-dependent, continues to increase over time, and can lead to mortality 7 days post-injection. Likewise, the migratory locust, *Locusta migratoria* was observed to be highly responsive to injected dsRNAs, exhibiting a dose-dependent response leading to target gene suppression and lethality, but was not responsive to environmental RNAi (Luo et al., 2013). A survey of the *Locusta* genome revealed putative orthologs for almost all of the genes implicated in dsRNA uptake in *Drosophila melanogaster* (Saleh et al., 2006), another species unresponsive to environmental RNAi.

4. BARRIERS TO DELIVERY AND EFFICACY IN RECALCITRANT SPECIES

Many recent studies have focused on the optimization of RNAi efficiency in recalcitrant species via selection of optimal target genes (Li et al.,

2013a; Wang et al., 2011; Yao et al., 2013; Zhang et al., 2013b). Contrary to this view, the results of dsRNA screens with sensitive coleopteran species yielding high hit rates among "housekeeping" gene targets using low dietary concentrations of dsRNA support the argument that biological barriers to RNAi, not target gene selection, are the problem. Likewise, the need to provide dsRNAs targeting obviously essential genes (e.g. the V-ATPase subunit genes) at dietary concentrations more than 1000-fold higher than are required for an RNAi phenotype in sensitive coleopteran species suggests that there are great inefficiencies in triggering and sustaining an RNAi response in recalcitrant species. It seems unlikely that the selection of alternative gene targets will close this efficacy gap. The absence of RNAi-induced lethal phenotypes in recalcitrant species may simply reflect the degree to which gene silencing actually occurs. While many studies report target gene silencing upon ingestion of dsRNA, the level of transcript knockdown seldom exceeds 60% and silencing is frequently transient (Huvenne and Smagghe, 2010; Li et al., 2013b). In the case of sensitive coleopteran species, target gene knockdown is frequently ≥90% (Baum et al., 2007; Bolognesi et al., 2012; Rangasamy and Siegfried, 2012; Zhu et al., 2011). The typical twofold reduction in mRNA transcript levels observed in many species may fall well within the range of normal variation observed in biological systems and, in any case, may not manifest immediately during the course of a feeding assay since the target protein itself needs to turnover first.

If we set aside the question of target gene selection and dsRNA design, the barriers to efficient RNAi in recalcitrant insects can be pictured as discreet steps in the mechanism of action of RNAi. DsRNAs, whether applied topically to an insect diet or expressed in transgenic plants, must be able to survive long enough to be taken up by insect cells. The insect midgut with its brush border membrane topography (Chapter 1) represents a large surface area that facilitates the absorption of nutrients and is the presumed site of entry for dsRNAs in most studies. It is not clear to what extent the peritrophic matrix in lepidopteran- and coleopteran midguts (Hegedus et al., 2009; Lehane, 1997) or the perimicrovillar membrane in hemipteran midguts (Silva et al., 2004) represents a physical barrier to dsRNA delivery. Upon safe arrival at the gut membrane surface, dsRNAs must be efficiently taken up by epithelial cells and delivered to the intracellular RNAi machinery. If the route of cellular uptake involves endocytosis, then release or escape from the endosome becomes a critical step for delivery to the cytoplasm (Varkouhi et al., 2011). The core RNAi machinery, being agnostic with respect to the source of the dsRNA, processes the dsRNA to generate siRNA–RISC complexes that can trigger a silencing event, provided transcripts of the gene target are present in that host cell. It is likely that expression of the RNAi core machinery, including Dicer and Argonaute proteins, varies among insect cell types and species, contributing to differences in the efficiency of cell-autonomous RNAi. Factors impacting RNA amplification and/or systemic movement of the silencing signal from the site of cellular entry represent additional barriers to RNAi efficiency and the generation of lethal phenotypes.

The alkaline nature of the lepidopteran midgut (Dow, 1992) could certainly represent one barrier to RNA delivery. Across insect orders, the presence of midgut and salivary nucleases that can rapidly degrade dsRNA molecules likely constitutes a more significant barrier (Arimatsu et al., 2007a,b; Christiaens et al., 2014; Furusawa et al., 1993; Liu et al., 2013; Luo et al., 2013; Rodríguez-Cabrera et al., 2010; Terenius et al., 2011; Wynant et al., 2014). In the case of hemipteran species such as the tarnished plant bug, *L. lineolaris*, which engage in extra-oral digestion of plant material prior to ingestion, the presence of dsRNAses in salivary secretions represents a significant barrier to dsRNA survival (Allen and Walker, 2012). Likewise, Christiaens et al. (2014) provided evidence for dsRNA degradation by dsRNAses in both pea aphid salivary secretions and haemolymph. No less than four different midgut-expressed dsRNAses were observed in the desert locust, *S. gregaria* (Wynant et al., 2014), a species that is recalcitrant to ingested dsRNA but exhibits a potent systemic RNAi response to injected dsRNA (Wynant et al., 2012). Knockdown experiments implicated at least one of these nucleases in dsRNA degradation in the midgut. Likewise, Luo et al. (2013) reported the presence of potent dsRNAses in midgut fluid from the migratory locust, *L. migratoria*, and concluded that rapid degradation of dsRNA in the midgut explains the recalcitrant nature of this species to dietary dsRNA. Given the sensitivity of both locust species to injected dsRNAs, it would be worth understanding the extent to which dsRNAses are also expressed in the haemolymph. Finally, the importance of a gut environment hostile to dsRNA is supported by our own observation that nucleases found in gut extracts from *S. frugiperda*, a lepidopteran species, are far more active in degrading dsRNAs than are nucleases found in corn rootworm gut extracts (Fig. 5.1), particularly at an alkaline pH more typical of the lepidopteran midgut.

Inefficient uptake by midgut epithelial cells represents the next barrier to efficacious RNAi-mediated insect control. The mechanism of dsRNA uptake by insect gut epithelial cells has not been defined, even for highly sensitive species such as the WCR. In the nematode, *C. elegans*, several genes

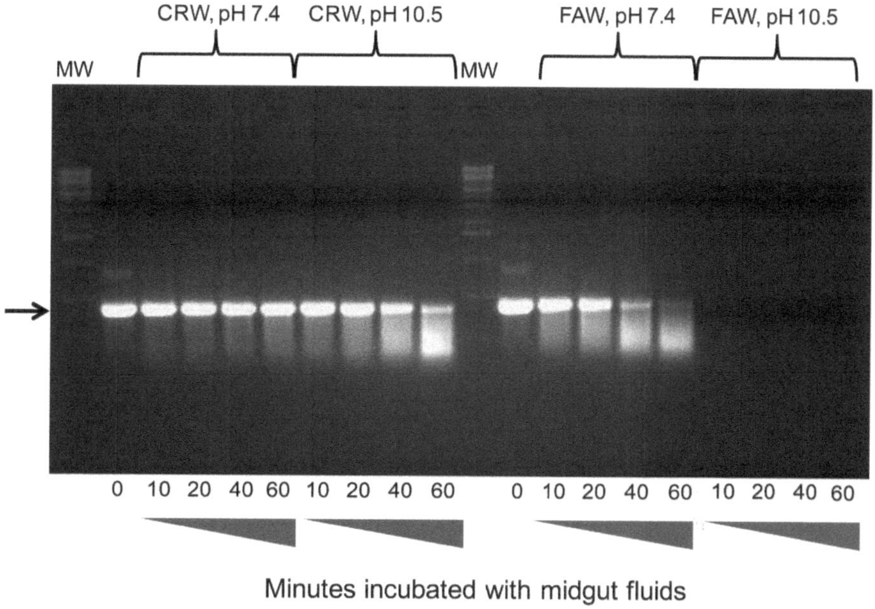

Figure 5.1 dsRNAs are rapidly degraded when treated with midgut extracts from the lepidopteran species, *Spodoptera frugiperda*, but remain intact when treated with midgut extracts from the western corn rootworm, *Diabrotica virgifera* virgifera. A 400-bp V-ATPase A dsRNA (10 μg, see arrow) was incubated at 23 °C with 50 μg of total protein extracted from isolated midgut tissue. Incubations were carried out in 100 μL volumes in either 25 mM Tris HCl (pH 7.4) or 25 mM sodium carbonate (pH 10.5). Aliquots (20 μL) were drawn at the times indicated, the digestions were quenched by ethanol precipitation, and the recovered dsRNAs were resolved by agarose gel electrophoresis. MW = Invitrogen molecular weight standards.

required for environmental RNAi and systemic spread have been identified. The *Sid-1* gene is required for systemic movement of the silencing signal between cells and is also required for environmental RNAi by gut epithelial cells. The *Sid-1* gene encodes a transmembrane protein that acts as a dsRNA-selective and dsRNA-gated channel to passively traffic dsRNAs between cells (Feinberg and Hunter, 2003; Jose et al., 2009; Shih and Hunter, 2011; Shih et al., 2009; Winston et al., 2002). The *Sid-2* gene encodes a membrane protein exclusively localized to intestinal cells that is required for dsRNA uptake from the intestinal lumen (McEwan et al., 2012; Winston et al., 2007). Environmental RNAi therefore requires coordination between the SID-1 and SID-2 proteins. A model has been proposed whereby dsRNAs are imported from the intestinal lumen by SID-2 via endocytosis and released from the internalized vesicles via passive movement through the SID-1 channel (McEwan et al., 2012; Whangbo and Hunter, 2008). While putative *Sid-1*-like genes have been identified in some insects, there are no reports of *Sid-2*-like genes in insect species whose genomes have been completely sequenced (e.g. Tomoyasu et al., 2008; Xu and Han, 2008; Zha et al., 2011).

In *Drosophila*, an insect lacking *Sid-1*-like genes, evidence has been presented for an energy-dependent endocytic mechanism of dsRNA uptake by cultured (S2) cells, apparently mediated by pattern-recognition scavenger receptors (Saleh et al., 2006; Ulvila et al., 2006). A genomic screen identified 23 genes required for RNAi in S2 cells including many components of the endocytic pathway involved in vesicle trafficking and protein transport or sorting. Subsequently, orthologs of these genes were selected for RNAi screens in *C. elegans*, resulting in the identification of 10 genes whose silencing disrupted the RNAi response triggered by feeding (Saleh et al., 2006). As noted by Hinas et al. (2012), this screen could not distinguish among knockouts that affected dsRNA uptake by intestinal cells, systemic RNAi, or the autonomous cellular RNAi machinery. In a subsequent study, mutations in several of the endocytic pathway genes required for cellular dsRNA uptake in *Drosophila* rendered flies that were hypersensitive to viral infection and unable to mount a systemic antiviral response (Saleh et al., 2009). Further support for the role of endocytic pathways in systemic RNAi proceeded from the characterization of *sid-3* and *sid-5* mutants in *C. elegans*, mutants that are defective in systemic RNAi but are otherwise RNAi competent. *Sid-3* encodes a conserved tyrosine kinase, homologs of which are known to associate with endocytic vesicles, and is required for the import of dsRNA into cells (Jose et al., 2012). *Sid-5* encodes an endosome-associated protein required for the export of dsRNAs from cells (Hinas et al., 2012). Three other mutants identified in RNAi screens, *rsd-2*, *rsd-3*, and *rsd-6*, are insensitive to ingested dsRNA targeting germline-expressed genes but remain sensitive to injected dsRNA (Tijsterman et al., 2004). Other mutants identified in this screen, *rsd-8* and *rsd-4*, were subsequently shown to be alleles of *sid-1* and *sid-2*, respectively, the former already identified as required for systemic RNAi. While the details are not yet understood, we expect that systemic RNAi in insects will prove to be equally complex and that *Sid-1*-like genes will not prove to be the sole determinant of systemic RNAi. Indeed, it is not entirely clear that the *Sid-1*-like genes identified to date in insects are even involved in systemic RNAi. A phylogenetic analysis of Sid-1-like proteins in *Tribolium* suggests that these proteins may not be orthologous to Sid-1, but rather to the *C. elegans* Tag-130 protein that is not involved in systemic RNAi (Tomoyasu et al., 2008). More recently, Xu et al. (2013) provided evidence through injection studies that the single *Sid-1*-like gene found in *N. lugens* is required for systemic RNAi.

With respect to environmental RNAi, we observe that immortalized insect cells scarcely resemble the differentiated cell types found in an insect midgut. The mechanism of dsRNA uptake by *Drosophila* S2 cells (Saleh et al., 2006; Ulvila et al., 2006) may or may not inform us regarding the mechanism of RNA uptake by gut epithelial cells, particularly since *Drosophila* is not sensitive to ingested dsRNAs in the absence of a transfection agent (Whyard et al, 2009). An alternative endocytic uptake mechanism may be operational in insect species that respond to bacterial-encapsulated dsRNAs in diet (Tian et al., 2009; Xiong et al., 2013; Zhang et al., 2013c; Zhu et al., 2012). In the future, the selection of insects resistant to orally toxic dsRNAs could provide an experimental system to identify genes involved in dsRNA uptake. While *Tribolium* may be the best model for understanding systemic silencing in insects, *Diabrotica* (corn rootworm) or *L. decemlineata* (CPB) may be the emerging model systems for understanding the mechanism of efficient environmental RNAi.

The presence of dsRNAses in the insect hemolymph, noted in some lepidopteran species, could prove to be a formidable barrier to systemic RNAi. A comparative study of the RNAi-sensitive cockroach, *Blattella germanica*, and the RNAi-recalcitrant lepidopteran, *M. sexta*, revealed that dsRNAs are rapidly degraded in the haemolymph fluids from *M. sexta* but not in the haemolymph fluids of *B. germanica* (Garbutt, 2011; Garbutt et al., 2013). Putative orthologs of the *Bombyx mori* and *M. sexta* dsRNAses were also identified in *Spodoptera littoralis* and *S. frugiperda*, two other lepidopteran species insensitive to environmental RNAi. Although not comprising an extensive dataset, these findings nevertheless prompted the suggestion that the capacity to rapidly degrade dsRNA in the hemolymph could represent an evolutionary adaptation to high viral exposure since lepidopteran-specific viruses are abundant and lepidopterans appear particularly prone to viral diseases among holometabolic species (Garbutt, 2011). More recently, Swevers et al. (2013) expanded on this hypothesis to discuss the possible role of persistent virus infections on RNAi efficiency, noting that viruses could interfere with RNAi in insects via number of strategies including expression of RNAi suppressor genes, production of RNA decoys, and manipulation of host gene expression. One attractive feature of this hypothesis is that it might explain why different research groups report such disparate results working with lepidopteran species (Terenius et al., 2011).

5. COMMERCIAL DEVELOPMENT OF RNAi ACTIVES

5.1. Next-generation rootworm-resistant corn

The exceptional sensitivity of corn rootworm larvae to environmental RNAi and the economic importance of this pest complex for corn production in the United States provided an excellent opportunity to develop a biotechnology-based strategy for corn rootworm management. Replicated field trials with transgenic corn plants expressing a corn rootworm V-ATPase dsRNA showed significant root protection (Fig. 5.2) similar to the pot assay results reported by Baum et al (2007).

While these plants showed significant reduction in root damage due to corn rootworm feeding, the damage observed was still above the commercially acceptable threshold of 0.25 for root damage ratings. Genetic crosses between the V-ATPase dsRNA-expressing transgenic event and a corn line expressing the corn rootworm-active Bt protein Cry3Bb1 served to combine these traits. Corn hybrids expressing both traits gave better root protection than did hybrids expressing either Cry3Bb1 or dsRNA alone (Fig. 5.2). The benefit of the RNAi-based trait became even more apparent when adult emergence was assessed. Corn plants expressing the V-ATPase dsRNA

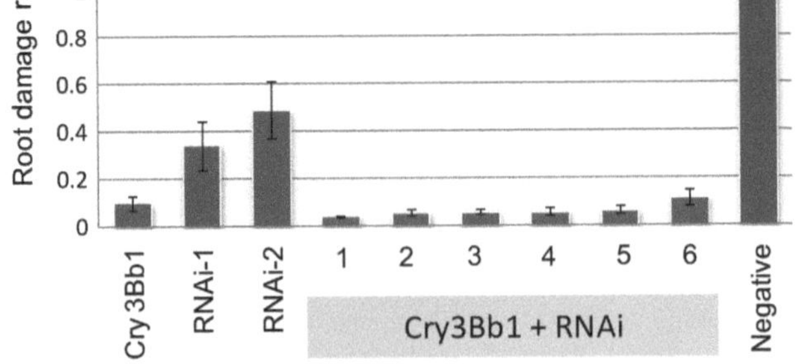

Figure 5.2 Transgenic plants expressing both the Cry3Bb1 protein and corn rootworm (CRW)-specific dsRNAs show excellent protection from rootworm feeding damage. Root damage ratings (0–3 Iowa scale) are averaged across nine locations from four states (IA, IL, IN, NE). Standard error bars are shown.

exhibited superior suppression of adult corn rootworm emergence in growth chamber assays (Fig. 5.3).

The identification of numerous efficacious targets through insect feeding assays (Baum et al., 2007) enabled a rational approach to the disruption of different cellular pathways including, in the case of the *Snf7* ortholog (Bolognesi et al., 2012), pathways involved in protein trafficking. Ramaseshadri et al. (2013) used histology and immuno-staining to show that knockdown of *Snf7* gene expression in corn rootworm larvae leads to cellular perturbations that are consistent with a gene encoding an essential component of the ESCORT (Endosomal Sorting Complex Required for Transport) pathway essential for protein trafficking within cells. Further electron microscopy studies (Koči et al., 2014) with WCR midguts have confirmed that loss of *Snf7* function causes a progressive degeneration of midgut enterocytes resulting in cell sloughing and lysis. These authors conclude that the severe cellular defects caused by loss of Snf7 in midgut enterocytes is the cause of death for rootworm larvae fed *Snf7* dsRNA. This RNAi-mediated disruption of cellular function represents a very different cause of insect mortality than that attributed to Bt insecticidal proteins, most of which bind to specific midgut receptors and create ion channels and pores to disrupt midgut function in susceptible species (Vachon et al., 2012; Chapter 2). Accordingly, a single transgenic corn line expressing both a corn rootworm-active Bt protein and an efficacious dsRNA (e.g. the *Snf7* dsRNA) can possess two very different modes of action for the control of

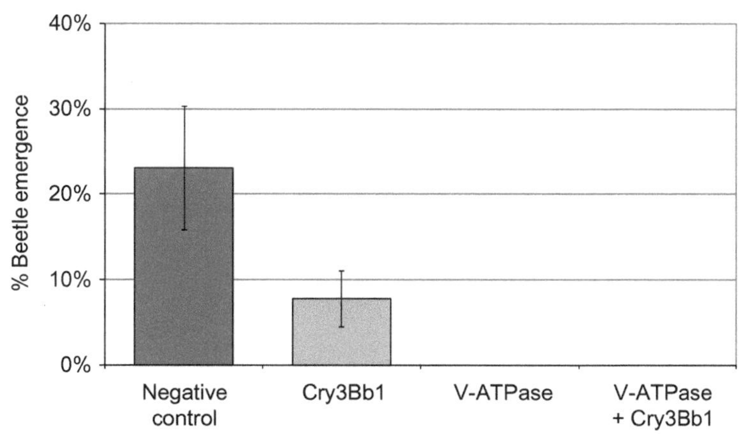

Figure 5.3 Transgenic plants expressing the CRW V-ATPase A dsRNA exhibit superior control of CRW adult beetle emergence in growth chamber assays. Standard error bars are shown.

corn rootworm. Commercial introduction of the *Snf7* RNAi trait in corn hybrids is expected by the end of this decade (Kupferschmidt, 2013).

5.2. Topical application

Given the recalcitrant nature of most lepidopteran and hemipteran pests towards environmental RNAi, the future of RNAi-based insect pest management strategies may depend on the development of topical formulations that facilitate delivery into insect cells. Significant progress has been made in the development of RNA delivery agents for RNAi human therapeutics (Zhou et al., 2013), a field that has wrestled with the challenges of RNA stability, endocytosis, and endosomal escape to deliver siRNA payloads. Examples of delivery agents being explored in the pharmaceutical arena are illustrated in Fig. 5.4.

These include cationic lipids, cationic dendrimers, cyclodextrin polymers, various forms of polethyleneimine, mesoporous silica, cell-internalizing peptides, and nucleic acid aptamers. Transfection agents have been shown to aid in the delivery of dsRNAs to a recalcitrant insect species,

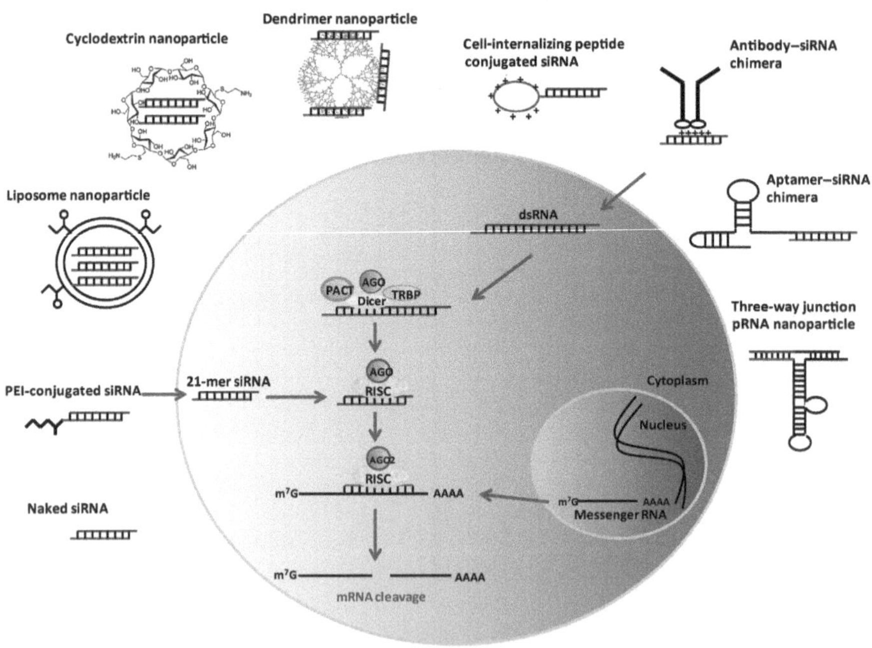

Figure 5.4 The mechanism and delivery strategies for RNA interference. RNAi therapeutics (e.g. siRNA) can be internalized into cells via different delivery vehicles. *Figure reproduced with permission from Zhou et al. (2013).* (See the colour plate.)

in this case *Drosophila* (Whyard et al., 2009). Zhang et al. (2010) demonstrated that chitosan–dsRNA complexes were stabilized in gel-based diets for delivery to mosquito larvae, but the actual impact of chitosan on dsRNA delivery to or uptake by mosquito gut cells was not evaluated in this study. More recently, He et al. (2013) reported that fluorescent nanoparticles (FNPs) comprised of a perylene-3,4,9,10-tetracarboxydiimide chromophore core, a polyphenylene dendrimer inner shell, and a cationic polymer outer shell can serve as a delivery vehicle for both DNA and dsRNA. FNPs coated with a dsRNA targeting the chitinase-like *CHT10* gene were fed at a dietary concentration of 250 ppm to the lepidopteran species, *O. furnacalis*, and shown to be more effective than dsRNA alone in arresting larval development.

A formulation of chemically modified AChE2 siRNA complexed with chitosan was shown to be toxic to DBM larvae in leaf disc assays and in a field trial (Gong et al., 2013). Larval mortality (%) in the field appeared to peak at 5 days post-treatment with the siRNA formulation (50, 100, and 200 ppm treatments), perhaps reflecting foliar degradation of the siRNA active or alternatively an increase in larval infestation leading to a higher recovery of survivors. The high sensitivity of CPB larvae to ingested dsRNAs provides an opportunity to explore the development of dsRNA formulations for use as foliar sprays in insect control. Zhu et al. (2011) evaluated the use of *E. coli*-expressed dsRNA to control CPB, demonstrating consistent target gene knockdown and high mortality of CPB larvae when fed potato leaves treated with either bacterial-expressed dsRNA or *in vitro*-synthesized dsRNA.

The cost of goods weighs heavily on the development of RNAi spray formulations, including the cost of dsRNA production as well as any requisite delivery agents. The first outdoor trial with an RNAi product was conducted by Hunter et al. (2010) to assess the efficacy of a dsRNA product (Remebee-IAPV) in reducing Israeli acute paralytic disease in honey bees, one of many viral diseases that afflict honey bee colonies worldwide. In this case, dsRNA was provided to honey bees in a sugar solution with the aim of blocking virus replication and transmission. Formulations that improve RNA delivery and insecticidal activity while enhancing environmental stability could enable a more favourable cost of goods (through lower use rates) and the development of economically viable topical applications for insect control. Factors that likely impact the environmental stability of dsRNA are not so different from those impacting the stability of synthetic insecticides and include UV damage from exposure to sunlight, microbial degradation (e.g. on foliar surfaces), and exposure to plant exudates. It is uncertain how formulations that improve the environmental stability of dsRNAs will affect bioavailability and efficacy. Likewise, it is too early in the development of formulation technologies to determine whether approaches taken in the pharmaceutical field can be applied to the development of effective crop protection products.

6. SAFETY CONSIDERATIONS

Safety assessments for the use of RNAi strategies in insect pest management include both the evaluation of environmental safety and food and feed safety. The impact of dsRNAs on non-target organisms, including beneficial insects, and the persistence of applied dsRNAs in the environment, are also addressed during the registration of Bt-based bioinsecticides and insect-protected crops. The current safety assessment approach to biotechnology-derived plants is appropriate for assessing the safety of plants produced using RNA-mediated gene regulation (Codex, 2003; EFSA, 2006; FDA, 1994; Parrott et al., 2010; Petrick et al., 2013; Redenbaugh et al., 1992), and the current environmental risk assessment approach can be applied to characterize risk for RNA-based technologies (Auer and Frederick, 2009; ILSI-CERA, 2011).

RNAi technology has the potential for high specificity towards target pest organisms, even more so than Bt-based technology (Bachman et al., 2013; Baum et al., 2007; Whyard et al., 2009). Larval feeding assays with orthologous CPB- and WCR dsRNAs demonstrated that the accumulation of base pair mismatches between the dsRNA sequence and target gene sequence results in reduced insecticidal activity (Baum et al., 2007). For example, a V-ATPase A dsRNA from WCR sharing 83% sequence identity with the CPB V-ATPase A coding region showed >10-fold lower activity against CPB larvae than did the orthologous CPB V-ATPase A dsRNA. Note that this residual activity towards CPB was expected because the WCR dsRNA sequence still retained stretches of >21 bp sequence identity with the CPB V-ATPase coding region, a result that is entirely consistent with the work of Bolognesi et al. (2012) and Bachman et al. (2013). The case for sequence specificity is further supported by the work of Whyard et al. (2009), who demonstrated that four *Drosophila* species could be selectively controlled using dsRNAs that target the divergent 3′ untranslated region of the γ*Tub23C* tubulin gene, a region lacking 21 bp matches among the four species. Other factors, including the barriers to RNAi described in this review, clearly play a role in determining sensitivity to dsRNA and assessing risk to non-target organisms. The aggregate of these factors provides outcomes such as those reported by Bachman et al. (2013) showing that a *Snf7* dsRNA from WCR has no effect when tested on a wide range of insect species. "Indirect" WCR feeding assays with *Snf7* orthologs from other coleopteran species further demonstrated that the spectrum of activity for the WCR *Snf7* dsRNA is narrow and only evident in a subset of beetles within the Galerucinae subfamily of Chrysomelidae (>90% identity with the WCR *Snf7* 240 bp dsRNA) that contain a shared sequence length of ≥21 bp within the *Snf7* coding sequence (Bachman et al., 2013). In summary, the careful selection of dsRNA sequences with no 21 bp match to orthologous genes further reduces the risk of unintended effects on non-target species that may be sensitive to ingested dsRNAs. With respect to RNAi-based strategies for corn rootworm control, however, any assessment of risk towards non-target insects or arachnids should start with the acknowledgement that non-coleopteran species will likely be at least three orders of magnitude less sensitive to environmental RNAi than the target pest itself (Table 5.1).

Regarding food safety, the natural occurrence of long dsRNAs and small RNAs in plants and other foods provides a very long history of safe use by humans (Heisel et al., 2008; Ivashuta et al., 2009; Jensen et al., 2013; Parrott et al., 2010; Petrick et al., 2013), even though sequence complementarity exists between small- and long dsRNAs in crops and human genes (Ivashuta et al., 2009; Jensen et al., 2013). The long list of biological barriers to oral activity of dietary small RNAs and longer dsRNAs in mammals and other vertebrates has been summarized (Petrick et al., 2013). Vertebrate digestive systems display common structural and functional features (Stevens and Hume, 2004); therefore, the same digestive barriers that greatly limit the potential for oral activity of ingested RNA in mammals are likely to be present in lower vertebrates. These barriers include salivary RNAses, harsh acidic conditions in the stomach that lead to depurination and denaturation of nucleic acids, nucleases in the lumen of the gastrointestinal tract, lytic enzymes from pancreatic secretions in the duodenum and nucleases in the blood (Houck and Berman, 1958; Loretz et al., 2006; O'Neill et al., 2011; Park et al., 2006; Petrick et al., 2013). Cellular membranes of the gut epithelium provide a physical barrier to uptake of high molecular weight, hydrophilic compounds like siRNAs (Akhtar, 2009; Jain, 2008). Recent studies employing small RNA sequencing and/or quantitative PCR support the view that dietary dsRNAs have extremely low bioavailability in mammals, at a level that is orders of magnitude below that needed for biological activity (Cottrill and Chan, 2014; Dickinson et al., 2013; Snow et al., 2013; Witwer and Hirschi, 2014; Witwer et al., 2013). Likewise, a recently reported survey of numerous animal small RNA datasets from public sources has not revealed evidence for any major plant-derived miRNA accumulation in animal samples (Zhang et al., 2012b), consistent with the lack of success in oral delivery of RNA-based therapeutics experienced by the pharmaceutical industry (Petrick et al., 2013). Current pharma approaches employ local and systemic delivery of RNA therapeutics through methods that bypass the oral route (Zhou et al., 2013). Based on the history of safe consumption of dsRNA in the diet, the extensive barriers to ingested RNA, and the lack of any appreciable dsRNA uptake from the diet, dietary dsRNA is not anticipated to represent any hazard or risk to humans, mammals, or other vertebrates.

7. INSECT RESISTANCE MANAGEMENT

In order to reduce the intensity of selection for resistance, insecticides can be applied judiciously when established economic thresholds of pest infestation have been crossed, tank mixed with insecticides providing a different MOA, or rotated with those insecticides during the growing season to alter the selective pressure on insect populations. The strategy of tank-mixing insecticides is fundamentally the same as stacking or pyramiding insect control traits with different MOAs in plants. RNAi-based insecticides would fit well with IPM strategies that rely on the use of multiple control strategies, including synthetic and biological insecticides, and that leverage host plant resistance traits and natural predators to provide economic control of insect pests. The selectivity and slow knockdown anticipated for RNAi insecticidal actives might encourage their use in the context of an IPM framework, perhaps as early-season applications to suppress target pest populations while allowing beneficial insects and predators to thrive.

For transgenic delivery of RNAi-based insect control traits, the path towards resistance management is well worn. Strategies were modelled and deployed in the United States and elsewhere to delay the emergence of insect resistance to insect-protected crops expressing Bt insecticidal proteins (Gassmann et al., 2009; Huang et al., 2011; Tabashnik, 2008). The high dose/refuge strategy for insect-protected crops relies on the out-crossing of rare resistant individuals with sensitive individuals in the non-transgenic refuge, but is only effective if the resistance phenotype is recessive and the dose sufficiently high to kill heterozygous progeny. In recent years, the stacking of insect resistance traits deploying different MOAs (Roush, 1998) to delay resistance development has become the standard for new commercial transgenic plant offerings. Examples include BollGard® II cotton expressing the Cry1Ac and Cry2Ab proteins for lepidopteran control, Genuity® VT Double PRO® corn expressing the Cry1A.105 and Cry2Ab proteins for lepidopteran control, and Genuity SmartStax® corn expressing the Cry3Bb1 and Cry34Ab–Cry35Ab proteins for corn rootworm control, and the Cry1A.105, Cry1Fa, and Cry2Ab proteins for broad-spectrum lepidopteran control (www.monsanto.com).

To ensure their durability, RNAi-based insect resistance traits should also be stacked with traits employing a different MOA. This is where the work in *C. elegans* provides additional value: the isolation of multiple mutants impaired in environmental and/or systemic RNAi in the nematode provides a cautionary tale for those of us in the field of insect pest management. The risk is not that target genes will be selected for single nucleotide polymorphism variants that evade RNAi processing (large dsRNAs can presumably mitigate this risk; Bachman et al., 2013; Bolognesi et al., 2012) but that, for instance, up-regulation of nucleases or defects in dsRNA uptake, processing, or systemic spread will compromise the triggering and spread of the RNAi response in the pest organism. The observation that most species of *Caenorhabditis* are not sensitive to environmental RNAi (Nuez and Félix, 2012; Whangbo and Hunter, 2008) and that several grassland nematode species are not sensitive to RNAi altogether (Wheeler et al., 2012) further suggests that this capability is under rapid evolution in nematodes. In the case of next-generation rootworm-resistant corn, the stacking of the *Snf7* RNAi trait with Bt genes encoding two rootworm insecticidal toxins with distinct MOAs could provide strong protection for all three traits. Likewise, one can envision the stacking of RNAi traits for lepidopteran control with any number of Bt genes encoding highly efficacious lepidopteran-active proteins. In contrast, the deployment of RNAi traits for hemipteran control is problematic because there are no viable biotech traits that confer effective protection from hemipteran feeding damage and consequently no traits to stack with. Progress has been made in the development of engineered Bt proteins for control of Lygus species in cotton (Baum et al., 2013) but, in general, Bt proteins tend to be ineffective against hemipteran pests. Transgenic plants expressing lectins have been shown to impact the growth, development, and fecundity of hemipteran pests, but regulatory approval of such traits is not assured given their apparent MOA in binding complex glycans that are also found along the intestinal tract of mammalian herbivores (Vandenborre et al., 2009). Without complementary traits for resistance management, the commercialization of durable RNAi-based hemipteran control traits may prove elusive.

8. CONCLUDING REMARKS

The mechanism of gene suppression manifested as RNAi provides an MOA unique among insecticidal agents. In addition, the potential for highly tailored insecticidal specificity afforded in part by the mechanism of RNAi differentiates it from other insect pest management strategies. Successful development of RNAi actives, either expressed in plants or applied topically as "biological" insecticide sprays, would provide growers with an important tool for sustainable insect pest management.

Despite the exceptional sensitivity of corn rootworm larvae to ingested *V-ATPase-* or *Snf7* dsRNAs, transgenic corn plants expressing these dsRNAs do not provide complete protection from rootworm feeding damage in the field, presumably because of the slow speed-to-kill typical of dsRNA actives. Transgenic corn plants combining the *Snf7* RNAi trait with the Cry3Bb1 gene, however, do provide superior control of corn rootworm larvae and emerging adults in the field. Accordingly, stacking the *Snf7* RNAi trait with suitable Bt traits in corn offers the best opportunity for efficacious and durable control of corn rootworm species. Looking towards the future, it seems likely that the slow speed-to-kill exhibited by insecticidal dsRNAs will affect how these agents are deployed commercially.

It is now firmly established that insect species differ greatly in their response to environmental RNA. Most notably, the lepidopteran and hemipteran species studied to date are far less sensitive to ingested dsRNAs than are the responsive coleopteran species exemplified by the corn rootworm. This differential sensitivity to environmental RNAi helps inform the environmental safety assessment of RNAi for management of coleopteran pests. It will also dictate the approach taken to exploit the phenomenon of RNAi for insect pest management. In order to achieve parity with the WCR *Snf7* dsRNA active, for example, dsRNAs for lepidopteran- or hemipteran pests will need at least a 1000-fold improvement in oral activity. Given the recalcitrant nature of most lepidopteran and hemipteran species to environmental RNAi and the minimal impact target gene selection will likely have on closing this efficacy gap, a clear path towards development of RNAi-based plant traits for lepidopteran- and hemipteran pest management is not yet apparent. For example, while transgenic plants expressing dsRNAs have been shown to impact the growth and survival of a few lepidopteran or hemipteran species in controlled environment tests, the effects are largely sub-lethal and have not been confirmed with field efficacy data demonstrating economic control of the pest species. Given this apparent "ineffective dose" activity towards recalcitrant species, it would be difficult to justify trait commercialization in light of the significant development and regulatory costs involved (McDougall, 2011; Prado et al., 2014). In the case of hemipteran species, the general lack of complementary (e.g. Bt) traits for use in insect resistance management strategies provides an additional hurdle towards commercial development of transgenic crops employing an RNAi-based trait.

For these reasons, the development of topically applied formulations that facilitate dsRNA delivery into insect cells provides a sensible alternative to the transgenic plant approach and may enable RNAi-based insect pest management for otherwise recalcitrant species. The success of this strategy depends on a detailed understanding of the barriers to efficient environmental RNAi, the development of cost-effective, stable formulations that overcome those barriers, and efficient robust systems for dsRNA production. In some insect species, the absence of an efficient systemic RNAi mechanism may mute any efficiency gained through the use of RNA delivery agents that enable environmental RNAi. Despite these uncertainties, initial studies with formulated dsRNAs suggest this is a promising approach.

ACKNOWLEDGEMENTS

The authors wish to thank Greg Heck, Sergey Ivashuta, Steve Levine, Bill Moar, Jay Petrick, Gerrit Segers, and Greg Watson for their thorough review of the manuscript.

REFERENCES

Akhtar, S., 2009. Oral delivery of siRNA and antisense oligonucleotides. J. Drug Target. 17, 491–495.

Allen, M.L., Walker III, W.B., 2012. Saliva of *Lygus lineolaris* digests double stranded ribonucleic acids. J. Insect Physiol. 58 (3), 391–396.

Araujo, R.N., Santos, A., Pinto, F.S., Gontijo, N.F., Lehane, M.J., Pereira, M.H., 2006. RNA interference of the salivary gland nitrophorin 2 in the triatomine bug Rhodnius prolixus Hemiptera, Reduviidae by dsRNA ingestion or injection. Insect Biochem. Mol. Biol. 36 (9), 683–693.

Arimatsu, Y., Furuno, T., Sugimura, Y., Togoh, M., Ishihara, R., Tokizane, M., Kotani, E., Hayashi, Y., Furusawa, T., 2007a. Purification and properties of double-stranded RNA degrading nuclease, dsRNase, from the digestive juice of the silkworm, *Bombyx mori*. J. Insect Biotechnol. Sericol. 76, 57–62.

Arimatsu, Y., Kotani, E., Sugimura, Y., Furusawa, T., 2007b. Molecular characterization of a cDNA encoding extracellular dsRNase and its expression in the silkworm, *Bombyx mori*. Insect Biochem. Mol. Biol. 37, 176–183.

Aronstein, K., Oppert, B., Lorenzen, M., 2011. RNAi in agriculturally-important arthropods. In: Grabowski, P.P. (Ed.), RNA Processing. In Tech, Shanghai, China, pp. 157–180.

Auer, C., Frederick, R., 2009. Crop improvement using small RNAs: applications and predictive ecological risk assessments. Trends Biotechnol. 27, 644–651.

Bachman, P., Bolognesi, R., Moar, W.J., Mueller, G.M., Paradise, M.S., Ramaseshadri, P., Tan, J., Uffman, J.P., Warren, J., Wiggins, B.E., Levine, S.L., 2013. Characterization of the spectrum of insecticidal activity of a double-stranded RNA with targeted activity against western corn rootworm *Diabrotica virgifera virgifera* LeConte. Transgenic Res. 22 (6), 1207–1222.

Barnard, A.-C., Nijhof, A., Fick, W., Stutzer, C., Maritz-Olivier, C., 2012. RNAi in arthropods: insight into the machinery and applications for understanding the pathogen-vector interface. Genes 3 (4), 702–741.

Bartel, D.P., 2009. MicroRNAs: target recognition and regulatory functions. Cell 136 (2), 215–233.

Baum, J.A., Bogaert, T., Clinton, W., Heck, G.R., Feldmann, P., Ilagan, O., Johnson, S., Plaetinck, G., Munyikwa, T., Pleau, M., Vaughn, T., Roberts, J., 2007. Control of coleopteran insect pests through RNA interference. Nat. Biotechnol. 25 (11), 1322–1326.

Baum, J.A., CaJacob, C.A., Feldmann, P., Heck, G.R., Nooren, I., Plaetinck, G., Maddelein, W., Vaughn, T., 2011. Methods for control of insect infestation in plants and compositions thereof. US Patent No. 7,943,819.

Baum, J.A., Evdokimov, A.G., Moshiri, F., Rydel, T.J., Sturman, E.J., von Rechenberg, M., Vu, H., Wollacott, A.M., Zheng, M., 2013. Proteins toxic to hemipteran insect species. US Patent Application No. 20130269060 A1.

Bautista, M.A.M., Miyata, T., Miura, K., Tanaka, T., 2009. RNA interference-mediated knockdown of a cytochrome P450, CYP6BG1, from the diamondback moth, *Plutella xylostella*, reduces larval resistance to permethrin. Insect Biochem. Mol. Biol. 39 (1), 38–46.

Belles, X., 2010. Beyond Drosophila, RNAi *in vivo* and functional genomics in insects. Annu. Rev. Entomol. 55, 111–128.

Bolognesi, R., Ramaseshadri, P., Anderson, J., Bachman, P., Clinton, W., Flannagan, R., Ilagan, O., Lawrence, C., Levine, S., Moar, W., Mueller, G., 2012. Characterizing the mechanism of action of double-stranded RNA activity against western corn rootworm *Diabrotica virgifera virgifera* LeConte. PLoS One 7 (10), e47534.

Boutros, M., Kiger, A.A., Armknecht, S., Kerr, K., Hild, M., Koch, B., Haas, S.A., Paro, R., Perrimon, N., 2004. Genome-wide RNAi analysis of growth and viability in Drosophila cells. Science 303 (5659), 832–835.

Boutros, M., Ahringer, J., 2008. The art and design of genetic screens: RNA interference. Nat. Rev. Genet. 9 (7), 554–566.

Burand, J.P., Hunter, W.B., 2013. RNAi, future in insect management. J. Invertebr. Pathol. 112 (Suppl. 1), S68–S74.

Campbell, E.M., Budge, G.E., Bowman, A.S., 2010. Gene-knockdown in the honey bee mite *Varroa destructor* by a non-invasive approach: studies on a glutathione S-transferase. Parasit. Vectors 3, 73.

Chen, J., Zhang, D., Yao, Q., Zhang, J., Dong, X., Tian, H., Zhang, W., 2010. Feeding-based RNA interference of a trehalose phosphate synthase gene in the brown planthopper, *Nilaparvata lugens*. Insect Mol. Biol. 19 (6), 777–786.

Christiaens, O., Sweveres, L., Smagghe, G., 2014. DsRNA degradation in the pea aphid (*Acyrthosiphon pisum*) associated with lack of response in RNAi feeding and injection assay. Peptides 53, 307–314. http://dx.doi.org/10.1016/j.peptides.2013.12.014.

Clemens, J.C., Worby, C.A., Simonson-Leff, N., Muda, M., Maehama, T., Hemmings, B.A., Dixon, J.E., 2000. Use of double-stranded RNA interference in Drosophila cell lines to dissect signal transduction pathways. Proc. Natl. Acad. Sci. U. S. A. 97 (12), 6499–6503.
Codex, 2003. Guideline for the conduct of food safety assessment of foods derived from recombinant-DNA plants (CAC/GL 45-2003). Monsanto Technical, Rome, Italy, pp. 7–26.
Cottrill, K.A., Chan, S.Y., 2014. Diet-derived microRNAs: separating the dream from reality. microRNA Diagn. Ther. 1, 46–57.
Coy, M.R., Sanscrainte, N.D., Chalaire, K.C., Inberg, A., Maayan, I., Glick, E., Paldi, N., Becnel, J.J., 2012. Gene silencing in adult *Aedes aegypti* mosquitoes through oral delivery of double-stranded RNA. J. Appl. Entomol. 136 (10), 741–748.
Dickinson, B., Zhang, Y., Petrick, J.S., Heck, G., Ivashuta, S., Marshall, W.S., 2013. Lack of detectable oral bioavailability of plant microRNAs after feeding in mice. Nat. Biotechnol. 31 (11), 965–967.
Dietzl, G., Chen, D., Schnorrer, F., Su, K.C., Barinova, Y., Fellner, M., Gasser, B., Kinsey, K., Oppel, S., Scheiblauer, S., Couto, A., Marra, V., Keleman, K., Dickson, B.J., 2007. A genome-wide transgenic RNAi library for conditional gene inactivation in *Drosophila*. Nature 448, 151–156.
Djebali, S., Davis, C.A., Merkel, A., Dobin, A., Lassmann, T., et al., 2012. Landscape of transcription in human cells. Nature 489, 101–108. http://dx.doi.org/10.1038/nature11233.
Dow, J., 1992. pH gradients in lepidopteran midgut. J. Exp. Biol. 172 (1), 355–375.
EFSA, 2006. Opinion of the Scientific Panel on Genetically Modified Organisms on a request from the Commission related to the notification (Reference C/SE/96/3501) for the placing on the market of genetically modified potato EH92-527-1 with altered starch composition, for cultivation and production of starch, under Part C of Directive 2001/18/EC from BASF Plant Science. EFSA J. 323, 1–20.
Elbashir, S.M., Lendeckel, W., Tuschl, T., 2001. RNA interference is mediated by 21- and 22-nucleotide RNAs. Genes Dev. 15 (2), 188–200.
FDA, 1994. Biotechnology of food. Monsanto technical report BG 94-4.
Feinberg, E.H., Hunter, C.P., 2003. Transport of dsRNA into cells by the transmembrane protein SID-1. Science 301, 1545–1547.
Figueira-Mansur, J., Ferreira-Pereira, A., Mansur, J.F., Franco, T.A., Alvarenga, E.S.L., Sorgine, M.H.F., Neves, B.C., Melo, A.C.A., Leal, W.S., Masuda, H., Moreira, M.F., 2013. Silencing of P-glycoprotein increases mortality in temephos-treated *Aedes aegypti* larvae. Insect Mol. Biol. 22 (6), 648–658.
Fire, A., Xu, S., Montgomery, M.K., Kostas, S.A., Driver, S.E., Mello, C.C., 1998. Potent and specific genetic interference by double-stranded RNA in *Caenorhabditis elegans*. Nature 391, 806–811.
Furusawa, T., Takayama, E., Ishihara, R., Hayashi, Y., 1993. Double-stranded ribonuclease activity in the digestive juice and midgut of the silkworm, *Bombyx mori*. Comp. Biochem. Physiol. A Mol. Integr. Physiol. 104, 795–801.
Garbian, Y., Maori, E., Kalev, H., Shafir, S., Sela, I., 2012. Bidirectional transfer of RNAi between honey bee and *Varroa destructor*: Varroa gene silencing reduces varroa population. PLoS Pathog. 8 (12), e1003035.
Garbutt, J.S., 2011. RNA Interference in Insects, Persistence and Uptake of Double-Stranded RNA and Activation of RNAi Genes. (Ph.D. thesis). University of Bath.
Garbutt, J.S., Bellés, X., Richards, E.H., Reynolds, S.E., 2013. Persistence of double-stranded RNA in insect hemolymph as a potential determiner of RNA interference success, Evidence from *Manduca sexta* and *Blattella germanica*. J. Insect Physiol. 59 (2), 171–178.
Gassmann, A.J., Onstad, D.W., Pittendrigh, B.R., 2009. Evolutionary analysis of herbivorous insects in natural and agricultural environments. Pest Manag. Sci. 65 (11), 1174–1181.

Georghiou, G.P., Lagunes-Tejeda, A., Food and Agriculture Organization of the United Nations, 1991. The Occurrence of Resistance to Pesticides in Arthropods: An Index of Cases Reported Through 1989. Georghiou, G.P., Lagunes-Tejeda, A., FAO, Rome.

Gong, L., Yang, X., Zhang, B., Zhong, G., Hu, M., 2011. Silencing of Rieske iron-sulfur protein using chemically synthesised siRNA as a potential biopesticide against *Plutella xylostella*. Pest Manag. Sci. 67 (5), 514–520.

Gong, L., Chen, Y., Hu, Z., Hu, M., 2013. Testing insecticidal activity of novel chemically synthesized sirna against *Plutella xylostella* under laboratory and field conditions. PLoS One 8 (5), e62990.

Griebler, M., Westerlund, S.A., Hoffmann, K.H., Meyering-Vos, M., 2008. RNA interference with the allatoregulating neuropeptide genes from the fall armyworm *Spodoptera frugiperda* and its effects on the JH titer in the hemolymph. J. Insect Physiol. 54 (6), 997–1007.

Gu, L., Knipple, D.C., 2013. Recent advances in RNA interference research in insects: Implications for future insect pest management strategies. Crop. Prot. 45, 36–40.

Gu, J., Liu, M., Deng, Y., Peng, H., Chen, X., 2011. Development of an efficient recombinant mosquito densovirus-mediated RNA interference system and its preliminary application in mosquito control. PLoS One 6 (6), e21329.

Hannon, G.J., 2002. RNA interference. Nature 418, 244–251.

He, B., Chu, Y., Yin, M., Müllen, K., An, C., Shen, J., 2013. Fluorescent nanoparticle delivered dsRNA toward genetic control of insect pests. Adv. Mater. 25 (33), 4580–4584.

Hegedus, D., Erlandson, M., Gillott, C., Toprak, U., 2009. New insights into peritrophic matrix synthesis, architecture, and function. Annu. Rev. Entomol. 54 (1), 285–302.

Heisel, S.E., Zhang, Y., Allen, E., Guo, L., Reynolds, T.L., Yang, X., Kovalic, D., Roberts, J.K., 2008. Characterization of unique small RNA populations from rice grain. PLoS One 3 (8), e2871.

Hinas, A., Wright, A.J., Hunter, C.P., 2012. SID-5 Is an endosome-associated protein required for efficient systemic RNAi in *C. elegans*. Curr. Biol. 22 (20), 1938–1943.

Honeybee Genome Sequencing Consortium, 2006. Insights into social insects from the genome of the honeybee *Apis mellifera*. Nature 443, 931–949.

Houck, J.C., Berman, L.B., 1958. Serum ribonuclease activity. J. Appl. Physiol. 12, 473–476.

Huang, F., Andow, D.A., Buschman, L.L., 2011. Success of the high-dose/refuge resistance management strategy after 15 years of Bt crop use in North America. Entomol. Exp. Appl. 140 (1), 1–16.

Hunter, W., Ellis, J., van Engelsdorp, D., Hayes, J., Westervelt, D., Glick, E., Williams, M., Sela, I., Maori, E., Pettis, J., Cox-Foster, D., Paldi, N., 2010. Large-scale field application of RNAi technology reducing Israeli acute paralysis virus disease in honey bees (*Apis mellifera*, Hymenoptera: Apidae). PLoS Pathog. 6 (12), e1001160.

Huvenne, H., Smagghe, G., 2010. Mechanisms of dsRNA uptake in insects and potential of RNAi for pest control: a review. J. Insect Physiol. 56 (3), 227–235.

ILSI-CERA, 2011. Problem Formulation for the Environmental Risk Assessment of RNAi Plants. Center for Environmental Risk Assessment, International Life Sciences Institute Research Foundation, Washington, D.C.

International Silkworm Genome Consortium, 2008. The genome of a lepidopteran model insect, the silkworm *Bombyx mori*. Insect Biochem. Mol. Biol. 38, 1036–1045.

Ivashuta, S.I., Petrick, J.S., Heisel, S.E., Zhang, Y., Guo, L., Reynolds, T.L., Rice, J.F., Allen, E., Roberts, J.K., 2009. Endogenous small RNAs in grain: semi-quantification and sequence homology to human and animal genes. Food Chem. Toxicol. 47 (2), 353–360.

Jain, K.K., 2008. Stability and delivery of RNA via the gastrointestinal tract. Curr. Drug Deliv. 5, 27–31.

James, C., 2013. Global status of commercialized biotech/GM crops: 2013. ISAAA brief no. 46, ISAAA, Ithaca, NY.

Jensen, P.D., Zhang, Y., Wiggins, B.E., Petrick, J.S., Zhu, J., Kerstetter, R.A., Heck, G.R., Ivashuta, S.I., 2013. Computational sequence analysis of predicted long dsRNA transcriptomes of major crops reveals sequence complementarity with human genes. GM Crops and Food: Biotechnology in Agriculture and the Food Chain 4, 90–97.

Jose, A.M., Smith, J.J., Hunter, C.P., 2009. Export of RNA silencing from *C. elegans* tissues does not require the RNA channel SID-1. Proc. Natl. Acad. Sci. U. S. A. 106 (7), 2283–2288.

Jose, A.M., Kim, Y.A., Leal-Ekman, S., Hunter, C.P., 2012. Conserved tyrosine kinase promotes the import of silencing RNA into *Caenorhabditis elegans* cells. Proc. Natl. Acad. Sci. U. S. A. 109 (36), 14520–14525.

Kennerdell, J.R., Carthew, R.W., 1998. Use of dsRNA-mediated genetic interference to demonstrate that *frizzled* and *frizzled 2* act in the wingless pathway. Cell 95, 1017–1026.

Khajuria, C., Buschman, L.L., Chen, M.-S., Muthukrishnan, S., Zhu, K.Y., 2010. A gut-specific chitinase gene essential for regulation of chitin content of peritrophic matrix and growth of *Ostrinia nubilalis* larvae. Insect Biochem. Mol. Biol. 40 (8), 621–629.

Khajuria, C., Li, H., Narva, K., Rangasamy, M., Siegfried, B., 2013. Effectiveness of dsRNA versus siRNA in RNAi mediated gene knock-down in western corn rootworm (*Diabrotica virgifera virgifera*). In: Program and Abstracts, 46th Annual Meeting of the Society for Invertebrate Pathology Conference on Invertebrate Pathology and Microbial Control. Society for Invertebrate Pathology. Pittsburgh, PA.

Koči, J., Ramaseshadri, P., Bolognesi, R., Segers, G., Flannagan, R., Park, Y., 2014. Ultrastructural changes caused by Snf7 RNAi in larval enterocytes of western corn rootworm *Diabrotica virgifera virgifera* Le Conte. PLoS One 9 (1), e83985.

Koizumi, K., Higashida, H., Yoo, S., Islam, M.S., Ivanov, A.I., Guo, V., Pozzi, P., Yu, S.H., Rovescalli, A.C., Tang, D., Nirenberg, M., 2007. RNA interference screen to identify genes required for *Drosophila* embryonic nervous system development. Proc. Natl. Acad. Sci. U. S. A. 104, 5626–5631.

Kontogiannatos, D., Swevers, L., Maenaka, K., Park, E.Y., Iatrou, K., Kourti, A., 2013. Functional characterization of a juvenile hormone esterase related gene in the moth *Sesamia nonagrioides* through RNA interference. PLoS One 8 (9), e73834.

Kumar, M., Gupta, G.P., Rajam, M.V., 2009. Silencing of acetylcholinesterase gene of *Helicoverpa armigera* by siRNA affects larval growth and its life cycle. J. Insect Physiol. 55 (3), 273–278.

Kumar, A., Wang, S., Ou, R., Samrakandi, M., Beerntsen, B.T., Sayre, R.T., 2013. Development of an RNAi based microalgal larvicide to control mosquitoes. MalariaWorld J. 4, 6.

Kupferschmidt, K., 2013. A lethal dose of RNA. Science 341, 732–733.

Kwon, D.H., Park, J.H., Lee, S.H., 2013. Screening of lethal genes for feeding RNAi by leaf disc-mediated systematic delivery of dsRNA in *Tetranychus urticae*. Pest. Biochem. Physiol. 105 (1), 69–75.

Lee, J.T., 2012. Epigenetic regulation by long noncoding RNAs. Science 338, 1435–1439.

Lehane, M.J., 1997. Peritrophic matrix structure and function. Annu. Rev. Entomol. 42, 525–550.

Li, J., Chen, Q., Lin, Y., Jiang, T., Wu, G., Hua, H., 2011a. RNA interference in *Nilaparvata lugens* (Homoptera, Delphacidae) based on dsRNA ingestion. Pest Manag. Sci. 67 (7), 852–859.

Li, X., Zhang, M., Zhang, H., 2011b. RNA interference of four genes in adult *Bactrocera dorsalis* by feeding their dsRNAs. PLoS One 6 (3), e17788.

Li, H., Jiang, W., Zhang, Z., Xing, Y., Li, F., 2013a. Transcriptome analysis and screening for potential target genes for RNAi-mediated pest control of the beet armyworm, *Spodoptera exigua*. PLoS One 8 (6), e65931.

Li, J., Wang, X.-P., Wang, M.-Q., Ma, W.-H., Hua, H.-X., 2013b. Advances in the use of the RNA interference technique in Hemiptera. Insect Sci. 20 (1), 31–39.

Liu, J., Smagghe, G., Swevers, L., 2013. Transcriptional response of BmToll9-1 and RNAi machinery genes to exogenous dsRNA in the midgut of *Bombyx mori*. J. Insect Physiol. 59 (6), 646–654.

Loretz, B., Foger, F., Werle, M., Bernkop-Schnurch, A., 2006. Oral gene delivery: strategies to improve stability of pDNA towards intestinal digestion. J. Drug Target. 14, 311–319.

Luo, Y., Wang, X., Yu, D., Chen, B., Kang, L., 2013. Differential responses of migratory locusts to systemic RNA interference via double-stranded RNA injection and feeding. Insect Mol. Biol. 22 (5), 574–583.

Manzano-Roman, R., Oleaga, A., Perez-Sanchez, R., Siles-Lucas, M., Rollinson, D., Hay, S.I., 2012. Gene silencing in parasites: current status and future prospects. Adv. Parasitol. 78, 1–55.

Mao, J., Zeng, F., 2012. Feeding-based RNA interference of a gap gene is lethal to the pea aphid, *Acyrthosiphon pisum*. PLoS One 7 (11), e48718.

Mao, J., Zeng, F., 2014. Plant-mediated RNAi of a gap gene-enhanced tobacco tolerance against the *Myzus persicae*. Transgenic Res. 23 (1), 145–152.

Mao, Y.-B., Cai, W.-J., Wang, J.-W., Hong, G.-J., Tao, X.-Y., Wang, L.-J., Huang, Y.-P., Chen, X.-Y., 2007. Silencing a cotton bollworm P450 monooxygenase gene by plant-mediated RNAi impairs larval tolerance of gossypol. Nat. Biotechnol. 25 (11), 1307–1313.

Mao, Y.-B., Tao, X.-Y., Xue, X.-Y., Wang, L.-J., Chen, X.-Y., 2011. Cotton plants expressing CYP6AE14 double-stranded RNA show enhanced resistance to bollworms. Transgenic Res. 20 (3), 665–673.

Mao, Y.-B., Xue, X.-Y., Tao, X.-Y., Yang, C.-Q., Wang, L.-J., Chen, X.-Y., 2013. Cysteine protease enhances plant-mediated bollworm RNA interference. Plant Mol. Biol. 83 (1–2), 119–129.

McDougall, P., 2011. The cost and time involved in the discovery, development and authorisation of a new plant biotechnology derived trait, a consultancy study for Crop Life International. http://www.croplife.org/PhillipsMcDougallStudy.

McEwan, D.L., Weisman, A.S., Hunter, C.P., 2012. Uptake of extracellular double-stranded RNA by SID-2. Mol. Cell 47 (5), 746–754.

Meyering-Vos, M., Müller, A., 2007. RNA interference suggests sulfakinins as satiety effectors in the cricket *Gryllus bimaculatus*. J. Insect Physiol. 53 (8), 840–848.

Miller, S.C., Brown, S.J., Tomoyasu, Y., 2008. Larval RNAi in Drosophila? Dev. Genes Evol. 218, 505–510.

Miller, S.C., Miyata, K., Brown, S.J., Tomoyasu, Y., 2012. Dissecting systemic RNA interference in the red flour beetle *Tribolium castaneum*, parameters affecting the efficiency of RNAi. PLoS One 7 (10), e47431.

Mutti, N.S., Park, Y., Reese, J.C., Reeck, G.R., 2006. RNAi knockdown of a salivary transcript leading to lethality in the pea aphid, *Acyrthosiphon pisum*. J. Insect Sci. 6, 38.

Mutti, N.S., Louis, J., Pappan, L.K., Pappan, K., Begum, K., Chen, M.-S., Park, Y., Dittmer, N., Marshall, J., Reese, J.C., Reeck, G.R., 2008. A protein from the salivary glands of the pea aphid, *Acyrthosiphon pisum*, is essential in feeding on a host plant. Proc. Natl. Acad. Sci. U. S. A. 105 (29), 9965–9969.

Nayak, A., Tassetto, M., Kunitomi, M., Andino, R., 2013. RNA interference-mediated intrinsic antiviral immunity in invertebrates. In: Cullen, B.R. (Ed.), Intrinsic Immunity, Current Topics in Microbiology and Immunology. In: 371, Springer-Verlag, Berlin, Heidelberg, pp. 183–200.

Nuez, I., Félix, M.-A., 2012. Evolution of susceptibility to ingested double-stranded RNAs in *Caenorhabditis* nematodes. PLoS One 7 (1), e29811.

Nunes, F.M.F., Simões, Z.L.P., 2009. A non-invasive method for silencing gene transcription in honeybees maintained under natural conditions. Insect Biochem. Mol. Biol. 39 (2), 157–160.

O'Neill, M.J., Bourre, L., Melgar, S., O'Driscoll, C.M., 2011. Intestinal delivery of non-viral gene therapeutics: physiological barriers and preclinical models. Drug Discov. Today 16, 203–218.

Paim, R.M.M., Araujo, R.N., Lehane, M.J., Gontijo, N.F., Pereira, M.H., 2013. Application of RNA interference in triatomine (Hemiptera, Reduviidae) studies. Insect Sci. 20 (1), 40–52.

Park, N.J., Li, Y., Yu, T., Brinkman, B.M., Wong, D.T., 2006. Characterization of RNA in saliva. Clin. Chem. 52, 988–994.

Parrott, W., Chassy, B., Ligon, J., Meyer, L., Petrick, J., Zhou, J., Herman, R., Delaney, B., Levine, M., 2010. Application of food and feed safety assessment principles to evaluate transgenic approaches to gene modulation in crops. Food Chem. Toxicol. 48, 1773–1790.

Petrick, J.S., Brower-Toland, B., Jackson, A.L., Kier, L.D., 2013. Safety assessment of food and feed from biotechnology-derived crops employing RNA-mediated gene regulation to achieve desired traits: a scientific review. Regul. Toxicol. Pharmacol. 66, 167–176.

Pilot, F., Philippe, J.M., Lemmers, C., Chauvin, J.P., Lecuit, T., 2006. Developmental control of nuclear morphogenesis and anchoring by *charleston*, identified in a functional genomic screen of *Drosophila* cellularisation. Development 133, 711–723.

Pitino, M., Coleman, A.D., Maffei, M.E., Ridout, C.J., Hogenhout, S.A., 2011. Silencing of aphid genes by dsRNA feeding from plants. PLoS One 6 (10), e25709.

Prado, J.R., Segers, G., Voelker, T., Carson, D., Dobert, R., Phillips, J., Cook, K., Cornejo, C., Monken, J., Grapes, L., Reynolds, T., Martino-Catt, S., 2014. Genetically engineered crops: from idea to product. Annu. Rev. Plant Biol. 65, 769–790.

Pridgeon, J.W., Zhao, L., Becnel, J.J., Strickman, D.A., Clark, G.G., Linthicum, K.J., 2008. Topically applied *AaeIAP1* double-stranded RNA kills female adults of *Aedes aegypti*. J. Med. Entomol. 45, 414–420.

Rajagopal, R., Sivakumar, S., Agrawal, N., Malhotra, P., Bhatnagar, R.K., 2002. Silencing of midgut aminopeptidase N of *Spodoptera litura* by double-stranded RNA establishes its role as *Bacillus thuringiensis* toxin receptor. J. Biol. Chem. 277 (49), 46849–46851.

Ramaseshadri, P., Segers, G., Flannagan, R., Wiggins, E., Clinton, W., Ilagan, O., McNulty, B., Clark, T., Bolognesi, R., 2013. Physiological and cellular responses caused by RNAi- mediated suppression of *Snf7* orthologue in western corn rootworm *Diabrotica virgifera virgifera* larvae. PLoS One 8 (1), e54270.

Rangasamy, M., Siegfried, B.D., 2012. Validation of RNA interference in western corn rootworm *Diabrotica virgifera virgifera* LeConte (Coleoptera, Chrysomelidae) adults. Pest Manag. Sci. 68 (4), 587–591.

Redenbaugh, K., Hiatt, W., Martineau, B., Kramer, M., Sheehy, R., Sanders, R., Houck, C., Emlay, D., 1992. Safety Assessment of Genetically Engineered Fruits and Vegetables: A Case Study of the Flavr Savr^{TM} Tomato. CRC Press, Inc., Boca Raton, FL

Rodríguez-Cabrera, L., Trujillo-Bacallao, D., Borrás-Hidalgo, O., Wright, D.J., Ayra-Pardo, C., 2010. RNAi-mediated knockdown of a *Spodoptera frugiperda* trypsin-like serine-protease gene reduces susceptibility to a *Bacillus thuringiensis* Cry1Ca1 protoxin. Environ. Microbiol. 12 (11), 2894–2903.

Roignant, J.Y., Carre, C., Mugat, B., Szymczak, D., Lepesant, J.A., Antoniewski, C., 2003. Absence of transitive and systemic pathways allows cell-specific and isoform-specific RNAi in *Drosophila*. RNA 9 (3), 299–308.

Roush, R.T., 1998. Two-toxin strategies for management of insecticidal transgenic crops: can pyramiding succeed where pesticide mixtures have not? Philos. Trans. R. Soc. Lond. Ser. B Biol. Sci. 353 (1376), 1777–1786.
Saleh, M.-C., van Rij, R.P., Hekele, A., Gillis, A., Foley, E., O'Farrell, P.H., Andino, R., 2006. The endocytic pathway mediates cell entry of dsRNA to induce RNAi silencing. Nat. Cell Biol. 8 (8), 793–802.
Saleh, M.C., Tassetto, M., van Rij, R.P., Goic, B., Gausson, V., Berry, B., Jacquier, C., Antoniewski, C., Andino, R., 2009. Antiviral immunity in *Drosophila* requires systemic RNA interference spread. Nature 458 (7236), 346–350.
Shakesby, A.J., Wallace, I.S., Isaacs, H.V., Pritchard, J., Roberts, D.M., Douglas, A.E., 2009. A water-specific aquaporin involved in aphid osmoregulation. Insect Biochem. Mol. Biol. 39 (1), 1–10.
Shih, J.D., Hunter, C.P., 2011. SID-1 is a dsRNA-selective dsRNA-gated channel. RNA 17, 1057–1065.
Shih, J.D., Fitzgerald, M.C., Sutherlin, M., Hunter, C.P., 2009. The SID-1 double-stranded RNA transporter is not selective for dsRNA length. RNA 15, 384–390.
Silva, C.P., Silva, J.R., Vasconcelos, F.F., Petretski, M.D.A., DaMatta, R.A., Ribeiro, A.F., Terra, W.R., 2004. Occurrence of midgut perimicrovillar membranes in paraneopteran insect orders with comments on their function and evolutionary significance. Arthropod Struct. Dev. 33 (2), 139–148.
Singh, A.D., Wong, S., Ryan, C.P., Whyard, S., 2013. Oral delivery of double-stranded RNA in larvae of the yellow fever mosquito, *Aedes aegypti*: implications for pest mosquito control. J. Insect Sci. 13, 69.
Snow, J.W., Hale, A.E., Isaacs, S.K., Baggish, A.L., Chan, S.Y., 2013. Ineffective delivery of diet-derived microRNAs to recipient animal organisms. RNA Biol. 10 (7), 1107–1116.
Stevens, C.E., Hume, I.D., 2004. Comparative Physiology of the Vertebrate Digestive System, second ed. Cambridge University Press, Cambridge.
Storer, N.P., Babcock, J.M., Schlenz, M., Meade, T., Thompson, G.D., Bing, J.W., Huckaba, R.M., 2010. Discovery and characterization of field resistance to Bt maize: *Spodoptera frugiperda* (Lepidoptera: Noctuidae) in Puerto Rico. J. Econ. Entomol. 103 (4), 1031–1038.
Surakasi, V.P., Mohamed, A.A.M., Kim, Y., 2011. RNA interference of β1 integrin subunit impairs development and immune responses of the beet armyworm, *Spodoptera exigua*. J. Insect Physiol. 57 (11), 1537–1544.
Swevers, L., Vanden Broeck, J., Smagghe, G., 2013. The possible impact of persistent virus infection on the function of the RNAi machinery in insects: a hypothesis. Front. Physiol. 4http://dx.doi.org/10.3389/fphys.2013.00319.
Tabashnik, B.E., 2008. Delaying insect resistance to transgenic crops. Proc. Natl. Acad. Sci. U. S. A. 105 (49), 19029–19030.
Tang, T., Zhao, C., Feng, X., Liu, X., Qiu, L., 2012. Knockdown of several components of cytochrome P450 enzyme systems by RNA interference enhances the susceptibility of *Helicoverpa armigera* to fenvalerate. Pest Manag. Sci. 68 (11), 1501–1511.
Terenius, O., Papanicolaou, A., Garbutt, J.S., Eleftherianos, I., Huvenne, H., Kanginakudru, S., Albrechtsen, M., An, C., Aymeric, J.L., Barthel, A., Bebas, P., Bitra, K., Bravo, A., Chevalier, F., Collinge, D.P., Crava, C.M., de Maagd, R.A., Duvic, B., Erlandson, M., Faye, I., Felfoldi, G., Fujiwara, H., Futahashi, R., Gandhe, A.S., Gatehouse, H.S., Gatehouse, L.N., Giebultowicz, J.M., Gomez, I., Grimmelikhuijzen, C.J., Groot, A.T., Hauser, F., Heckel, D.G., Hegedus, D.D., Hrycaj, S., Huang, L., Hull, J.J., Iatrou, K., Iga, M., Kanost, M.R., Kotwica, J., Li, C., Li, J., Liu, J., Lundmark, M., Matsumoto, S., Meyering-Vos, M., Millichap, P.J., Monteiro, A., Mrinal, N., Niimi, T., Nowara, D., Ohnishi, A., Oostra, V., Ozaki, K., Papakonstantinou, M., Popadic, A., Rajam, M.V., Saenko, S., Simpson, R.M.,

Soberon, M., Strand, M.R., Tomita, S., Toprak, U., Wang, P., Wee, C.W., Whyard, S., Zhang, W., Nagaraju, J., Ffrench-Constant, R.H., Herrero, S., Gordon, K., Swevers, L., Smagghe, G., 2011. RNA interference in Lepidoptera: an overview of successful and unsuccessful studies and implications for experimental design. J. Insect Physiol. 57 (2), 231–245.

The International Aphid Genomics Consortium, 2010. Genome sequence of the pea aphid *Acyrthosiphon pisum*. PLoS Biol. 8 (2), e1000313.

Tian, H., Peng, H., Yao, Q., Chen, H., Xie, Q., Tang, B., Zhang, W., 2009. Developmental control of a lepidopteran pest *Spodoptera exigua* by ingestion of bacteria expressing dsRNA of a non-midgut gene. PLoS One 4 (7), e6225.

Tijsterman, M., May, R.C., Simmer, F., Okihara, K.L., Plasterk, R.H.A., 2004. Genes required for systemic RNA interference in *Caenorhabditis elegans*. Curr. Biol. 14, 111–116.

Timmons, L., Fire, A., 1998. Specific interference by ingested dsRNA. Nature 395, 854.

Timmons, L., Court, D.L., Fire, A., 2001. Ingestion of bacterially expressed dsRNAs can produce specific and potent genetic interference in *Caenorhabditis elegans*. Gene 263, 103–112.

Tomari, Y., Zamore, P.D., 2005. Perspective, machines for RNAi. Genes Dev. 19 (5), 517–529.

Tomoyasu, Y., Miller, S.C., Tomita, S., Schoppmeier, M., Grossmann, D., Bucher, G., 2008. Exploring systemic RNA interference in insects, a genome-wide survey for RNAi genes in *Tribolium*. Genome Biol. 9 (1), R10.

Tribolium Genome Sequence Consortium, 2008. The genome of the model beetle and pest *Tribolium castaneum*. Nature 452, 949–955.

Turner, C.T., Davy, M.W., MacDiarmid, R.M., Plummer, K.M., Birch, N.P., Newcomb, R.D., 2006. RNA interference in the light brown apple moth, *Epiphyas postvittana* Walker induced by double-stranded RNA feeding. Insect Mol. Biol. 15 (3), 383–391.

Ulvila, J., Parikka, M., Kleino, A., Sormunen, R., Ezekowitz, R.A., Kocks, C., Rämet, M., 2006. Double-stranded RNA is internalized by scavenger receptor-mediated endocytosis in *Drosophila* S2 cells. J. Biol. Chem. 281 (20), 14370–14375.

Upadhyay, S.K., Chandrashekar, K., Thakur, N., Verma, P.C., Borgio, J.F., Singh, P.K., Tuli, R., 2011. RNA interference for the control of whiteflies (*Bemisia tabaci*) by oral route. J. Biosci. 36, 153–161.

Vachon, V., Laprade, R., Schwartz, J.-L., 2012. Current models of the mode of action of *Bacillus thuringiensis* insecticidal crystal proteins: a critical review. J. Invertebr. Pathol. 111, 1–12.

Vandenborre, G., Van Damme, E.J.M., Smagghe, G., 2009. Natural products: plant lectins as important tools in controlling pest insects. In: Ishaaya, I., Horowitz, A.R. (Eds.), Biorational Control of Arthropod Pests. Springer, Netherlands, pp. 163–187.

Vander Meer, R.K., Choi, M.Y., 2013. Formicidae (ant) control using double-stranded RNA constructs. US Patent No. 8,575,328.

Varkouhi, A.K., Scholte, M., Storm, G., Haisma, H.J., 2011. Endosomal escape pathways for delivery of biologicals. J. Control. Release 151 (3), 220–228.

Walker, W.B., Allen, M.L., 2011. RNA interference-mediated knockdown of IAP in *Lygus lineolaris* induces mortality in adult and pre-adult life stages. Entomol. Exp. Appl. 138, 83–92.

Walshe, D.P., Lehane, S.M., Lehane, M.J., Haines, L.R., 2009. Prolonged gene knockdown in the tsetse fly *Glossina* by feeding double stranded RNA. Insect Mol. Biol. 18 (1), 11–19.

Wang, Y., Zhang, H., Li, H.-C., Miao, X.-X., 2011. Second-generation sequencing supply an effective way to screen RNAi targets in large scale for potential application in pest insect control. PLoS One 6 (4), e18644.

Wang, J., Wu, M., Wang, B., Han, Z., 2013. Comparison of the RNA interference effects triggered by dsRNA and siRNA in *Tribolium castaneum*. Pest Manag. Sci. 69, 781–786.

Whangbo, J.S., Hunter, C.P., 2008. Environmental RNA interference. Trends Genet. 24 (6), 297–305.

Wheeler, D., Darby, B.J., Todd, T.C., Herman, M.A., 2012. Several grassland soil nematode species are insensitive to RNA-mediated interference. J. Nematol. 44, 92–101.

Whyard, S., Singh, A.D., Wong, S., 2009. Ingested double-stranded RNAs can act as species-specific insecticides. Insect Biochem. Mol. Biol. 39 (11), 824–832.

Winston, W.M., Molodowitch, C., Hunter, C.P., 2002. Systemic RNAi in *C. elegans* requires the putative transmembrane protein SID-1. Science 295, 2456–2459.

Winston, W.M., Sutherlin, M., Wright, A.J., Feinberg, E.H., Hunter, C.P., 2007. *Caenorhabditis elegans* SID-2 is required for environmental RNA interference. Proc. Natl. Acad. Sci. U. S. A. 104, 10565–10570.

Witwer, K.W., Hirschi, K.D., 2014. Transfer and functional consequences of dietary microRNAs invertebrates: concepts in search of corroboration. Bioessays 36http://dx.doi.org/10.1002/bies.201300150.

Witwer, K.W., McAlexander, M.A., Queen, S.E., Adams, R.J., 2013. Real-time quantitative PCR and droplet digital PCR for plant miRNAs in mammalian blood provide little evidence for general uptake of dietary miRNAs. RNA Biol. 10 (7), 1080–1086.

Wuriyanghan, H., Rosa, C., Falk, B.W., 2011. Oral delivery of double-stranded RNAs and siRNAs induces RNAi effects in the potato/tomato psyllid, *Bactericerca cockerelli*. PLoS One 6 (11), e27736.

Wynant, N., Verlinden, H., Breugelmans, B., Simonet, G., Vanden Broeck, J., 2012. Tissue-dependence and sensitivity of the systemic RNA interference response in the desert locust, *Schistocerca gregaria*. Insect Biochem. Mol. Biol. 42 (12), 911–917.

Wynant, N., Santos, D., Verdonck, R., Spit, J., Van Wielendaele, P., Vanden Broeck, J., 2014. Identification, functional characterization and phylogenetic analysis of double stranded RNA degrading enzymes present in the gut of the desert locust, *Schistocerca gregaria*. Insect Biochem. Mol. Biol.. http://dx.doi.org/10.1016/j.ibmb.2013.12.008.

Xiong, Y., Zeng, H., Zhang, Y., Xu, D., Qiu, D., 2013. Silencing the *HaHR3* gene by transgenic plant-mediated RNAi to disrupt *Helicoverpa armigera* development. Int. J. Biol. Sci. 9 (4), 370–381.

Xu, W., Han, Z., 2008. Cloning and phylogenetic analysis of *sid-1*-like genes from aphids. J. Insect Sci. 8, 30.

Xu, H.J., Chen, T., Ma, X.F., Xue, J., Pan, P.L., Zhang, X.C., Cheng, J.A., Zhang, C.X., 2013. Genome-wide screening for components of small interfering RNA (siRNA) and micro-RNA (miRNA) pathways in the brown planthopper, *Nilaparvata lugens* (Hemiptera: Delphacidae). Insect Mol. Biol. 22 (6), 635–647. http://dx.doi.org/10.1111/imb.12051.

Yang, J., Han, Z.-j., 2014. Efficiency of different methods for dsRNA delivery in cotton bollworm (*Helicoverpa armigera*). J. Integr. Agric. 13 (1), 115–123.

Yao, J., Rotenberg, D., Afsharifar, A., Barandoc-Alviar, K., Whitfield, A.E., 2013. Development of RNAi methods for *Peregrinus maidis*, the corn planthopper. PLoS One 8 (8), e370243.

Yu, N., Christiaens, O., Liu, J., Niu, J., Cappelle, K., Caccia, S., Huvenne, H., Smagghe, G., 2013. Delivery of dsRNA for RNAi in insects, an overview and future directions. Insect Sci. 20, 4–14.

Zha, W., Peng, X., Chen, R., Du, B., Zhu, L., He, G., 2011. Knockdown of midgut genes by dsRNA-transgenic plant-mediated RNA interference in the hemipteran insect *Nilaparvata lugens*. PLoS One 6 (5), e20504.

Zhang, X., Zhang, J., Zhu, K.Y., 2010. Chitosan/double-stranded RNA nanoparticle-mediated RNA interference to silence chitin synthase genes through larval feeding in the African malaria mosquito *Anopheles gambiae*. Insect Mol. Biol. 19 (5), 683–693.
Zhang, X., Zhang, J., Zhu, K., 2012a. Advances and prospects of RNAi technologies in insect pest management. In: Liu, T., Kang, L. (Eds.), Recent Advances in Entomological Research. Springer, Berlin Heidelberg, pp. 347–358.
Zhang, Y., Wiggins, B.E., Lawrence, C., Petrick, J., Ivashuta, S., Heck, G., 2012b. Analysis of plant-derived miRNAs in animal small RNA datasets. BMC Genomics 13, 381.
Zhang, Y.-l., Zhang, S.-z., Kulye, M., Wu, S.-r., Yu, N.-t., Wang, J.-h., Zeng, H.-m., Liu, Z.-x., 2012c. Silencing of molt-regulating transcription factor gene, *CiHR3*, affects growth and development of sugarcane stem borer, *Chilo infuscatellus*. J. Insect Sci. 12 (91), 1–12.
Zhang, H., Li, H.-C., Miao, X.-X., 2013a. Feasibility, limitation and possible solutions of RNAi-based technology for insect pest control. Insect Sci. 20, 15–30.
Zhang, M., Zhou, Y., Wang, H., Jones, H.D., Gao, Q., Wang, D., Ma, Y., Xia, L., 2013b. Identifying potential RNAi targets in grain aphid *Sitobion avenae* F. based on transcriptome profiling of its alimentary canal after feeding on wheat plants. BMC Genomics 14, 560.
Zhang, X., Liu, X., Ma, J., Zhao, J., 2013c. Silencing of cytochrome P450 CYP6B6 gene of cotton bollworm (*Helicoverpa armigera*) by RNAi. Bull. Entomol. Res. 103 (5), 584–591.
Zhao, L., Chen, J., 2013. Double stranded RNA constructs to control ants. US Patent Application Publication No. 2013/0078212.
Zhao, Y., Yang, G., Wang-Pruski, G., You, M., 2008. *Phyllotreta striolata* (Coleoptera, Chrysomelidae): arginine kinase cloning and RNAi-based pest control. Eur. J. Biochem. 105, 815–822.
Zhou, X., Wheeler, M.M., Oi, F.M., Scharf, M.E., 2008. RNA interference in the termite *Reticulitermes flavipes* through ingestion of double-stranded RNA. Insect Biochem. Mol. Biol. 38 (8), 805–815.
Zhou, J., Shum, K.-T., Burnett, J., Rossi, J., 2013. Nanoparticle-based delivery of RNAi therapeutics, progress and challenges. Pharmaceuticals 61, 85–107.
Zhu, F., Xu, J., Palli, R., Ferguson, J., Palli, S.R., 2011. Ingested RNA interference for managing the populations of the Colorado potato beetle, *Leptinotarsa decemlineata*. Pest Manag. Sci. 67 (2), 175–182.
Zhu, J.-Q., Liu, S., Ma, Y., Zhang, J.-Q., Qi, H.-S., Wei, Z.-J., Yao, Q., Zhang, W.-Q., Li, S., 2012. Improvement of pest resistance in transgenic tobacco plants expressing dsRNA of an insect-associated gene EcR. PLoS One 7 (6), e38572.

CHAPTER SIX

Detection and Mechanisms of Resistance Evolved in Insects to Cry Toxins from *Bacillus thuringiensis*

Yidong Wu
*College of Plant Protection, Nanjing Agricultural University, Nanjing, China

Contents

Abstract

Transgenic crops producing *Bacillus thuringiensis* Cry toxins (*Bt* crops) have been planted globally to control some key pests. Benefited from implementation of proactive resistance management strategies such as the refuge strategy and the pyramid strategy in many countries, most of the target pests of *Bt* crops have been sustainably and effectively controlled for nearly 20 years. However, several cases of field-evolved resistance to *Bt* corn and *Bt* cotton have been documented, causing reduced field efficacy. Evolution of resistance by target pests is a real threat to the continued success of *Bt* crops. It is crucial to employ sensitive detection methods to monitor the evolution of the resistance in the target insects and thereby adapt resistance management strategies

Advances in Insect Physiology, Volume 47
ISSN 0065-2806
http://dx.doi.org/10.1016/B978-0-12-800197-4.00006-3

proactively to delay resistance evolution. Recent progress and considerations were reviewed on the four resistance detection approaches: concentration-response assay, F_1 screen, F_2 screen and DNA screen. Diverse genetic options for target pests to cope with *Bt* crops are challenging the efforts to understand mechanisms of resistance and to design rational resistance management strategies. Understanding the molecular genetic basis of *Bt* resistance is essential for developing sensitive resistance detection methods and intelligent resistance management strategies. Clarifying resistance mechanisms have facilitated and advanced our understanding on modes of action of *Bt* Cry toxins. The most recent advances on resistance mechanisms to *Bt* toxins are summarized in both laboratory-selected strains and field-evolved populations. According to current knowledge of *Bt* resistance mechanisms, some implications for resistance management and directions for future research are suggested.

1. INTRODUCTION

Bacillus thuringiensis (*Bt*) is a gram-positive bacterium that is characterized by producing parasporal crystal proteins with insecticidal activity (Cry toxins) during sporulation (Knowles, 1994; Schnepf et al., 1998). *Bt* sprays have been widely used as a bioinsecticide in agriculture, forestry and mosquito control for several decades (Sanahuja et al., 2011). Transgenic crops expressing genes encoding *Bt* toxins (*Bt* crops) were commercialized for the first time in 1996 in the United States; since then, the cumulative total of more than 570 million ha of *Bt* crops (mainly corn and cotton) has been adopted globally (James, 2013; Tabashnik et al., 2013). Intensive planting of *Bt* crops inevitably creates strong selection pressure on the target insect pests, thereby the resistance to these *Bt* crops evolved by target insect pests is considered a major threat to the durability of *Bt* crops (Carrière et al., 2010; Gould, 1998; Huang et al., 2011; Tabashnik, 1994).

As a proactive effort to preserve the value and benefits of *Bt* crops, the refuge strategy has been widely used to delay development of insect resistance to *Bt* crops (Tabashnik et al., 2004). In theory, three key conditions should be met in order to maximize the effectiveness of the refuge strategy: recessive inheritance of resistance phenotype, low frequency of initial resistance alleles and abundant susceptible insects provided by non-*Bt* host plants nearby (Tabashnik et al., 2013). In practice, target insects have variable and unpredictable inheritance patterns of the resistance trait and the initial frequencies of resistance alleles within geographical populations. So, it is crucial to employ sensitive detection methods to monitor the dynamic evolution of the resistance alleles over space and time for the target insects and thereby adapt resistance management strategies proactively to delay resistance evolution for a given pest species–*Bt* crop combination. This review summarizes recent progress and considerations on different resistance detection approaches.

The genetic capacity of insect populations to evolve resistance to *Bt* toxins has been well demonstrated in many species within several insect orders. Numerous *Bt*-resistant strains have been developed by laboratory selection (Bauer, 1995; Ferré and Van Rie, 2002; Ferré et al., 2008). Other insects such as diamondback moth, *Plutella xylostella*, and cabbage looper, *Trichoplusia ni*, have evolved resistance to *Bt* sprays in the open field and green houses, respectively (Janmaat and Myers, 2003; Tabashnik et al., 1990). Recently, field-evolved resistance to *Bt* corn in three pests (maize stem borer, *Busseola fusca*; western corn rootworm, *Diabrotica virgifera virgifera*; and fall armyworm, *Spodoptera frugiperda*) and to *Bt* cotton in two pests (American bollworm, *Helicoverpa zea*; pink bollworm, *Pectinophora gossypiella*) has been documented and reported to cause reduced field control efficacy (Tabashnik et al., 2013, 2014). Therefore, understanding the molecular genetic basis of *Bt* resistance is essential for developing sensitive resistance detection methods and intelligent resistance management strategies (Ferré and Van Rie, 2002; Heckel, 1994, 2012; Heckel et al., 2007). This review will also focus to describe the most recent advances on resistance mechanisms to *Bt* toxins in both laboratory-selected strains and field-evolved populations. Given current knowledge of *Bt* resistance mechanisms, some implications for resistance management and directions for future research are suggested.

2. DETECTION METHODS AND CURRENT STATUS OF INSECT RESISTANCE TO *Bt* CROPS

2.1. Definition of resistance

The WHO Expert Committee on Insecticides defined insecticide resistance as "the development of an ability in a strain of insects to tolerate doses of toxicants which would proved to be lethal to the majority of individuals of the normal population from the same species" (WHO, 1957). This definition was the established operational definition for years and has been adopted widely. However, this definition is considered not complete because it is not applicable to individual insects, and it is difficult to define an unselected or susceptible "normal population" (Sawicki, 1987). Crow (1960) proposed a more general definition as "Insecticide resistance marks a genetic change in response to selection". Crow's definition is applicable to both single insects and whole populations, and the severity of resistance can range from no effect on field control to field failure of an insecticide application. Brent (1986) defined resistance as "any heritable decrease in sensitivity to a chemical within a pest population". Brent (1986) emphasized that insecticide resistance can cause complete loss of action of an agrochemical or may have little practical significance. Sawicki (1987) modified Crow's definition to "Insecticide resistance marks a genetic change in response to selection by toxicants that may impair control in the field". In the definitions mentioned above, both a change in susceptibility to the insecticide detectable with bioassays in the laboratory and field control failure are considered resistance. In other words, resistance to an insecticide need not result in loss of insect pest control.

Tabashnik et al. (2009a, 2014) defined the term "field-evolved resistance" as "a genetically based decrease in susceptibility of a population to a pesticide caused by exposure to the pesticide in the field". The term "field-evolved resistance" applies to resistance in both pest and beneficial organisms and does not necessarily imply loss of economic efficacy in the field (Tabashnik et al., 2009a). This general definition favours proactive detection and management of resistance.

Another definition of resistance proposed by the Insecticide Resistance Action Committee (IRAC) is "a heritable change in the sensitivity of a pest population that is reflected in the repeated failure of a product to achieve the expected level of control when used according to the label recommendation for that pest species" (http://www.irac-online.org/about/resistance). This definition emphasizes a causal relationship between resistance and field control failure. Agrochemical industry intends to use this definition to validate and confirm resistance as the cause of observed losses in field efficacy.

There are debates on the term, "field-evolved resistance", in defining insect resistance to *Bt* crops. A general definition of "field-evolved resistance" considers that the primary goal of monitoring resistance to *Bt* crops is to detect evolution of resistance early enough to enable proactive management of resistant insects, and field control problems associated with field-evolved resistance vary from none to severe (Tabashnik et al., 2008, 2013). A narrow definition of "field-evolved resistance" regards decreased field efficacy and/or survival on *Bt* plants as the decisive criteria for defining a case of field-evolved resistance (Moar et al., 2008; Sumerford et al., 2013). To avoid this confusion, Tabashnik et al. (2014) defined "practical resistance" as field-evolved resistance that reduces the efficacy of a pesticide and has practical consequences for pest control.

2.2. Resistance detection methods

The major objectives of resistance detection and monitoring should include (1) determining baseline susceptibility and initial resistance allele frequency of target pests to pesticides; (2) detecting changes in resistance frequency to gain early warning of a resistance problem; (3) determining the effectiveness of the implemented resistance management strategies; and (4) documenting losses of insect control in the field associated with resistance (Brent, 1986; Caprio and Sumerford, 2007). The primary objective of resistance monitoring to *Bt* crops is to detect field-evolved resistance early enough to enable proactive management before field failure occurs (Tabashnik et al., 2009a, 2013).

Resistance detection means identifying a significant change in the susceptibility of a pest population to pesticides. Effective resistance detection techniques are essential for useful early warning systems and in defining the extent and severity of resistance. This information is also critical for evaluating resistance management strategies. There are a number of methods developed for detection of *Bt* resistance in field insect populations (Caprio and Sumerford, 2007; Huang, 2006). The four methods reviewed here for *Bt* resistance detection are divided into two types: (1) the concentration–response and diagnostic concentration assays used for determining resistance intensity and the frequency of resistant individuals and (2) F_1 screens, F_2 screens and DNA screens used for detecting resistance allele frequency. These two types of detection approaches are complementary and mutually confirmatory, and the combination of these two types of detection methods has proved to be especially effective for *Bt* resistance monitoring in *Helicoverpa armigera* from China (Zhang et al., 2011, 2012a, 2013).

2.2.1 Concentration–response and diagnostic concentration assays

Concentration–response assays are commonly used to estimate the toxin concentration that kills 50% of a pest population (LC_{50}). Resistance ratios (RRs) are then measured by comparing the LC_{50} values obtained with insect strains derived from populations exposed to the toxin against those of susceptible laboratory strains (Tabashnik et al., 2008, 2009a). An RR greater than 10 is more likely to indicate heritable decrease in susceptibility and higher RRs provide stronger evidence of resistance (Tabashnik, 1994, Tabashnik et al., 2009a).

The frequency of resistant individuals can be measured as the proportion of a field population surviving at an appropriate diagnostic concentration which kills all or nearly all individuals of the susceptible population, but few or no resistant individuals. Resistance allele frequency can be readily estimated if the genetic basis of resistance is known (Tabashnik et al., 2009a; Zhang et al., 2012a). Compared with the concentration–response assay, the diagnostic concentration test is more efficient for detecting an incipient resistance outbreak (Roush and Miller, 1986).

When designing and implementing an efficient detection programme for *Bt* resistance based on the concentration–response and/or diagnostic concentration assays, several key factors should be considered:

(1) Establishment of baseline susceptibility data and the calibration of a diagnostic concentration. It is important to survey the baseline susceptibility of target pest populations sampled from geographical areas prior to wide commercial adoption of a *Bt* crop. The baseline data can reveal the range of geographical variations of unexposed field populations which will then be necessary for defining susceptibility changes relating to exposure to *Bt* crops. Similarly, it is necessary to calibrate the diagnostic concentrations against a range both of unexposed field populations and of susceptible laboratory strains (Forrester et al., 1993; Khakame et al., 2013; Wang et al., 2010). Compared with the appropriate baseline comparators considering natural susceptibility variations, a statistically significant increase in survival at a diagnostic concentration and/or in LC_{50} values or LC_{90} will provide initial evidence of resistance evolution in the field.

(2) Maintaining a susceptible laboratory strain for comparison across years and regions. For valid comparison with the susceptibility of field populations, it is necessary to choose a reference strain to be kept in the laboratory, which should be neither extremely susceptible nor unusually tolerant to the *Bt* toxin concerned. Concurrent comparative susceptibility data for a specific *Bt* toxin from the susceptible strain are essential for comparison with field population susceptibility across years and regions.

(3) Sampling and testing. The scale of sampling depends on the aims and types of resistance detection as well as on the mobility and population structure of the target pest. Each collection needs to be representative of the local field population. It is suggested that hundreds of insects should be collected from many locations instead of thousands of insects from a few locations (ffrench-Constant and Roush, 1990). Considering costs of time and labour, it is often best that an extensive survey be done first, followed by a more intensive monitoring of selected locations, which have intensive adoption of *Bt* crops and possibly a higher risk of resistance evolution.

Comprehensive resistance monitoring requires evaluation of insect field populations on *Bt* crops as well as from other sources including non-*Bt* plants. Sampling and testing of target pest insects surviving on or near *Bt* crops is essential for early detection of field-evolved resistance (Tabashnik et al., 2008, 2013).

Field-collected insects are pooled in large groups for mating in the laboratory to generate field-derived strains for bioassays which are normally carried out on the damaging life stages, usually larval or nymphal. Both diet incorporation and diet-overlay methods are commonly used to determine responses of the target pest to a series of concentrations and/or a diagnostic concentration of *Bt* toxin. Although the diet incorporation method is closer to exposure of target pests to *Bt* plants, diet-overlay assays are simpler and require less *Bt* toxin (Sivasupramaniam et al., 2007). In addition, precautions should be taken to avoid high temperatures that would degrade the *Bt* toxin and give misleading results.

(4) *Bt* protein source and forms. Provision and tracking of a reliable *Bt* protein source is extremely important in interpreting data across laboratories and making valid comparison over time (Sivasupramaniam et al., 2007). In the United States and some other countries, *Bt* proteins required for resistance monitoring programmes are mainly supplied by industry. Centralized supplies of *Bt* proteins can make sure reliable and consistent quality throughout monitoring programmes. It is always important to use a tester strain (a susceptible laboratory strain) of a target pest to check the efficacy of new protein sources, or new batches from the same source. Mortality in the laboratory susceptible strain should remain constant at a given toxin dose, and conversion factors should be used when comparing results using proteins of different toxicity.

The toxin form to be used for a monitoring programme needs to be as similar as possible to the form ingested by the target pest when feeding on a *Bt* crop. Adoption of the most appropriate form of a *Bt* protein for resistance selection in the laboratory or resistance monitoring in the field has been an issue for debate (Anilkumar et al., 2008). A partially activated form of Cry1Ac toxin (C-terminal removed, but N-terminal retained) is expressed in transgenic cotton (Perlak et al., 1990), but a formulation called MVPII has been widely used for monitoring resistance of cotton pests to Cry1Ac (Sivasupramaniam et al., 2007). MVPII contains a hybrid protoxin that is identical to Cry1Ac in its active region and is encapsulated in *Pseudomonas fluorescens* (Tabashnik et al., 2002). One resistant strain of cotton bollworm,

H. armigera, and one resistant strain of *H. zea*, both pests that are targeted by *Bt* cotton, were ca. 10 times more resistant to Cry1Ac toxin than to Cry1Ac protoxin. The SCD-r1 strain of *H. armigera* had ca. 500-fold resistance to Cry1Ac toxin, but only 39-fold resistance to Cry1Ac protoxin (Tabashnik et al., 2011; Xu et al., 2005). Similarly, the AR strain of *H. zea* had more than 100-fold resistance to Cry1Ac toxin, but only 10-fold to Cry1Ac protoxin (Anilkumar et al., 2008). For *Bt* resistance monitoring of these two cotton pests, it could be better to use Cry1Ac toxin for early detection of resistance.

2.2.2 F_1 screen

The principles of complementation tests are as follows (Zhang et al., 2012a): If two resistant strains are crossed, each with recessive alleles for resistance at separate loci, allelic complementation will restore susceptibility (the wild-type phenotype) in the progeny. However, if the recessive resistance alleles occur at the same locus in different strains, the progeny from the cross between strains will be resistant because they will inherit resistance alleles at the same locus from both parents. Based on the principle of complementation tests, Gould et al. (1997) developed an F_1 screen method to estimate the frequency of Cry1Ac-resistance alleles in field populations of the tobacco budworm, *Heliothis virescens*. Over 2000 single-pair families were made among field-collected male moths and virgin female moths from the laboratory-selected YHD2 strain of *H. virescens*, which has a high level of Cry1Ac resistance conferred by a recessive gene (producing a truncated cadherin). F_1 offspring from each of 1025 families were screened with a discriminating concentration of Cry1Ac that could distinguish heterozygous from homozygous resistant individuals. Four of these families produced 30–42% offspring with resistance to Cry1Ac, indicating that in each of the four field-collected males was heterozygous and carried a recessive resistance allele. By the F_1 screen, the initial frequency of alleles for Cry1Ac resistance was able to be estimated as 0.0015 in field populations of *H. virescens* (Gould et al., 1997).

The F_1 screen method has been used to determine initial *Bt* resistance allele frequencies in field populations of several pests other than *H. virescens*. A total of 286 single-pair crosses were made between field-derived adults of the poplar leaf beetle, *Chrysomela tremulae*, and adults of a laboratory strain with recessive resistance to Cry3Aa. F_1 neonates from each of the 176 single-pair families with enough F_1 offspring were screened with leaf discs from a *Bt* poplar line. Three of the 176 families screened produced resistant F_1 larvae, and the estimated frequency of resistance allele was 0.0113 (Wenes et al., 2006). In the sugarcane borer, *Diatraea saccharalis* (F.), a total of 364 single-pair matings were made between field-collected individuals with those from a laboratory strain with recessive resistance to Cry1Ab. F_1 neonates from each of 256 single-pair families with sufficient F_1 offspring were screened with *Bt* maize leaf tissue. One of the 256 families screened produced resistant F_1 larvae, and the estimated frequency of resistance allele was 0.0028 for the combined five populations collected from Louisiana and Texas of the United States (Yue et al., 2008).

The F_1 screen method has been confirmed to be effective for detecting early shifts in the resistance frequency of target pests after exposure to *Bt* crops. F_1 screens were conducted in Australia to monitor Cry2Ab resistance allele frequencies in field populations of the Australian bollworm, *Helicoverpa punctigera*, sampled from both *Bt* cotton (Bollgard II, pyramided with Cry1Ac and Cry2Ab) cropping areas and non-cropping areas in Australia (Downes et al., 2010a). Cry2Ab resistance allele frequency of populations of *H. punctigera* from cropping areas increased from 0.015 in 2007/2008 to 0.048 in 2008/2009 cotton season. The Cry2Ab resistance allele frequency for cropping populations collected in 2008/2009 is eightfold higher than for non-cropping populations sampled in 2009 (Downes et al., 2010a). *Bt* cotton expressing only Cry1Ac protein has been intensively planted in northern China for more than a decade, but planting of this cotton has been limited in northwestern China (Zhang et al. 2011). F_1 screens conducted by Zhang et al. (2012a) with 593 *H. armigera* males caught in the field during 2009–2010 show that the proportion of males in which resistance to Cry1Ac was detected three times higher for northern China field populations (0.16) than for a field population from northwestern China (0.057).

The F_1 screen method is designed to detect rare recessive resistance alleles in field-derived individuals that occur at the same locus as the resistance alleles in the resistant laboratory strain. In theory, the F_1 screen can also detect non-recessive resistance alleles at any locus. Zhang et al. (2012a) extended the F_1 screen method for detection of both recessive and non-recessive alleles from field populations of *H. armigera*. Figure 6.1 is a schematic diagram of the F_1 screen programme for resistance alleles in *H. armigera*. With this approach, resistance alleles detected by F_1 screen in northern China were composed of 84% recessive cadherin alleles and 16% non-recessive alleles at either a cadherin locus or another locus. In contrast, for northwestern China, all of the resistance alleles were recessive cadherin alleles (Zhang et al., 2012a).

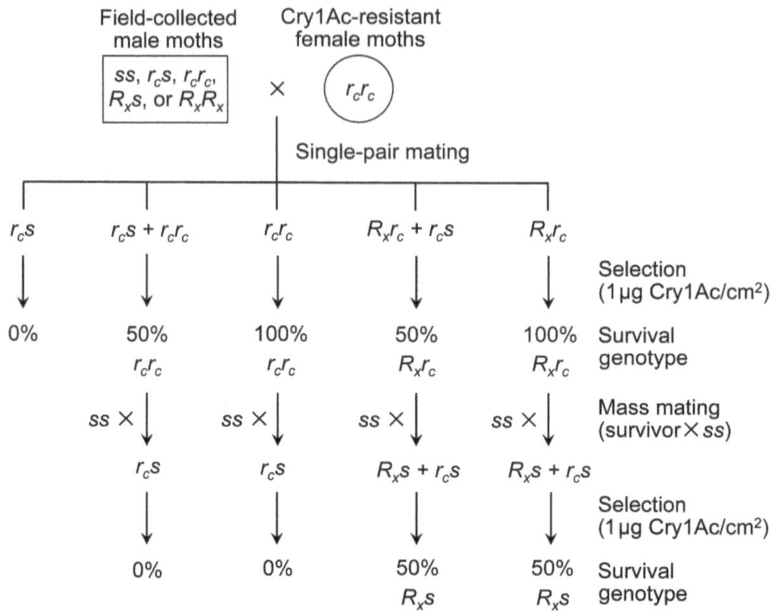

Figure 6.1 F_1 screen for resistance alleles in *Helicoverpa armigera* (Zhang et al., 2012a). Field-caught male moths were crossed individually to homozygous female moths with a recessive resistance allele at the cadherin locus (r_cr_c) from a laboratory strain of *H. armigera*. F_1 offspring from single-pair families were tested at a diagnostic concentration of Cry1Ac. Expected survival of the F_1 progeny in these bioassays depends on the genotype of the field-caught male: 0% for a susceptible homozygote (*ss*), 50% for individuals with one susceptible allele and either any recessive resistance allele at the cadherin locus (r_c) or a dominant resistance allele at any locus (R_x) (genotypes r_cs or R_xs) and 100% for individuals with the genotypes r_cr_c or R_xR_x. To determine the dominance of the resistance alleles detected with the F_1 screen, survivors of the F_1 screen are crossed with a susceptible strain (*ss*) and the progeny tested at the diagnostic concentration. Expected survival of the progeny of these F_1 survivors × *ss* crosses is 0% if the resistance allele is recessive (progeny are r_cs) and 50% if the allele is dominant (half of progeny are R_xs and half are *ss*). It should be noted that F_1 screen does not detect recessive alleles at loci other than the same resistance locus as in the tester laboratory strain.

The prerequisite for the success of an F_1 screen is to possess a laboratory strain which has recessive resistance conferred by a single locus. If the resistance genes are well characterized, such as cadherin mutations in the three cotton pests (*H. virescens*, *P. gossypiella* and *H. armigera*), the F_1 screen can be used to recover resistance alleles from field populations. With F_1 screens and molecular cloning, 15 cadherin alleles (r_1–r_{15}) associated with Cry1Ac resistance were discovered from field populations of *H. armigera* from China (Yang et al., 2007; Zhang et al., 2012a,b; Zhao et al., 2010).

2.2.3 F_2 screen

Andow and Alstad (1998) developed the F_2 screen technique to detect recessive alleles in field populations. The principle of an F_2 screen is that F_1 offspring of each mated female collected from the field are sib-mated to produce F_2 iso-female families. Each mated female carries four haploid genomes, two from her own and two from her mate. If the field-collected female or her mate carries one recessive resistance allele (*r*), the F_2 offspring of the iso-female family will have approximately 6.25% survival (*rr*) when screened with a diagnostic concentration of *Bt* toxin or with *Bt* plants. Compared with the diagnostic concentration assay, the F_2 screen extends the sensitivity to detect recessive resistance traits by more than an order of magnitude. Compared with the F_1 screen, the F_2 screen can detect resistance alleles at any locus rather than just the same locus where resistance alleles occur in the laboratory strains (Andow and Alstad, 1998).

Because of its robustness, the F_2 screen has been employed to detect resistance allele frequency to *Bt* toxins for at least eight lepidopteran pests: European corn borer, *Ostrinia nubilalis* (Andow et al., 1998, 2000; Bourguet et al., 2003; Siegfried et al., 2014; Stodola et al., 2006), rice yellow stem borer, *Scirpophaga incertulas* (Bentur et al., 2000), southwestern corn borer, *Diatraea grandiosella* (Huang et al., 2007a), Mediterranean corn borer, *Sesamia nonagrioides* (Andreadis et al., 2007), sugarcane borer, *D. saccharalis* (Huang et al., 2007b, 2008, 2009), *H. armigera* (Liu et al., 2010; Mahon et al., 2007, 2012; Xu et al., 2009; Zhang et al., 2012a), *H. punctigera* (Downes et al. 2009, 2010a; Mahon et al. 2012) and *H. virescens* (Blanco et al., 2009) and one coleopteran pest: *C. tremulae* (Génissel et al., 2003).

Because the F_2 screen involves intensive input of time and resources, it has been used only to determine initial resistance allele frequency for most monitoring plans. However, F_2 screen has been practically used as the mainstay of the routine monitoring of *Bt* resistance in two cotton pests, *H. armigera* and *H. punctigera* in Australia (Downes and Mahon, 2012a,b). From 2002/2003 until 2010/2011, F_2 screens were used to test 1222 iso-female lines (4888 alleles) of *H. armigera* and 1558 (6232 alleles) iso-female lines of *H. punctigera* for Cry1Ac resistance. Two *H. armigera* and three *H. punctigera* lines were positive for a resistance allele to Cry1Ac. Based on data pooled since 2002/2003, the estimated Cry1Ac resistance allele frequency for *H. armigera* is 0.0006 (95% CI between 0.0001 and 0.002) and for *H. punctigera* is 0.0006 (95% CI between 0.0002 and 0.001). During the same period, F_2 screens were used to test 1303 iso-female lines (5212 alleles) of *H. armigera* and 1642 iso-female lines (6566 alleles) of *H. punctigera* for Cry2Ab resistance. Forty-seven *H. armigera* and 33 *H. punctigera* lines were positive for a resistance allele to Cry2Ab. Based on data pooled since 2002/2003, the estimated Cry2Ab resistance allele frequency for *H. armigera* is 0.009 (95% CI between 0.007 and 0.012) and for *H. punctigera* is 0.005 (95% CI between 0.004 and 0.007) (Downes and Mahon, 2012a,b). The F_2 screen data clearly showed that there has been a statistically significant increase in the frequency of Cry2Ab resistance alleles in *H. punctigera* since the widespread adoption of Bollgard II in 2004/2005 in Australia. The F_2 screen therefore fulfilled the intended function of resistance monitoring for providing an early warning of resistance evolution (Downes et al., 2010a). In China, the F_2 screen data of 363 field-derived iso-female families of *H. armigera* showed that the frequency of Cry1Ac resistance was three times higher for a field population from northern China (0.052), where there is intensive *Bt* cotton planting than for field populations from northwestern China (0.016), where there is limited *Bt* cotton planting (Zhang et al., 2012a).

If the resistance of the isolated strain from an F_2 screen is recessive, it can be used to perform F_1 screens. From the positive lines with a major resistance allele detected by F_2 screens, strains with recessive *Bt* resistance at a single locus have been successfully isolated from *C. tremulae*, *D. saccharalis*, *H. armigera* and *H. punctigera*. The R#60 strain of *C. tremulae* resistant to Cry3Aa, the *Bt*–RR strain of *D. saccharalis* resistant to Cry1Ab, the SP15 strain of *H. armigera* resistant to Cry2Ab and the Hp4-13 strain of *H. punctigera* resistant to Cry2Ab were subsequently employed in performing F_1 screens (Augustin et al., 2004; Downes et al. 2010b; Huang et al., 2007b; Mahon et al., 2007).

If resistance is controlled by a single locus in field populations of a target pest, F_1 screens and F_2 screens should detect similar frequencies of resistance alleles. If there are different resistance genes at more than one locus, F_2 screens are expected to detect higher resistance frequencies than F_1 screens because F_1 screens detect only the type of resistance present in the resistant tester strain. It is very interesting to note that *Bt* resistance frequencies are generally higher when estimated by F_1 screen than by F_2 screen in several pests (Table 6.1). For *C. tremulae* and *D. saccharalis*, the discrepancy in frequencies identified using the two methodologies might be explained by the fact that the field populations were sampled from different locations and in different years. For *H. armigera* and *H. punctigera* from Australia, F_1 screens and F_2 screens were conducted concurrently on same batches of field populations, but resistance frequencies to Cry2Ab were 3–5.5 times higher estimated by F_1 screens than by F_2 screens. Mahon et al. (2010) hypothesized

Table 6.1 Comparison of *Bt* resistance allele frequencies estimated from the F_1 screens and F_2 screens

Insect species	*Bt* toxin	Location	Year	Frequency estimated by F_2 screen (95% CI)	Frequency estimated by F_1 screen (95% CI)	References
Chrysomela tremulae	Cry3Aa	La Chesnaye, France	1999–2001	0.0037 (0.00045–0.008)	NA	Génissel et al. (2003)
		Bar-le-Duc, France	2003	NA	0.0113 (0.0031–0.247)	Wenes et al. (2006)
Diatraea saccharalis	Cry1Ab	Winnsboro, LA, USA	2004	0.0012	NA	Huang et al. (2007b)
	Cry1Ab	Franklin, Tensas, East Carroll, and Rapides, LA, USA	2006	NA	0.0030 (0.0004–0.0084)	Yue et al. (2008)
Ostrinia nubilalis	Cry1F	Saunders, NE, USA	2003–2005	NA	0.0286	Siegfried et al. (2014)
	Cry1F	Saunders, NE, USA	2006–2008	NA	0.0253	Siegfried et al. (2014)
	Cry1F	Saunders, NE, USA Story, IA, USA	2008–2009	0.0093–0.0142	NA	Siegfried et al. (2014)
Helicoverpa armigera	Cry2Ab	Cotton regions, Estern Australia	2007–2008	0.006 (0.0002–0.013)	0.033 (0.021–0.047)	Mahon et al. (2010)
	Cry1Ac	Anyang, Henan, China	2009	0.052 (0.034–0.076)	0.091 (0.066–0.123)	Zhang et al. (2012a)
	Cry1Ac	Shawan, Xinjiang, China	2010	0.019 (0.0084–0.041)	0.029 (0.012–0.064)	Zhang et al. (2012a)
Helicoverpa punctigera	Cry2Ab	Cotton regions, Estern Australia	2007–2008	0.005 (0.002–0.010)	0.015 (0.003–0.036)	Downes et al. (2010a)
	Cry2Ab	Cotton regions, Estern Australia	2008–2009	0.012 (0.006–0.020)	0.048 (0.033–0.065)	Downes et al. (2010a)

that some "resistance alleles" are homozygous lethal if autozygous (as generated in F_2 tests) but not as allozygous homozygotes (as generated in F_1 tests). The hypothesis was extended to accommodate the possibility that alleles at linked loci may be homozygous lethal. However, neither of two tests of the hypothesis carried out to date provided evidence that any alleles that confer resistance are associated with severe fitness costs (Mahon et al., 2010), which does not support the hypothesis.

In China, diverse cadherin mutants associated with Cry1Ac resistance in *H. armigera* were detected from F_1 screens and F_2 screens (Yang et al., 2007; Zhang et al., 2012a; Zhao et al., 2010). A number of resistant strains with autozygous genotypes of cadherin showed various resistance intensities to Cry1Ac, ranging from 31- to 530-fold (Zhang et al., 2012a). Under a diagnostic concentration of Cry1Ac (1 μg/cm^2 diet surface), the AY148 strain (r_9r_9) had 46% survival, the SCD-r1 strain (r_1r_1) had 92% survival and F_1 offspring (r_1r_9) from AY148 × SCD-r1 had 81% survival (Zhang et al., 2012a). If one of the field parents of an iso-female line in an F_2 screen carries a copy of the r_9 allele, the F_2 offspring will have only 2.9% survival (6.25% × 0.46) under the diagnostic concentration, which will be scored negative. In contrast, if one of the field males carrying a copy of r_9 allele (r_9s) is crossed with a female from SCD-r1 (r_1r_1) in an F_1 screen, the F_1 offspring (r_1r_9) will have 41% survival (81% × 0.5) under the diagnostic concentration, which will be scored positive. In F_2 screens of Cry1Ac resistance alleles in two populations of *H. armigera* from China, the parents of iso-female lines were scored as having at least one resistance allele if F_2 progeny survival was >3% instead of >6.25% (Zhang et al., 2012a). Thus, Cry1Ac resistance allele frequencies estimated by F_1 screens were similar to that by F_2 screens (Zhang et al., 2012a). So, variability in resistance intensities among different autozygotes (such as r_ar_a and r_br_b) and/or allozygotes (r_ar_b) might cause discrepancy in results of F_1 screens and F_2 screens.

2.2.4 DNA screen

Bioassay-based screens (including concentration-response assays, F_1 screens and F_2 screens) have been used as primary tools to detect and monitor *Bt* resistance, but these approaches need huge manpower, material resources and time for sampling insects from the field and rearing their progeny in laboratory conditions. Compared with bioassay-based screens, DNA screens can save labour and time, for it can detect resistance genotypes on individual insects at any life stages. Further, DNA screens are efficient for detecting recessive resistance alleles at low frequencies (Zhang et al., 2013). In other words, DNA screen can only pick up resistance data for which genetic basis is already well characterized.

DNA-based screening has become applicable to *Bt* resistance since cadherin mutations were confirmed to cause Cry1Ac resistance in three lepidopteran pests of cotton (Gahan et al., 2001; Morin et al., 2003; Xu et al., 2005). In *P. gossypiella*, >8000 insects collected from cotton fields in Arizona, California and Texas during 2001–2009 were screened with a DNA-based diagnostic PCR for three resistance alleles from laboratory-selected strains, and none of these three resistance alleles were detected (Tabashnik et al., 2006, 2010). Similarly, the cadherin resistance allele from a laboratory-selected strain of *H. virescens* was not detected in >7000 field-collected individuals (Gahan et al., 2007). These results indicate extremely low frequencies of the cadherin resistance alleles that could be detected, but they do not exclude the presence of resistance alleles at either the cadherin locus or other loci that could not be detected by the PCR methods used. In both cases, however, bioassay data show that resistance remained rare in field populations (Blanco et al., 2009; Tabashnik et al., 2010).

In *H. armigera*, an allele (r_{15}) with a 55aa deletion in the intracellular domain of cadherin (*HaCad*) was identified to confer non-recessive resistance to Cry1Ac (Zhang et al., 2012b). Subsequently, a DNA-based PCR method was developed to screen for the r_{15} allele in field populations collected from the main cotton planting areas of China in 2011 and 2012 (Zhang et al., 2013). Three heterozygous r_{15} alleles were detected from 562 moths collected from northern China (with intensive *Bt* cotton planting), and the frequency of r_{15} allele was estimated to be 0.0027. However, no r_{15} allele was detected from 314 moths collected from Xinjiang (with limited *Bt* cotton use). Although all the r_{15} alleles have the same deletion in the cDNA sequence, at least four different indels causing loss of exon 32 have been detected in the genomic DNA sequences flanking exon 32 of *HaCad*. Thus, designing a new method to detect the resistance alleles such as r_{15} at cDNA level can avoid underestimating their frequency.

DNA screening is still in its infancy for *Bt* resistance monitoring. There are two major limiting factors to be resolved yet. The first is the fact that diverse mechanisms of resistance may exist in field populations (Zhang et al., 2012a), challenging the accurate prediction of different or novel resistance mechanisms evolved in the field. Selection for resistance in the laboratory is a routine approach to predict likely mechanisms, but the resistance may not be representative of resistance found in the field. F_2 screens might be applicable for isolating a collection of resistant strains representing the important resistance types in the field populations, molecular mechanisms for which can be further characterized in the laboratory. The second limiting factor is multiple mutations occurring at a single resistance locus complicating DNA screens. At least 15 cadherin alleles have been identified associated with Cry1Ac resistance in *H. armigera* (Xu et al., 2005; Yang et al., 2006, 2007; Zhang et al., 2012a,b; Zhao et al., 2010). Even the single r_{15} cadherin allele has been shown to have at least four origins, each of which has a different indel in the genomic DNA sequence (Zhang et al., 2013). So, the use of DNA-based diagnostics to screen for specific resistance-associated alleles is prone to underestimating resistance gene frequencies as new mutations within the same gene may be overlooked.

2.3. Current status of field-evolved resistance to *Bt* crops

The area planted with *Bt* crops worldwide increased from 1.1 million ha in 1996 to 75 million ha in 2013 (James, 2013). *Bt* corn accounted for 67% of corn planted in the United States during 2012 and *Bt* cotton accounted for 79–95% of cotton planted in Australia, China, India and the United States during 2010–2012 (Tabashnik et al., 2013). Considering the strong adaptation capacity of insects to insecticides, large scale and intensive planting of *Bt* crops will inevitably result in evolution of resistance in the field by the target pests to *Bt* crops.

Monitoring data on insect resistance to *Bt* crops were recently reviewed by Tabashnik et al. (2013) from 77 studies conducted in eight countries. Whereas most previous assessments characterized pest populations only as resistant or not, the new analysis by Tabashnik et al. (2013, 2014) introduced a series of four levels of field-evolved resistance to *Bt* crops, ranging from "incipient resistance", to the most serious cases of resistance "practical resistance". Among the 13 cases examined with field-evolved resistance (Table 6.2), 3 cases were categorized as "incipient resistance" (a statistically significant increase in resistance occurred, but <1% of the individuals were resistant), 4 cases as "early warning" of resistance (with 1–6% of individuals with resistant phenotypes), 1 case as >50% resistant individuals and reduction of the expected efficacy, and 5 cases as "practical resistance" (>50% resistant individuals and reduced efficacy reported).

Analysis of the available data from resistance monitoring suggests that the high-dose/refuge strategy has helped to delay pest resistance to *Bt* crops (Huang et al., 2011; Tabashnik et al., 2013). For the high-dose/refuge strategy to be effective, there are three conditions to be met: (1) high dose. The

Table 6.2 Status of field-evolved resistance to *Bt* crops in nine species of major insect pests (Tabashnik et al., 2013, 2014)

Pest	Crop	Toxin	Country	References
Incipient resistance (<1% resistant individuals)				
H. armigera	Cotton	Cry1Ac	Australia	Downes and Mahon (2012b)
H. armigera	Cotton	Cry2Ab	Australia	Downes and Mahon (2012b)
H. punctigera	Cotton	Cry2Ab	Australia	Downes et al. (2010a)
Early warning (1–6% resistant individuals)				
D. saccharalis	Corn	Cry1Ab	United States	Huang et al. (2012)
H. armigera	Cotton	Cry1Ac	China	Zhang et al. (2011, 2012a) and Jin et al. (2013)
O. furnacalis	Corn	Cry1Ab	Philippines	Alcantara et al. (2011)
P. gossypiella	Cotton	Cry1Ac	China	Wan et al. (2012)
>50% resistant individuals and reduced efficacy expected				
H. zea	Cotton	Cry2Ab	United States	Ali and Luttrell (2007) and Tabashnik et al. (2009a, 2013)
Practical resistance (>50% resistant individuals and reduced efficacy reported)				
B. fusca	Corn	Cry1Ab	South Africa	Van Rensburg (2007) and Kruger et al. (2011)
D.v. virgifera	Corn	Cry3Bb	United States	Gassmann et al. (2011, 2012)
H. zea	Cotton	Cry1Ac	United States	Luttrell et al. (2004), Ali et al. (2006) and Tabashnik et al. (2008)
P. gossypiella	Cotton	Cry1Ac	India	Dhurua and Gujar (2011)
S. frugiperda	Corn	Cry1F	United States	Storer et al. (2010, 2012)

concentration of toxins produced by *Bt* plants should be high enough to kill all or nearly all heterozygotes, rendering resistance as a recessive trait; (2) abundant non-*Bt* refuges to provide susceptible insects; and (3) low initial resistance allele frequency. Among the five cases of practical resistance (Table 6.2; Tabashnik et al., 2013), at least one of the key requirements for the high-dose/refuge strategy was violated. For *B. fusca*, *D.v. virgifera*, *H. zea* and *S. frugiperda*, the high-dose standard is not met. For

P. gossepiella, the first-generation *Bt* cotton (producing Cry1Ac) meets the high-dose standard, but *P. gossepiella* resistance to *Bt* cotton evolved quickly in western India, though not in the southwestern United States. Although both countries had regulations requiring refuges of non-*Bt* cotton to be planted near *Bt* cotton, poor compliance with the refuge regulations by Indian cotton farmers may play a key role in the rapid evolution of resistance and reduced field control in India (Tabashnik et al., 2013).

3. RESISTANCE MECHANISMS

Mode of action of *Bt* Cry toxins has been intensively studied and frequently reviewed (Bravo et al., 2011; Ferré and Van Rie, 2002; Griffitts and Aroian, 2005; Ibrahim et al., 2010; Knowles, 1994; Pardo-López et al., 2013; Pigott and Ellar, 2007; Soberón et al., 2009; Vachon et al., 2012; Chapter 2). Although many details of mode of action of *Bt* Cry toxins are far from understood, the major steps (crystal solubilization, proteolytic activation, receptor binding, membrane insertion and pore formation) are generally agreed on (Knowles, 1994; Schnepf et al., 1998). Insects could develop resistance to Cry toxins due to alteration at any step of the sequential procession of intoxication, and as many as 10 potential *Bt* resistance mechanisms were proposed (Heckel, 1994). For brevity, major steps of mode of action, potential and observed resistance mechanisms are represented in Fig. 6.2. This review focuses on resistance mechanism involving alterations in receptor binding and proteolytic activation of Cry toxins in lepidopteran insects. Other mechanisms such as more efficient repair of damaged midgut cells and elevated immune responses in *Bt*-resistant strains are not included.

3.1. Mode of action of *Bt* Cry toxins

3.1.1 Structure of Bt *Cry toxins*

B. thuringiensis bacteria can produce various kinds of proteins with insecticidal activity. The major insecticidal toxins include the δ-endotoxins produced as parasporal crystals during sporulation growth phase of *Bt*, and the vegetative insecticidal proteins (VIP proteins) synthesized during the vegetative phase of growth and secreted into the medium. The δ-endotoxins are mainly composed of two families, Cry and Cyt, which are significantly different in primary sequence. Cry toxins are insecticidal to the insects from Lepidoptera, Coleoptera, Hymenoptera and Diptera orders, and also to nematodes. Cyt toxins, characterized for possessing cytolytic activity, are specifically toxic to the Dipteran insects.

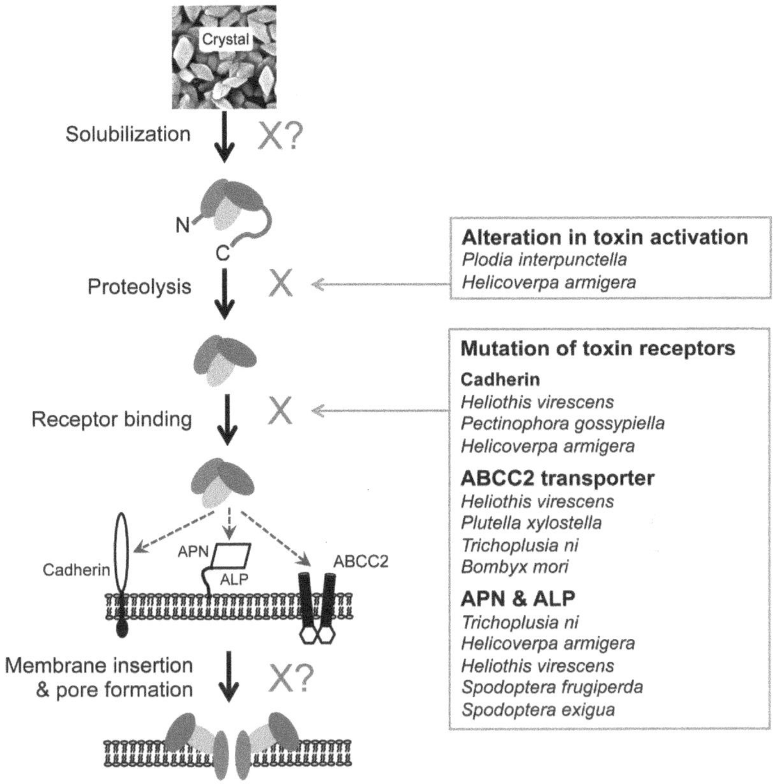

Figure 6.2 Schematic diagram of the steps of Cry toxin action and possible resistance mechanisms. Receptor mutations and/or proteolytic alterations have been identified in a number of *Bt*-resistant lepidopterans, whereas resistance mechanisms involving solubilization failure or abnormal membrane insertion/pore formation have not been reported yet. (See the colour plate.)

Since identification of the first *cry* genes (Schnepf and Whiteley, 1981), more than 700 *cry* genes divided into 72 groups have been reported (Crickmore et al., 2014). Cry toxins are grouped in three families that are not related in structure: Bin-like, Mtx-like and 3d-Cry (Pardo-López et al., 2013). Although the sequence identity is low between different groups of 3d-Cry proteins, the three-dimensional structure is highly conserved. The structure of 3d-Cry toxins is composed of three structural domains (Li et al., 1991). Domain I consists of a bundle of seven α-helices with a central helix α-5 surrounded by six amphipathic helices. Domain I is the most conserved domain among Cry toxins and is responsible for pore formation in the cell membrane. Domain II is composed of three antiparallel β-sheets packed around a hydrophobic core to form a "β-prism". Domain II is the most divergent domain among Cry toxins and is involved in defining receptor binding specificity. Domain III is a β-sandwich of two antiparallel β-sheets that form a "jelly-roll" topology. Domain III is implicated in both receptor binding and pore formation in the cell membrane (de Maagd et al., 2001; Schwartz et al., 1997). An extensive discussion of these toxins, its solubilization and activation is presented in Chapter 2.

3.1.2 Receptors for Bt *Cry toxins*

A variety of proteins have been identified and characterized as receptors or putative receptors for Cry toxins. These include cadherins (Chen et al., 2009a; Contreras et al., 2013; Fabrick et al., 2009; Flannagan et al., 2005; Gahan et al., 2001; Hua et al., 2008, 2013; Morin et al., 2003; Nagamatsu et al., 1998; Ren et al., 2013; Sayed et al., 2007; Vadlamudi et al., 1995; Xu et al., 2005), aminopeptidases (APNs; Chen et al., 2009b; Crava et al., 2013; Gill et al., 1995; Gill and Ellar, 2002; Knight et al., 1994, 1995; Luo et al., 1997; Oltean et al., 1999; Rajagopal et al., 2002; Sivakumar et al., 2007), alkaline phosphatases (ALPs; Hua et al., 2009; Jiménez et al., 2012; Martins et al., 2010; McNall and Adang, 2003; Ning et al., 2010; Perera et al., 2009; Zúñiga-Navarrete et al., 2013) and ATP-binding cassette (ABC) transporters (Atsumi et al., 2012; Gahan et al., 2010; Tanaka et al., 2013). Please refer to Chapter 2 for additional details on the receptor proteins.

3.1.3 Models for mode of action

Crystal proteins ingested by the larvae are solubilized in the insect midgut to release protoxins, and protoxins are then proteolytically cleaved to active toxins. These activated toxins bind to specific receptors at the surface of midgut epithelial cells and insert into the membrane to form non-selective pores which are permeable to ions and small molecules. The presence of such pores in the plasma membrane leads to osmotic lysis and death of the cells (Carroll and Ellar, 1993; Knowles, 1994; Knowles and Ellar, 1987; also refer to Chapter 2 for more details). Although the mode of action of Cry toxins described above is widely accepted, many details in the process of intoxication are still lacking or controversial. Two models have been proposed to depict details of the mode of action of Cry toxins: the sequential binding model (Bravo et al., 2004) and the signalling pathway model (Zhang et al., 2005, 2006). These two models were critically reviewed in detail by Vachon et al. (2012).

The sequential binding model comprises several steps: the monomeric toxin binds to ALP and APN receptors in a low-affinity interaction to locate the toxin in close proximity to the membrane surface; the monomeric toxin binds to the cadherin receptor in a high-affinity interaction to induce proteolytic cleavage of the N-terminal end containing helix α-1 of domain I, and removal of helix α-1 enables the rest of the toxin to form a toxin prepore oligomer; the oligomeric prepore structure binds to ALP and APN receptors with high affinity; and this binding is conducive to the insertion of the prepore oligomer into the membrane, creating permeable pores in the apical membrane of midgut cells (Pardo-López et al., 2013). Recently, additional evidence supporting the sequential binding model indicate that pore formation is an important step in the mechanism of action of Cry proteins and that protoxin binding to cadherin receptor may have a functional role inducing an alternative oligomer formation different from the oligomer induced by the activated toxin (Gómez et al., 2014). Upon identification of ABCC2 as a novel receptor for Cry toxins, Heckel (2012) proposed that ABCC2 may be involved in facilitating the insertion of the prepore structure into the membrane as an extension of the sequential binding model.

The signalling pathway model proposed that the toxicity of Cry proteins is due to the activation of a Mg^{2+}-dependent signal cascade pathway that is triggered by the interaction of the monomeric toxin with the cadherin receptor. Binding of Cry1Ab monomer to cadherin activates a guanine nucleotide-binding protein (G protein), which in turn activates an adenylcyclase promoting the production of intracellular cAMP. The increased cAMP levels activate protein kinase A which activates an intracellular pathway resulting in cell death (Zhang et al., 2005, 2006). A recent study by Zhang et al. (2012b) demonstrated that a deletion in the intracellular domain of cadherin is genetically linked with non-recessive resistance to Cry1Ac in *H. armigera*, providing evidence that the cytoplasmic domain of cadherin is involved in mediating toxicity. This evidence is consistent with the vital role of the intracellular region of cadherin proposed by the cell signalling model.

Based on work dealing with Cry1A receptors and resistance mechanisms in *H. virescens*, a combination of the above two models has also been proposed. In the combined model, osmotic cell lysis due to pore formation and programmed cell death are not mutually exclusive, and both events are possibly involved and have a direct effect in toxicity (Jurat-Fuentes and Adang, 2006).

3.2. Alterations in proteolytic processing of Cry toxins in resistant insects

Proper activation of *Bt* Cry protoxins is critical for mediating toxicity to target insects, and improper processing of a protoxin (insufficient activation or increased degradation) can result in insect resistance to the protoxin (Oppert, 1999). The 198r strain of Indian meal moth, *Plodia interpunctella* was selected with *B. thuringiensis* subsp. *entomocidus* HD198 resulting in high levels of resistance to Cry1Ab protoxin (Herrero et al., 2001). A major gut protease (T1) that activates Cry1A protoxins was found to be absent in the gut extracts from the resistant 198r strain. Subsequently, bulked segregant analyses revealed that the absence of T1 was genetically linked to resistance to HD198 *Bt*-strain (Oppert et al., 1997). A laboratory-selected strain of *H. armigera* (designated Akola-R) exhibited 72-fold resistance to Cry1Ac protoxin. The larval midgut juice from Akola-R strain did not activate Cry1Ac protoxin properly, producing a mixture of 95 and 68 kDa Cry1Ac polypeptides instead of the biologically active 65 kDa toxin core. N-terminal sequencing of these 95 and 68 kDa polypeptides produced by gut juices of resistant insects revealed an intact N-terminus (Rajagopal et al., 2009). Considering removal of N-terminal of Cry1Ac is an essential early step in the mode of action (Bravo et al., 2002), retention of the N-terminal of Cry1Ac protoxin is expected to impair Cry1Ac toxicity. Further studies by Rajagopal et al. (2009) identified a proteinase (HaSP2) responsible for Cry1Ac activation, and downregulation of HaSP2 was claimed to cause improper processing of the protoxin. But the deduced amino acid sequence of *HaSP2* lacked two (Asp102 and Ser195) of the three conserved residues (His57, Asp102 and Ser195) important for the catalytic triad, suggesting that HaSP2 may be a serine protease homologue without proteinase activity. The function of HaSP2 on Cry1Ac protoxin activation needs to be further investigated.

The 198r strain of *P. interpunctella* showed 264-fold resistance to Cry1Ab protoxin but only 25-fold to Cry1Ab toxin (Herrero et al., 2001). The Akola-R strain of *H. armigera* showed 72-fold resistance to Cry1Ac protoxin, but no resistance to Cry1Ac toxin (Rajagopal et al., 2009). Much higher levels of resistance to a protoxin than to its toxin form in a *Bt*-resistant strain may be a good indication of the involvement of proteinases in *Bt* resistance.

3.3. Modifications of Cry toxin receptors in resistant insects

3.3.1 Cadherin

The YHD2 strain of *H. virescens* is a laboratory-selected strain that is about 10,000-fold resistant to Cry1Ac toxin (Gould et al., 1995). A single major gene (BtR-4) is responsible for 40–80% of Cry1Ac resistance levels in YHD2 and was mapped to linkage group 9 (LG9; Heckel et al., 1997). At that time, two types of *Bt* receptors were identified: APNs and cadherins (Knight et al., 1994; Vadlamudi et al., 1995). While two APNs were mapped on other linkage groups, a cadherin gene (*HvCad*) was mapped on the same linkage group (LG9) as BtR-4. Further genetic and molecular analysis revealed that the *HvCad* in the YHD2 strain was inactivated by a retrotransposon, introducing a premature stop codon and preventing the translation of the full-length protein (Gahan et al., 2001). After the first identification of cadherin as a *Bt* resistance gene in *H. virescens*, linkage between Cry1Ac resistance and the cadherin gene were tested in another two cotton pests: *P. gossypiella* and *H. armigera* (Morin et al., 2003; Xu et al., 2005).

In *P. gossypiella*, three mutated cadherin gene alleles (r_1, r_2 and r_3) were genetically linked to resistance to Cry1Ac in the AZP-R strain (Morin et al., 2003), and a fourth resistance allele (r_4) was identified from the Bt4R strain (Fabrick and Tabashnik, 2012). The r_1 allele has a deletion mutation of 24 bp that results in two amino acid substitutions and the omission of eight amino acids. The r_2 allele has a 202-bp deletion introducing a premature stop codon, the r_3 allele has a 126-bp deletion that eliminates 42 amino acids and the r_4 allele has a 15-bp deletion that causes loss of 5 amino acids. All four deletions were located upstream of the Cry1Ac-binding region of the cadherin protein (Fabrick and Tabashnik, 2012; Morin et al., 2003). Recently, eight novel alleles (r_5–r_{12}) associated with Cry1Ac resistance were revealed from two field populations of *P. gossypiella* from western India. For seven of the eight alleles, each produced two or more different transcript isoforms by alternative splicing (Fabrick et al., 2014).

In *H. armigera*, a deletion between exons 8 and 25 of *HaCad* (r_1 allele) created a premature stop codon that was genetically linked to Cry1Ac resistance in the laboratory-selected GYBT strain (Xu et al. 2005; Yang et al., 2006). Through F_1 screens and F_2 screens of field populations of *H. armigera* from northern China, where *Bt* cotton expressing Cry1Ac has been intensively adopted, a total of 15 resistance alleles of *HaCad* (r_1–r_{15}) were identified (Yang et al., 2007; Zhang et al., 2012a; Zhao et al., 2010). Unlike most cadherin mutants previously identified, which have mutations at the ectodomain and confer recessive resistance, the r_{15} allele of *HaCad* showed a 55 amino acids deletion in the intracellular domain of cadherin and it was reported to cause non-recessive resistance to Cry1Ac in *H. armigera* (Zhang et al., 2012b). Although all the r_{15} alleles have the same deletion in the cDNA sequence, at least four different indels causing loss of exon 32 have been detected in the genomic DNA sequences flanking exon 32 of *HaCad*. Multiple origins of the r_{15} alleles illustrate parallel genotypic adaptation of *H. armigera* to the selection pressure of *Bt* cotton (Zhang et al., 2013).

Unlike mutations that occurred directly in the cadherin gene, reduced expression levels of cadherin (*DsCad*) were reported to be associated with Cry1Ab resistance in the Cry1Ab-RR strain of *D. saccharalis* (Yang et al., 2011). Whereas cDNA sequences of *DsCad* were identical between the resistant and susceptible strains, the transcript levels of *DsCad* was significantly lower in Cry1Ab-RR. The role of DsCad protein on Cry1Ab toxicity was further confirmed by RNAi analysis, which demonstrated that knockdown of *DsCad* expression correlated with increased tolerance to Cry1Ab in the susceptible strain (Yang et al., 2011).

So far, a total of 28 cadherin alleles genetically linked to Cry1Ac resistance were identified from three lepidopteran pests: 1 from *H. virescens*, 12 from *P. gossypiella* and 15 from *H. armigera* (Fabrick and Tabashnik, 2012; Fabrick et al., 2014; Gahan et al., 2001; Morin et al., 2003; Xu et al., 2005; Yang et al., 2006, 2007; Zhang et al., 2012a,b; Zhao et al., 2010). These 28 resistance alleles can be grouped into four types (illustrated in Fig. 6.3): (1) Truncation caused by premature stop codons at the extracellular domain are expected to create truncated proteins that lose the transmembrane anchor region and cannot interact with Cry1Ac. (2) Deletions from the Cry1Ac-binding region or somewhere else at the extracellular domain are expected to damage normal interaction between cadherin and Cry1Ac. (3) Deletion at the intracellular domain (such as r_{15} of *HaCad*) does not affect toxin binding, and this mutation is suggested to affect post-binding events (Zhang et al., 2012b). Evidence of involvement of the intracellular cadherin-domain in the Cry1Ac toxicity suggest that both pore formation and cell signalling pathways contribute to the efficacy of *Bt* toxins (Zhang et al., 2012b). (4) Single amino acid mutations in the Cry1Ac-binding region of the cadherin receptor from *H. virescens* were demonstrated to affect its toxin-binding ability to Cry1Ac (Xie et al., 2005). Resistance alleles r_{10}–r_{14} of *HaCad* were suggested to confer resistance through amino acid substitutions, although exact amino acid substitutions have not been functionally confirmed yet (Zhang et al., 2012a).

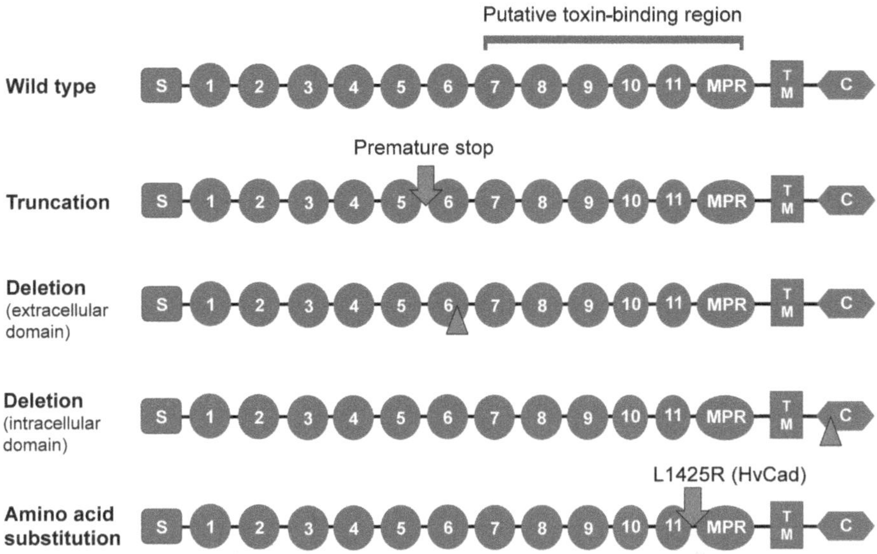

Figure 6.3 Types of cadherin mutations associated with Cry1Ac resistance in three cotton pests (*Heliothis virescens, Pectinophora gossypiella* and *Helicoverpa armigera*). (1) *Truncation*: a premature stop codon is present at any sites on the extracellular domain of cadherin. (2) *Deletion in the extracellular domain*: a stretch of amino acid residues on the extracellular domain, present within or ahead of the toxin-binding region. (3) *Deletion in the intracellular domain*: lack of 55 amino acid residues in the r_{15} allele of *HaCad*, which confers non-recessive resistance to Cry1Ac (Zhang et al., 2012b). (4) *Amino acid substitution*: the L1425R mutant of HvCad of *H. virescens* can decrease binding to Cry1Ac and has the potential to confer resistance (Xie et al., 2005). The r_{10}–r_{14} alleles of *HaCad* in *H. armigera* were suggested to confer Cry1Ac resistance by amino acid substitutions (Zhang et al., 2012a). (See the colour plate.)

3.3.2 Aminopeptidase

Mutations or altered expression of APN has been suggested or confirmed to confer high levels of resistance to Cry1 toxins in three lepidopterans: *S. exigua*, *T. ni* and *H. armigera* (Herrero et al., 2005; Tiewsiri and Wang, 2011; Zhang et al., 2009).

Lack of APN1 expression was detected in a Cry1Ca-resistant strain of *S. exigua* by using suppression subtractive hybridization. Northern blot analysis confirmed APN1 was not expressed in the resistant strain, whereas other three APNs (APN2, APN3 and APN4) had no difference in expression between the resistant and susceptible strains (Herrero et al., 2005). These data suggest that the lack of APN1 expression plays a role in Cry1Ca resistance in *S. exigua*, although a linkage between Cry1Ca resistance and the lack of APN1 expression needs to be determined. Cry1Ac resistance in the GLEN-Cry1Ac-BCS of *T. ni* was introgressed into a susceptible strain

(Benzon) by eight generations of backcrossing. A proteomic screen of midgut BBMV proteins from the two isogenic strains revealed that Cry1Ac resistance is associated with differential alteration of two midgut APNs, down-regulation of APN1 and up-regulation of APN6, conferred by a *trans*-regulatory mechanism. Genetic analysis showed that only down-regulation of APN1 was linked to Cry1Ac resistance (Tiewsiri and Wang, 2011), supporting the importance of APN1 on Cry1Ac resistance and on the mode of action of Cry1Ac toxin in *T. ni*.

Interestingly, a deletion mutation of *HaAPN1* (at amino acids 938–1004) was suggested to confer Cry1Ac resistance in the BtR strain of *H. armigera* (Zhang et al., 2009). A peptide fragment of HaAPN1 with the deletion expressed in *Escherichia coli* cells lost binding with Cry1Ac, whereas the peptide fragment without the deletion bound to Cry1Ac (Zhang et al., 2009). Both the wild type and the short type of *HaAPN1* (i.e. heterozygote) were present in many individuals from the resistant BtR strain, suggesting that other resistance mechanism(s) may exist in this resistant strain. Genetic analysis showed Cry1Ac resistance in BtR is incompletely recessive and polygenic (Liang et al., 2008), which confirms that there are other resistance mechanisms in this resistant strain besides the *HaAPN1* mutation.

3.3.3 Alkaline phosphatase

Reduced expression of ALP protein and mRNA has been associated with Cry1 resistance in some strains of three Noctuid species (*H. virescens*, *H. armigera* and *S. frugiperda*) with different resistance phenotypes to Cry1 toxins (Jurat-Fuentes and Adang, 2004; Jurat-Fuentes et al., 2011). In the YHD3 strain of *H. virescens*, two genes at different loci (*HvCad* and *HvABCC2*) have been identified to confer extremely high levels of resistance to Cry1Ac (Gahan et al., 2001, 2010). Although reduced binding with Cry1Ac mediated by HvALP was observed in YHD3 (Jurat-Fuentes and Adang, 2004), a direct correlation between HvALP levels and resistance is still lacking (Jurat-Fuentes et al., 2011). Two field-derived strains of *S. frugiperda* (456 and 512) with extremely high levels of resistance to Cry1Fa showed three- to fourfold reduction in ALP activity and protein levels, but no differences in APN activity in midgut BBMVs when compared with two susceptible strains (Jurat-Fuentes et al., 2011). In the Cry1Ac-selected AR1 strain of *H. zea*, a 10-fold increase in specific ALP activity in the midgut lumen was suggested as a resistance mechanism. The soluble form of ALP in midgut lumen may sequester some of Cry1Ac toxins and, therefore, affect toxin binding with other receptors (Caccia et al., 2012). Further studies are needed to verify if alteration of ALP expression in resistant strains is just an adaptive consequence or directly impact toxin binding resulting in insect resistance.

3.3.4 ABCC2

In the YHD2 strain of *H. virescens*, a mutation in the cadherin gene *HvCad* was identified as a major mechanism conferring high levels of resistance to Cry1Ac toxin (Gahan et al., 2001). YHD2 lost binding with Cry1Aa, but not with Cry1Ab and Cry1Ac (Lee et al., 1995). The YHD2 strain was further selected with Cry1Ac to produce the YHD3 strain, which had much higher resistance levels to Cry1Ac than YHD2 and loss binding with all three Cry1A toxins (Jurat-Fuentes and Adang, 2004). A second resistance gene (*HvABCC2*) contributing to the additional resistance and lost of binding with Cry1Ab and Cry1Ac in YHD3 was then identified by genetic mapping and positional cloning (Gahan et al., 2010). A 22-bp deletion in exon 2 of *HvABCC2* in YHD3 was predicted to result in a truncated 99-residue protein instead of the full-length protein composed of 1339 amino acids. The inactivation mutation of *HvABCC2* was correlated with both higher resistance levels and the loss of Cry1Ab and Cry1Ac binding to midgut membranes (Gahan et al., 2010).

Resistance to Cry1Ac spray formulations has also evolved in field populations of *P. xylostella* and *T. ni* (Janmaat and Myers, 2003; Tabashnik et al., 1990). Genetic mapping demonstrated that field-evolved resistance to Cry1Ac in the NO-QA strain of *P. xylostella* and the GLEN-Cry1Ac-BCS strain of *T. ni* was also genetically linked with *ABCC2* gene, named *PxABCC2* and *TnABCC2*, respectively (Baxter et al., 2011). In the NO-QA strain of *P. xylostella*, a 10-residue deletion in the middle of transmembrane helix 12 of *PxABCC2* was predicted to be located in the extracellular region of the second NBD domain (Baxter et al., 2011). However, the mutation of *TnABCC2* has not yet been identified in the GLEN-Cry1Ac-BCS strain of *T. ni*.

In the C2 strain of *B. mori*, a candidate locus for recessive resistance to Cry1Ab was mapped to a region of 82 kb (containing *BmABCC2*) on chromosome 15. Comparisons of *BmABCC2* sequences among 10 susceptible and 7 resistant strains revealed a common tyrosine insertion in resistant alleles which is located in an outer loop of the predicted transmembrane structure. Introduction of a wild-type allele of *BmABCC2* into a resistant strain by using germline transformation restored susceptibility to Cry1Ab in the resistant strain, which provides very strong evidence for the role of this single tyrosine insertion of *BmABCC2* in conferring Cry1Ab resistance (Atsumi et al., 2012). As mentioned above, the additional tyrosine at position 234 is located in one of the extracellular loops of a transmembrane domain of *BmABCC2*, and it was proposed that this loop is involved in binding to the toxin in the midgut lumen. Expression of *BmABCC2* with and without Tyr234 in Sf9 cells showed that insertion of Tyr234 caused reduced Cry1Ab cytotoxicity (Tanaka et al., 2013). Reduced binding of Cry1Ab was associated with this insertion in Sf9 cells (Tanaka et al., 2013), but not in brush border membrane vesicles from larval midguts (Atsumi et al., 2012).

Diversified resistance mutations in the same *ABCC2* gene of at least four lepidopteran species, from single amino acid insertions to deletion of the entire protein, strongly suggest that this ABC transporter plays a crucial role in the toxin mode of action as a functional receptor (Heckel, 2012; Tanaka et al., 2013). According to the existing evidence, it is concluded that the ABCC2 protein is a functional receptor for Cry1Ab and Cry1Ac but not for Cry1Aa, and mutations in this gene confer high levels of resistance to Cry1Ab and Cry1Ac with recessiveness.

4. GENETIC DIVERSITY OF RESISTANCE AND IMPLICATIONS FOR RESISTANCE MANAGEMENT

4.1. Laboratory-selected and field-evolved resistance

A prominent hypothesis about insecticide resistance is that selection in the laboratory favours a polygenic response, whereas selection in the field favours a monogenic response (McKenzie and Batterham, 1994; Roush and McKenzie, 1987). The idea is that the starting laboratory populations for selection are usually small and not likely to contain rare, major resistance mutations; thus, laboratory selection is prone to pick up pre-existing genes of minor effect (ffrench-Constant, 2013). However, for conventional insecticides, a review of empirical data and simulation results did not support this hypothesis (Groeters and Tabashnik, 2000). Likewise, this hypothesis is not supported by the available data on *Bt* resistance. First, several strains with high levels of resistance to Cry1Ac conferred by major genes (*cadherin* or *ABCC2*) were selected for each of the three lepidopterans: *H. virescens*, *P. gossypiella* and *H. armgiera* (Gahan et al., 2001, 2010; Morin et al., 2003; Xu et al., 2005; Zhang et al., 2012a,b). Second, the correspondence between the genetic basis of laboratory-selected resistance and field-selected resistance to Cry1Ac has been established in two cases. In northern China, the r_1 resistance allele of cadherin of *H. armigera* was first identified in a laboratory-selected strain collected from Hebei Province in 2001 (Xu et al., 2005) and the r_1 allele was also detected in three resistant strains isolated in 2009 from the field-selected Anyang population in Henan Province of China (Zhang et al., 2012a). Mutations in the same cadherin gene are associated with *P. gossypiella* resistance to Cry1Ac in laboratory-selected strains from Arizona of the United States and field-selected populations from India (Fabrick et al., 2014). It shows that laboratory-selected strains can be a valuable tool for finding loci that are important in field-evolved resistance to *Bt* crops.

In view of diverse and complex genetic basis of insect resistance to *Bt* proteins that have been identified (as summarized in Table 6.3), we can say that the use of laboratory selection to predict the exact types of field-selected resistance is facing challenges and should be taken with caution. Screen of F_1 and F_2 progeny from several geographic field-derived populations for *Bt* resistance in the laboratory will provide better prediction of resistance mechanism than selection from a single field population or from a susceptible strain kept in the laboratory for a long period.

Since resistance to Cry toxin may result from multiple mutational events at a single resistance locus (cadherin) such as have been revealed in field-selected populations of two cotton pests *H. armigera* from China and *P. gossypiella* from India (Fabrick et al., 2014; Yang et al., 2007; Zhang et al., 2012a; Zhao et al., 2010), the F_1 screen will be a practical and reliable method to recover resistance alleles at a known resistance locus from field populations.

The F_2 screen is an efficient approach to isolate a collection of strains representing the important types of resistance mechanisms that are present in field populations of the target pests. Large-scale F_2 screens have been successfully employed to monitor resistance frequency and also to isolate different *Bt*-resistant strains from geographical field populations of *H. armigera* in China (Zhang et al., 2012a) and *H. armigera* and *H. punctigera* in Australia (Downes et al., 2009, 2010a; Mahon et al., 2007, 2008, 2012). Eight strains of *H. armigera* resistant to Cry1Ac were isolated from an F_2 screen of field populations in China, and resistance types of these eight strains were determined through an approach involving crossing with a susceptible or resistant strain, bioassay and sequencing of the cadherin gene (Fig. 6.4). Among these eight resistant strains, six strains had recessive resistance alleles at a cadherin locus, one strain had a non-recessive resistance allele at the cadherin locus and the other strain had non-recessive resistance but not linked with the

Table 6.3 Mechanisms of resistance to Cry1 toxins in lepidopterans

Strain	Toxin	RR[a]	Altered proteolytic processing	Binding site modification				References
				Cadherin	ABCC2	APN	ALP	
H. virescens								
YFO	Cry1Ac	>78		△[b]				Gahan et al. (2010)
YEE	Cry1Ac	>780			△			Gahan et al. (2010)
YHD3/ YHD2-B	Cry1Ac	73,700		△	△		△	Gahan et al. (2010) and Jurat-Fuentes et al. (2011)
H. armigera								
GYBT	Cry1Ac	564		△				Xu et al. (2005)
SCD-r1	Cry1Ac	438		△				Yang et al. (2009)
Xj-r15	Cry1Ac	140		△				Zhang et al. (2012b)
BtR	MVP/Cry1Ac[c]	2971				△	△	Zhang et al. (2009) and Jurat-Fuentes et al. (2011)
Akola-R	Cry1Ac[c]	72	△					Rajagopal et al. (2009)
H. zea								
AR1	Cry1Ac	~100					△	Caccia et al. (2012)
P. gossypiella								
AZP-R	Cry1Ac	3100		△				Morin et al. (2003)
AGJ, KMP	Cry1Ac	NA[d]		△				Fabrick et al. (2014)

P. xylostella							
NO-QA/ NO-QAGE	MVP/Cry1Ac[c]	>6800		△			Baxter et al. (2011) and Tabashnik et al. (1993)
T. ni							
GLEN-Cry1Ac-BCS	Cry1Ac	2344		△	△		Tiewsiri and Wang (2011) and Baxter et al. (2011)
B. mori							
C2	Cry1Ab	>2860		△			Atsumi et al. (2012)
S. frugiperda							
456, 512	Cry1Fa	7720–10,000				△	Jurat-Fuentes et al. (2011) and Blanco et al. (2010)
P. interpunctella							
198-r	Cry1Ab[c]	264	△				Oppert et al. (1997) and Herrero et al. (2001)
S. exigua							
Resistant	Cry1Ca	850			△		Herrero et al. (2005) and Moar et al. (1995)

[a]RR at LC_{50}.
[b]Presence of a specific resistance mechanism.
[c]The proteins used for bioassays are protoxins.
[d]Survivors collected from *Bt* cotton in western India.

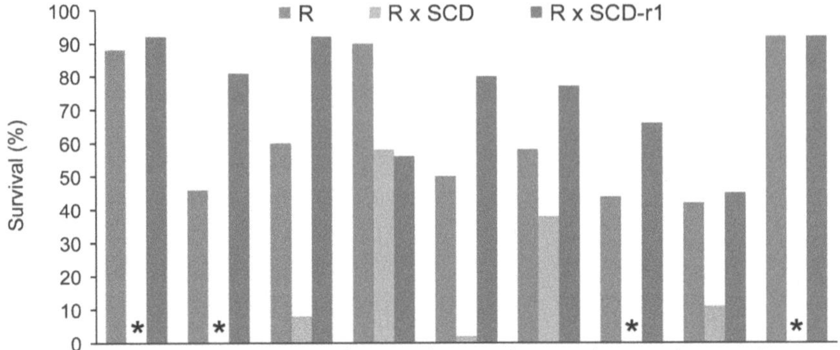

Strain	AY9	AY148	AY335	AY423	AY440	AY441	SC23	SW34	SCD-r1
R allele	*HaCad* r_1	*HaCad* r_9	*HaCad* r_{10}	Non-HaCad	*HaCad* r_{11}	*HaCad* r_{12}	*HaCad* r_{13}	*HaCad* r_{14}	*HaCad* r_1
Dominance	Rec	Rec	Rec	Dom	Rec	Dom	Rec	Rec	Rec

Figure 6.4 Diverse genetic basis of Cry1Ac resistance in field populations of *Helicoverpa armigera* from China (Zhang et al., 2012a). Survival at the diagnostic concentration of Cry1Ac of nine resistant (R) strains (red bars; eight strains isolated from the F_2 screen and the laboratory-selected strain SCD-r1 with the r_1 allele of *HaCad* at the cadherin locus), and progeny from crosses between each resistant strain and either the susceptible SCD strain (green bars) or the resistant SCD-r1 strain (blue bars). Asterisks indicate 0% survival for progeny from crosses between the SCD strains. Rec, recessive; Dom, dominant. (See the colour plate.)

cadherin locus (Zhang et al., 2012a). In contrast with diverse genetic basis of Cry1Ac resistance in *H. armigera* from China, five Cry2Ab-resistant strains of *H. armigera*, which were isolated from an F_2 screen of field populations in Australia, were all recessive and shared a common resistance locus (Mahon et al., 2008).

4.2. Resistance dominance and the refuge strategy

The refuge strategy has been the primary approach to delay evolution of pest resistance to *Bt* crops and has been mandatory in the United States, Australia and elsewhere. The refuge is expected to produce relative abundant susceptible target insects that will be able to mate with rare survivors from *Bt* crops. If resistance is recessive, the progeny from such matings will be killed by *Bt* crops and the evolution of resistance will be substantially delayed (Tabashnik et al., 2009a, 2013). The refuge strategy will be most effective if *Bt* crops express consistently a high dose of *Bt* proteins to render resistance a functionally recessive trait and thus kill almost all of the heterozygous progeny.

In contrast, effectiveness of the refuge strategy will be largely compromised if resistance is non-recessive or partial recessive and *Bt* crops do not meet the high-dose standard for the target pest.

Selection for resistance to *Bt* Cry toxins in the laboratory has shown that the levels of insect resistance to Cry proteins can vary from complete recessiveness to complete dominance (Ferré et al., 2008). Even for a single species *H. armigera*, resistance to Cry1Ac is completely recessive in SCD-r1 strain (r_1 allele, with a truncated cadherin) and completely dominant in AY2 and QX7 strains at the diagnostic concentration of Cry1Ac (Jin et al., 2013). Even further, different alleles at the same locus of cadherin in *H. armigera* can confer different levels of dominance, r_1 allele is recessive and r_{15} allele (with a 55aa deletion in the cytoplasmic domain of cadherin) is non-recessive (Zhang et al., 2012b).

Similar to results from laboratory selection, dominance of field-evolved resistance also ranges from being recessive to dominant. Resistance to Cry1F corn is recessive in *S. frugiperda* from Puerto Rico (Storer et al., 2010). Based on results from laboratory-selected resistant strains, resistance to Cry1Ac cotton is expected to be recessive in *P. gossypiella* from India (Fabrick et al., 2014). In contrast, resistance to *Bt* corn (Cry1Ab) is dominant in *B. fusca* from South Africa (Campagne et al., 2013), and resistance to Cry3Bb corn is expected to be non-recessive in *D.v. virgifera* from the United States (Devos et al., 2013; Petzold-Maxwell et al., 2012).

Results from evidence of both the laboratory-selected and field-evolved resistance to Cry toxins, the diversity of evolved resistance dominance to *Bt* toxins is high. It is implying that target pests could evolve dominant resistance alleles that could cope with the refuge strategy. Compared with the progress on understanding of the molecular mechanisms mediating recessive resistance to *Bt* toxins, our understanding on mechanisms of dominant resistance is limited. It deserves more attention because it is tougher for us to deal with such mechanism of resistance to Cry proteins.

4.3. Cross-resistance and the pyramid strategy

Cross-resistance means resistance to a particular pesticide that results in resistance to other pesticides through a common resistance mechanism. Cross-resistance is usually present among pesticides sharing similar binding target sites or similar detoxifying pathways. For example, selection with Cry1Ac in *H. armigera* caused cross-resistance to Cry1Aa and Cry1Ab, which is conferred by cadherin mutations (Xu et al., 2005). Multiple resistance is defined as resistance to more than one class of pesticides in a single pest through different resistance mechanisms. Multiple resistance is generally caused by sequential or mosaic uses of different classes of pesticides. For example, a strain of *P. xylostella* (NO-95C), selected in the field with *Bt* formulations (containing Cry1A and Cry1C toxins) and further selected with Cry1C in the laboratory, developed high levels of resistance to Cry1A and moderate levels of resistance to Cry1C (Liu and Tabashnik, 1997; Liu et al., 1996). Bioassays of progeny from split broods of single-pair families demonstrated that resistance to Cry1C segregates independently of resistance to Cry1Ab in NO-95C (Liu and Tabashnik, 1997). Resistance to Cry1A and Cry1C in the NO-95C strain of *P. xylostella* is a typical phenomenon of multiple resistance.

The "pyramid" strategy for *Bt* resistance management uses crops that produces two or more dissimilar toxins that kill the same target pest (Roush, 1998). Toxins to be used for pyramiding should have no cross-resistance, so that the risk to evolve a single mechanism of resistance against both toxins could be dramatically reduced. Mathematical models and empirical studies have suggested resistance to pyramided two-gene plants that show no cross-resistance can be significantly delayed as compared with resistance to single-gene plants adopted sequentially or in mosaics (Roush, 1998; Zhao et al., 2003).

However, target pests can possibly evolve a single gene that can overcome both *Bt* genes to be used in the pyramid, even if they have different binding sites. The most widely used pyramid is the second-generation *Bt* cotton that produces Cry1Ac and Cry2Ab. No cross-resistance between these two toxins is presumed because they bind to different larval midgut target sites. However, several strains of *H. virescens* selected with Cry1Ac in the laboratory caused significant cross-resistance to Cry2Aa (Gould et al., 1992, 1995; Jurat-Fuentes et al., 2003). In addition, laboratory selection of a strain of *P. gossypiella* with Cry2Ab resulted in 420-fold cross-resistance to Cry1Ac as well as 240-fold resistance to Cry2Ab (Tabashnik et al., 2009b). Also, two Cry1Ac-resistant strains of *H. armigera* isolated from field-selected populations from northern China had low, but significant cross-resistance to Cry2Ab (4–6-fold; Jin et al., 2013), and a strain of *H. zea* selected for resistance to Cry1Ac increased survival on *Bt* cotton expressing Cry1Ac and Cry2Ab (Brévaulta et al., 2013). All these results show that cross-resistance occurs between Cry1Ac and Cry2Ab in at least three key cotton pests. Even though there are no shared binding sites between two Cry proteins, there are several potential resistance mechanisms that could result in cross-resistance in the target insects. For example, if different classes of *Bt* toxins are activated by the same proteases, alterations of the proteases could cause cross-resistance. If a novel proteolytic mechanism is evolved to degrade the activated toxin core, it could cause cross-resistance. Pore formation is believed to be general and an important step in the mechanism of action of many Cry proteins, thus any change affecting pore formation or pore function may result in broad cross-resistance (Heckel, 1994). With increasing deployment of the pyramid *Bt* cotton and corn crops, more work is needed to identify potential molecular basis of cross-resistance between two distinct toxins such as Cry1Ac and Cry2Ab. In the future, the toxins for stacking varieties should be as disparate as possible, showing highly different mechanism of actions such as a *Bt* toxin and a non-*Bt* factor, to minimize the chance that mutations in a single gene could confer resistance to both factors.

5. CONCLUSIONS

Large-scale adoption of *Bt* crops have brought both ecological benefits for environment and economic benefits for farmers. In the foreseeable future, adoption of *Bt* crops will keep increasing globally. Benefited from implementation of proactive resistance management strategies such as the refuge strategy and the pyramid strategy in many countries, most of the target pests of *Bt* crops have been sustainably and effectively controlled for nearly 20 years. However, several cases of field-evolved resistance to *Bt* corn and *Bt* cotton have been documented causing reduced field efficacy. Evolution of resistance by target pests is a real threat to the continued success of *Bt* crops.

The genetic capacity of insect populations to evolve resistance to Cry toxins has been well demonstrated by both laboratory-selected and field-evolved resistance in many species within several insect orders. Diverse genetic options for target pests to cope with *Bt* crops are challenging the efforts to understand mechanisms of resistance and to design rational resistance management strategies. Genetic mapping approach has been proved to be a powerful tool and it is anticipated that it will continue to be essential in dissecting the complex genetic basis of *Bt* resistance. Clarifying resistance mechanisms can facilitate and advance our understanding on modes of action of Cry toxins.

Proactive evaluation of the inheritance and initial frequency of resistance are useful for assessing the risk of resistance problems in the close future.

Although laboratory selection can provide useful information on characters of *Bt* resistance, employing the F_1 screens and F_2 screens to isolate various resistance alleles from geographical field populations and to track evolution trajectory of resistance are greatly encouraged. Detection of early resistance is critical for the establishment of an adaptive and intelligent resistance management strategy.

Intensive research on *Bt* resistance focuses primarily on resistance to Cry1A toxins in lepidopteran insects. More work is needed to cover resistance in other insect orders and resistance to *Bt* toxins from other Cry proteins and families. Resistance mechanisms and resistance evolution for several cases of field-evolved resistance deserve further studies. Retrospective analysis of these cases can provide better understanding of resistance mechanisms to Cry proteins to help improve *Bt* resistance management practices.

ACKNOWLEDGEMENTS

I am grateful to Yihua Yang, Derek Russell, Alejandra Bravo, Mario Soberon, Bruce Tabashnik and the editors of this volume for reading the text and making helpful comments. This work was supported in part by grants from the Ministry of Agriculture of China (Grant no. 2014ZX08012-004) and the NSFC of China (Grant no. 31272382).

REFERENCES

Alcantara, E., Estrada, A., Alpuerto, V., Head, G., 2011. Monitoring Cry1Ab susceptibility in Asian corn borer (Lepidoptera: Crambidae) on Bt corn in the Philippines. Crop. Prot. 30, 554–559.

Ali, M.I., Luttrell, R.G., 2007. Susceptibility of bollworm and tobacco budworm (Noctuidae) to Cry2Ab2 insecticidal protein. J. Econ. Entomol. 100, 921–931.

Ali, M.I., Luttrell, R.G., Young III, S.Y., 2006. Susceptibility of *Helicoverpa zea* and *Heliothis virescens* (Lepidoptera: Noctuidae) populations to Cry1Ac insecticidal protein. J. Econ. Entomol. 99, 164–175.

Andow, D.A., Alstad, D.N., 1998. F_2 screen for rare resistance alleles. J. Econ. Entomol. 91, 572–578.

Andow, D.A., Alstad, D.N., Pang, Y.-H., Bolin, P.C., Hutchinson, W.D., 1998. Using an F_2 screen to search for resistance alleles to *Bacillus thuringiensis* toxin in European corn borer (Lepidoptera: Crambidae). J. Econ. Entomol. 91, 579–584.

Andow, D.A., Olson, D.M., Hellmich, R.L., Alstad, D.N., Hutchison, W.D., 2000. Frequency of resistance to *Bacillus thuringiensis* toxin Cry1Ab in an Iowa population of European corn borer (Lepidoptera: Crambidae). J. Econ. Entomol. 93, 26–30.

Andreadis, S.S., Alvarez-Alfageme, F., Sánchez-Ramos, I., Stodola, T.J., Andow, D.A., Milonas, P.G., Savopoulou-Soultani, M., Castánera, P., 2007. Frequency of resistance to *Bacillus thuringiensis* toxin CrylAb in Greek and Spanish population of *Sesamia nonagrioides* (Lepidoptera: Noctuidae). J. Econ. Entomol. 100, 195–201.

Anilkumar, K.J., Rodrigo-Simón, A., Ferré, J., Pusztai-Carey, M., Sivasupramaniam, S., Moar, W.J., 2008. Production and characterization of *Bacillus thuringiensis* Cry1Ac-resistant cotton bollworm *Helicoverpa zea* (Boddie). Appl. Environ. Microbiol. 74, 462–469.

Atsumi, S., Miyamoto, K., Yamamoto, K., Narukawa, J., Kawai, S., Sezutsu, H., Kobayashi, I., Uchino, K., Tamura, T., Mita, K., Kadono-Okuda, K., Wada, S., Kanda, K., Goldsmith, M.R., Noda, H., 2012. Single amino acid mutation in an ATP-binding cassette transporter gene causes resistance to Bt toxin Cry1Ab in the silkworm, *Bombyx mori*. Proc. Natl. Acad. Sci. U.S.A. 109, E1591–E1598.

Augustin, S., Courtin, C., Réjasse, A., Lorme, P., Génissel, A., Bourguet, D., 2004. Genetics of resistance to transgenic *Bacillus thuringiensis* poplars in *Chrysomela tremulae* (Coleoptera: Chrysomelidae). J. Econ. Entomol. 97, 1058–1064.

Bauer, L.S., 1995. Resistance: a threat to the insecticidal crystal proteins of *Bacillus thuringiensis*. Fla. Entomol. 78, 414–443.

Baxter, S.W., Badenes-Perez, F.R., Morrison, A., Vogel, H., Crickmore, N., Kain, W., Wang, P., Heckel, D.G., Jiggins, C.D., 2011. Parallel evolution of *Bacillus thuringiensis* toxin resistance in lepidoptera. Genetics 189, 675–679.

Bentur, J.S., Andow, D.A., Cohen, M.B., Romena, A.M., Gould, F., 2000. Frequency of alleles conferring resistance to a *Bacillus thuringiensis* toxin in a Philippine populations of *Scirpophaga incertulas* (Lepidoptera: Pyralidae). J. Econ. Entomol. 93, 1515–1521.

Blanco, C.A., Andow, D.A., Abel, C.A., Sumerford, D.V., Hernandez, G., López, J.D., Adams, L., Groot, A., Leonard, R., Parker, R., Payne, G., Perera, O.P., Terán-Vargas, A.P., Azuara-Dominquez, A., 2009. *Bacillus thuringiensis* Cry1Ac resistance frequency in tobacco budworm (Lepidoptera: Noctuidae). J. Econ. Entomol. 102, 381–387.

Blanco, C.A., Portilla, M., Jurat-Fuentes, J.L., Sánchez, J.F., Viteri, D., Vega-Aquino, P., Terán-Vargas, A.P., Azuara-Domínquez, A., López, J.D., Arias, R., Zhu, Y.C., Lugo-Barera, D., Jackson, R., 2010. Susceptibility of isofamilies of *Spodoptera frugiperda* (Lepidoptera: Noctuidae) to Cry1Ac and Cry1Fa proteins of *Bacillus thuringiensis*. Southwest. Entomol. 35, 409–415.

Bourguet, D., Chaufaux, J., Séguin, M., Buisson, C., Hinton, J.L., Stodola, T.J., Porter, P., Cronholm, G., Buschman, L.L., Andow, D.A., 2003. Frequency of alleles conferring resistance to Bt maize in French and US corn belt populations of the European corn borer, *Ostrinia nubilalis*. Theor. Appl. Genet. 106, 1225–1233.

Bravo, A., Sanchez, J., Kouskoura, T., Crickmore, N., 2002. N-terminal activation is an essential early step in the mechanism of action of the *Bacillus thuringiensis* Cry1Ac insecticidal toxin. J. Biol. Chem. 277, 23985–23987.

Bravo, A., Gómez, I., Conde, J., Muñoz-Garay, C., Sánchez, J., Miranda, R., Zhuang, M., Gill, S.S., Soberón, M., 2004. Oligomerization triggers binding of a Bacillus thuringiensis Cry1Ab pore-forming toxin to aminopeptidase N receptor leading to insertion into membrane microdomains. Biochim. Biophys. Acta 1667, 38–46.

Bravo, A., Likitvivatanavong, S., Gill, S.S., Soberón, M., 2011. *Bacillus thuringiensis*: a story of a successful bioinsecticide. Insect Biochem. Mol. Biol. 41, 423–431.

Brent, K.J., 1986. Detection and monitoring of resistant forms: an overview. In: National Research Council, (Ed.), Pesticide Resistance: Strategies and Tactics for Management. National Academy Press, Washington, DC, pp. 298–312.

Brévaulta, T., Heubergerb, S., Zhang, M., Ellers-Kirkb, C., Ni, X., Massond, L., Li, X., Tabashnik, B.E., Carrière, Y., 2013. Potential shortfall of pyramided transgenic cotton for insect resistance management. Proc. Natl. Acad. Sci. U.S.A. 110, 5806–5811.

Caccia, S., Moar, W.J., Chandrashekhar, J., Oppert, C., Anilkumar, K.J., Jurat-Fuentes, J.L., Ferré, J., 2012. Association of Cry1Ac toxin resistance in *Helicoverpa zea* (Boddie) with increased alkaline phosphatase levels in the midgut lumen. Appl. Environ. Microbiol. 78, 5690–5698.

Campagne, P., Kruger, M., Pasquet, R., Le Ru, B., Van den Berg, J., 2013. Dominant inheritance of field-evolved resistance to Bt corn in *Busseola fusca*. PLoS One 8, e69675.

Caprio, M.A., Sumerford, D.V., 2007. Evaluating transgenic plants for suitability in pest and resistance management problems. In: Lacey, L.A., Kaya, H.K. (Eds.), Field Manual of Techniques in Invertebrate Pathology. Springer, Netherlands, pp. 769–789.

Carrière, Y., Crowder, D.W., Tabashnik, B.E., 2010. Evolutionary ecology of adaptation to Bt crops. Evol. Appl. 3, 561–573.

Carroll, J., Ellar, D.J., 1993. An analysis of *Bacillus thuringiensis* δ-endotoxin action on insect-midgut-membrane permeability using a light-scattering assay. Eur. J. Biochem. 214, 771–778.

Chen, J., Aimanova, K.G., Fernandez, L.E., Bravo, A., Soberón, M., Gill, S.S., 2009a. *Aedes aegypti* cadherin serves as a putative receptor of the Cry11Aa toxin from *Bacillus thuringiensis* subsp. *israelensis*. Biochem. J. 424, 191–200.

Chen, J., Aimanova, K.G., Pan, S., Gill, S.S., 2009b. Identification and characterization of *Aedes aegypti* aminopeptidase N as a putative receptor of *Bacillus thuringiensis* Cry11A toxin. Insect Biochem. Mol. Biol. 39, 688–696.

Contreras, E., Schoppmeier, M., Real, M.D., Rausell, C., 2013. Sodium solute symporter and cadherin proteins act as *Bacillus thuringiensis* Cry3Ba toxin functional receptors in *Tribolium castaneum*. J. Biol. Chem. 288, 18013–18021.

Crava, C.M., Bel, Y., Jakubowska, A.K., Ferré, J., Escriche, B., 2013. Midgut aminopeptidase N isoforms from *Ostrinia nubilalis*: activity characterization and differential binding to Cry1Ab and Cry1Fa proteins from *Bacillus thuringiensis*. Insect Biochem. Mol. Biol. 43, 924–935.

Crickmore, N., Baum, J., Bravo, A., Lereclus, D., Narva, K., Sampson, K., Schnepf, E., Sun, M., Zeigler, D.R., 2014. *Bacillus thuringiensis* toxin nomenclature. http://www.btnomenclature.info/.

Crow, J.F., 1960. Genetics of insecticide resistance: general considerations. Misc. Publ. Entomol. Soc. Am. 2, 69–74.

de Maagd, R.A., Bravo, A., Crickmore, N., 2001. How *Bacillus thuringiensis* has evolved specific toxins to colonize the insect world. Trends Genet. 17, 193–199.

Devos, Y., Meihls, L.N., Kiss, J., Hibbard, B.E., 2013. Resistance evolution to the first generation of genetically modified *Diabrotica*-active Bt-maize events by western corn rootworm: management and monitoring considerations. Transgenic Res. 22, 269–299.

Dhurua, S., Gujar, G.T., 2011. Field-evolved resistance to Bt toxin Cry1Ac in the pink bollworm, *Pectinophora gossypiella* (Saunders) (Lepidoptera: Gelechiidae), from India. Pest Manag. Sci. 67, 898–903.

Downes, S., Mahon, R., 2012a. Evolution, ecology and management of resistance in *Helicoverpa* spp. to Bt cotton in Australia. J. Invertebr. Pathol. 110, 281–286.

Downes, S., Mahon, R., 2012b. Successes and challenges of managing resistance in *Helicoverpa armigera* to Bt cotton in Australia. GM Crops Food 3, 228–234.

Downes, S., Parker, T.L., Mahon, R.J., 2009. Frequency of alleles conferring resistance to the *Bacillus thuringiensis* toxins Cry1Ac and Cry2Ab in Australian populations of *Helicoverpa punctigera* (Lepidoptera: Noctuidae) from 2002 to 2006. J. Econ. Entomol. 102, 733–742.

Downes, S., Parker, T., Mahon, R., 2010a. Incipient resistance of *Helicoverpa punctigera* to the Cry2Ab Bt toxin in Bollgard II(r) cotton. PLoS One 5, e12567.

Downes, S., Parker, T.L., Mahon, R.J., 2010b. Characteristics of resistance to *Bacillus thuringiensis* toxin Cry2Ab in a strain of *Helicoverpa punctigera* (Lepidoptera: Noctuidae) isolated from a field population. J. Econ. Entomol. 103, 2147–2154.

Fabrick, J.A., Tabashnik, B.E., 2012. Similar genetic basis of resistance to Bt toxin Cry1Ac in boll-selected and diet-selected strains of pink bollworm. PLoS One 7, e35658.

Fabrick, J., Oppert, C., Lorenzen, M.D., Morris, K., Oppert, B., Jurat-Fuentes, J.L., 2009. A novel *Tenebrio molitor* cadherin is a functional receptor for *Bacillus thuringiensis* Cry3Aa toxin. J. Biol. Chem. 284, 18401–18410.

Fabrick, J.A., Ponnuraj, J., Singh, A., Tanwar, R.K., Unnithan, G.C., Yelich, A.J., Li, X.C., Carrière, Y., Tabashnik, B.E., 2014. Alternative splicing and highly variable cadherin transcripts associated with field-evolved resistance of pink bollworm to Bt cotton in India. PLoS One 9, e97900.

Ferré, J., Van Rie, J., 2002. Biochemistry and genetics of insect resistance to *Bacillus thuringiensis*. Annu. Rev. Entomol. 47, 501–533.

Ferré, J., Van Rie, J., MacIntosh, S.C., 2008. Insecticidal genetically modified crops and insect resistance management (IRM). In: Romeis, J., Shelton, A.M., Kennedy, G.G. (Eds.), Integration of Insect-Resistant Genetically Modified Crops Within IPM Programs. Springer, Netherlands, pp. 41–85.

ffrench-Constant, R.H., 2013. The molecular genetics of insecticide resistance. Genetics 194, 807–815.

ffrench-Constant, R.H., Roush, R.T., 1990. Resistance detection and documentation: the relative roles of pesticidal and biochemical assays. In: Roush, R.T., Tabashnik, B.E. (Eds.), Pesticide Resistance in Arthropods. Chapman and Hall, New York, pp. 4–38.

Flannagan, R.D., Yu, C.G., Mathis, J.P., Meyer, T.E., Shi, X., Siqueira, H.A., Siegfried, B.D., 2005. Identification, cloning and expression of a Cry1Ab cadherin receptor from European corn borer, *Ostrinia nubilalis* (Hübner) (Lepidoptera: Crambidae). Insect Biochem. Mol. Biol. 35, 33–40.

Forrester, N.W., Cahill, M., Bird, L.J., Layland, J.K., 1993. Management of pyrethroid and endosulfan resistance in *Helicoverpa armigera* (Lepidoptera: Noctuidae) in Australia. Bull. Entomol. Res. Suppl. Ser. 1, 1–132.

Gahan, L.J., Gould, F., Heckel, D.G., 2001. Identification of a gene associated with Bt resistance in *Heliothis virescens*. Science 293, 857–860.

Gahan, L.J., Gould, F., López Jr., J.D., Micinski, S., Heckel, D.G., 2007. A polymerase chain reaction screen of field populations of *Heliothis virescens* for a retrotransposon insertion conferring resistance to *Bacillus thuringiensis* toxin. J. Econ. Entomol. 100, 187–194.

Gahan, L.J., Pauchet, Y., Vogel, H., Heckel, D.G., 2010. An ABC transporter mutation is correlated with insect resistance to *Bacillus thuringiensis* Cry1Ac toxin. PLoS Genet. 6, e1001248.

Gassmann, A.J., Petzold-Maxwell, J.L., Keweshan, R.S., Dunbar, M.W., 2011. Field-evolved resistance to Bt maize by western corn rootworm. PLoS One 6, e22629.

Gassmann, A.J., Petzold-Maxwell, J.L., Keweshan, R.S., Dunbar, M.W., 2012. Western corn rootworm and Bt maize: challenges of pest resistance in the field. GM Crops Food 3, 235–244.

Génissel, A., Augustin, S., Courtin, C., Pilate, G., Lorme, P., Bourguet, D., 2003. Initial frequency of alleles conferring resistance to *Bacillus thuringiensis* poplar in a field population of *Chrysomela tremulae*. Proc. Biol. Sci. 270, 791–797.

Gill, S.S., Cowles, E.A., Francis, V., 1995. Identification, isolation, and cloning of a Bacillus thuringiensis CryIAc toxin-binding protein from the midgut of the lepidopteran insect Heliothis virescens. J. Biol. Chem. 270, 27277–27282.

Gill, M., Ellar, D., 2002. Transgenic *Drosophila* reveals a functional *in vivo* receptor for the *Bacillus thuringiensis* toxin Cry1Ac1. Insect Mol. Biol. 11, 619–625.

Gómez, I., Sánchez, J., Muñoz-Garay, C., Matus, V., Gill, S.S., Soberón, M., Bravo, A., 2014. *Bacillus thuringiensis* Cry1A toxins are versatile proteins with multiple modes of action: two distinct pre-pores are involved in toxicity. Biochem. J. 459, 383–396.

Gould, F., 1998. Sustainability of transgenic insecticidal cultivars: integrating pest genetics and ecology. Annu. Rev. Entomol. 43, 701–726.

Gould, F., Martinez-Ramirez, A., Anderson, A., Ferre, J., Silva, F.J., Moar, W.J., 1992. Broad-spectrum resistance to *Bacillus thuringiensis* toxins in *Heliothis virescens*. Proc. Natl. Acad. Sci. U.S.A. 89, 7986–7990.

Gould, F., Anderson, A., Reynolds, A., Bumgarner, L., Moar, W.J., 1995. Selection and genetic analysis of a *Heliothis virescens* (Lepidoptera: Noctuidae) strain with high levels of resistance to *Bacillus thuringiensis* toxins. J. Econ. Entomol. 88, 1545–1559.

Gould, F., Anderson, A., Jones, A., Sumerford, D., Heckel, D.G., Lopez, J., Micinski, S., Leonard, R., Laster, M., 1997. Initial frequency of alleles for resistance to *Bacillus thuringiensis* toxins in field populations of *Heliothis virescens*. Proc. Natl. Acad. Sci. U.S.A. 94, 3519–3523.

Griffitts, J.S., Aroian, R.V., 2005. Many roads to resistance: how invertebrates adapt to Bt toxins. Bioessays 27, 614–624.

Groeters, F.R., Tabashnik, B.E., 2000. Roles of selection intensity, major genes, and minor genes in evolution of insecticide resistance. J. Econ. Entomol. 93, 1580–1587.

Heckel, D.G., 1994. The complex genetic basis of resistance to *Bacillus thuringiensis* toxin in insects. Biocontrol Sci. Technol. 4, 405–417.

Heckel, D.G., 2012. Learning the ABCs of Bt: ABC transporters and insect resistance to *Bacillus thuringiensis* provide clues to a crucial step in toxin mode of action. Pestic. Biochem. Physiol. 104, 103–110.

Heckel, D.G., Gahan, L.C., Gould, F., Anderson, A., 1997. Identification of a linkage group with a major effect on resistance to *Bacillus thuringiensis* Cry1Ac endotoxin in the tobacco budworm (Lepidoptera: Noctuidae). J. Econ. Entomol. 90, 75–86.

Heckel, D.G., Gahan, L.J., Baxter, S.W., Zhao, J.Z., Shelton, A.M., Gould, F., Tabashnik, B.E., 2007. The diversity of Bt resistance genes in species of Lepidoptera. J. Invertebr. Pathol. 95, 192–197.

Herrero, S., Oppert, B., Ferré, J., 2001. Different mechanisms of resistance to *Bacillus thuringiensis* toxins in the indianmeal moth. Appl. Environ. Microbiol. 67, 1085–1089.

Herrero, S., Gechev, T., Bakker, P.L., Moar, W.J., de Maagd, R.A., 2005. *Bacillus thuringiensis* Cry1Ca-resistant *Spodoptera exigua* lacks expression of one of four aminopeptidase N genes. BMC Genomics 6, 96.

Hua, G., Zhang, R., Abdullah, M.A., Adang, M.J., 2008. *Anopheles gambiae* cadherin AgCad1 binds the Cry4Ba toxin of *Bacillus thuringiensis israelensis* and a fragment of AgCad1 synergizes toxicity. Biochemistry 47, 5101–5110.

Hua, G., Zhang, R., Bayyareddy, K., Adang, M.J., 2009. *Anopheles gambiae* alkaline phosphatase is a functional receptor of *Bacillus thuringiensis jegathesan* Cry11Ba toxin. Biochemistry 48, 9785–9793.

Hua, G., Zhang, Q., Zhang, R., Abdullah, A.M., Linser, P.J., Adang, M.J., 2013. AgCad2 cadherin in *Anopheles gambiae* larvae is a putative receptor of Cry11Ba toxin of *Bacillus thuringiensis* subsp. *jegathesan*. Insect Biochem. Mol. Biol. 43, 153–161.

Huang, F., 2006. Detection and monitoring of insect resistance to transgenic Bt crops. Insect Sci. 13, 73–84.

Huang, F., Leonard, B.R., Cook, D.R., Lee, D.R., Andow, D.A., Baldwin, J.L., Tindall, K.V., Wu, X., 2007a. Frequency of alleles conferring resistance to *Bacillus thuringiensis* maize in Louisiana populations of the southwestern corn borer. Entomol. Exp. Appl. 122, 53–58.

Huang, F.N., Leonard, B.R., Andow, D.A., 2007b. F_2 screen for resistance to a *Bacillus thuringiensis*-maize hybrid in the sugarcane borer (Lepidoptera: Crambidae). Bull. Entomol. Res. 97, 437–444.

Huang, F.N., Leonard, B.R., Moore, S.H., Cook, D.R., Baldwin, J., Tindall, K.V., Lee, D.R., 2008. Allele frequency of resistance to *Bacillus thuringiensis* Cry1Ab corn in Louisiana populations of sugarcane borer (Lepidoptera: Crambidae). J. Econ. Entomol. 101, 492–498.

Huang, F., Parker, R., Leonard, R., Yong, Y., Liu, J., 2009. Frequency of resistance alleles to *Bacillus thuringiensis*-corn in Texas populations of the sugarcane borer, *Diatraea saccharalis* (F.) (Lepidoptera: Crambidae). Crop. Prot. 28, 174–180.

Huang, F., Andow, D.A., Buschman, L.L., 2011. Success of the high-dose/refuge resistance management strategy after 15 years of Bt crop use in North America. Entomol. Exp. Appl. 140, 1–16.

Huang, F., Ghimire, M.N., Leonard, B.R., Daves, C., Levy, R., Baldwin, J., 2012. Extended monitoring of resistance to *Bacillus thuringiensis* Cry1Ab maize in *Diatraea saccharalis* (Lepidoptera: Crambidae). GM Crops Food 3, 245–254.

Ibrahim, M.A., Griko, N., Junker, M., Bulla, L.A., 2010. *Bacillus thuringiensis*: a genomics and proteomics perspective. Bioeng. Bugs 1, 31–50.

James, C., 2013. Global Status of Commercialized Biotech/GMCrops: 2013: ISAAA Brief No. 46. International Service for the Acquisition of Agri-biotech Applications (ISAAA), Ithaca, NY.

Janmaat, A.F., Myers, J., 2003. Rapid evolution and the cost of resistance to *Bacillus thuringiensis* in greenhouse populations of cabbage loopers, *Trichoplusia ni*. Proc. Biol. Sci. 270, 2263–2270.

Jiménez, A.I., Reyes, E.Z., Cancino-Rodezno, A., Bedoya-Pérez, L.P., Caballero-Flores, G.G., Muriel-Millan, L., Likitvivatanavong, F.S., Gill, S.S., Bravo, A., Soberón, M., 2012. *Aedes aegypti* alkaline phosphatase ALP1 is a functional receptor of *Bacillus thuringiensis* Cry4Ba and Cry11Aa toxins. Insect Biochem. Mol. Biol. 42, 683–689.

Jin, L., Yin, W., Zhang, L., Yang, Y., Tabashnik, B.E., Wu, Y., 2013. Dominant resistance to Bt cotton and minor cross-resistance to Bt toxin Cry2Ab in cotton bollworm from China. Evol. Appl. 6, 1222–1235.

Jurat-Fuentes, J.L., Adang, M.J., 2004. Characterization of a Cry1Ac-receptor alkaline phosphatase in susceptible and resistant *Heliothis virescens* larvae. Eur. J. Biochem. 271, 3127–3135.

Jurat-Fuentes, J.L., Adang, M.J., 2006. The *Heliothis virescens* cadherin protein expressed in *Drosophila* S2 cells functions as a receptor for *Bacillus thuringiensis* Cry1A but not Cry1Fa toxins. Biochemistry 45, 9688–9695.

Jurat-Fuentes, J.L., Gould, F.L., Adang, M.J., 2003. Dual resistance to *Bacillus thuringiensis* Cry1Ac and Cry2Aa toxins in *Heliothis virescens* suggests multiple mechanisms of resistance. Appl. Environ. Microbiol. 69, 5898–5906.

Jurat-Fuentes, J.L., Karumbaiah, L., Jakka, S.R.K., Ning, C.M., Liu, C.X., Wu, K.M., Jackson, J., Gould, F., Blanco, C., Portilla, M., Perera, O., Adang, M., 2011. Reduced levels of membrane-bound alkaline phosphatase are common to lepidopteran strains resistant to Cry toxins from *Bacillus thuringiensis*. PLoS One 6, e17606.

Khakame, S.K., Wang, X., Wu, Y., 2013. Baseline toxicity of metaflumizone and lack of cross resistance between indoxacarb and metaflumizone in diamondback moth (Lepidoptera: Plutellidae). J. Econ. Entomol. 106, 1423–1429.

Knight, P.J., Crickmore, N., Ellar, D.J., 1994. The receptor for *Bacillus thuringiensis* CrylA(c) delta-endotoxin in the brush border membrane of the lepidopteran *Manduca sexta* is aminopeptidase N. Mol. Microbiol. 11, 429–436.

Knight, P.J., Knowles, B.H., Ellar, D.J., 1995. Molecular cloning of an insect aminopeptidase-N that serves as a receptor for *Bacillus thuringiensis* CryIA(c) toxin. J. Biol. Chem. 270, 17765–17770.

Knowles, B.H., 1994. Mechanism of action of *Bacillus thuringiensis* insecticidal delta-endotoxins. Adv. Insect Physiol. 24, 275–308.

Knowles, B.H., Ellar, D.J., 1987. Colloid-osmotic lysis is a general feature of the mechanism of action of *Bacillus thuringiensis* δ-endotoxins with different insect specificity. Biochim. Biophys. Acta 924, 507–518.

Kruger, M., Van Rensburg, J.B.J., Van Den Berg, J., 2011. Resistance to Bt maize in *Busseola fusca* (Lepidoptera: Noctuidae) from Vaalharts, South Africa. Environ. Entomol. 40, 477–483.

Lee, M.K., Rajamohan, F., Gould, F., Dean, D.H., 1995. Resistance to *Bacillus thuringiensis* Cry1A delta-endotoxins in a laboratory-selected *Heliothis virescens* strain is related to receptor alteration. Appl. Environ. Microbiol. 61, 3836–3842.

Li, J., Carrol, J., Ellar, D.J., 1991. Crystal structure of insecticidal delta-endotoxin from *Bacillus thuringiensis* at 2.5 Å resolution. Nature 353, 815–821.

Liang, G.M., Wu, K.M., Yu, H.K., Li, K.K., Feng, X., Guo, Y.Y., 2008. Changes of inheritance mode and fitness in *Helicoverpa armigera* (Hübner) (Lepidoptera: Noctuidae) along with its resistance evolution to Cry1Ac toxin. J. Invertebr. Pathol. 97, 142–149.

Liu, Y.B., Tabashnik, B.E., 1997. Inheritance of resistance to the *Bacillus thuringiensis* toxin Cry1C in the diamondback moth. Appl. Environ. Microbiol. 63, 2218–2223.

Liu, Y.B., Tabashnik, B.E., Pusztai-Carey, M., 1996. Field-evolved resistance to *Bacillus thuringiensis* toxin Cry1C in diamondback moth (Lepidoptera: Plutellidae). J. Econ. Entomol. 89, 798–804.

Liu, F., Xu, Z., Zhu, Y.C., Huang, F., Wang, Y., Li, H.L., Li, H., Gao, C., Zhou, W., Shen, J., 2010. Evidence of field-evolved resistance to Cry1Ac-expressing Bt cotton in *Helicoverpa armigera* (Lepidoptera: Noctuidae) in northern China. Pest Manag. Sci. 66, 155–161.

Luo, K., Sangadala, S., Masson, L., Mazza, A., Brousseau, R., Adang, M.J., 1997. The Heliothis virescens 170-kDa aminopeptidase functions as 'Receptor A' by mediating specific *Bacillus thuringiensis* Cry1A δ-endotoxin binding and pore formation. Insect Biochem. Mol. Biol. 27, 735–743.

Luttrell, R.G., Ali, I., Allen, K.C., Young III, S.Y., Szalanski, A., Williams, K., Lorenz, G., Parker Jr., C.D., Blanco, C., 2004. Resistance to Bt in Arkansas populations of cotton bollworm. In: Richter, D.A. (Ed.), Proceedings, 2004 Beltwide Cotton Conferences, 5–9 January 2004, San Antonio, TX. National Cotton Council of America, Memphis, TN, pp. 1373–1383.

Mahon, R.J., Olsen, K.M., Downes, S., Addison, S., 2007. Frequency of alleles conferring resistance to the Bt toxins Cry1Ac and Cry2Ab in Australian populations of *Helicoverpa armigera*. J. Econ. Entomol. 100, 1844–1853.

Mahon, R.J., Olsen, K.M., Downes, S., 2008. Isolations of Cry2Ab resistance in Australian populations of *Helicoverpa armigera* (Lepidoptera: Noctuidae) are allelic. J. Econ. Entomol. 101, 909–914.

Mahon, R.J., Downes, S., James, W., Parker, T., 2010. Why do F_1 screens estimate higher frequencies of Cry2Ab resistance in *Helicoverpa armigera* (Lepidoptera: Noctuidae) than do F_2 screens? J. Econ. Entomol. 103, 472–481.

Mahon, R.J., Downes, S.J., James, B., 2012. Vip3A resistance alleles exist at high levels in Australian targets before release of cotton expressing this toxin. PLoS One 7, e39192.

Martins, E.S., Monnerat, R.G., Queiroz, P.R., Dumas, V.F., Braz, S.V., de Souza Aguiar, R.W., Gomes, A.C., Sanchez, J., Bravo, A., Ribeiro, B.M., 2010. Midgut GPI-anchored proteins with alkaline phosphatase activity from the cotton boll weevil (*Anthonomus grandis*) are putative receptors for the Cry1B protein of *Bacillus thuringiensis*. Insect Biochem. Mol. Biol. 40, 138–145.

McKenzie, J.A., Batterham, P., 1994. The genetic, molecular and phenotypic consequences of selection for insecticide resistance. Trends Ecol. Evol. 9, 166–169.

McNall, R.J., Adang, M.J., 2003. Identification of novel *Bacillus thuringiensis* Cry1Ac binding proteins in *Manduca sexta* midgut through proteomic analysis. Insect Biochem. Mol. Biol. 33, 999–1010.

Moar, W.J., Pusztai-Carey, M., Van Faassen, H., Bosch, D., Frutos, R., Rang, C., Luo, K., Adang, M.J., 1995. Development of *Bacillus thuringiensis* CryIC resistance by *Spodoptera exigua* (Hübner) (Lepidoptera: Noctuidae). Appl. Environ. Microbiol. 61, 2086–2092.

Moar, W., Roush, R., Shelton, A., Ferré, J., MacIntosh, S., Leonard, B.R., Abel, C., 2008. Field-evolved resistance to Bt toxins. Nat. Biotechnol. 26, 1072–1074.

Morin, S., Biggs, R.W., Sisterson, M.S., Shriver, L., Ellers-Kirk, C., Higginson, D., Holley, D., Gahan, L.J., Heckel, D.G., Carrière, Y., 2003. Three cadherin alleles associated with resistance to *Bacillus thuringiensis* in pink bollworm. Proc. Natl. Acad. Sci. U.S.A. 100, 5004–5009.

Nagamatsu, Y., Toda, S., Koike, T., Miyoshi, Y., Shigematsu, S., Kogure, M., 1998. Cloning, sequencing, and expression of the *Bombyx mori* receptor for *Bacillus thuringiensis* insecticidal CryIA(a) toxin. Biosci. Biotechnol. Biochem. 62, 727–734.

Ning, C.M., Wu, K.M., Liu, C.X., Gao, Y.L., Jurat-Fuentes, J.L., Gao, J.L., 2010. Characterization of a Cry1Ac toxin-binding alkaline phosphatase in the midgut from *Helicoverpa armigera* (Hübner) larvae. J. Insect Physiol. 56, 666–672.
Oltean, D.I., Pullikuth, A.K., Lee, H.-K., Gill, S.S., 1999. Partial purification and characterization of Bacillus thuringiensis Cry1A toxin receptor A from Heliothis virescens and cloning of the corresponding cDNA Appl. Environ. Microbiol. 65, 4760–4766.
Oppert, B., 1999. Protease interactions with *Bacillus thuringiensis* insecticidal toxins. Arch. Insect Biochem. Physiol. 42, 1–12.
Oppert, B., Kramer, K.J., Beeman, R.W., Johnson, D., McGaughey, W.H., 1997. Proteinase-mediated insect resistance to *Bacillus thuringiensis* toxins. J. Biol. Chem. 272, 23473–23476.
Pardo-López, L., Soberón, M., Bravo, A., 2013. *Bacillus thuringiensis* insecticidal three-domain Cry toxins: mode of action, insect resistance and consequences for crop protection. FEMS Microbiol. Rev. 37, 3–22.
Perera, O.P., Willis, J.D., Adang, M.J., Jurat-Fuentes, J.L., 2009. Cloning and characterization of the Cry1Ac-binding alkaline phosphatase (HvALP) from *Heliothis virescens*. Insect Biochem. Mol. Biol. 39, 294–302.
Perlak, F.J., Deaton, R.W., Armstrong, T.A., Fuchs, R.L., Sims, S.R., Greenplate, J.T., Fischhoff, D.A., 1990. Insect resistant cotton plants. Nat. Biotechnol. 8, 939–943.
Petzold-Maxwell, J.L., Cibils-Stewart, X., Wade French, B., Gassmann, A.J., 2012. Adaptation by western corn rootworm (Coleoptera: Chrysomelidae) to Bt maize: inheritance, fitness costs, and feeding preference. J. Econ. Entomol. 105, 1407–1418.
Pigott, C.R., Ellar, D.J., 2007. Role of receptors in *Bacillus thuringiensis* crystal toxin activity. Microbiol. Mol. Biol. Rev. 71, 255–281.
Rajagopal, R., Sivakumar, S., Agrawal, N., Malhotra, P., Bhatnagar, R.K., 2002. Silencing of midgut aminopeptidase N of *Spodoptera litura* by double stranded RNA establishes its role as *Bacillus thuringiensis* toxin receptor. J. Biol. Chem. 277, 46849–46851.
Rajagopal, R., Arora, N., Sivakumar, S., Rao, N.G., Nimbalkar, S.A., Bhatnagar, R.K., 2009. Resistance of *Helicoverpa armigera* to Cry1Ac toxin from *Bacillus thuringiensis* is due to improper processing of the protoxin. Biochem. J. 419, 309–316.
Ren, X.L., Chen, R.R., Zhang, Y., Ma, Y., Cui, J.J., Han, Z.J., Mu, L.L., Li, G.Q., 2013. A *Spodoptera exigua* cadherin serves as a putative receptor for *Bacillus thuringiensis* Cry1Ca toxin and shows differential enhancement of Cry1Ca and Cry1Ac toxicity. Appl. Environ. Microbiol. 79, 5576–5583.
Roush, R.T., 1998. Two-toxin strategies for management of insecticidal transgenic crops: can pyramiding succeed where pesticide mixtures have not? Philos. Trans. R. Soc. B 353, 1777–1786.
Roush, R.T., McKenzie, J.A., 1987. Ecological genetics of insecticide and acaricide resistance. Annu. Rev. Entomol. 32, 361–380.
Roush, R.T., Miller, G.L., 1986. Considerations for design of insecticide resistance monitoring. J. Econ. Entomol. 79, 293–298.
Sanahuja, G., Banakar, R., Twyman, R.M., Capell, T., Christou, P., 2011. *Bacillus thuringiensis*: a century of research, development and commercial applications. Plant Biotechnol. J. 9, 283–300.
Sawicki, R., 1987. Definition, detection and documentation of insecticide resistance. In: Ford, M.G., Holloman, D.W., Khambay, B.P.S., Sawicki, R.M. (Eds.), Combating Resistance to Xenobiotics: Biological and Chemical Approaches. Ellis Horwood Ltd., Chichester, UK, pp. 105–117
Sayed, A., Nekl, E.R., Siqueira, H.A., Wang, H.C., ffrench-Constant, R.H., Bagley, M., Siegfried, B.D., 2007. A novel cadherin-like gene from western corn rootworm, *Diabrotica virgifera virgifera* (Coleoptera: Chrysomelidae), larval midgut tissue. Insect Mol. Biol. 16, 591–600.

Schnepf, H.E., Whiteley, H.R., 1981. Cloning and expression of the *Bacillus thuringiensis* crystal protein gene in *Escherichia coli*. Proc. Natl. Acad. Sci. U.S.A. 78, 2893–2897.
Schnepf, E., Crickmore, N., Van Rie, J., Lereclus, D., Feitelson, J., Zeigler, D.R., Dean, D.H., 1998. *Bacillus thuringiensis* and its pesticidal crystal proteins. Microbiol. Mol. Biol. Rev. 62, 775–806.
Schwartz, J.L., Potvin, L., Chen, X.J., Brousseau, R., Laprade, R., Dean, D.H., 1997. Single-site mutations in the conserved alternating arginine region affect ionic channels formed by CryIAa, a *Bacillus thuringiensis* toxin. Appl. Environ. Microbiol. 63, 3978–3984.
Siegfried, B.D., Rangasamy, M., Wang, H., Spencer, T., Haridas, C.V., Tenhumberg, B., Sumerford, D.V., Storer, N.P., 2014. Estimating the frequency of Cry1F resistance in field populations of the European corn borer (Lepidoptera: Crambidae). Pest Manag. Sci. 70, 725–733.
Sivakumar, S., Rajagopal, R., Venkatesh, G.R., Srivastava, A., Bhatnagar, R.K., 2007. Knockdown of aminopeptidase-N from *Helicoverpa armigera* larvae and in transfected Sf21 cells by RNA interference reveals its functional interaction with *Bacillus thuringiensis* insecticidal protein Cry1Ac. J. Biol. Chem. 282, 7312–7319.
Sivasupramaniam, S., Head, G.P., English, L., Li, Y.J., Vaughn, T.T., 2007. A global approach to resistance monitoring. J. Invertebr. Pathol. 95, 224–226.
Soberón, M., Gill, S.S., Bravo, A., 2009. Signaling versus punching hole: how do *Bacillus thuringiensis* toxins kill insect midgut cells? Cell. Mol. Life Sci. 66, 1337–1349.
Stodola, T.J., Andow, D.A., Hyden, A.R., Hinton, J.L., Roark, J.J., Buschman, L.L., Porter, P., Cronholm, G.B., 2006. Frequency of resistance to *Bacillus thuringiensis* toxin Cry1Ab in southern United States corn belt population of European corn borer (Lepidoptera: Crambidae). J. Econ. Entomol. 99, 502–507.
Storer, N.P., Babcock, J.M., Schlenz, M., Meade, T., Thompson, G.D., Bing, J.W., Huckaba, R.M., 2010. Discovery and characterization of field resistance to Bt maize: *Spodoptera frugiperda* (Lepidoptera: Noctuidae) in Puerto Rico. J. Econ. Entomol. 103, 1031–1038.
Storer, N.P., Kubiszak, M.E., King, J.E., Thompson, G.D., Santos, A.C., 2012. Status of resistance to Bt maize in *Spodoptera frugiperda*: lessons from Puerto Rico. J. Invertebr. Pathol. 110, 294–300.
Sumerford, D.V., Head, G.P., Shelton, A., Greenplate, J., Moar, W., 2013. Field-evolved resistance: assessing the problem and moving forward. J. Econ. Entomol. 106, 1525–1534.
Tabashnik, B.E., 1994. Evolution of resistance to *Bacillus thuringiensis*. Annu. Rev. Entomol. 39, 47–79.
Tabashnik, B.E., Cushing, N.L., Finson, N., Johson, M.W., 1990. Field development of resistance to *Bacillus thuringiensis* in diamondback moth (Lepidoptera: Plutellidae). J. Econ. Entomol. 83, 1671–1676.
Tabashnik, B.E., Finson, N., Johnson, M.W., Moar, W.J., 1993. Resistance to toxins from *Bacillus thuringiensis* subsp. *kurstaki* causes minimal cross-resistance to *B. thuringiensis* subsp. *aizawai* in the diamondback moth (Lepidoptera: Plutellidae). Appl. Environ. Microbiol. 59, 1332–1335.
Tabashnik, B.E., Liu, Y.B., Dennehy, T.J., Sims, M.A., Sisterson, M.S., Biggs, R.W., Carrière, Y., 2002. Inheritance of resistance to Bt toxin Cry1Ac in a field-derived strain of pink bollworm (Lepidoptera: Gelechiidae). J. Econ. Entomol. 95, 1018–1026.
Tabashnik, B.E., Gould, F., Carrière, Y., 2004. Delaying evolution of insect resistance to transgenic crops by decreasing dominance and heritability. J. Evol. Biol. 17, 904–912.
Tabashnik, B.E., Fabrick, J.A., Henderson, S., Biggs, R.W., Yafuso, C.M., Nyboer, M.E., Manhardt, N.M., Coughlin, L.A., Sollome, J., Carrière, Y., Dennehy, T.J., Morin, S., 2006. DNA screening reveals pink bollworm resistance to Bt cotton remains rare after a decade of exposure. J. Econ. Entomol. 99, 1525–1530.

Tabashnik, B.E., Gassmann, A.J., Crowder, D.W., Carrière, Y., 2008. Insect resistance to Bt crops: evidence versus theory. Nat. Biotechnol. 26, 199–202.
Tabashnik, B.E., Van Rensburg, J.B.J., Carrière, Y., 2009a. Field-evolved insect resistance to Bt crops: definition, theory, and data. J. Econ. Entomol. 102, 2011–2025.
Tabashnik, B.E., Unnithan, G.C., Masson, L., Crowder, D.W., Li, X., Carrière, Y., 2009b. Asymmetrical cross-resistance between *Bacillus thuringiensis* toxins Cry1Ac and Cry2Ab in pink bollworm. Proc. Natl. Acad. Sci. U.S.A. 106, 11889–11894.
Tabashnik, B.E., Sisterson, M.S., Ellsworth, P.C., Dennehy, T.J., Antilla, L., Liesner, L., Whitlow, M., Staten, R.T., Fabrick, J.A., Unnithan, G.C., Yelich, A.J., Ellers-Kirk, C., Harpold, V.S., Li, X., Carriere, Y., 2010. Suppressing resistance to Bt cotton with sterile insect releases. Nat. Biotechnol. 28, 1304–1307.
Tabashnik, B.E., Huang, F.N., Ghimire, M.N., Leonard, B.R., Siegfried, B.D., Rangasamy, M., Yang, Y.J., Wu, Y.D., Gahan, L.J., Heckel, D.G., Bravo, A., Soberón, M., 2011. Efficacy of genetically modified Bt toxins against insects with different genetic mechanisms of resistance. Nat. Biotechnol. 29, 1128–1131.
Tabashnik, B.E., Brévault, T., Carrière, Y., 2013. Insect resistance to Bt crops: lessons from the first billion acres. Nat. Biotechnol. 31, 510–521.
Tabashnik, B.E., Mota-Sanchez, D., Whalon, M.E., Hollingworth, R.M., Carrière, Y., 2014. Defining terms for proactive management of resistance to Bt crops and pesticides. J. Econ. Entomol. 107, 496–507.
Tanaka, S., Miyamoto, K., Noda, H., Jurat-Fuentes, J.L., Yoshizawa, Y., Endo, H., Sato, R., 2013. The ATP-binding cassette transporter subfamily C member 2 in *Bombyx mori* larvae is a functional receptor for Cry toxins from *Bacillus thuringiensis*. FEBS J. 280, 1782–1794.
Tiewsiri, K., Wang, P., 2011. Differential alteration of two aminopeptidases N associated with resistance to *Bacillus thuringiensis* toxin Cry1Ac in cabbage looper. Proc. Natl. Acad. Sci. U.S.A. 108, 14037–14042.
Vachon, V., Laprade, R., Schwartz, J.L., 2012. Current models of the mode of action of *Bacillus thuringiensis* insecticidal crystal proteins: a critical review. J. Invertebr. Pathol. 111, 1–12.
Vadlamudi, R.K., Weber, E., Ji, I., Ji, T.H., Bulla Jr., L.A., 1995. Cloning and expression of a receptor for an insecticidal toxin of *Bacillus thuringiensis*. J. Biol. Chem. 270, 5490–5494.
Van Rensburg, J.B.J., 2007. First report of field resistance by stem borer, *Busseola fusca* (Fuller) to Bt-transgenic maize. S. Afr. J. Plant Soil 24, 147–151.
Wan, P., Huang, Y., Wu, H., Huang, M., Cong, S., Tabashnik, B.E., Wu, K., 2012. Increased frequency of pink bollworm resistance to Bt toxin Cry1Ac in China. PLoS One 7, e29975.
Wang, X., Li, X., Shen, A., Wu, Y., 2010. Baseline susceptibility of the diamondback moth (Lepidoptera: Plutellidae) to chlorantraniliprole in China. J. Econ. Entomol. 103, 843–848.
Wenes, A.L., Bourguet, D., Andow, D.A., Courtin, C., Carré, G., Lorme, P., Sanchez, L., Augustin, S., 2006. Frequency and fitness cost of resistance to *Bacillus thuringiensis* in *Chrysomela tremulae* (Coleoptera: Chrysomelidae). Heredity 97, 127–134.
WHO, 1957. Expert Committee on Malaria, Seventh Report: WHO Technical Report Series No. 125. World Health Organization, Geneva, Switzerland.
Xie, R., Zhuang, M., Ross, L.S., Gomez, I., Oltean, D.I., Bravo, A., Soberon, M., Gill, S.S., 2005. Single amino acid mutations in the cadherin receptor from *Heliothis virescens* affect its toxin binding ability to Cry1A toxins. J. Biol. Chem. 280, 8416–8425.
Xu, X., Yu, L., Wu, Y., 2005. Disruption of a cadherin gene associated with resistance to Cry1Ac δ-endotoxin of *Bacillus thuringiensis* in *Helicoverpa armigera*. Appl. Environ. Microbiol. 71, 948–954.

Xu, Z., Liu, F., Chen, J., Huang, F., Andow, D.A., Wang, Y., Zhu, Y.C., Shen, J., 2009. Using an F_2 screen to monitor frequency of resistance alleles to Bt cotton in field populations of *Helicoverpa armigera* (Hübner) (Lepidoptera: Noctuidae). Pest Manag. Sci. 65, 391–397.

Yang, Y.J., Chen, H., Wu, S., Xu, X., Wu, Y., 2006. Identification and molecular detection of a deletion mutation responsible for a truncated cadherin of *Helicoverpa armigera*. Insect Biochem. Mol. Biol. 36, 735–740.

Yang, Y.J., Chen, H., Wu, Y., Yang, Y.-H., Wu, S., 2007. Mutated cadherin alleles from a field population of *Helicoverpa armigera* confer resistance to *Bacillus thuringiensis* toxin Cry1Ac. Appl. Environ. Microbiol. 73, 6939–6944.

Yang, Y.H., Yang, Y.J., Gao, W.Y., Guo, J.J., Wu, Y.H., Wu, Y.D., 2009. Introgression of a disrupted cadherin gene enables susceptible *Helicoverpa armigera* to obtain resistance to *Bacillus thuringiensis* toxin Cry1Ac. Bull. Entomol. Res. 99, 175–181.

Yang, Y., Zhu, Y.C., Ottea, J., Husseneder, C., Leonard, B.R., Abel, C., Luttrell, R., Huang, F., 2011. Down regulation of a gene for cadherin, but not alkaline phosphatase, associated with Cry1Ab resistance in the sugarcane borer *Diatraea saccharalis*. PLoS One 6, e25783.

Yue, B., Huang, F., Leonard, B.R., Moore, S., Parker, R., Andow, D.A., Cook, D., Emfinger, K., Lee, D.R., 2008. Verifying an F_1 screen for identification and quantification of rare *Bacillus thuringiensis* resistance alleles in field populations of the sugarcane borer, *Diatraea saccharalis*. Entomol. Exp. Appl. 129, 172–180.

Zhang, X., Candas, M., Griko, N.B., Rose-Young, L., Bulla, L.A., 2005. Cytotoxicity of *Bacillus thuringiensis* Cry1Ab toxin depends on specific binding of the toxin to the cadherin receptor BT-R1 expressed in insect cells. Cell Death Differ. 12, 1407–1416.

Zhang, X., Candas, M., Griko, N.B., Taussig, R., Bulla, L.A., 2006. A mechanism of cell death involving an adenylyl cyclase/PKA signaling pathway is induced by the Cry1Ab toxin of *Bacillus thuringiensis*. Proc. Natl. Acad. Sci. U.S.A. 103, 9897–9902.

Zhang, S., Cheng, H., Gao, Y., Wang, G., Liang, G., Wu, K., 2009. Mutation of an aminopeptidase N gene is associated with *Helicoverpa armigera* resistance to *Bacillus thuringiensis* Cry1Ac toxin. Insect Biochem. Mol. Biol. 39, 421–429.

Zhang, H., Yin, W., Zhao, J., Jin, L., Yang, Y., Wu, S., Tabashnik, B.E., Wu, Y., 2011. Early warning of cotton bollworm resistance associated with intensive planting of Bt cotton in China. PLoS One 6, e22874.

Zhang, H., Tian, W., Zhao, J., Jin, L., Yang, J., Liu, C., Yang, Y., Wu, S., Wu, K., Cui, J., Tabashnik, B.E., Wu, Y., 2012a. Diverse genetic basis of field-evolved resistance to Bt cotton in cotton bollworm from China. Proc. Natl. Acad. Sci. U.S.A. 109, 10275–10280.

Zhang, H., Wu, S., Yang, Y., Tabashnik, B.E., Wu, Y., 2012b. Non-recessive Bt toxin resistance conferred by an intracellular cadherin mutation in field-selected populations of cotton bollworm. PLoS One 7, e53418.

Zhang, H., Tang, M., Yang, F., Yang, Y., Wu, Y., 2013. DNA-based screening for an intracellular cadherin mutation conferring non-recessive Cry1Ac resistance in field populations of *Helicoverpa armigera*. Pestic. Biochem. Physiol. 107, 148–152.

Zhao, J.Z., Cao, J., Li, Y., Collins, H.L., Roush, R.T., Earle, E.D., Shelton, A.M., 2003. Transgenic plants expressing two *Bacillus thuringiensis* toxins delay insect resistance evolution. Nat. Biotechnol. 21, 1493–1497.

Zhao, J., Jin, L., Yang, Y., Wu, Y., 2010. Diverse cadherin mutations conferring resistance to *Bacillus thuringiensis* toxin Cry1Ac in *Helicoverpa armigera*. Insect Biochem. Mol. Biol. 40, 113–118.

Zúñiga-Navarrete, F., Gómez, I., Peña, G., Bravo, A., Soberón, M., 2013. A *Tenebrio molitor* GPI-anchored alkaline phosphatase is involved in binding of *Bacillus thuringiensis* Cry3Aa to brush border membrane vesicles. Peptides 41, 81–86.

CHAPTER SEVEN

Photorhabdus Toxins

Richard H. ffrench-Constant, Andrea J. Dowling
Biosciences, University of Exeter, Cornwall, United Kingdom

Contents

Advances in Insect Physiology, Volume 47
ISSN 0065-2806
http://dx.doi.org/10.1016/B978-0-12-800197-4.00007-5

Abstract

The last 10 years has seen an explosion in our knowledge of the diversity of toxins produced by the insect pathogenic bacteria *Photorhabdus* and *Xenorhabdus*. Here we review new data on the structure, mode of action and biophysics of *Photorhabdus* toxins and place them into context with similar toxins made in other bacteria such as those of the genus *Yersinia* and *Serratia*. We look in detail at structure–function studies of the Toxin complexes or 'Tcs' and suggest that these might constitute a new mode of bacterial secretion, here termed the Toxin complex secretion system or 'TC-SS'. We examine current data on the Makes Caterpillars Floppy or 'Mcf' toxins both from insect pathogenic bacteria and also the related Mcf-like Fit toxin from *Pseudomonas*. We also review data on the *Photorhabdus* Virulence Cassettes or 'PVC's' and their equivalent phage-like secretion systems in other bacteria. Again these appear to be a totally novel secretion system, termed PVC-SS, derived from tailed myophages. Finally, we review the sheer diversity of other candidate toxins identified from *Photorhabdus* and *Xenorhabdus* genomes, such as PaTox and Photox, and from simple gain-of-function screens devised to identify novel effectors when expressed in recombinant *Escherichia coli*, termed 'Rapid Virulence Annotation' or 'RVA'. Taken together, available data support the early contention that *Photorhabdus* genomes do indeed encode more candidate toxins than those of other bacteria and suggest that this unique genus demands more time from serious toxinologists than has been devoted to date.

1. *PHOTORHABDUS* LIFESTYLES, RELATIVES AND GENOMES

Before we can fully appreciate the stunning diversity of toxins made by *Photorhabdus* bacteria, and indeed their potential utility, we first need to understand their lifestyles and how they are related to other entomopathogenic bacteria such as those from the related genus *Xenorhabdus*.

1.1. The life cycle of *Photorhabdus temperata* and *Photorhabdus luminescens*

Photorhabdus are γ-proteobacteria that live in a bizarre 'symbiosis of pathogens' with nematodes of the genus *Heterhorhabditis* and their life cycle has been extensively reviewed previously (Forst and Nealson, 1996; Forst et al., 1997; Waterfield et al., 2009). While the bacteria appear to act as benign symbionts of the nematode gut, once regurgitated into the open blood system (hemocoel) of their insect hosts they switch to becoming potent pathogens that rapidly overcome the insect immune system and then kill the insect (Forst et al., 1997). The insect cadaver is then bioconverted into a realm of future bacteria and their nematode partners in a life cycle involving exquisitely precise interactions with nematodes of different stages (Somvanshi et al., 2012). The transmission of *Photorhabdus* to the infective juvenile nematode progeny requires bacterial colonization of maternal rectal glands and hatching of the progeny *within* the mother, in a process termed *endotokia matricida* (Somvanshi et al., 2012). This is a stunning piece of natural history that effectively reclones the bacterium prior to its repackaging in the infective juvenile. However for the purposes of this review, which focuses on insecticidal toxins, it is really only the biology of insect infection that demands our immediate attention.

The infective juvenile nematodes are the vectors that actively seek out and penetrate their insect hosts. They then individually regurgitate 50–250 bacterial cells from their guts directly into the insect hemocoel (Ciche and Ensign, 2003). Unfortunately from then on, the precise path of the insect infection and the stages at which different toxins are expressed is still largely a 'black box' (ffrench-Constant et al., 2003). Part of this problem results from the difficulty of relating bacteria grown in aerobic, nutrient-rich broth in the laboratory to the different stages of insect infection and eventual death. In fact, only a very few studies have attempted to correlate the production of putative anti-insect virulence factors with their production in infected insects themselves (Daborn et al., 2001). In reality, therefore, we remain with a series of isolated observations about different classes of toxins and a central failure to relate even the most detailed biochemistry to any aspect of infection itself (ffrench-Constant et al., 2003). While this continues to be a depressing concept for the biologist, we have been lucky enough to be able to identify and isolate a plethora of insecticidal toxins from *Photorhabdus* and these will, therefore, be the focus of this review.

1.2. The unusual life cycle of *Photorhabdus asymbiotica*

Most attempts to develop a family tree for *Photorhabdus* divide the genus into three species groups *P. temperata*, *P. luminescens* and *P. asymbiotica* (Peat et al., 2010). The strictly insect-associated life cycle briefly described above is undertaken by bacteria from the two groups termed *P. temperata* and *P. luminescens*. These two groups seem to lack plasmids and have been widely used with their nematode vectors in biological control. However, the third group *P. asymbiotica*, which carry plasmids (Wilkinson et al., 2010), seems to have evolved a recent and more sinister side to its life cycle. As its specific name suggests, *P. asymbiotica* was originally named as such as it did not appear to have a nematode vector, it was therefore literally termed 'a-symbiotic'.

However over time, a number of cases of human infection by *P. asymbiotica* have been reported both from North America and from Australia (Gerrard et al., 2004), and a vector nematode was eventually discovered after one particular Australian patient had been digging in the soil with his bare hands (Gerrard et al., 2006). While the number of cases of *P. asymbitioca* infection can hardly be described as an epidemic, there are several factors we should probably take into account in relation to both the evolution of human virulence and also the continued use of these bacteria in biocontrol. (1) At this stage, we cannot determine if the infection of humans is due solely to inappropriate infection of patients by infective juvenile nematodes or if there are accompanying changes in virulence factor expression, for example, gains or losses of specific genes and/or changes in their genetic regulation. (2) Because of the relative awareness/watchfulness of clinicians in America and Australia, we cannot be sure that *P. asymbiotica* infections are also occurring at low frequency in other countries and that they are simply not being diagnosed as such. (3) What is clear is that all bacteria associated with human infection seem to carry one or more plasmids and, so far, those used in crop protection do not. So it would seem prudent to avoid the use of plasmid-carrying bacteria in crop protection.

Why is all this relevant to a comprehensive review on insecticidal toxins? First, the relationship between bacteria–invertebrate and bacteria–vertebrate interactions is clearly important from the point of view of disease emergence, specifically, when the bacterial vector starts to change its host specificity. Second, if any of the insecticidal toxins discovered from the system are seen to have any potential utility in crop protection, it may be useful to have some understanding of their role in virulence in order to make informed decisions about their likely mode of action and safety.

1.3. *Xenorhabdus* and comparative genomics

In a remarkable example of parallel evolution, *Xenorhabdus* bacteria have a very similar lifestyle to *Photorhabdus*. *Xenorhabdus* are also vectored by symbiotic entomopathogenic nematodes, this time from the family Steinernematidae. The closer relationship between *Xenorhabdus* and *Photorhabdus* than between either of their nematode hosts may suggest that their recent common ancestor may have been able to colonize both *Steinernema* and *Heterorhabditis* nematodes (Chaston et al., 2011). *Xenorhabdus* bacteria are also subdivided into several species groups, specifically *Xenorhabdus nematophila* and *Xenorhabdus bovienii*. Given the relatively small size of both *Photorhabdus* and *Xenorhabdus* genomes

(4–5 Mb) and recent rapid advances in next-generation sequencing, we are beginning to see a steady stream of finished and draft genome sequences appearing in the genome databases. As early as the year 2000, a simple sample sequence (2000 random reads) from the genome of *P. luminescens* strain W14 began to demonstrate that these bacteria were a potential gold mine for novel bacterial effectors (ffrench-Constant et al., 2000). The first full genome to be sequenced was that of *P. luminescens* strain TT01 which was 5,688,987 bp in length and contained 4839 predicted protein-coding genes (Duchaud et al., 2003). At the time in 2003, the authors claimed that the TT01 genome encoded more predicted toxins than any other bacterial genomes sequenced. Some 10 years later in 2014, this claim may still hold true as *Photorhabdus* genomes are literally stuffed with genes encoding proven or putative toxins. Now we also have a finished genome for the North American *P. asymbiotica* ATCC43949 (Wilkinson et al., 2009) and a draft genome of the Australian *P. asymbiotica* strain Kingscliff (Wilkinson et al., 2010), both of which can infect both insects and man. The genome of *P. asymbiotica* will hopefully, therefore, not only be a source of novel insecticidal toxins but will help us understand why the vector nematode sometimes infects man rather than insects and indeed, having done so, how the *P. asymbiotica* bacteria themselves persist in the face of the human immune system. Several genomes for the closely related *Xenorhabdus* bacteria are also now available and the comparative genomics of different *Photorhabdus* and *Xenorhabdus* stains has proved to be a powerful tool in the identification of novel insecticidal toxins. The type strain of *X. nematophila* ATCC 19061, a North American strain isolated from the nematode symbiont *S. carpocapsae*, has been fully sequenced (Chaston et al., 2011), and more recently, a draft genome of a French strain *X. nematophila* strain F1 has also been completed (Lanois et al., 2013). For *X. bovienii*, we also now have the complete sequence of strain SS-2004 (Chaston et al., 2011). The utility of all these genome sequences to the hunt for *Photorhabdus* toxins will become clear when we examine how they can be mined both *in silico* and biologically for novel effectors.

2. THE TOXIN COMPLEXES

Much of the original interest in *Photorhabdus* as a source of novel insecticidal toxins was centred on the discovery that the cell-free supernatant of certain strains was orally active to the model insect, Tobacco (tomato) hornworm (hawkmoth), *Manduca sexta* (Bowen and Ensign, 1998). It is worth noting at the outset that the delivery of the *Photorhabdus* bacteria directly into the insect's hemocoel does not make it obvious as to why *any Photorhabdus* toxins are indeed active on the insect midgut, unless they are somehow able to exert their effects from either side of the gut (gut lumen or insect hemocoel). This section, therefore, describes the discovery, purification, cloning and structure of the toxins conferring this oral activity, namely the Toxin complexes or Tcs, and finally attempts reconcile the mode of action of these important toxins with their potential role in infection.

2.1. Tc discovery, gene cloning and ABC nomenclature

The oral activity of the cell-free bacterial supernatant of *Photorhabdus* was first demonstrated in *P. luminescens* strain W14 (Bowen and Ensign, 1998). The toxicity was associated with large protein complexes with an estimated molecular weight of >1,000,000 which migrate as single complexes on a native gel but could be resolved into a range of different polypeptides (30–200 kDa) under denaturing conditions (Bowen and Ensign, 1998). Four such million molecular weight complexes were isolated from strain W14 and were termed Toxin complex (Tc) A, B, C and D or TcA, TcB, TcC and TcD (Bowen et al., 1998). By raising a polyclonal antibody against a mixture of all four complexes, the loci encoding each complex were identified from an expression library and the corresponding loci were termed the *toxin complex* (*tc*) encoding loci *tca*, *tcb*, *tcc* and *tcd*, encoding complexes A, B, C and D, respectively (Bowen et al., 1998). At the time, this seemed like a remarkably convenient and appropriate nomenclature, with four genomic *tc* loci encoding four final Tc complexes. However, this nomenclature was to be undone by two further findings. First, each *tc* locus was usually split into two or three separate open reading frames (Bowen et al., 1998). Thus, individual open reading frames in the *tca* locus had to be termed *tcaA*, *tcaB* and *tcaC*. Second, it rapidly became clear from the growing number of genome sequences that there was no limit to the number of potential *tc* loci in any one genome and that we were simply going to run out of enough letters and/or numbers to name each locus sensibly.

The growing confusion around the rapidly growing *tc* nomenclature came hand in hand with a more detailed understanding of what minimum set of genes was required to reconstitute recombinant oral toxicity in *Escherichia coli*. Specifically, expression of all three A, B and C subunit-encoding genes (e.g. *tcaA*, *tcaB* and *tcaC*) was necessary to reconstitute recombinant oral toxicity (Waterfield et al., 2001a; Yang and Waterfield, 2013). Moreover, for reasons we will return to later, expression of *tcaB* and *tcaC* (encoding B and C) in the same bacterial cytoplasm was also essential (Waterfield et al., 2001a), whereas the *tcaA* open reading frame (encoding A) could be expressed alone and then mixed with the products of *tcaB*/*tcaC* (B and C) to reconstitute full oral toxicity (Waterfield et al., 2001a). To put it most simply, a mixture of A, B and C toxin subunits was required and, for some mysterious reason, B and C needed to be made together in the same bacterial cytoplasm. This 'mechanistic' explanation of the A, B and C subunits also provided an instant cure to the growing nomenclature problem as the massive list of gene names derived for different *tc* loci within the same genome or the genomes of different bacteria, such as *Xenorhabdus*, *Serratia entomophila*, *Yersinia entomophaga* and *Yersinia pestis*, could all be resolved down to whether they encoded an A, B or C subunit of these 'ABC' toxins.

2.2. Diversity of *tc*-like genes from other bacteria

As discussed above, following the cloning of the *tc* genes from *P. luminescens*, a number of *tc*-like genes were cloned from a range of other bacteria. These were all given a range of very different names resulting in considerable confusion in the resulting rather complex literature. For example, the *tc*-like genes from *S. entomophila* are found on the 'amber disease-associated plasmid' or *pADAP* which is responsible for amber disease in the New Zealand grass grub, *Costelytra zealandica*. These were termed '*Serratia entomophila pathogenicity*' or '*sep*' genes A, B and C, specifically *sepA*, *sepB* and *sepC* (Hurst et al., 2000). Insertion of a transposon into either of the three genes abolished the gut clearance and cessation of feeding associated with amber disease (Hurst et al., 2000) and of course, knowing what we now know from the *Photorhabdus tc* genes above, it is now clear that they simply encode the three different A, B and C subunits of the *S. entomophila* Tc-like toxin. Similarly another bacterium, *Y. entomophaga* strain MH96, that kills the New Zealand grass grub (Hurst et al., 2011) also carries *tc*-like genes. Here the three A, B and C subunit-encoding genes are found in a 32-kb pathogenicity island together with two chitinase genes (Hurst et al., 2011), whose functions will be discussed below. In a more biochemical approach, Tc-like toxins were also purified from *X. nematophila* strain PMF1296 (Morgan et al., 2001; Sergeant et al., 2003). Here they were termed '*Xenorhabdus particulate toxins*' or '*xpt*' and again genes encoding each of the A, B and C subunits are found, namely *xptA1*, *xptB1* and *xptC1* (Morgan et al., 2001), although to add further to the growing confusion the B and C subunit-encoding genes are in fact in a different order due to the split nature of the first gene in the operon. Interestingly, and importantly, in some species of bacteria genes encoding B and C subunits are found as gene fusions. In other words, both the B and C subunits are encoded together in the same polypeptide. In fact, Dow AgroSciences still holds patents that protect the fusion of B and C encoding genes for their expression in transgenic plants (see Section 2.6 on the use of Tc genes in crop protection below). This clearly suggests that the B and C subunits must have some very close association in formation of the final ABC complex, a point to which we will return below.

2.3. Structure, function and biophysics of the Tc ABC complexes

2.3.1 'A' subunit assembly

While the list of genes encoding different *tc*-like homologs from different bacteria continued to grow, a few laboratories began to focus on resolving the structure of the Toxin complexes in order to begin to understand how these huge structures actually conferred their toxicity. The first stoichiometric and biophysical studies of the Tcs were from *Xenorhabdus*. Here, using a combination of biophysical techniques including electron microscopy, Lee and co-workers suggested that the A subunits formed a large (~1.15 MDa) bottle-shaped tetramer with an open central channel (Lee et al., 2007). This A subunit tetramer was inferred to bind to the insect brush border membrane and then co-associate with B and C subunits. Four years later, this A subunit tetramer was again shown to bind to insect brush border membranes and to cause pore formation in black lipid membranes (Sheets et al., 2011). This confirmed that it was an assembly of the A subunits that were responsible for recognition and attachment to the insect gut but left the role of the B and C subunits unclear. Further work on the *Xenorhabdus* Tc subunits also showed that the B and C subunits bound together in a 1:1 stoichiometry and that B/C then binds irreversibly to the A subunit assembly, giving a final A, B, C stoichiometry of 4:1:1 (Sheets et al., 2011). Excitingly, these studies also revealed that the A and B/C subunits from *Photorhabdus* and *Xenorhabdus* were functionally interchangeable, suggesting that the A attachment vehicle (four A subunits as a tetramer) could be combined with B/C components from other bacterial species. In other words, the Tc toxins could be 'mixed and matched' (ffrench-Constant et al., 2007) and it did not matter where individual component subunits came from as long as they were combined in the correct manner (B and C subunits made in the same bacterial cytoplasm and then combined with A subunits). Similar studies of the

ABC complex from *Y. entomophaga* appeared to show, in contrast, that the A subunits form a pentamer, rather than a tetramer (Landsberg et al., 2011). Again in an apparent departure from the *Xenorhabdus* toxin, the *Yersinia* ABC complex is also decorated with chitinase enzymes that confer chitinase activity on the purified native complex. This chitinase activity may be responsible for the disappearance of the peritrophic membrane in the guts of toxin-treated Diamondback moth, *Plutella xylostella*, caterpillars (Landsberg et al., 2011) which could represent the first step in allowing the mature toxin access to the brush border membrane.

2.3.2 Low-resolution structure of the Tc ABC holotoxin

At this point, we are left with a picture whereby the A subunits combine to form either a pentamer (as suggested by the *Yersinia* data) or a tetramer (as suggested by the *Xenorhabdus* data) with an open central channel. This cage-like structure then attaches to the brush border membrane where it exerts its toxic effect. However, the precise role of the integral channel and its interaction with the B and C subunits of the toxin had to await further detailed analysis of the *Photorhabdus* toxin at 6.3 Å resolution by cryo-electron microscopy and single particle analysis (Gatsogiannis et al., 2013). This very elegant work showed that the A subunits (formed from TcdA1-encoded A subunits) formed a syringe-like pentamer with a molecular mass of ~1.14 MDa and a long funnel-like central pore that is capable of conducting ions. These channels are cation selective due to a ring of negative charges displayed at the mouth of the channel. The structure of the Tc channel is also very different from that of other AB-type toxins as the binding components of anthrax, C2 and iota toxins all form β-hairpins whereas the cryo-EM structure of the *Photorhabdus* A subunit suggests it is composed entirely of α-helices with a minimal channel diameter of ~1.5 nm (Gatsogiannis et al., 2013). Further, the A subunit-mediated channel is reconstituted in a fully oriented manner when added to only one side of the lipid bilayer and its conductance can be blocked when either the B or the C subunits are added to the same side of the membrane (Lang et al., 2013). This suggests that B and C co-assemble with the A assembly in a very specific manner and that their interaction blocks the pore formed by the A subunits themselves. But what do the B and C subunits actually do and how are they involved in toxicity?

The precise role of the C subunits of the Tcs was an intriguing mystery due to the unusual structure of the *tccC*-like genes that encode them. These genes clearly encoded proteins with three separate domains (Waterfield et al., 2001b). The central 'core' of a *tccC* gene encodes a highly conserved motif with similarity to sequences previously termed *rearrangement hotspots* or *rhs* elements in *E. coli*. The N-terminal sequence is also well conserved but the C-terminal tail is extremely variable. This variability of the C-termini of *tccC*-like genes is found across all *tccC* genes in one genome or indeed across all *tccC*-like genes in any bacterial genome. Critically, the laboratory of Klaus Aktories has recently shown that two of these C subunits (termed TccC3 and TccC5) are ADP-ribosyltransferases that selectively modify unusual amino acids with actin and Rho proteins, respectively (Gatsogiannis et al., 2013; Lang et al., 2010, 2011). Specifically, TccC3 ADP-ribosylates threonine-148 in actin, resulting in actin polymerization, and TccC5 ADP-ribosylates Rho at glutamine-61/63, inducing Rho activation (Gatsogiannis et al., 2013), and the net effect of these two enzymatic activities is to cause extensive polymerization and clustering of actin in target cells. Intriguingly, given that the active site of the ribosyltransferase is encoded within the C-termini of the C subunits confirms the hypothesis that the C-termini of the C subunits encode a range of different potential effectors (Waterfield et al., 2001b). The *tccC* genes that encode the C subunits are, therefore, a potential gold mine for the identification of further enzymatic activities and translocated effectors. For example, the C-terminal end of one C subunit from *Y. pestis* (termed YipA) carries a consensus sequence for protein tyrosine phosphatase (Spinner et al., 2012), suggesting one possible further mode of enzymatic action. All of these data suggest a model whereby the A subunits of the Tc toxins come together to form a syringe-like injection mechanism that somehow translocates the C-terminus of the C subunit into the target cell where it exerts its toxic effect. This hypothesis is consistent with data showing rearrangement of the A subunits upon membrane insertion, suggesting that they do indeed act like a 'syringe' for insertion of the toxic moiety into the host cell. However, the precise interaction of the B and C subunits and how it leads to the release of the C-terminus of the C subunit awaited further detailed structural studies, this time again from the *Yersinia* toxin.

2.3.3 Encapsulation and auto-proteolysis of the C subunit

The precise interaction of the B and C subunits had to await the crystallization of the *Y. entomophaga* B/C in complex and structural determination to a resolution of 2.5 Å. This crystal structure showed that the B/C complex forms a large hollow structure that encapsulates the cytotoxic C-terminus of the C subunit like the 'shell of an egg' (Busby et al., 2013), thus presumably protecting that host bacterium from any of the toxic effects that the C subunit might exert on the host itself. The shell of the 'egg' has a β-propeller domain at one end which it uses to attach to the A subunit in the mature native ABC complex. Moreover, when the C subunit is folded in complex with B, it auto-proteolyses to release its toxic C-terminus for translocation into the target cell (Busby et al., 2013). This correlates with the earlier observation that the *Photorhabdus* C subunit, specifically TccC3, is cleaved at its hyper-variable sequence but only in the presence of the B subunit, in this case TcdB2. This superb piece of work in *Y. entomophaga* finally explains why the central core of the *tccC*-like genes is so conserved, as it encodes the precise site at which auto-proteolysis occurs and it must therefore always remain invariant for the C subunit cleavage to work. The structure of the C subunit is also important as it is the first structure determined for proteins containing 'Rearrangement hotspot' (Rhs) repeats, similar structures to which are found in the eukaryotic tyrosine–aspartate YD-repeat-containing family, including the teneurins which are type II integral membrane proteins that help establish neuronal cell connections during development (Hong et al., 2012). This may suggest that the Tc B/C-mediated encapsulation device is used in other systems for protein encapsulation and delivery. Indeed, recent work has suggested that Gram-negative Rhs proteins and the distantly related wall-associated protein A (WapA) from Gram-positive bacteria mediate intercellular competition in bacteria. Specifically, Rhs and WapA carry polymorphic C-terminal toxin domains that are deployed to inhibit the growth of neighbouring bacterial cells (Kung et al., 2012). Similarly, RhsT from *Pseudomonas aeruginosa* PSE9 is translocated into mouse phagocytic cells and mice survive infection with rhsT mutants, showing that RhsT is a real virulence factor (Kung et al., 2012). These striking findings show that YD-repeat proteins in bacteria are involved in contact-dependent growth inhibition and/or virulence, and suggests that B/C-like YD-repeat containing 'egg shell' structures could play a more general role in protecting host bacterial cells from their delivered bacterial toxins.

2.3.4 Insights from the high-resolution structure of the ABC holotoxin

During the final drafting of this chapter, Meusch et al. published a high-resolution structure of the complete *P. luminescens* ABC Tc holotoxin at a resolution of 4.0 Å (Meusch et al., 2014), which now allows us to begin to put all of the pieces of this complex puzzle together into a novel picture. To add to the already confusing nomenclature, they call the Tc holotoxin

'PCT3' as it is composed of TcdA1, TcdB2 and TccC3 polypeptides, whereby presumably 'PCT5' would contain TccC5. However, here again we will try and ignore the over-complicated nomenclature and simply refer to the A, B and C subunits of the Tc holotoxin. This exquisite holotoxin structure reveals how the A subunits come together to form four receptor-binding sites and a neuraminidase-like region. The neuraminidase-like region is structurally homologous to neuraminidases and closes the bottom of the pentameric Tc shell. Neuraminidases are exosialidase enzymes that cleaves α-ketosidic linkage between the sialic (*N*-acetylneuraminic) acid and an adjacent sugar residue. The precise role of this domain in the Tc toxins is unclear. However, the influenza virus neuraminidase may take part in fusion of the viral and host cell membranes (Wagner et al., 2000), suggesting that it may somehow be involved in docking of the Tc A subunits onto the host membrane. The four receptor-binding domains have an immunoglobulin-like β-sandwich fold structurally similar to the receptor-binding domains of the well-studied diphtheria and anthrax toxins. However, it is notable that the Tc toxins have four such receptor-like domains compared to other toxins that have only one or two (Choe et al., 1992; Petosa et al., 1997). Meusch et al. (2014) argue that the very closely spaced (only 8 Å apart) amino- and carboxy-terminal regions of the two insertions that form the receptor-binding domains suggest that the inserted domains were added to an originally much simpler shell (Meusch et al., 2014). The authors also argue that the variability of this region between different Tc toxins might begin to explain their putative host specificity. It is, however, worth reminding ourselves that in *Photorhabdus* at least, host specificity (choice of insect host) is determined by the nematode vector and not the bacterium. An alternative role for the Tc receptor variability found in different toxins in the same bacterium is that the different Tcs formed might target different tissues within the same insect.

To understand how the Tc toxin changes from its pre-pore to pore state, Meusch et al. (2014) built a molecular model by fitting the crystal structure of the A pre-pore into the cryo-EM structure. This analysis shows in incredible detail how formation of the pore is associated with dramatic conformational changes that open the shell and shift the central channel by 12 nm. This dramatic change in structure of the of the A pore complex is facilitated by three defined hinge regions and results in a large-scale re-organization of the four receptor domains (labelled 'Receptor-binding domain A, B, C and D' in Fig. 7.1), suggesting a potential alteration in their binding geometry and/or affinity.

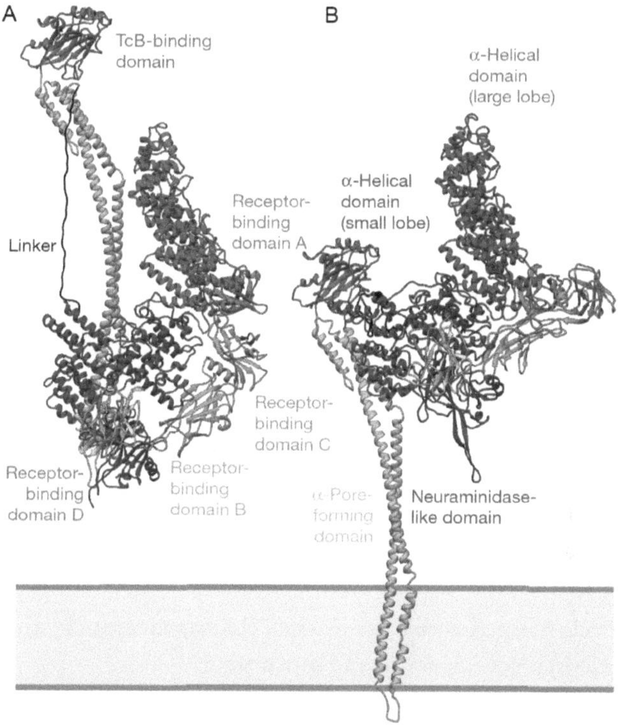

Figure 7.1 Ribbon diagram showing the TcA protomer in (A) pre-pore state and (B) in pore-forming state. Receptor regions A–D and neuramidase-like region are labelled, along with the Tcb-binding domain, linker, pore-forming domain and α-helical domains (the small and large lobes). Note the dramatic changes that occur as the TcA protomer switches from pre-pore to pore-forming states, specifically, the collapse of the elastic linker, the extension of the pore-forming domain and the re-organization of the receptor domains A–D. *Reprinted from Meusch et al. (2014) with permission from MacMillan.* (See the colour plate.)

How is the energy provided for such a dramatic change of conformation? The answer comes from the linker (labelled 'Entropic spring' in Fig. 7.2) which connects that A subunit shell and the channel in the pre-pore state. This linker has an unusually elongated structure with a length of 113 Å covered by only 48 residues. The authors therefore suggest that this linker acts like a stretched 'elastic band' or 'entropic spring' that is stretched in the pre-pore state and can only achieve its preferred 'relaxed' state when the shell is opened and the channel extended. This begins to explain how the channel is driven through the complex for entry into the host target membrane. The concept behind this entropic spring mechanism can be seen very clearly in Fig. 7.2 as the spring contracts to send the pore shooting down into the

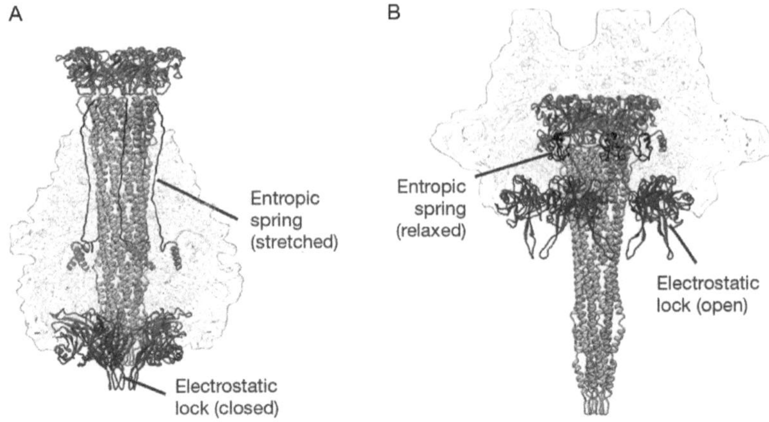

Figure 7.2 Another view of the structure of (A) Tca pre-pore and (B) pore-forming complex. Note the entropic spring or linker (black), release of which provides the elastic energy to drive the pore-forming channel down towards the membrane. The neuramidase-like domain is shown in blue (see text for discussion). *Reprinted from Meusch et al. (2014) with permission from MacMillan.* (See the colour plate.)

membrane following the release of the 'electrostatic lock' that otherwise keeps the delivery vehicle in its pre-pore state.

A question remains as to how the C subunit toxin is delivered down this dramatically extended pore? To answer this question, Meusch et al. also solved the crystal structure of the B and C subunits in complex (Meusch et al., 2014). To get around the fact that B and C subunits need to be made together in the same bacterial cytoplasm, as discussed above, they made a B/C fusion protein (by fusing the genes encoding TcdB2 and TccC3). Although the *P. luminescens* B/C complex only shares 56% amino acid identity to the equivalent structure in the *Y. entomophaga* B/C complex (termed YenB-YenC2), the eggshell-like structure formed by the B subunit and the manner in which it encases the toxic C-terminus of the C subunit are almost identical. Similarly, as suggested by the work in the *Yersinia* Tc toxin, the C subunit acts as an aspartyl autoprotease with aspartates 651 and 674 forming the catalytic dyad. This means that the toxic C-terminal of the C subunit is again cleaved with the shell provided by the B subunit. This does not explain how the C subunit is translocated down the very narrow channel so dramatically extended by the A subunits and how it gets into the channel in the first place?

The inner surface of the B/C-formed 'shell' is positively charged and contains large hydrophobic patches. Meusch et al. (2014) suggest that the inside of the B/C shell creates a 'hostile milieu for protein folding'

(Meusch et al., 2014). They propose that this hostile environment might be able to directly unfold native proteins and suggest that the lack of resolution of the cleaved ADP-ribosyltransferase domain residing in the shell is consistent with it being either unfolded or in static disorder. This leaves us with a picture of the unfolded ADP-ribosyltransferase domain sitting within the B/C shell and awaiting translocation into the delivery channel formed by the A subunits. So how does it get into the channel? The β-propeller of the B subunit interacts strongly with the highly conserved central funnel formed by the A subunits, as shown in Fig. 7.3A–C. The narrow gate formed by this β-propeller is open in the holotoxin, suggesting that it forms the entrance to the delivery channel, as depicted in Fig. 7.3B. The open β-propeller pore is also hydrophobic and the authors suggest it therefore plays a similar role to the Φ-clamp in anthrax toxin which protects hydrophobic patches in the translocated protein.

The narrow passage of the β-propeller suggests that the ADP-ribosyltransferase must be unfolded before it passes this gate and critically the authors found extra density (labelled in red in Fig. 7.4B) within the translocation channel after holotoxin formation, confirming the idea that the unfolded ADP-ribosyltransferase passes through the β-propeller gate and enters the translocation channel before membrane permeation. Finally, before we leave this description of a stunning new delivery mechanism, it is worth returning to the unexplained diversity of potential toxins encoded by the C-termini of different TccC-like proteins. Meusch et al. (2014) point out that the C-terminal (hyper-variable) regions of TccC3 and TccC5 are both cationic with isoelectric points of 9.68 and 8.65, respectively (Meusch et al., 2014). This suggests that translocation of the ADP-ribosyltransferase domains of TccC3 and TccC5, and indeed equivalent domains from other unstudied TccC-like proteins are pH-independent. This is in stark contrast to the pH gradient-dependent, unidirectional, translocation of anthrax lethal factor across the cation-selective anthrax translocation pore (Brown et al., 2011). This also suggests that the other cleaved toxic C-termini of other C subunits should also be cationic (have similar isoelectric points) to take advantage of the same delivery mechanism.

2.3.5 Potential Tc chaperones, release factors and delivery co-factors

Finally, before we leave the detailed study of Tc structure and its mode of action, it is worth noting that there are several unresolved questions relating to both toxin complex assembly and toxin delivery itself. Specifically, what are the roles of molecular chaperones in assembling the ABC complex itself?

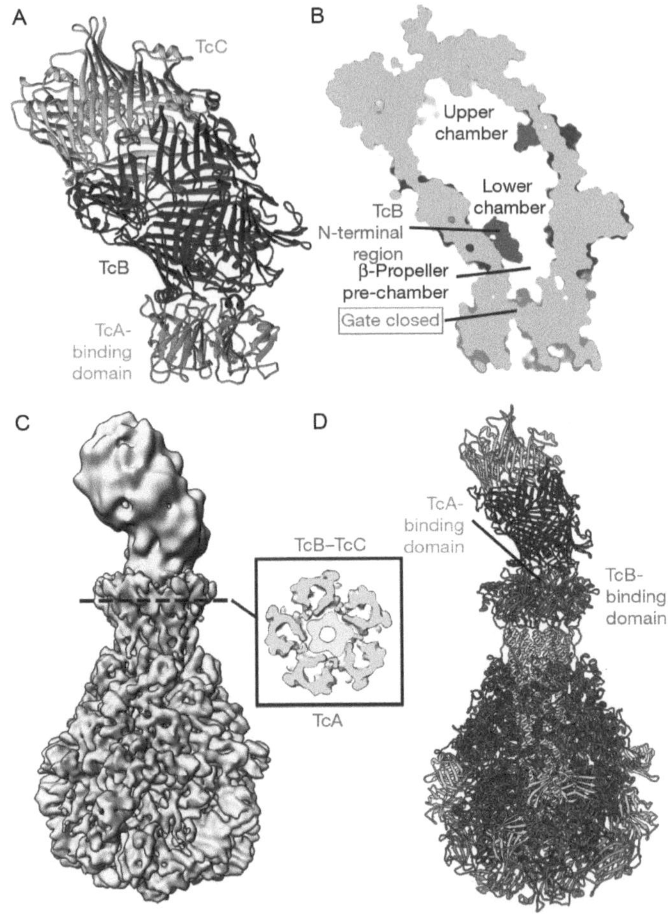

Figure 7.3 Interaction of the B/C (TcB/TcC) subunits with the A (TcA) subunit assembly. (A) The B subunit (labelled TcB in blue) and the C subunit (TcC in yellow) interact to form a large 'cocoon-' or 'egg-shell'-shaped structure. The B subunit also comprises the A subunit-binding domain (labelled Tca-binding domain in light blue) which folds into an asymmetric six-bladed β-propeller. (B) A vertical section through the egg-shell-like structure formed by B/C to show the upper and lower chambers and the pre-chamber formed by the β-propeller. Note that the gate out of the B/C chamber is closed. Subunit domains are coloured: Tca-binding domain (light blue), TcB (dark blue), TcC (yellow) and Tcb N-terminal region (red). (C) Cryo-EM structure of the ABC holotoxin. Subunits and domains are labelled: TcA (grey) and TcB/TcC (yellow). The inset labelled TcB–TcC shows that the β-propeller gate is open here. (D) Final molecular model of the ABC holotoxin formed by fitting the crystal structure of the A subunits and the crystal structure of B/C into the overall cryo-EM structure. *Reprinted from Meusch et al. (2014) with permission from MacMillan.* (See the colour plate.)

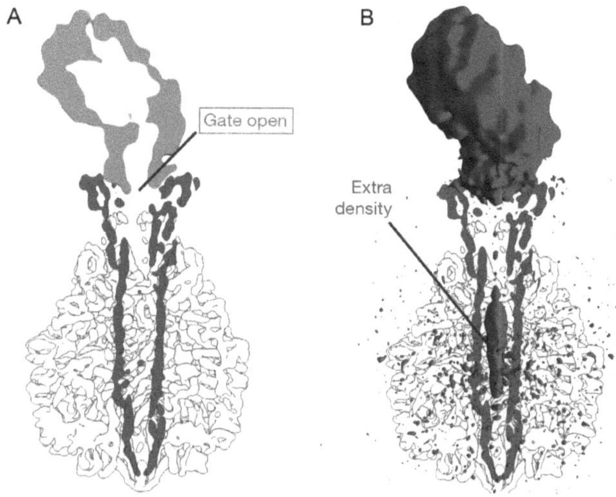

Figure 7.4 Vertical sections through the density of the ABC holotoxin channel. (A) Section showing the continuous channel between the egg-shell-like structure of B/C in yellow and the channel of the A subunits in grey. (B) Difference map between the cryo-EM structure of the ABC holotoxin and the cryo-EM structure of the A subunits alone. The difference map is overlaid in red. Note the extra density present in the channel which may correspond to the delivered ADP-ribosyltransferase domain of the C subunit (see text above for discussion). *Reprinted from Meusch et al. (2014) with permission from MacMillan.* (See the colour plate.)

How are such large proteins extruded through such a small pore and exactly how are they unfolded and re-folded? How are the mature holotoxins released from the bacterial outer membrane to be found free in the bacterial supernatant? In relation to the first question, it is interesting that the ABC complexes of *Yersinia* appear to be decorated with chitinases, whereas no similar companions have been noted for Tc holotoxins from *Photorhabdus* or *Xenorhabdus*. There are clearly several potential and simple explanations for the lack of these co-inhabitants in *Photorhabdus* and *Xenorhabdus*. (1) It may simply be that the operons studied in *Photorhabdus* and *Xenorhabdus* lack nearby genes encoding chitinases. There is, therefore, the formal possibility that chitinases will form partners in ABC complexes if encoded within, or nearby, the *tc*-encoding locus itself. (2) The potential work of the chitinases as chaperones (i.e. the presence of chitinases is required for correct folding and assembly of the mature ABC toxin) might be performed by other partners in other bacteria. To this end, it is noteworthy that we recovered GroEL-related amino acid sequences from our original purification of native *P. luminescens* W14 Toxin complex A (D. Bowen, M. Blackburn and R.H. ffrench-Constant, unpublished data). It is, therefore, possible that this well-known molecular chaperone does indeed help fold and assemble the massive ABC toxin itself.

Little work has been done to look at the localization of the mature Tc holotoxins on the bacteria themselves. Using a polyclonal antibody raised against the complete Tca holotoxin in *P. luminescens* W14, we showed that the mature Tc particles are displayed on the outer membrane of the host bacterium (ffrench-Constant et al., 2003). In some members of the *P. luminescens* clade, the Tc complexes are released into the bacterial supernatant when grown in culture, resulting in an oral toxicity to caterpillars of the cell-free supernatant. In other members of this clade, the Tc toxins are not released from the outer membrane when grown in culture in the laboratory (Waterfield et al., 2001a). The release of the mature holotoxin from the bacterial outer membrane is mediated by the *pdl1*-encoded lipase, a gene that lies within the *tcd* locus of *P. luminescens* W14 but is lacking from the equivalent region of the *P. luminescens* TT01 genome (Yang and Waterfield, 2013). These data are therefore consistent with the hypothesis that the PDL1 lipase releases Tc complexes from the outer membrane of strain W14 but not from strain TT01. The precise role of such release in the biology of insect infection is not shown and presumably for *most* strains of *Photorhabdus* the Tc holotoxins must remain displayed on the bacterial outer membrane, presumably for use only in a contact-dependent manner.

To determine if factors within the host cell itself can also influence Tc delivery, perhaps by refolding the extruded C subunit toxin, Lang et al. (2014) looked at the effects of different drugs which block different co-factors in the target cell. They discovered that both the cellular uptake and resultant toxicity of the two C subunits TccC3 and TccC5 could be blocked by radiciol, cyclosporine A and FK506 or tacrolimus (Lang et al., 2014). The blocking effect of these drugs implicates the molecular chaperone heat shock protein (Hsp) 90 and the peptidyl prolyl *cis/trans* isomerases (PPIases) of the cyclophilin and FKBP families in the uptake or refolding of the ADP-ribosylating C subunits. Specifically, given the tiny size of the pore formed by the A subunits (~1.5 nm minimal diameter), the authors speculate that these putative chaperones (such as Hsp90 and FKBP51) might play a role in refolding translocated proteins after passage through the pore in an unfolded state (Lang et al., 2014). In summary, there is clearly still much we need to learn both about how these enormous ABC complexes are made and secreted by bacteria and also how they exert their toxic action on the host target cell.

2.4. The Tcs as 'polymorphic' toxins

Before we leave the structure of this fascinating group of toxins, it is worth looking at the Tcs in their broadest context; in other words, recognizing that they belong to a large and diverse group of toxins recently defined as 'polymorphic toxin systems' (Zhang et al., 2012). Polymorphic toxin systems are characterized by an amazing diversity of C-terminal toxin domains which appear to have been generated by recombination with standalone toxin-coding cassettes. They are found across all bacterial lineages and are delivered by seven distinct secretory systems including type II, V, VI, VII and the PVCs (see Section 3), PrsW (a putative site 1 protease)-dependent and MuF (a component of the phage prohead)-phage-capsid-like systems (Zhang et al., 2012). These have a general domain template of (a) a trafficking domain, (b) a series of repeats, (c) a pre-toxin domain, (d) a releasing auto-peptidase domain and (e) a toxin domain. Note that the TcB and TcC proteins share some of the characteristics of this system. Specifically, TcB proteins have an N-terminal *Salmonella* plasmid-borne virulence factor B-like trafficking domain, followed by an integrin-like β-propeller, whereas the TcC proteins carry multiple Rearrangement Hot Spots repeats followed by a variable C-terminal toxin domain (as discussed above).

2.5. The role of the Tcs in the biology of infection

Despite our detailed knowledge about the structure, mode of action and even biophysics of the Tcs, it is worth considering if any of this new knowledge really tells us anything about their potential role in infection. For example, it is notable that the bacteria that carry *tc* gene homologs have very different lifestyles and only some of them currently have known associations with insects, or other invertebrates, at any stage within their lifecycles. The most notable bacterium outside of *Photorhabdus* and *Xenorhabdus* that carries copies of *tc*-like genes is none other than *Bacillus thuringiensis* itself (Blackburn et al., 2011). These *tc*-like genes in *B. thuringiensis* may be important in increasing the potency of this well-known bacterium against insects particularly when it comes to oral routes of infection. In contrast, other bacteria which carry *tc*-like genes, such as *Pseudomonas syringae* pv. *tomato* (a plant pathogen), *Fibrobacter succinogenes* (a cellulolytic bacterium found in ruminants) and *Treponema denticola* (associated with periodontal disease), currently have no known associations with insects. Of course, the simplest hypothesis is indeed that the presence of *tc*-like genes within a bacterial genome somehow *predicts* an interaction with insects at some stage in the bacteria's lifecycle. However, the natural history of most bacteria is far from complete and we are therefore only left to speculate on the role of *tc*-like genes based on the bacteria for which some insect-associated natural history is available. For example, in the well-studied insect pathogen *S. entomophila*, the *pADAP* plasmid carrying A, B and C subunit-encoding *tc*-like genes (*sepA*, *sepB* and *sepC*) clearly does indeed cause amber disease in grass grubs (Tan et al., 2006). Similarly, three *tc*-like genes are also found in *Y. entomophaga*, suggesting that the resultant ABC toxin could have effects on the *C. zealandica* midgut (Hurst et al., 2011). In fact, examination of the midgut histopathology of *C. zealandica* treated with purified *Yersinia* Tc toxins (Marshall et al., 2012) is strikingly similar to the effects seen on feeding purified native *Photorhabdus* holotoxin 'A' to *M. sexta* caterpillars (Blackburn et al., 1998). Specifically, first the peritrophic membrane that surrounds the food in the gut is destroyed and then the microvilli of the midgut itself are disrupted. While such studies make for elegant cross sections of the insect midgut, they may be failing to tell us much about how the Tc toxins are really attaching to and destroying the midgut cells. Thus, it is also noteworthy that the midgut histopathology of both Bt toxins and indeed cholesterol oxidase (Purcell et al., 1993) are little different from that of the Tc toxins. This histopathology may therefore simply reflect a common pattern of midgut self-destruction, whatever the proximal toxin or toxic insult. However, taken together, the available data suggest that different insect pathogens do indeed use the Tcs to disrupt the insect midgut, presumably either from the gut lumen itself (*Y. entomophaga* or *B. thuringiensis*) or from the hemocoel side (*Photorhabdus* and *Xenorhabdus*). Indeed, some pathogens that carry *tc*-like genes, such as *Y. entomophaga*, have been shown to be able to pass from the gut lumen into the hemocoel itself (Marshall et al., 2012), suggesting the *tc*-like genes may somehow facilitate this passage across the gut.

Insects can also act as reservoirs or vectors of human diseases. To this end, the role of *tc*-like genes in the plague bacillus *Y. pestis* has, therefore, attracted considerable recent attention. In this case, *Y. pestis* is normally maintained by flea–rodent enzootic cycles but is occasionally transferred to man through the bite of an infected flea. Originally when the first genomes of *Y. pestis* were sequenced, these *tc*-like genes were found disrupted in some strains leading to the original hypothesis that their loss may somehow be involved in increased pathogenicity to man. More recently, the genes encoding the A and B subunits (termed *yitAB* and *yitC*) and two C subunits (the *tccC* homologues *yipA* and *yipB*) have been shown to be largely intact in the genomes of most other sequenced *Y. pestis* strains, suggesting that they may in fact play an undescribed role in interacting with either the flea vector or incidental human host (Spinner et al., 2012). The currently available data on exactly what role the Tc toxins might perform in the unusual lifecycle of the plague bacillus are however potentially conflicting. If we ignore the suggestion that *Y. pestis* Tcs are type III secreted (Gendlina et al., 2007), which now seems hard to believe given that they effectively form their own secretion machinery, and concentrate on data available for *tc* gene expression, we can see several interesting points. First, *Y. pestis tc* genes (*yitAB* and *yitC*) are up-regulated in J774A.1 macrophages. Second, repression of *tc* genes via knock-out of their *yitR* regulator allowed for increased phagocytosis of *Y. pestis* isolated directly from fleas (Vadyvaloo et al., 2010). Suggesting that the *tc* genes may be playing some role in inhibiting phagocytosis of the flea-borne bacteria by the macrophages. Third expression of *yitR*, the regulator of *Y. pestis tc* expression, is up-regulated in the flea itself (Vadyvaloo et al., 2010) but the Tc proteins themselves do not play a detectable role in flea infection or the ability to produce a transmissible infection (Spinner et al., 2012). Fourth, the *Y. pestis* Tc proteins (A subunit protein, YitA and C subunit protein, YipA) themselves are highly expressed in the flea but not at the same temperature (22 °C) in the laboratory (Spinner et al., 2012). Fifth, Tc (YitA and YipA) production is higher at lower temperature (<22 °C) and is cut-off at 37 °C, the temperature the bacterium would reach after being released into a human, but the proteins persist for over 9 h (Spinner et al., 2012). Finally, *Y. pestis* Tc proteins are localized to the outer membrane of the bacterium (Spinner et al., 2012), like the Tcs of *Photorhabdus*. Taking all of this available evidence together, this suggests that Tc proteins are made within infected fleas and then displayed on the outer membrane of *Y. pestis* as it is introduced into mammalian host upon flea feeding. This role would be consistent with them resisting phagocytosis via rodent or human phagocytes after their release from the flea. Such a role is consistent with them playing a role in pathogenicity against mammalian hosts but potentially via the nature of the delivered C subunit toxin (i.e. in preventing phagocytosis) rather than via the loss of insect-specific *tc* loci, as originally suggested by the sequencing of the first *Y. pestis* genome.

2.6. The potential role of Tcs in crop protection

Aside from these major gaps in our understanding of how *tc*-like genes contribute to lifestyle changes in the bacteria in which they are found, much of the original interest around the *tc* genes was for their potential inclusion in transgenic crops, specifically as alternatives to the widely deployed toxins from Bt.

Unfortunately our detailed knowledge of the structure and function of the Tcs has not come hand in hand with their successful expression in transgenic crops, despite their obvious potential utility as insecticidal proteins. Here we will not provide a detailed breakdown of the patent literature on this area, which is considerable, but it is worth considering why the expression of these genes in plants is so technically demanding. First, before it was understood that all three A, B and C subunits needed to be expressed in plants, considerable effort was placed into making plants that only expressed A subunits, with the underlying assumption that they would kill insects. For example, as a model system the A subunit from *P. luminescens* W14 was expressed in transgenic *Arabidopsis* and its toxicity to neonate *M. sexta* was examined (Liu et al., 2003). We now clearly know that there are a number of reasons as to why we would not expect this experiment to work. First, *M. sexta* (Tobacco or Tomato hornworm) does not feed on this plant by choice, so while *Arabidopsis* is a useful model for testing the success of expressing a large gene (the intact *tcdA* mRNA expressed was 8 kb) and resulting protein (TcdA at 283 kDa) in plants, we would not expect the feeding results to be very meaningful. Second, and most importantly, apart from the potential for pore formation, it is hard to see why expression of the A subunits alone would be toxic in the absence of B and C subunits, specifically the delivered C toxin, unless binding and pore formation by A subunits alone are indeed enough to trigger oral toxicity. However, despite all this, bioassays of 234 T_0 lines transformed with construct pDAB7026 (*osm-tcdA*) identified nine transformed lines that caused 100% mortality in neonate *M. sexta* infested on transgenic *Arabidopsis* (Liu et al., 2003). These data are therefore consistent with the hypothesis that binding and potential pore formation by A subunits alone are enough to trigger insect mortality, at least in this model insect. However, full recombinant toxicity clearly requires full assembly of the ABC holotoxin which to our knowledge has not been successfully achieved in a transgenic plant. The simplest reason for this is that expression of all three relatively large genes encoding A, B and C subunits (or expression of A with a B/C fusion) is still beyond the limits of current plant transformation technology or that the plant is simply not able to perform all of assembly/chaperone functions required to assemble a functional Tc holotoxin.

3. *PHOTORHABDUS* VIRULENCE CASSETTES

The recognition that the Tc toxins form part of the super-family of polymorphic toxin systems also helps us understand the next toxin delivery vehicle from *Photorhabdus* that we will discuss, termed the phage-like 'PVCs. Belonging to a group of phage-tail-like particles termed 'tailocins', PVCs are a fascinating structure with a potentially very interesting biology.

3.1. Discovery and organization of PVCs

Following the sequencing of the genomes of both *P. luminescens* TT01 and *P. asymbiotica* ATCC43949, a large number of pro-phage-like loci were noted in each genome and termed the PVCs. The number and location of these phage-like loci differed between the different sequenced genomes. For example, in *P. luminescens* four PVC encoding loci were found in a tandem array between a type IV DNA conjugation pilus operon and a *mukB* replicon partitioning gene. However at the equivalent position in *P. asymbiotica* genome, only one PVC locus was found and in *P. temperata* K122, the PVC loci have been replaced by a Mu-like phage (Yang et al., 2006). This suggests that the PVC loci are not only phage-like in sequence but also that they can readily move around and between bacterial genomes, as also supported by the presence of the PVC-like anti-feeding pro-phage (Afp) locus on the virulence plasmid *pADAP* in *S. entomophila* (as discussed in Section 3.2). The PVC loci themselves have a specific overall structure with 12–15 open reading frames encoding a conserved 'PVC unit' followed by 2–4 loci encoding putative effector proteins. The conserved PVC unit encodes caudate phage-derived proteins, such as the tail sheath protein and gp19, which both form the tail tubule, gp25 which forms the baseplate and a distinct clade of AAA + ATPases (AAA representing ATPases associated with diverse cellular activities) that are related to CDC48 (Yang et al., 2006). This overall picture suggests that the PVC-Secretion System (PVC-SS) is similar to the type VI secretion system in being derived from the tails of pro-phage but differs in its associated AAA + ATPase, which in the type VI system is a member of the ClpB clade (ClpV). We are therefore left with a picture that both the type VI-SS and PVC-SS have both independently evolved from pro-phage tails using different AAA + ATPases to recycle the injection apparatus after deployment (Zhang et al., 2012).

The putative PVC encoded effectors appeared to correspond to proteins with similarity to a range of different toxins including TccC, from the Toxin complexes (discussed above); part of Makes Caterpillars Floppy 1 (Mcf1) toxin (discussed in Section 4); LopT (a homologue of the *Yersinia* cysteine protease YopT) an effector delivered by the Type III secretion system in other bacteria and the *Photorhabdus* necrosis factor (Pnf; a homolog of

Cnf). The open reading frames encoding these putative effectors are also regularly flanked by transposons, suggesting that the different effector genes may be swapped between different PVC loci. This array of putative toxins encoded at the end of each PVC locus, and the potential to swap these toxin-encoding genes around, suggests that the PVCs also belong to the polymorphic toxin system super-group (Zhang et al., 2012), as discussed above for the Tc-encoding genes. Zhang and co-workers have noted that several toxin domains are secreted by the PVC-SS across most major bacterial lineages and even the euryarchaea. Moreover, proteins secreted via the PVC-SS carry a highly conserved metallopeptidase domain (identified by the motif HExxHxxQ-E) N-terminal to the toxin domain itself. So again we see a system, like the TccC's, whereby a 'constant' N-terminal peptidase domain can recombine to gather different variable C-terminal toxin domains (Zhang et al., 2012) and again this suggests that these peptidase domains are likely to act as auto-proteolytic domains that release the toxin during or after its secretion by the PVC-SS. The C-terminal toxins found to be associated with these PVC metalloproteases span an incredible diversity of nucleases, nucleic acid deaminases, peptidases, pore-forming domains and several other enzymatic domains (Zhang et al., 2012).

PVCs can be readily expressed in recombinant *E. coli* and expression of the proteins corresponding to the first three open reading frames of the *P. asymbiotica* PVC carrying *Photorhabdus* necrosis factor (PaPVCpnf) was confirmed via 2D gel analysis and protein sequencing. Transmission electron microscopy (TEM) of the recombinant PVCs revealed a phage-tail-like particle very similar to R-type pyocins (Yang et al., 2006). The PVCs are ~30 nm wide but show a huge variety in length (up to 800 nm long) depending on whether they are found in a contracted (needle within sheath) or extended state (needle protruding from sheath). Interestingly, polyclonal antibodies raised against the putative effector proteins do seem to label proteins extruded from the PVC when anti-effector antibody-conjugated gold particles are viewed under TEM (N.R. Waterfield and R.H. ffrench-Constant, unpublished data). The candidate effector proteins also cause extensive rearrangement of the actin cytoskeleton when expressed directly inside transfected NIH-3T3 cells, consistent with them being *bona fide* effectors. Finally, injection of larvae of the wax moth, *Galleria mellonella*, with recombinant PaPVCpnf killed the larvae and caused dramatic rearrangement of the actin cytoskeleton of hemocytes recovered from injected larvae (Yang et al., 2006). Such effects disappeared when transposons were inserted into the open reading frames encoding the putative effector proteins (Yang et al.,

2006). Taken together, these rather preliminary data support a hypothesis whereby the PVCs are phage-tail-like delivery vehicles that export candidate effectors (encoded at the end of each of the individual PVC encoding loci) into their target cells. However at this point, we have no idea as to what cells are targeted and indeed what the full range of potential effectors might be, as their predicted amino acid sequences yield few clues as to their likely modes of action.

3.2. The role of PVC-like structures in other bacteria

To get some idea of how a phage-tail-like PVC structure might be playing a role in the delivery of bacterial effectors to eukaryotic cells, we need to look at the overall diversity of phage-like structures made by bacteria. Phages with contractile tails, or 'myophages', infect a wide range of bacteria and are important for our discussions here as they are related to several other structures made by bacteria, including *Photorhabdus*. First, the important type VI secretion system of bacteria is evolutionarily related to myophage tail structures and shares several common components (Leiman et al., 2009). Second, the bacteria killing 'bacteriocins', such as the R-type pyocins of *P. aeruginosa*, are also closely related to phage (Nakayama et al., 2000). These phage-tail-like bacteriocins have a contractile sheath, inner tube, baseplate components and tail fibres, but lack a DNA-filled head and therefore cannot replicate on their own. Third, the bacteriocin-like structures made by the insect killing *S. entomophila*, where they are termed the Afp, and the metamorphosis inducing *Pseudoalteromonas luteoviolacea*, where they are termed 'Metamorphosis-associated contractile structures' or 'Macs' (Shikuma et al., 2014), are also very similar in their overall morphology to the PVCs of *Photorhabdus*. But what does this tell us about the potential function of the PVCs themselves? To address this question, we need to look in more detail at the structure and biology of the Afp and the Macs and ask how they mediate their specific bacteria–animal interactions.

In *S. entomophila*, the genes encoding the Afp are found on the same pADAP plasmid as those (*sepABC*) encoding the Tc-like ABC toxin. The Afp toxin delivery system causes the initial cessation in feeding of the New Zealand grass grub (*C. zealandica*) larva, and the Tc-like toxin then causes clearance of the gut leading to amber colouration in the dying grub (Hurst et al., 2000). Eighteen open reading frames in the pADAP plasmid are associated with the anti-feeding activity and these appear to encode structures similar to different components of the sheathed tail of bacteriophage

T4 and also some components of the type VI secretion system. Using cryo-EM and electron tomography of negatively stained Afp particles, Heymann et al. (2013) have recently shown that the Afp sheath is similar to that of the T4 phage tail but with a different sub-domain arrangement of the polymerizing sheath proteins. The central channel, which presumably delivers the putative effectors encoded at the end of the PVC-like locus, is similar in diameter and axial width to the Hcp1 hexamer of the *P. aeruginosa* type VI secretion system. This tube extends through the baseplate into a 'needle', resembling the 'puncture device' of the T4 phage tail. Interestingly, the tube itself appears to contain either the delivered toxin or potentially a 'ruler' protein which may 'measure' the length of the tube while under assembly. Taken together, the available Afp data support a hypothesis whereby one of two different candidate effectors, also encoded on pADAP, is delivered by the modified phage tail into target cells in the insect midgut. At the time of writing, however, the exact receptors for the Afp particle, how it binds to the midgut cell and also exactly how the toxin is delivered all remain a mystery.

More recently, another stunning piece of bacterial natural history has been described around a second PVC-like structure, the Mac, of the marine bacterium *P. luteoviolacea*. Larvae of the serpulid polychaete tubeworm, *Hydroides elegans*, require contact with surface-bound *P. luteoviolacea* bacteria in order to undergo metamorphosis. This has recently been shown to be determined by an array of PVC-like phage-tail-like structures that the bacterium makes. These Mac arrays are formed from ~100 contractile structures with outward-facing baseplates, linked by tail fibres and a dynamic hexagonal net (Shikuma et al., 2014). These arrays are synthesized intracellularly, are released by cell lysis and then expand extracellularly into an ordered 'multi-Mac' array. Previously, Huang and co-authors had speculated that the Macs might be involved in transporting specific molecules either into the biofilm matrix or into larval cells directly that then leads to larval metamorphosis (Huang et al., 2012). The presentation of the Macs in an ordered array may therefore suggest that their 'firing' or delivery is synchronized or that the array somehow guarantees the correct dosage of delivered 'effector' to trigger metamorphosis when contacted by a settling worm larva. Either way, this incredible piece of biology supports the idea that a PVC-like structure can deliver an unknown bacterial effector into a marine worm larva that is then triggered to undergo metamorphosis. This piece of biology reminds us that the relationship between bacteria, phage and eukaryotes is very ancient and that bacterial effectors that we view simply as 'toxins' due to their effects on isolated cells may in fact be effectors that modulate exquisite aspects of bacterium–animal interactions. Thus in the case of the PVCs expressed and released by *Photorhabdus* in the infected insect, these phage-like delivery vehicles might even be interacting with the infective (dauer) juvenile nematode hosts and triggering their development (metamorphosis) into reproductive hermaphrodites.

How do PVC-like secretion systems recognize their target cells? To begin to come up with some hypotheses to answer this question, its notable that some tailed phage also carry virion proteins that show peptidoglycan hydrolytic activity (Moak and Molineux, 2004), suggesting that these are used by the phage to penetrate the peptidoglycan cell wall during the delivery of DNA. Phage P2 gpX also carries LysM domains which have been implicated in binding to both peptidoglycan and chitin, the latter of which would clearly be important in the recognition of insect tissue. In conclusion, and at this point in time, we are clearly not certain as to how the released *Photorhabdus* PVCs bind to or indeed penetrate host cell walls but, given the above, it is interesting to speculate that they may be redeploying mechanisms evolved to get into prokaryotic cells (Gram-positive and Gram-negative bacteria) in their interactions with eukaryotic cells.

3.3. Implications for PVC biology

To conclude this section, the PVC-like 'tailocins' are modified phage tails that appear to be capable of delivering candidate effectors into target eukaryotic cells. While the nature of the delivered effectors, and exactly how they enter the target cell, remains unclear, what is clear is that PVC-like particles seem to have clearly defined phenotypes (either death, cessation of feeding or the triggering of metamorphosis). This seems to suggest that the PVCs are binding to defined receptors in their target cells or tissues and that they can 'recognize' specific structures (e.g. to destroy midgut cells) or indeed specific species (e.g. the settling *Hydroides* tubeworm larva). We can, therefore, speculate that the PVCs have defined receptors capable of recognizing displayed sugars or proteins (like most phage) and that the effectors they deliver are likely to interfere with deep physiological processes that alter the developmental trajectory of the target tissue. In this context, it is fascinating that *Photorhabdus* genomes carry so many PVC loci and it raises the interesting possibility that the effectors they deliver may be involved both in destruction of the insect midgut (like the Afp particle from *Serratia*) or indeed in triggering developmental changes in their nematode hosts (by analogy with the role of Macs in triggering tubeworm metamorphosis). This latter possibility is particularly interesting given that *Photorhabdus* must somehow trigger reproductive development in its nematode host following its delivery from the infective juvenile, which is equivalent to the 'dauer' juvenile of *Caenorhabditis elegans* and is therefore effectively a 'resting' stage.

4. THE MCF TOXINS

The Mcf toxins are another group of large insecticidal toxins that have close relatives in other bacteria. Here we will look at their discovery in *Photorhabdus*, their close relatives in *Pseudomonas* and what little we know about their potential mode(s) of action. We will also begin to examine how the discovery of the Mcf toxin led to a series of very simple but very effective gain-of-function screens carried out in recombinant *E. coli.*

4.1. Discovery and mode of action of Mcf1

Due to the relatively small genome size (4–5 Mb) of *Photorhabdus*, it is easy to envisage screening a relatively small number of recombinant *E. coli* clones with large insert sizes ~30–50 kb. Although standard laboratory cloning strains of *E. coli* lack specific secretion machineries, such as the well-studied type III secretion syringe, if we are simply interested in looking for novel toxins capable of their own export then a simple gain-of-function screen becomes easy. To this end, we arrayed 50 kb cosmids from *P. luminescens* W14 into 96 well plates and then simply grew up individual clones and injected them into individual *M. sexta* caterpillars (Daborn et al., 2002). On the face of it, the caterpillar immune system, which has both cellular (hemocytes) and microbial (anti-microbial peptides) components, is another good reason why such a screen should not work. For example, without any mechanism of evading their own phagocytosis, all the recombinant clones might be successfully engulfed by the *Manduca* hemocytes. However to our surprise, we discovered that 1 in every 300 or so clones killed the injected caterpillars within 24 h. These caterpillars not only died but within 12 hours they exhibited a significant loss of body turgor, in other words they became 'floppy' (Daborn et al., 2002). We bombarded this positive clone with 126 transposons and then sequenced the complete 33 kb insert using the transposons to sequence from. On retesting all these mutants, we found that 39 insertions negated both the lethal and the floppy phenotypes and that all 39 insertions were in a single long open reading frame (8.8 kb) termed the *makes caterpillars floppy* gene or *mcf*. When sub-cloned alone, this gene still had the ability to kill caterpillars and also caused apoptotic cell death in both isolated insect hemocytes and also the midgut of injected caterpillars, suggesting that Mcf may be promoting apoptosis (Daborn et al., 2002).

The *mcf* gene predicts a 2929-amino acid protein with a molecular weight of 324 kDa. Homology searches suggested that this large protein was divided into three overall domains (Daborn et al., 2002). First, the N-terminal domain carries a domain similar to a Bcl2-homology 3-like (BH3-like) domain (labelled as BH3 in Fig. 7.5), whose putative function we will return to. Second, the central domain is similar to *Clostridium difficile* toxin B over a region important for toxin translocation (labelled as 'Cytotoxin B-like' in Fig. 7.5). Third, the C-terminus carries a region similar to repeats in a 'Repeats in toxins' (RTX)-like toxin from *Actinobacillus pleuropneumoniae* (labelled as 'RTX' in Fig. 7.5).

While none of these homologies are particularly striking, we wanted to test the possibility that Mcf was promoting apoptosis via its similarity to a BH3-domain-only protein. BH3-domain-only proteins are important in the homeostatic balance of Bcl2 proteins in the outer membrane of the mitochondrion, and overexpression of BH3-domain-only proteins in mammals causes mitochondrially mediated apoptosis (Budd, 2001). To test this hypothesis, we looked at the effects of recombinant Mcf toxin on mammalian cell lines. We found that Mcf-treated cells showed apoptotic nuclear

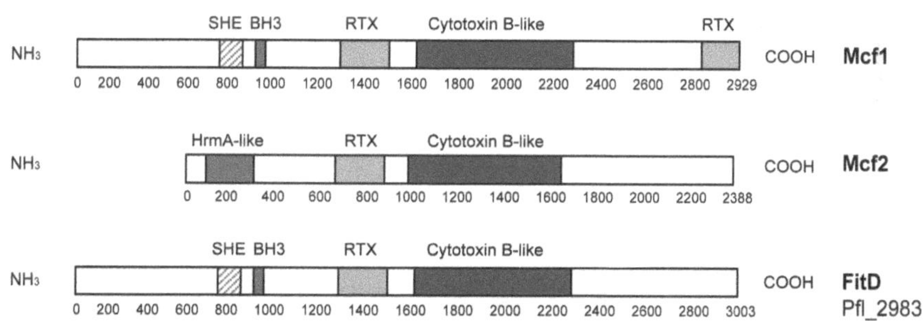

Figure 7.5 Diagram showing the inter-relationship between the different domains in Mcf-like toxins. The *Photorhabdus* toxin Mcf1 carries similarities to the predicted MCF1/SHE proteolysis domain (domain labelled SHE), a BH3-like domain (BH3) and two regions that carry similarity to 'Repeats in toxins' (RTX) toxins. The central domain is similar to *Clostridium difficile* toxin B over a region important for toxin translocation (labelled as Cytotoxin B-like). The *Photorhabdus* toxin Mcf2 has a different N-terminal domain which carries similarity to the type III delivered plant effector HrmA (domain labelled HrmA-like). Finally, the Mcf-like FitD toxin carries the SHE, BH3 and one of the RTX-like domains but lacks similarity to RTX-like toxins at its C-terminus (see text for further discussion). (See the colour plate.)

morphology, activation of the effector caspase-3 and DNA laddering after 6 h and the presence of cleaved PARP after 16 h (Dowling et al., 2004). All of these cellular phenotypes could be prevented by incubation of the treated cells with the apoptosis inhibitor zVAD-fmk. Further, transfection of cells with only the N-terminal 1280 amino acids, which contains the BH3-like domain, also triggered cell destruction (Dowling et al., 2004). Taken together, these data are consistent with the hypothesis that Mcf causes apoptosis and that the domain responsible is within the N-terminal 1280 amino acids of the protein.

To test the hypothesis that the BH3-like domain was indeed responsible for the apoptotic activity of Mcf, we target the domain using site-directed mutagenesis. A double mutant within the BH3 domain dramatically decreases apoptosis in treated cells, again consistent with the hypothesis that Mcf is acting via its BH3-like domain on the mitochondrion (Dowling et al., 2004). Topical treatment of cells with the Mcf toxin also directly alters mitochondrial membrane potential, causes the translocation of the Bcl2-associated X protein to the mitochondrion and triggers the release of cytochrome *c*. Further, and perhaps most convincingly, cells over-expressing Bcl-x_L, a member of the Bcl-2 family that is anti-apoptotic, are resistant to Mcf-mediated apoptosis (Dowling et al., 2004). All of these data are consistent with the idea that Mcf mimics a BH3-domain-only protein and thereby interferes with the balance of Bcl-2-containing proteins in the outer membrane of the mitochondrion, thus promoting mitochondrially mediated apoptosis.

4.2. Diversity of Mcf-like toxins

4.2.1 The Mcf toxin 2

Following the sample sequencing of *P. luminescens* strain W14, we also discovered a second *mcf*-like gene in the W14 genome which we termed *mcf2* (Waterfield et al., 2003). The *mcf2* gene predicts another large toxin with a similar three domain structure. However this time, the N-terminus of the predicted Mcf2 shows similarity to the type III secreted plant 'avirulence' gene (*HrmA*) from the plant pathogenic bacterium *P. syringae* (see Fig. 7.5 for location of HrmA-like region), the transient expression of which in tobacco cells results in cell death. This suggested to us that the N-terminus of Mcf-like toxins encodes the toxic activity and that this activity differs between Mcf-like homologs. To test the role of the Mcf2 N-terminus, we tagged it with a c-Myc antibody tag and showed that it was translocated to the nucleus of transfected mammalian cells which were subsequently destroyed by an uncharacterized mechanism (Waterfield et al., 2003).

4.2.2 *The Mcf-like 'Fit' toxin from* Pseudomonas

In parallel work, an Mcf-like toxin was also found in *Pseudomonas fluorescens* CHA0 (recently renamed as *Pseudomonas protegens*), a bacterium found in the plant rhizosphere that can protect plants against fungi. This insecticidal toxin, termed '*P. fluorescens* insecticidal toxin' or 'Fit', is also encoded by a single long open reading frame (*fitD*) which lies within the *fitABCDEFGH* operon (Pechy-Tarr et al., 2008). The sub-cloned *fit* gene produced the same floppy phenotype in injected *Manduca* larvae as the *mcf1* gene. Knock-out mutants of the *fit* gene also showed reduced toxicity to insects, although they still killed injected insects presumably due to a similar level of functional redundancy to that seen in single gene knock-outs in *Photorhabdus*. In other words, both *Photorhabdus* and *Pseudomonas* genomes encode so many genes with insecticidal activity that knock-out of one alone is not sufficient to eliminate toxicity in the resulting mutant. This is a point that has played a large part in our development of gain-of-function screens, and one which we will return to below. Similarly, other rhizosphere-inhabiting pseudomonads that naturally lack the *fit*-encoding locus also showed little or no insecticidal activity, suggesting that Fit is indeed like Mcf a dominant toxin. In conclusion, while the Fit toxin has a similar three-part domain structure to Mcf1 and Mcf2, and even similarity to the consensus BH3 domain within its N-terminus, its precise mode of action is currently unclear.

4.2.3 *The MCF1-SHE domain as a novel serine peptidase*

Before we leave our consideration of the primary structure of Mcf-like toxins, it is worth returning again to the work of Zhang and co-workers who have identified the MCF1-SHE domain as a possible serine peptidase shared both by members of the polymorphic toxin super-family and also by secreted effectors (Zhang et al., 2012). They initially identified this conserved domain as a region shared by certain polymorphic toxins and the PVC-SS toxins. However, iterative PSI-BLAST searches also revealed that a similar domain is present in Mcf1, FitD and the phytotoxin $HopT_{1-1}$ which is secreted by the type III-SS in *P. syringae*. The domain highlights a putative catalytic triad and suggests that the C-terminus of Mcf1 and FitD might be released by auto-proteolysis, in an analogous fashion to the release of the C-terminal 'bullet' from TccC. However, exactly how Mcf-like toxins enter cells (presumably via endocytosis) and how they are cleaved within their target cells (presumably within endosomes) remains unclear.

4.3. Studying Mcf-like toxins *in vivo*

Both Mcf1 and the Mcf-like Fit toxin are dominant toxins that appear to act quickly on their insect hosts. While it is possible to isolate hemocytes from toxin-injected insects and to study the histopathology of other tissues such as the midgut, these rather static end-point-mediated assays do not really tell us much about how the toxins are really acting *in vivo*. Moreover, they do not really tell us why such active toxins are not active on the host bacterium, or in the case of *Photorhabdus* why they do not kill the nematode vector as well. To this end, recent studies on Mcf-like toxins have focused on observations of both toxin activities *in vitro* and in real time and also on their precise genetic regulation in insects.

4.3.1 The Drosophila *embryo as a microcosm for toxin mode-of-action studies*

To look at how quickly Mcf1 acts on insect hemocytes, Vlisidou and co-workers have co-opted the *Drosophila* embryo system to look at toxin mode of action in real time (Vlisidou et al., 2009). The *Drosophila* embryo system was pioneered to look at the role of insect hemocytes in development and wound repair but the system has also recently been used to look at the effects of Mcf1 toxin on the hemocytes in real time. The advantage of this system is that labelled hemocytes can be visualized in real time and the response of these hemocytes to toxin action can be tested in *Drosophila* strains of differing genetic background. In other words, the role of potential toxin targets can be tested directly via genetic manipulation of the insect host. For this study, a strain of the insect and human pathogen *P. asymbiotica* was injected into *Drosophila* embryos and the response of the patrolling hemocytes recorded using confocal microscopy of the whole embryo. Vlisidou et al. (2009) found that injection of *P. asymbiotica* caused a dramatic 'freezing' of the hemocytes shortly following introduction of the bacteria into the hemocoel. This was in stark contrast to wild-type *E. coli* which was phagocytosed by the embryonic hemocytes in a Dscam-independent manner (Vlisidou et al., 2009). This freezing phenotype was seen both with injection of purified Mcf1 and also with recombinant *E. coli* expressing the *mcf1* gene, suggesting that the recombinant bacteria make enough Mcf1 toxin and that it is displayed/released in a biologically meaningful manner. This freezing phenotype is accompanied by a dramatic rearrangement of the actin cytoskeleton of affected hemocytes (Vlisidou et al., 2009). Importantly, genetic modulation of the different *Drosophila* strains injected with Mcf1 showed that the freezing phenotype is *shibire* dependent (*shibire* is a temperature-sensitive *Drosophila* dynamin mutant that results in the inability of endocytic vesicles to be separated from the parent membrane), suggesting that endocytosis of the Mcf1 toxin is required for toxicity, and that effects on the actin cytoskeleton could be modulated by dominant negative or constitutively active Rac expression. Taken together, these results not only elegantly illustrate how the *Drosophila* embryo can be used to test bacterial toxins in real time and in the presence of different host cell backgrounds but also show unexpected and very early effects of Mcf1 on the actin cytoskeleton. The interaction between the BH3-domain-related apoptosis triggered by Mcf1 and these early effects on the actin cytoskeleton should therefore form a productive area for future research on this fascinating toxin.

4.3.2 Regulation of Fit toxin expression in the insect host

Finally, the broad spectrum of activity of Mcf-like toxins (the effect of Mcf1 on a range of insect and mammalian cells) suggests that they must be very tightly regulated during insect infection; otherwise, toxin production might kill other unintended partners, such as the host nematodes in the case of *Photorhabdus*. To this end, recent studies on the regulation of the Mcf-like Fit toxin in *Pseudomonas* are extremely interesting. As described above, the *fitABCDEFGH* operon contains the *fitD* gene which encodes the Fit Mcf-like toxin. However, *fitF*, *fitG* and *fitH* appear to code for regulatory proteins. Until recently, the precise role of these candidate regulators was obscure; however, recent work by Pechy-Tarr et al. has shown that expression of Fit in culture can be activated by over-transcription of *fitG* or deletion of *fitH*, thus suggesting that the encoded proteins are Fit activators and repressors, respectively (Pechy-Tarr et al., 2013). More recently, the same group has shown that the histidine kinase encoded by *fitF* is the primary sensor that signals the presence of the appropriate conditions for making the toxin (Kupferschmied et al., 2014), which presumably activates a phosphorelay from the histidine kinase to the receiver and phosphotransfer domain of FitF. FitF is then thought to inactivate FitH via phosphorylation of its conserved aspartate, as replacement of this amino acid locks the protein in it repressing state. The periplasmic sensory domain of FitF is proposed to have evolved via domain shuffling from a common ancestor with DctB (a histidine kinase that regulates the uptake of C4-dicarboxylates in Proteobacteria), and the authors speculate that FitF is inhibited by plant-associated small molecules which then block expression of the insecticidal toxin when the bacterium is in the rhizosphere (Kupferschmied et al., 2014). Similarly, toxin expression may then be switched on in the insect by a combination of the absence of plant-associated molecules and/or the presence of signals within the insect. Either way, this begins to demonstrate how such a toxic toxin can be tightly shut-off in situations where its production could be detrimental but quickly turned on as the bacterium switches to an insect host. The equivalent genes for the control of Mcf1 in *Photorhabdus* are not known, but a number of such sensor kinases such as PhoQ are present in the genome and could play a similar role to limit toxin expression in the nematode host.

5. PATOX AND PHOTOX

The toxin classes we have discussed so far illustrate three very different approaches to toxin gene isolation from *Photorhabdus*. First, in the case of the *tc* genes, Tc proteins were purified, anti-Tc antibodies raised and then the Tc-encoding genes were recovered from expression libraries. Second, in the case of the PVCs, phage-like loci were expressed in recombinant *E. coli* and found to produce visible phage-tail-like structures. Third, in the case of Mcf, we have seen how gain-of-function screens can be used to identify novel toxins with killing caterpillars being the simple gain-of-function phenotype. Clearly, however, given the large amount of genomic sequence now available for different strains and species of *Photorhabdus*, one can also use bioinformatic tools to simply search for motifs found in other toxins. This seems to have been the approach used in the discovery of the next group of toxins we are going to talk about and it is the one that has probably been underutilized in the analysis of all the candidate toxins identified from *Photorhabdus* and *Xenorhabdus* genomics.

5.1. PaTox structure and function

Jank and co-workers (Jank et al., 2013) have recently used 'BLAST' sequence alignment tool to look for sequence motifs in the *P. asymbiotica* genome that are similar to those found in the glycosylating toxins of *Clostridium* and *Legionella*. These toxins are characterized by an aspartic acid-X-aspartic acid (or DxD) motif which is thought to be important for the coordination of a divalent cation and for nucleotide–sugar binding (Wiggins and Munro, 1998). Jank et al. (2013) discovered a gene in the genome encoding a putative glycosylating toxin which they termed *P. asymbiotica* Toxin or 'PaTox'. They then cloned the associated gene, purified the PaTox protein and tested its activity by injecting it into larvae of the Greater wax moth, *G. mellonella*. They found that the full-length toxin killed all of the injected larvae in 88 h. They also found that when they changed the DxD motif in PaTox to asparagine-X-asparagine (NxN), 61% of the larvae survived (Jank et al., 2013). The authors also tested the role of the putative glycosyltransferase domain (termed PaTox^{G}) in inhibiting phagocytosis by macrophages using protective antigen (the component of anthrax toxin responsible for binding and translocation) as a delivery system. Strikingly, PaTox^{G} strongly inhibited phagocytosis of labelled *E. coli*, an effect that was eliminated when the DxD motif was again changed to NxN. Similarly, the PaTox^{G} domain also caused disassembly of actin filaments in HeLa cells, suggesting that phagocytosis is blocked via interference with the actin cytoskeleton. Jank and others went on to solve the structure of PaTox using X-ray crystallography to a resolution of 1.8 Å (Jank et al., 2013). PaTox is extremely interesting as it effectively performs two functions and the crystal structure helps us understand how it performs both functions. First, it targets Rho proteins (like TccC5 which is delivered by the Toxin Complex Secretion System discussed above) by glycosylation of tyrosine (GlcNAcylates) at Y32. Second, it also attacks heterotrimeric G proteins via deamidation. Interestingly, *P. asymbiotica* does not carry a copy of TccC5, and therefore, the authors only detected tyrosine GlcNAcylation during *P. asymbiotica* infection of insect cells but not following infection with *P. luminescens*. This might suggest that PaTox is somehow unique to, or required by, *Photorhabdus* strains that can infect man but this hypothesis remains to be tested.

The overall structure of the toxin is a single-chain AB toxin with two catalytic domains within the C-terminus and a receptor-translocation domain at the N-terminus (Jank et al., 2013). Much like Mcf, its precise mechanism of entry into the cell is uncertain, but it has been speculated that it enters host cells from early endosomes (Jank et al., 2013). The crystal structure of PaTox^{G} shows a unique GT-A fold, which is the smallest found in currently known glycosyltransferase toxins, and lacks extending sub-domains. As these sub-domains are critical in distinguishing Rho from Ras as targets for enzymic attack, PaTox^{G} is proposed to target Rho in a highly specific manner. Downstream of the glycosyltransferase domain in the crystal structure is a deamidase with amino acid similarity to *Salmonella* SseI, and this PaTox domain is termed PaTox^{D} (Jank et al., 2013). *Salmonella* SseI is secreted into host cells by the SPI-2 type III secretion system and is thought to control the migration of host immune cells (McLaughlin et al., 2009). The SseI-like domain of PaTox activates G_i and $G_{q/11}$ via deamidation of a critical glutamine residue involved in the GTP hydrolysis of the G proteins. However, because PaTox appears to activate several different heterotrimeric Gα proteins, the downstream consequences of this activation are currently unclear. What is clear, however, is that the preferred substrate of Rho inhibition by glucosylation is the GTP-loaded form of Rho, suggesting that the two functions are closely linked. In conclusion, this elegant structure–function analysis of PaTox has shown that the same toxin (PaTox) causes both activation (via $PaTox^D$) and inactivation (via $PaTox^G$) of the same Rho protein target. Although glycosylation-induced inactivation appears to be the dominant activity, this suggests that PaTox is designed to precisely modulate the activity of Rho in a delicate balance between activation and inactivation. Perhaps more importantly, this begins to suggest that the other large toxins made by *Photorhabdus*, such as Mcf, may also carry more than one enzymatic activity. Indeed, this may thus begin to explain the complex phenotypes observed following the treatment of different cell types with Mcf.

5.2. Photox as a novel mART toxin

Another *Photorhabdus* toxin was identified using bioinformatic approaches in the mono-ADP-ribosyltransferase (mART) toxin termed '*Photorhabdus* toxin' or 'Photox'. The *plu0822* gene was identified as the putative gene encoding mART toxin protein using an array of parallel bioinformatic approaches that basically looked for the characteristic primary amino acid sequence and predicted folding pattern of the mART toxins (Visschedyk et al., 2010). Amino acid alignments over the three key regions (termed regions 1–3) of similarity provided strong support for the hypothesis that *plu0822* encodes a mART toxin. Specifically, in region 1, the catalytic Arg^{288} is preceded by a Tyr. In region 2, the Ser-Thr-Ser motif is preceded by aromatic and hydrophobic residues. Finally in region 3, both primary and secondary glutamate residues are found, namely Glu^{355} and Glu^{353}. The specific site of Photox modification of actin was analyzed *in vivo* using a yeast-based assay system and showed that Photox specifically ADP-ribosylates actin at Arg^{177}. Photox has an unusual domain structure with the N-terminal domain (residues 1–185) lacking any similarity to other toxin domains and being predicted to be disordered in structure. The authors speculate that this N-terminal domain may be involved in entry of Photox into the target cell, potentially using VgrG as a translocator during type VI secretion (Visschedyk et al., 2010). In conclusion, Photox is clearly a novel mART that ADP-ribosylates actin, but the exact point at which it is expressed during *Photorhabdus* infection and therefore its potential interaction with other toxins that alter actin assembly remains unclear.

6. BINARY TOXINS

Finally, we will briefly review the binary toxins that have been found in either *Photorhabdus* or *Xenorhabdus*. These toxins are clearly smaller and potentially easier to get to grips with than the large *Photorhabdus* toxins and toxin delivery systems discussed so far. They may, therefore, also be easier to clone and utilize in insect control.

6.1. The PirAB binary toxins

The products of genes (*plu4093-plu4092* and *plu4437-plu4436*) in *P. luminescens* TT01 show oral activity against mosquito larvae (Waterfield et al., 2005). Because the encoded proteins showed similarity to both delta-endotoxins from *B. thurigiensis* and also a developmentally regulated protein from the Colorado potato beetle, *Leptinotarsa decemlineata*, they were termed the *Photorhabdus* insect related or Pir toxins. The developmentally regulated beetle protein has been inferred to possess juvenile hormone esterase activity but the PirAB toxins themselves show no such activity (Waterfield et al., 2005). Instead, they show injectable activity against the wax moth, *G. mellonella*, but only when injected together and not when injected individually, suggesting that both pirA and pirB are necessary as a binary toxin to cause insecticidal activity. The mode of action of these interesting proteins is still unclear but they are clearly potentially useful as mosquito larvicides. Specifically, PirAB shows good activity against the larvae of both *Aedes aegypti* (the vector of Dengue fever) and *Aedes albopictus*, while not affecting the *Mesocyclops thermocyclopoides* predator (Ahantarig et al., 2009). *Photorhabdus* PirAB may, therefore, form a useful alternative to toxins from *B. thuringiensis* and *Bacillus sphaericus* in the control of mosquito larvae.

6.2. The XaxAB and YaxAB cytotoxins

The XaxAB cytotoxin was first purified from the supernatant of *X. nematophila* grown in laboratory culture and was shown to have activity both against insect hemocytes and mammalian red blood cells (Vigneux et al., 2007). Pairs of genes encoding similar proteins are found in the insect pathogens *P. luminescens* (where they would presumably be termed *paxAB*) and *Pseudomonas entomophila*, the plant pathogen *P. syringae* and the human pathogens *Yersinia enterocolitica* (see following discussion) and *Proteus mirabilis*. The XaxAB toxin appears to trigger apoptosis in treated HeLa cells, and pre-treatment of the cells with the pan-caspase inhibitor z-VAD-fmk inhibits XaxAB cytotoxicity. Interestingly, maximal haemolytic activity is observed with exactly equal concentrations of both XaxA and XaxB. Moreover, addition of XaxA *then* XaxB in that specific order is required for toxicity. This is interesting and hard to understand that in some AB toxins, the A subunit carries the enzymatic activity and the B subunit is required for toxin binding, in which case one might expect the binding subunit (B) to have to be added before the toxic A subunit (Barth et al., 2004). Most recently, the activity of the XaxAB-like toxin in *P. luminescens* TT01 (here termed PaxAB), encoded by *plu1961* and *plu1962*, has also been examined (Zhang et al., 2014). Again both PaxA and PaxB are required for insecticidal activity and in combination showed effects on a Spruce budworm, *Choristoneura fumiferana*, a midgut-derived cell line FPMI-CF-203/2.5 or CF-203. Finally, homologous proteins to *Xenorhabdus* XaxAB are also found in *Y. enterocolitica* and have been named 'YaxAB' (Wagner et al., 2013). Mutant *Y. enterocolitica* lacking *yaxAB* genes shows similar final levels of infection as wild-type bacteria, but the time course of infection is delayed and a different pathology results in the mouse spleen. Again, the two proteins YaxA and YaxB are required for cytotoxicity which appears to be related to osmotic lysis triggered by pore formation. Indeed, protein modelling work shows similarities between YaxA and pore-forming toxins from *Bacillus cereus* (HBL-B) and *E. coli* (HlyE). Taken together, these data are consistent with the hypothesis that YaxAB is a pore-forming toxin that plays an important role in the progression of mouse infection. Overall, this suggests that the XaxAB-like binary toxins demand further work on their specific mode of action.

7. CLASSICAL SECRETIONS SYSTEMS AND NOVEL SCREENS

7.1. Type III and other classical secretion systems

It is not the intention of this review to produce another long list of the incredible diversity of secretion machinery found in the genomes of *Photorhabdus* and *Xenorhabdus*. If the reader is specifically interested in toxin secretion, then the reader is referred to previous reviews which list all known secretion systems as having representatives in *P. luminescens* except for those of the type IV secretion system (Rodou et al., 2010). Many of these classical secretion systems, such as the molecular syringe encoded by the type III secretion system, are clearly important in *Photorhabdus* infection but generally they are more highly described in other more human pathogenic bacteria such as *Salmonella*. That is not to say that study of the *P. luminescens* type III secretion device does not demand attention. Thus, elegant early studies by Brugirard-Ricaud et al. showed that *Photorhabdus* LopT, which is homologous to the *Yersinia* cysteine-protease YopT, could be translocated into mammalian cells in an active form when heterologously expressed in *Yersinia* (Brugirard-Ricaud et al., 2005). Moreover, the *Photorhabdus lopT* gene itself only appears to be transcribed in interactions with insect hemocytes, and knock-out of the type III secretion system increased phagocytosis and reduced nodulation (the formation of a melanin case around invading bacteria by a body of hemocytes). It, therefore, appears that the delivered LopT effector is important in reducing the phagocytosis of *Photorhabdus* and also in reducing the capacity of the insect hemocytes to envelop the invading bacteria in a melanin nodule. Similarly, *Photorhabdus* type III secretion system also delivers the cycle-inhibiting factor, named Cif (Chavez et al., 2010). Cycle-inhibiting factors induce the arrest of the cell cycle at the G_2/M or G_1/S transition by catalyzing the deamidation of a specific glutamine residue (Gln40) in the Neural-precursor-cell-Expressed Developmentally Down-regulated 8 (NEDD8) protein and the related protein ubiquitin (Crow et al., 2012). This is interesting from an evolutionary perspective as type III effectors often act as proteases or phosphatases but here a cysteine-protease-like fold in Cif has been adapted to perform a deamidation reaction. The *P. luminescens* Cif blocks the cell cycle at the G_2/M transition and triggers apoptosis in treated cells. Cif is expressed in the hemolymph of infected insects but is not essential for anti-insect virulence. Both of these elegant studies therefore show that the type III secretion system is deployed during infection of the insect by *Photorhabdus* and that an array of other undocumented effectors probably also have subtle but critical effects on the resulting infection.

7.2. RVA-like screens for novel effectors

Given the brief discussion of the array of classical secretion systems present in *Photorhabdus*, it does not seem likely that expressing random fragments of *Photorhabdus* DNA in laboratory strains of *E. coli* which are not secretion proficient or indeed armed with any specific secretion machinery (such as the type III-SS) would be useful in producing biologically active toxins. However, we reasoned that if a toxin was sufficiently toxic (like Mcf1) or if a toxin effectively encoded its own secretion system (like the Tc secretion system or TC-SS), then it should be possible to perform simple gain-of-function screens to look for novel toxins active in recombinant *E. coli*. We, therefore, designed a suite of simple gain-of-function screens that we termed 'Rapid Virulence Annotation' or 'RVA' (Waterfield et al., 2008). The first RVA-like screen obviously led to the discovery of Mcf1 and simply involved injecting individual recombinant *E. coli* clones carrying large (30–50 kb) insert sizes. As discussed above, this rapidly led to the discovery of the dominant Mcf1 toxin which has then also served as a useful positive control to test for the quality and insert size of any screened library. Since *Photorhabdus* has to interact with, and suppress, a range of different organisms, including its host nematode and a range of saprophytes such as bacteria, nematodes and fungi that invade the infected insect corpse, we expanded these simple screens to look at different insects, the model nematode, *C. elegans*, amoebae (as models for phagocytosis), insect phagocytes and macrophages (mammalian phagocytes). These screens are not only highly effective in recovering loci that encode toxic proteins but also recovered polyketide synthase and non-ribosomal peptide synthase encoding loci that make a range of different small molecules and peptides (Waterfield et al., 2008). Again, it is not the point of this review to simply provide a shopping list of all the hits from these screens, which are reported elsewhere (Waterfield et al., 2008), but to simply point out that the screens have revealed a further wealth of potentially bioactive proteins and small molecules from *Photorhabdus* that demand further detailed investigation.

7.3. Clustering methods to look for novel effector phenotypes

Finally, one might question the validity of isolating hundreds of new recombinant clones without the ability to rapidly characterize their potential modes of action. To this end, we have used high content microscopy and cluster analysis of the resulting phenotypic data to assign candidate clones into their likely phenotypic clusters (Dowling and Hodgson, 2014). In other words, we have taken detailed measurements of everything from treated cell shape, cell density, nuclear morphology, actin clustering and mitochondrial permeability to provide a snapshot of the treated cell. By then clustering unknown effectors with effectors for which we have a known mode of action, we can determine whether any new candidate effectors fall into clusters with known toxins or if their phenotype(s) suggest a novel mode of action (Dowling and Hodgson, 2014).

8. PERSPECTIVES FOR THE FUTURE OF *PHOTORHABDUS* TOXINS

Since 1998 and the first descriptions of the orally active Toxin complexes, we have come a long way in our understanding of *Photorhabdus* toxins. The study of *Photorhabdus* has been important not only because of the promise of producing an orally active agent for insect-resistant transgenic crops, but perhaps more importantly because *Photorhabdus* has now led the way in highlighting new secretion systems such as the TC-SS and the PVC-SS. Where do we go from here? Clearly, one central rate-limiting step in the discovery of insecticidal toxins is the assumption that they must be orally active against insects in their native recovered state. In this context, since we now understand something about how *Photorhabdus* toxins are actually delivered to their target cells, this demands a more thorough investigation of the receptors on those target cells and, therefore, of the likely tissue specificity (insect vs. mammalian or insect gut vs. insect hemocyte for example) of any delivery device. In fact, Raunser and co-workers specifically imply at the end of their beautiful paper on the mode of action of the Tcs that the A subunit receptors could be specifically modified or adapted to act as a drug delivery syringe in medicine. This makes the critical point that any given toxin is not specifically insecticidal because of the nature of the toxin that it delivers, which may be equally active on the cellular components of both insects and man, but that it is the nature and specificity of the cell surface receptor that must dictate a specific interaction with the insect midgut. Clearly, any successful use of *Photorhabdus* toxins in agriculture must therefore demand a further detailed investigation of their receptors in the insect midgut.

ACKNOWLEDGEMENTS

We would like to thank all members of the Richard ffrench-Constant laboratory, past and present, for their work on *Photorhabdus*. We would also like to thank the BBSRC, Leverhulme Trust and Royal Society of London for research funding.

REFERENCES

Ahantarig, A., Chantawat, N., Waterfield, N.R., ffrench-Constant, R., Kittayapong, P., 2009. PirAB toxin from *Photorhabdus asymbiotica* as a larvicide against dengue vectors. Appl. Environ. Microbiol. 75, 4627–4629.

Barth, H., Aktories, K., Popoff, M.R., Stiles, B.G., 2004. Binary bacterial toxins: biochemistry, biology, and applications of common *Clostridium* and *Bacillus* proteins. Microbiol. Mol. Biol. Rev. 68, 373–402 (table of contents).

Blackburn, M., Golubeva, E., Bowen, D., ffrench-Constant, R.H., 1998. A novel insecticidal toxin from *Photorhabdus luminescens*, Toxin complex a (Tca), and its histopathological effects on the midgut of *Manduca sexta*. Appl. Environ. Microbiol. 64, 3036–3041.

Blackburn, M.B., Martin, P.A., Kuhar, D., Farrar Jr., R.R., Gundersen-Rindal, D.E., 2011. The occurrence of *Photorhabdus*-like Toxin complexes in *Bacillus thuringiensis*. PLoS One 6, e18122.

Bowen, D.J., Ensign, J.C., 1998. Purification and characterization of a high-molecular-weight insecticidal protein complex produced by the entomopathogenic bacterium *Photorhabdus luminescens*. Appl. Environ. Microbiol. 64, 3029–3035.

Bowen, D., Rocheleau, T.A., Blackburn, M., Andreev, O., Golubeva, E., Bhartia, R., ffrench-Constant, R.H., 1998. Insecticidal toxins from the bacterium *Photorhabdus luminescens*. Science 280, 2129–2132.

Brown, M.J., Thoren, K.L., Krantz, B.A., 2011. Charge requirements for proton gradient-driven translocation of anthrax toxin. J. Biol. Chem. 286, 23189–23199.

Brugirard-Ricaud, K., Duchaud, E., Givaudan, A., Girard, P.A., Kunst, F., Boemare, N., Brehelin, M., Zumbihl, R., 2005. Site-specific antiphagocytic function of the *Photorhabdus luminescens* type III secretion system during insect colonization. Cell. Microbiol. 7, 363–371.

Budd, R.C., 2001. Activation-induced cell death. Curr. Opin. Immunol. 13, 356–362.

Busby, J.N., Panjikar, S., Landsberg, M.J., Hurst, M.R., Lott, J.S., 2013. The BC component of ABC toxins is an RHS-repeat-containing protein encapsulation device. Nature 501, 547–550.

Chaston, J.M., Suen, G., Tucker, S.L., Andersen, A.W., Bhasin, A., Bode, E., Bode, H.B., Brachmann, A.O., Cowles, C.E., Cowles, K.N., Darby, C., De Leon, L., Drace, K., Du, Z., Givaudan, A., Herbert Tran, E.E., Jewell, K.A., Knack, J.J., Krasomil-Osterfeld, K.C., Kukor, R., Lanois, A., Latreille, P., Leimgruber, N.K., Lipke, C.M., Liu, R., Lu, X., Martens, E.C., Marri, P.R., Medigue, C., Menard, M.L., Miller, N.M., Morales-Soto, N., Norton, S., Ogier, J.C., Orchard, S.S., Park, D., Park, Y., Qurollo, B.A., Sugar, D.R., Richards, G.R., Rouy, Z., Slominski, B., Slominski, K., Snyder, H., Tjaden, B.C., Van Der Hoeven, R., Welch, R.D., Wheeler, C., Xiang, B., Barbazuk, B., Gaudriault, S., Goodner, B., Slater, S.C., Forst, S., Goldman, B.S., Goodrich-Blair, H., 2011. The entomopathogenic bacterial endosymbionts *Xenorhabdus* and *Photorhabdus*: convergent lifestyles from divergent genomes. PLoS One 6, e27909.

Chavez, C.V., Jubelin, G., Courties, G., Gomard, A., Ginibre, N., Pages, S., Taieb, F., Girard, P.A., Oswald, E., Givaudan, A., Zumbihl, R., Escoubas, J.M., 2010. The cyclomodulin Cif of *Photorhabdus luminescens* inhibits insect cell proliferation and triggers host cell death by apoptosis. Microbes Infect. 12, 1208–1218.

Choe, S., Bennett, M.J., Fujii, G., Curmi, P.M., Kantardjieff, K.A., Collier, R.J., Eisenberg, D., 1992. The crystal structure of diphtheria toxin. Nature 357, 216–222.

Ciche, T.A., Ensign, J.C., 2003. For the insect pathogen *Photorhabdus luminescens*, which end of a nematode is out? Appl. Environ. Microbiol. 69, 1890–1897.

Crow, A., Hughes, R.K., Taieb, F., Oswald, E., Banfield, M.J., 2012. The molecular basis of ubiquitin-like protein NEDD8 deamidation by the bacterial effector protein Cif. Proc. Natl. Acad. Sci. U.S.A. 109, E1830–E1838.

Daborn, P.J., Waterfield, N., Blight, M.A., ffrench-Constant, R.H., 2001. Measuring virulence factor expression by the pathogenic bacterium *Photorhabdus luminescens* in culture and during insect infection. J. Bacteriol. 183, 5834–5839.

Daborn, P.J., Waterfield, N., Silva, C.P., Au, C.P., Sharma, S., ffrench-Constant, R.H., 2002. A single *Photorhabdus* gene, makes caterpillars floppy (mcf), allows *Escherichia coli* to persist within and kill insects. Proc. Natl. Acad. Sci. U.S.A. 99, 10742–10747.

Dowling, A.J., Hodgson, D.J., 2014. An unbiased method for clustering bacterial effectors using host cellular phenotypes. Appl. Environ. Microbiol. 80, 1185–1196.

Dowling, A.J., Daborn, P.J., Waterfield, N.R., Wang, P., Streuli, C.H., ffrench-Constant, R.H., 2004. The insecticidal toxin Makes caterpillars floppy (Mcf) promotes apoptosis in mammalian cells. Cell. Microbiol. 6, 345–353.

Duchaud, E., Rusniok, C., Frangeul, L., Buchrieser, C., Givaudan, A., Taourit, S., Bocs, S., Boursaux-Eude, C., Chandler, M., Charles, J.F., Dassa, E., Derose, R., Derzelle, S., Freyssinet, G., Gaudriault, S., Medigue, C., Lanois, A., Powell, K., Siguier, P., Vincent, R., Wingate, V., Zouine, M., Glaser, P., Boemare, N., Danchin, A., Kunst, F., 2003. The genome sequence of the entomopathogenic bacterium *Photorhabdus luminescens*. Nat. Biotechnol. 21, 1307–1313.

ffrench-Constant, R.H., Waterfield, N., Burland, V., Perna, N.T., Daborn, P.J., Bowen, D., Blattner, F.R., 2000. A genomic sample sequence of the entomopathogenic bacterium *Photorhabdus luminescens* W14: potential implications for virulence. Appl. Environ. Microbiol. 66, 3310–3329.

ffrench-Constant, R., Waterfield, N., Daborn, P., Joyce, S., Bennett, H., Au, C., Dowling, A., Boundy, S., Reynolds, S., Clarke, D., 2003. *Photorhabdus*: towards a functional genomic analysis of a symbiont and pathogen. FEMS Microbiol. Rev. 26, 433–456.

ffrench-Constant, R.H., Dowling, A., Waterfield, N.R., 2007. Insecticidal toxins from *Photorhabdus* bacteria and their potential use in agriculture. Toxicon 49, 436–451.

Forst, S., Nealson, K., 1996. Molecular biology of the symbiotic-pathogenic bacteria *Xenorhabdus* spp. and *Photorhabdus* spp.. Microbiol. Rev. 60, 21–43.

Forst, S., Dowds, B., Boemare, N., Stackebrandt, E., 1997. *Xenorhabdus* and *Photorhabdus* spp.: bugs that kill bugs. Annu. Rev. Microbiol. 51, 47–72.

Gatsogiannis, C., Lang, A.E., Meusch, D., Pfaumann, V., Hofnagel, O., Benz, R., Aktories, K., Raunser, S., 2013. A syringe-like injection mechanism in *Photorhabdus luminescens* toxins. Nature 495, 520–523.

Gendlina, I., Held, K.G., Bartra, S.S., Gallis, B.M., Doneanu, C.E., Goodlett, D.R., Plano, G.V., Collins, C.M., 2007. Identification and type III-dependent secretion of the *Yersinia pestis* insecticidal-like proteins. Mol. Microbiol. 64, 1214–1227.

Gerrard, J., Waterfield, N., Vohra, R., ffrench-Constant, R., 2004. Human infection with *Photorhabdus asymbiotica*: an emerging bacterial pathogen. Microbes Infect. 6, 229–237.

Gerrard, J.G., Joyce, S.A., Clarke, D.J., ffrench-Constant, R.H., Nimmo, G.R., Looke, D.F., Feil, E.J., Pearce, L., Waterfield, N.R., 2006. Nematode symbiont for *Photorhabdus asymbiotica*. Emerg. Infect. Dis. 12, 1562–1564.

Heymann, J.B., Bartho, J.D., Rybakova, D., Venugopal, H.P., Winkler, D.C., Sen, A., Hurst, M.R., Mitra, A.K., 2013. Three-dimensional structure of the toxin-delivery particle antifeeding prophage of *Serratia entomophila*. J. Biol. Chem. 288, 25276–25284.

Hong, W., Mosca, T.J., Luo, L., 2012. Teneurins instruct synaptic partner matching in an olfactory map. Nature 484, 201–207.

Huang, Y., Callahan, S., Hadfield, M.G., 2012. Recruitment in the sea: bacterial genes required for inducing larval settlement in a polychaete worm. Sci. Rep. 2, 228.

Hurst, M.R., Glare, T.R., Jackson, T.A., Ronson, C.W., 2000. Plasmid-located pathogenicity determinants of *Serratia entomophila*, the causal agent of amber disease of grass grub, show similarity to the insecticidal toxins of Photorhabdus luminescens. J. Bacteriol. 182, 5127–5138.

Hurst, M.R., Jones, S.A., Binglin, T., Harper, L.A., Jackson, T.A., Glare, T.R., 2011. The main virulence determinant of *Yersinia entomophaga* MH96 is a broad-host-range toxin complex active against insects. J. Bacteriol. 193, 1966–1980.

Jank, T., Bogdanovic, X., Wirth, C., Haaf, E., Spoerner, M., Bohmer, K.E., Steinemann, M., Orth, J.H., Kalbitzer, H.R., Warscheid, B., Hunte, C., Aktories, K., 2013. A bacterial toxin catalyzing tyrosine glycosylation of Rho and deamidation of Gq and Gi proteins. Nat. Struct. Mol. Biol. 20, 1273–1280.

Kung, V.L., Khare, S., Stehlik, C., Bacon, E.M., Hughes, A.J., Hauser, A.R., 2012. An rhs gene of *Pseudomonas aeruginosa* encodes a virulence protein that activates the inflammasome. Proc. Natl. Acad. Sci. U.S.A. 109, 1275–1280.

Kupferschmied, P., Pechy-Tarr, M., Imperiali, N., Maurhofer, M., Keel, C., 2014. Domain shuffling in a sensor protein contributed to the evolution of insect pathogenicity in plant-beneficial *Pseudomonas protegens*. PLoS Pathog. 10, e1003964.

Landsberg, M.J., Jones, S.A., Rothnagel, R., Busby, J.N., Marshall, S.D., Simpson, R.M., Lott, J.S., Hankamer, B., Hurst, M.R., 2011. 3D structure of the *Yersinia entomophaga* toxin complex and implications for insecticidal activity. Proc. Natl. Acad. Sci. U.S.A. 108, 20544–20549.

Lang, A.E., Schmidt, G., Schlosser, A., Hey, T.D., Larrinua, I.M., Sheets, J.J., Mannherz, H.G., Aktories, K., 2010. *Photorhabdus luminescens* toxins ADP-ribosylate actin and RhoA to force actin clustering. Science 327, 1139–1142.

Lang, A.E., Schmidt, G., Sheets, J.J., Aktories, K., 2011. Targeting of the actin cytoskeleton by insecticidal toxins from *Photorhabdus luminescens*. Naunyn Schmiedebergs Arch. Pharmacol. 383, 227–235.

Lang, A.E., Konukiewitz, J., Aktories, K., Benz, R., 2013. TcdA1 of *Photorhabdus luminescens*: electrophysiological analysis of pore formation and effector binding. Biophys. J. 105, 376–384.

Lang, A.E., Ernst, K., Lee, H., Papatheodorou, P., Schwan, C., Barth, H., Aktories, K., 2014. The chaperone Hsp90 and PPIases of the cyclophilin and FKBP families facilitate membrane translocation of *Photorhabdus luminescens* ADP-ribosyltransferases. Cell. Microbiol. 16, 490–503.

Lanois, A., Ogier, J.C., Gouzy, J., Laroui, C., Rouy, Z., Givaudan, A., Gaudriault, S., 2013. Draft genome sequence and annotation of the entomopathogenic bacterium *Xenorhabdus nematophila* strain F1. Genome Announc. 1 e00342-13.

Lee, S.C., Stoilova-Mcphie, S., Baxter, L., Fulop, V., Henderson, J., Rodger, A., Roper, D.I., Scott, D.J., Smith, C.J., Morgan, J.A., 2007. Structural characterisation of the insecticidal toxin XptA1, reveals a 1.15 MDa tetramer with a cage-like structure. J. Mol. Biol. 366, 1558–1568.

Leiman, P.G., Basler, M., Ramagopal, U.A., Bonanno, J.B., Sauder, J.M., Pukatzki, S., Burley, S.K., Almo, S.C., Mekalanos, J.J., 2009. Type VI secretion apparatus and phage tail-associated protein complexes share a common evolutionary origin. Proc. Natl. Acad. Sci. U.S.A. 106, 4154–4159.

Liu, D., Burton, S., Glancy, T., Li, Z.S., Hampton, R., Meade, T., Merlo, D.J., 2003. Insect resistance conferred by 283-kDa *Photorhabdus luminescens* protein TcdA in Arabidopsis thaliana. Nat. Biotechnol. 21, 1222–1228.

Marshall, S.D., Hares, M.C., Jones, S.A., Harper, L.A., Vernon, J.R., Harland, D.P., Jackson, T.A., Hurst, M.R., 2012. Histopathological effects of the Yen-Tc toxin complex from *Yersinia entomophaga* MH96 (Enterobacteriaceae) on the *Costelytra zealandica* (Coleoptera: Scarabaeidae) larval midgut. Appl. Environ. Microbiol. 78, 4835–4847.

Mclaughlin, L.M., Govoni, G.R., Gerke, C., Gopinath, S., Peng, K., Laidlaw, G., Chien, Y.H., Jeong, H.W., Li, Z., Brown, M.D., Sacks, D.B., Monack, D., 2009. The *Salmonella* SPI2 effector SseI mediates long-term systemic infection by modulating host cell migration. PLoS Pathog. 5, e1000671.

Meusch, D., Gatsogiannis, C., Efremov, R.G., Lang, A.E., Hofnagel, O., Vetter, I.R., Aktories, K., Raunser, S., 2014. Mechanism of Tc toxin action revealed in molecular detail. Nature 508, 61–65.

Moak, M., Molineux, I.J., 2004. Peptidoglycan hydrolytic activities associated with bacteriophage virions. Mol. Microbiol. 51, 1169–1183.

Morgan, J.A., Sergeant, M., Ellis, D., Ousley, M., Jarrett, P., 2001. Sequence analysis of insecticidal genes from *Xenorhabdus nematophilus* PMFI296. Appl. Environ. Microbiol. 67, 2062–2069.
Nakayama, K., Takashima, K., Ishihara, H., Shinomiya, T., Kageyama, M., Kanaya, S., Ohnishi, M., Murata, T., Mori, H., Hayashi, T., 2000. The R-type pyocin of *Pseudomonas aeruginosa* is related to P2 phage, and the F-type is related to lambda phage. Mol. Microbiol. 38, 213–231.
Peat, S.M., ffrench-Constant, R.H., Waterfield, N.R., Marokhazi, J., Fodor, A., Adams, B.J., 2010. A robust phylogenetic framework for the bacterial genus *Photorhabdus* and its use in studying the evolution and maintenance of bioluminescence: a case for 16S, gyrB, and glnA. Mol. Phylogenet. Evol. 57, 728–740.
Pechy-Tarr, M., Bruck, D.J., Maurhofer, M., Fischer, E., Vogne, C., Henkels, M.D., Donahue, K.M., Grunder, J., Loper, J.E., Keel, C., 2008. Molecular analysis of a novel gene cluster encoding an insect toxin in plant-associated strains of *Pseudomonas fluorescens*. Environ. Microbiol. 10, 2368–2386.
Pechy-Tarr, M., Borel, N., Kupferschmied, P., Turner, V., Binggeli, O., Radovanovic, D., Maurhofer, M., Keel, C., 2013. Control and host-dependent activation of insect toxin expression in a root-associated biocontrol pseudomonad. Environ. Microbiol. 15, 736–750.
Petosa, C., Collier, R.J., Klimpel, K.R., Leppla, S.H., Liddington, R.C., 1997. Crystal structure of the anthrax toxin protective antigen. Nature 385, 833–838.
Purcell, J.P., Greenplate, J.T., Jennings, M.G., Ryerse, J.S., Pershing, J.C., Sims, S.R., Prinsen, M.J., Corbin, D.R., Tran, M., Sammons, R.D., et al., 1993. Cholesterol oxidase: a potent insecticidal protein active against boll weevil larvae. Biochem. Biophys. Res. Commun. 196, 1406–1413.
Rodou, A., Ankrah, D.O., Stathopoulos, C., 2010. Toxins and secretion systems of *Photorhabdus luminescens*. Toxins (Basel) 2, 1250–1264.
Sergeant, M., Jarrett, P., Ousley, M., Morgan, J.A., 2003. Interactions of insecticidal toxin gene products from *Xenorhabdus nematophilus* PMFI296. Appl. Environ. Microbiol. 69, 3344–3349.
Sheets, J.J., Hey, T.D., Fencil, K.J., Burton, S.L., Ni, W., Lang, A.E., Benz, R., Aktories, K., 2011. Insecticidal toxin complex proteins from *Xenorhabdus nematophilus*: structure and pore formation. J. Biol. Chem. 286, 22742–22749.
Shikuma, N.J., Pilhofer, M., Weiss, G.L., Hadfield, M.G., Jensen, G.J., Newman, D.K., 2014. Marine tubeworm metamorphosis induced by arrays of bacterial phage tail-like structures. Science 343, 529–533.
Somvanshi, V.S., Sloup, R.E., Crawford, J.M., Martin, A.R., Heidt, A.J., Kim, K.S., Clardy, J., Ciche, T.A., 2012. A single promoter inversion switches *Photorhabdus* between pathogenic and mutualistic states. Science 337, 88–93.
Spinner, J.L., Jarrett, C.O., Larock, D.L., Miller, S.I., Collins, C.M., Hinnebusch, B.J., 2012. *Yersinia pestis* insecticidal-like Toxin complex (Tc) family proteins: characterization of expression, subcellular localization, and potential role in infection of the flea vector. BMC Microbiol. 12, 296.
Tan, B., Jackson, T.A., Hurst, M.R., 2006. Virulence of *Serratia* strains against *Costelytra zealandica*. Appl. Environ. Microbiol. 72, 6417–6418.
Vadyvaloo, V., Jarrett, C., Sturdevant, D.E., Sebbane, F., Hinnebusch, B.J., 2010. Transit through the flea vector induces a pretransmission innate immunity resistance phenotype in *Yersinia pestis*. PLoS Pathog. 6, e1000783.
Vigneux, F., Zumbihl, R., Jubelin, G., Ribeiro, C., Poncet, J., Baghdiguian, S., Givaudan, A., Brehelin, M., 2007. The *xaxAB* genes encoding a new apoptotic toxin from the insect pathogen *Xenorhabdus nematophila* are present in plant and human pathogens. J. Biol. Chem. 282, 9571–9580.

Visschedyk, D.D., Perieteanu, A.A., Turgeon, Z.J., Fieldhouse, R.J., Dawson, J.F., Merrill, A.R., 2010. Photox, a novel actin-targeting mono-ADP-ribosyltransferase from *Photorhabdus luminescens*. J. Biol. Chem. 285, 13525–13534.

Vlisidou, I., Dowling, A.J., Evans, I.R., Waterfield, N., ffrench-Constant, R.H., Wood, W., 2009. *Drosophila* embryos as model systems for monitoring bacterial infection in real time. PLoS Pathog. 5, e1000518.

Wagner, R., Wolff, T., Herwig, A., Pleschka, S., Klenk, H.D., 2000. Interdependence of hemagglutinin glycosylation and neuraminidase as regulators of influenza virus growth: a study by reverse genetics. J. Virol. 74, 6316–6323.

Wagner, N.J., Lin, C.P., Borst, L.B., Miller, V.L., 2013. YaxAB, a *Yersinia enterocolitica* pore-forming toxin regulated by RovA. Infect. Immun. 81, 4208–4219.

Waterfield, N., Dowling, A., Sharma, S., Daborn, P.J., Potter, U., ffrench-Constant, R.H., 2001a. Oral toxicity of *Photorhabdus luminescens* W14 toxin complexes in *Escherichia coli*. Appl. Environ. Microbiol. 67, 5017–5024.

Waterfield, N.R., Bowen, D.J., Fetherston, J.D., Perry, R.D., ffrench-Constant, R.H., 2001b. The *tc* genes of *Photorhabdus*: a growing family. Trends Microbiol. 9, 185–191.

Waterfield, N.R., Daborn, P.J., Dowling, A.J., Yang, G., Hares, M., ffrench-Constant, R.H., 2003. The insecticidal toxin Makes caterpillars floppy 2 (Mcf2) shows similarity to HrmA, an avirulence protein from a plant pathogen. FEMS Microbiol. Lett. 229, 265–270.

Waterfield, N., Kamita, S.G., Hammock, B.D., ffrench-Constant, R., 2005. The *Photorhabdus* Pir toxins are similar to a developmentally regulated insect protein but show no juvenile hormone esterase activity. FEMS Microbiol. Lett. 245, 47–52.

Waterfield, N.R., Sanchez-Contreras, M., Eleftherianos, I., Dowling, A., Yang, G., Wilkinson, P., Parkhill, J., Thomson, N., Reynolds, S.E., Bode, H.B., Dorus, S., ffrench-Constant, R.H., 2008. Rapid Virulence Annotation (RVA): identification of virulence factors using a bacterial genome library and multiple invertebrate hosts. Proc. Natl. Acad. Sci. U.S.A. 105, 15967–15972.

Waterfield, N.R., Ciche, T., Clarke, D., 2009. *Photorhabdus* and a host of hosts. Annu. Rev. Microbiol. 63, 557–574.

Wiggins, C.A., Munro, S., 1998. Activity of the yeast MNN1 alpha-1,3-mannosyltransferase requires a motif conserved in many other families of glycosyltransferases. Proc. Natl. Acad. Sci. U.S.A. 95, 7945–7950.

Wilkinson, P., Waterfield, N.R., Crossman, L., Corton, C., Sanchez-Contreras, M., Vlisidou, I., Barron, A., Bignell, A., Clark, L., Ormond, D., Mayho, M., Bason, N., Smith, F., Simmonds, M., Churcher, C., Harris, D., Thompson, N.R., Quail, M., Parkhill, J., ffrench-Constant, R.H., 2009. Comparative genomics of the emerging human pathogen *Photorhabdus asymbiotica* with the insect pathogen *Photorhabdus luminescens*. BMC Genomics 10, 302.

Wilkinson, P., Paszkiewicz, K., Moorhouse, A., Szubert, J.M., Beatson, S., Gerrard, J., Waterfield, N.R., ffrench-Constant, R.H., 2010. New plasmids and putative virulence factors from the draft genome of an Australian clinical isolate of *Photorhabdus asymbiotica*. FEMS Microbiol. Lett. 309, 136–143.

Yang, G., Waterfield, N.R., 2013. The role of TcdB and TccC subunits in secretion of the *Photorhabdus* Tcd toxin complex. PLoS Pathog. 9, e1003644.

Yang, G., Dowling, A.J., Gerike, U., ffrench-Constant, R.H., Waterfield, N.R., 2006. *Photorhabdus* virulence cassettes confer injectable insecticidal activity against the wax moth. J. Bacteriol. 188, 2254–2261.

Zhang, D., De Souza, R.F., Anantharaman, V., Iyer, L.M., Aravind, L., 2012. Polymorphic toxin systems: comprehensive characterization of trafficking modes, processing, mechanisms of action, immunity and ecology using comparative genomics. Biol. Direct 7, 18.

Zhang, X., Hu, X., Li, Y., Ding, X., Yang, Q., Sun, Y., Yu, Z., Xia, L., Hu, S., 2014. XaxAB-like binary toxin from *Photorhabdus luminescens* exhibits both insecticidal activity and cytotoxicity. FEMS Microbiol. Lett. 350, 48–56.

CHAPTER EIGHT

Methods for Deployment of Spider Venom Peptides as Bioinsecticides

Volker Herzig*, Niraj S. Bende*, Md. Shohidul Alam*, H. William Tedford†, Robert M. Kennedy†, Glenn F. King*

*Division of Chemistry and Structural Biology, Institute for Molecular Bioscience, The University of Queensland, St. Lucia, Queensland, Australia
†Vestaron Corporation, Kalamazoo, Michigan, USA

Contents

Abstract

Despite intensive control measures, insect pests cause enormous damage to crops and stored grain. In addition, insects vector some of the world's most devastating human diseases, including malaria, dengue and Chagas disease. Chemical insecticides remain the dominant method of controlling insect pests but the arsenal of extant insecticides is rapidly diminishing due to the evolution of resistance in pest species and a more difficult regulatory environment due to heightened concern about the potential adverse impacts of chemical insecticides on human health and the environment. Along with predatory beetles, spiders are the most successful insect predators on the planet and

Advances in Insect Physiology, Volume 47
ISSN 0065-2806
http://dx.doi.org/10.1016/B978-0-12-800197-4.00008-7

their venoms contain a diverse array of small, disulfide-rich insecticidal peptides. In addition to being potent and highly selective for insects, they collectively offer very diverse pharmacology and should degrade to innocuous breakdown products in the field. However, their major disadvantage is a low level of intrinsic oral activity. Consequently, much research has been directed towards improving the oral activity of these peptide toxins or, alternatively, obviating the oral activity problem altogether by incorporating transgenes encoding the toxins into suitable biological delivery vehicles. Here, we discuss recent advances in both of these areas. We also discuss an approach that merges the advantages of small-molecule and peptide-based insecticides, namely using the pharmacophore of an insecticidal spider-venom peptide to rationally develop a small-molecule mimetic that has improved oral activity, but retains activity for a novel insecticide target.

1. INTRODUCTION

The human population is projected to reach to 9.3 billion by 2050 (Bloom, 2011). This expansion of the population will create enormous pressure on food production since there is limited scope for increasing the amount of cultivated land (Oerke, 2006). Despite intensive control measures, herbivorous insects reduce world crop yields by 10–14% (Oerke, 2006) and damage as much as 30% of stored grain (Boyer et al., 2012), and hence their control is a key factor in determining food productivity. Increases in human population density combined with global warming and a dramatic escalation in international travel has also placed considerable pressure on the public health sector to control the spread of insects that vector human diseases such as malaria, dengue and Chagas disease.

Unfortunately, the arsenal of chemical insecticides available to combat insect pests is in rapid decline. This is largely due to the fact that most chemical insecticides target a handful of enzymes, receptors and ion channels in the insect nervous system, and consequently their long-term use has led to widespread resistance in pest species (King and Hardy, 2013; Smith et al., 2013). Concerns about the adverse impact of certain classes of chemical insecticides on the environment and human health have also led to widespread de-registrations and use cancellations; for example, over the period 2005–2009, the U.S. Environmental Protection Agency (EPA) de-registered or limited the use of 169 insecticides, while only nine new insecticides were registered during the same time period (Windley et al., 2012). In simple terms, we have to combat an increasing insect pest problem with a rapidly diminishing chemical arsenal.

2. SPIDER VENOM PEPTIDES AS BIOINSECTICIDES

Spiders are among the most successful and abundant insect predators on the planet. They play a key role in keeping insect populations at bay in a wide range of ecosystems (Nyffeler and Sunderland, 2003). During more than 300 million years of co-evolution with insects, spiders not only evolved sophisticated ways of using silk for prey capture (Blackledge et al., 2009; Vollrath and Selden, 2007), but also acquired powerful venoms that rapidly paralyse insect prey. Spider venoms are complex chemical cocktails of salts, small organic molecules, peptides and proteins, but the dominant components in most spider venoms are small, disulfide-rich peptides (King and Hardy, 2013; Kuhn-Nentwig et al., 2011). Individual spider venoms can contain more than 1000 small peptides (Escoubas et al., 2006). These peptides simultaneously attack virtually every component of the synaptic machinery in the central or peripheral nervous system of envenomated prey, which is tantamount to "shock-and-awe" at the molecular level (Ikonomopoulou and King, 2013; King and Hardy, 2013).

Given that more than 44,500 extant species of spiders have been described to date (Platnick, 2013), then a conservative estimate of 250 peptides per venom indicates that as a group spider venoms are likely to contain >10 million bioactive peptides (King, 2011), the vast majority of which are expected to be insecticidal. Hence, spider venoms provide an almost inexhaustible source of insecticidal molecules. To date, more than 200 disulfide-rich insecticidal spider-venom peptides (ISVPs) have been sequenced, ranging in size from 3.3 to 9.0 kDa and containing three to six disulfide bonds (King and Hardy, 2013). Several of these peptides are devoid of vertebrate activity and sufficiently potent against pest insects (i.e. $LD_{50} < 1500$ pmol/g) and to warrant serious consideration as bioinsecticides (Table 8.1).

Almost all ISVPs isolated to date contain a structural motif known as an inhibitor cystine knot (King et al., 2002; Pallaghy et al., 1994) that provides these peptides with remarkable chemical and thermal stability as well as resistance to proteases (King and Hardy, 2013; Saez et al., 2010). Thus, in contrast with most other peptides and proteins, biological stability is not a concern with respect to use of ISVPs as bioinsecticides. ISVPs should be stable during long periods of storage as well as in the field, and they should degrade to innocuous breakdown products. Spider-venom peptides can also be produced using a range of heterologous expression systems that enable cost-effective large-scale production (Klint et al., 2013).

Table 8.1 Properties of ISVPs with potential for development as bioinsecticides[a]

Rational name	Synonyms	Species	ArachnoServer ID[b]	Mass (kDa)	Disulfide bonds	Molecular target	Phyletic specificity	LD_{50} (pmol/g)[c]	Oral activity	Vertebrate activity
μ-AGTX-Aa1d	μ-Agatoxin IV	*Agelenopsis aperta*	381	4.20	4	Na_V channel	D, L, O[d]	30 (D)	N.D.[e]	No
μ-CUTX-As1a	Aptotoxin 3; Aps III	*Apomastus schlingeri*	403	3.76	4	Na_V channel	D,[d] L	133 (L)	N.D.	N.D.
U$_1$-CUTX-As1c	Aptotoxin 6; Aps VI	*Apomastus schlingeri*	407	8.31	4	N.D.	L	2 (L)	N.D.	N.D.
U$_1$-NETX-Csp1a	Peptide A	*Calisoga* sp.	28	4.30	4	N.D.	L	551 (L)	N.D.	No
μ-DGTX-Dc1a	DTX9.2	*Diguetia canities*	352	6.39	4	Na_V channel	B,[d] D,[d] L	380 (L)	N.D.	No
κ-HXTX-Hv1c	J-ACTX-Hv1c	*Hadronyche versuta*	174	3.76	4	K_{Ca} channel	B, C,[d] D, L, O	91 (D)	Low[d]	No
ω-HXTX-Hv1a	V-1	*Hadronyche versuta*	193	4.05	3	Ca_V channel	B, C, D, H, L, O	77 (D)	Low[d]	No
ω-HXTX-Hv2a	–	*Hadronyche versuta*	204	4.48	3	Ca_V channel	B,[d] C,[d] D, O	160 (C)	N.D.	No
Γ-CNTX-Pn1a	Tx4(5-5); Pn4A	*Phoneutria nigriventer*	279	5.17	5	Glu receptor (NMDA subtype)	B, D, O	<100 (D)	N.D.	No

δ-CNTX-Pn1a	Tx4(6-1); PnTx4(6-1)	*Phoneutria nigriventer*	280	5.24	5	Na_V channel	B, D	36 (D)	N.D.	No
δ-AMATX-Pl1a	δ-palutoxin IT1	*Pireneitega luctuosa*	301	4.03	4	Na_V channel	D, H, L	2347 (L)	No	N.D.
U_1-PLTX-Pt1a	Plectoxin-V/VI	*Plectreurys tristis*	405	5.07	5	N.D.	D,[d] L	14 (L)	N.D.	N.D.
U_2-SGTX-Sf1b	Toxin SF12; Toxin F5.7	*Segestria florentina*	117	5.12	4	N.D.	L	1365 (L)	N.D.	No
U_1-TRTX-Sp1a	OAIP-1	*Selenotypus plumipes*	1495	3.81	3	N.D.	B, C, L	2 (C)	Moderate	N.D.
U_1-AGTX-Ta1a	TaITX-1	*Tegenaria agrestis*	349	5.68	3	N.D.	C, D,[d] L	780 (L)	N.D.	No

[a]Table is modified and updated from King and Hardy (2013) and includes only ISVPs reported to have no toxicity on vertebrates. Phyletic specificity refers to the insect orders in which activity has been reported: Blattaria (B), Coleoptera (C), Diptera (D), Hemiptera (H), Lepidoptera (L) and Orthoptera (O). Na_V, voltage-gated sodium channel; Ca_V, voltage-activated calcium channel; K_{Ca}, calcium-activated potassium channel.
[b]Refers to the toxin ID number in the ArachnoServer database. To navigate to the toxin card, use http://www.arachnoserver.org/toxincard.html?id=ID-number.
[c]LD_{50} value is given for the insect order (in parentheses) in which the peptide was found to be most potent.
[d]V. Herzig, N. Bende, and G.F. King (unpublished data).
[e]N.D., not determined.

The biggest potential disadvantage of ISVPs is that they are not orally or topically active. Spiders have solved the delivery problem by evolving fangs that act like hypodermic syringes to deliver their venoms into the haemolymph of prey, thus enabling ISVPs to access their targets in the insect nervous system. We have found that the LD_{50} values for ISVPs fed to insects are about 100-fold *higher* than when they are injected (Hardy et al., 2013; King and Hardy, 2013), presumably due to a slow rate of absorption in the insect gut, as observed previously for insect neuropeptides (Audsley et al., 2008) and disulfide-rich peptides from scorpion and snake venoms (Casartelli et al., 2005). Nevertheless, the low level of intrinsic oral activity of ISVPs is not restricted to insects, as other arthropod pests such as ticks were found to be sensitive to oral administration of ω-hexatoxin-Hv1a (hereafter Hv1a) (Mukherjee et al., 2006).

There is growing evidence that some peptides and proteins can access paracellular and transcellular transport routes in the insect midgut. Septate junctions in insects are leaky for molecules smaller than 5 kDa (Casartelli et al., 2005; Skaer et al., 1987), and this paracellular route might explain the observed uptake of insect neuropeptides and ISVPs (Audsley et al., 2007; Hardy et al., 2013). However, size alone is not the sole determinant of uptake rate, as the 0.9-kDa insect neuropeptide cydiastatin-4 is translocated across the midgut of the lepidopteran *Manduca sexta* at a rate 42 times *lower* than for Hv1a, a 4-kDa ISVP (V. Herzig, N. Audsley, and G.F. King, unpublished data). In addition to the paracellular route, the translocation of large proteins such as albumin across the insect midgut epithelium involves active energy-dependent processes such as transcytosis (Casartelli et al., 2005).

Even though our understanding of transport routes across the insect midgut is still in its infancy, it is clear that some peptides can pass undegraded through the midgut, consistent with the observed insecticidal effects of ISVPs after oral administration. Further research in this area might facilitate the rational introduction of chemical modifications and/or the use of formulations that will enhance the oral activity of ISVPs and facilitate their direct deployment as bioinsecticides. However, the remainder of this chapter will focus on alternative methods of delivering ISVPs to target insect species as well as the option of developing orally active small-molecule ISVP mimetics.

3. TRANSCYTOSIS OF SPIDER VENOM PEPTIDES ACROSS THE INSECT GUT EPITHELIUM VIA FUSION TO MOLECULAR TRANSPORT VEHICLES

ISVPs appear to be able to access the insect haemolymph via paracellular transport. Although this route of uptake is extremely inefficient, it could be enhanced by formulation with appropriate solvents and adjuvants. However, a more generic approach for improving the oral activity of these peptides might be to fuse them to a protein that is capable of being transcytosed across the insect gut epithelium. These fusion proteins would thus serve as molecular transporters that could ferry ISVPs across the insect gut.

This approach was first demonstrated via fusion of ISVPs to a plant lectin. Lectins are protease-resistant, carbohydrate-binding proteins that are widely distributed in animals, plants and micro-organisms. Insecticidal plant lectins play an important role in defence against herbivorous insects (Michiels et al., 2010). *Galanthus nivalis* agglutinin (GNA), a small 14-kDa mannose-specific lectin from the snowdrop plant, is taken up by receptor-mediated endocytosis in the insect midgut after binding to aminopeptidase N, a membrane-bound glycoprotein. Subsequent transcytosis leads to its accumulation in haemolymph, Malpighian tubules, fat bodies, ovarioles and the central nerve cord (Fitches et al., 2001, 2012) (Fig. 8.1). Thus, the transcellular uptake of GNA can be exploited as a mechanism for delivering ISVPs to the insect haemolymph and nervous system (reviewed in Bonning and Chougule, 2014). For example, the oral activity of both Hv1a and the ISVP U_2-SGTX-Sf1a (SFI1) was improved dramatically by fusion to the N-terminus of GNA. Second instar cabbage moths, *Mamestra brassicae*, fed on cabbage leaves coated with 0.2% Hv1a–GNA fusion protein suffered 85% mortality after 10 days, whereas control mortality was <20% (Fitches et al., 2012). Similarly, although neither GNA nor SFI1 was toxic when fed to fourth instar tomato moths, *Lacanobia oleracea*, the inclusion of a GNA–SFI1 fusion protein as 2.5% of dietary protein was insecticidal to first instar larvae, causing 100% mortality after 6 days, and to fifth instar larvae, causing 20% mortality and threefold reduced weight gain compared to controls (Fitches et al., 2004).

An alternative fusion-protein approach recently introduced for specific delivery of ISVPs to aphids (Bonning et al., 2014) relies on the fact that aphids and other sap-sucking hemipterans vector a large number of plant viruses. Luteoviruses are specifically vectored by aphids; the virus is ingested during feeding on an infected plant and then it persists in the aphid without replicating. Remarkably, the virus is able to move from the aphid gut into the haemocoel via transcytosis, which relies on specific recognition of the viral coat protein (CP) via gut receptors followed by receptor-mediated endocytosis (Bonning and Chougule, 2014). Thus, as for GNA, the virus CP can be used as a delivery vehicle for ISVPs. Hv1a has very little intrinsic oral activity against aphids (Pal et al., 2013), but it was recently demonstrated to be highly orally active against a wide range of aphids when fused to the

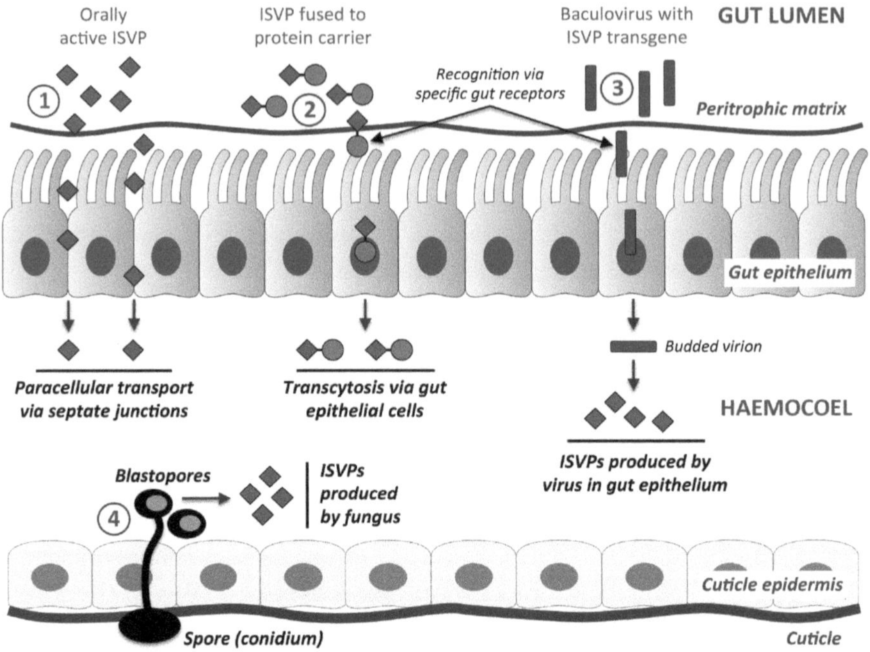

Figure 8.1 Summary of methods for delivering ISVPs to the insect haemocoel: (1) Some ISVPs appear to be able to access the haemocoel via paracellular uptake through leaky septate junctions. (2) ISVPs can be fused to carrier proteins such as snowdrop lectin (GNA) and the coat protein of pea enation mosaic virus, which are recognised by specific receptors on gut epithelial cells. These fusion proteins can be taken up via receptor-mediated endocytosis and delivered to the haemocoel by transcytosis. (3) Baculovirus can be engineered to encode an ISVP transgene. Upon recognition by specific gut receptors, budded virions infect gut epithelial cells and other body tissues and release ISVPs. (4) Entomopathogenic fungi such as *Metarhizium* and *Beauveria* can also be engineered to encode an ISVP transgene. Fungal spores (conidia) adhere to the insect cuticle, germinate and then invade the integument. The fungus then proliferates in the haemocoel in the form of wall-less blastopores, which produce and release ISVPs. *Adapted from Bonning and Chougule (2014).* (See the colour plate.)

C-terminus of the CP of pea enation mosaic virus (PEMV); the CP–Hv1a fusion protein was even orally active against the soybean aphid, *Aphis glycines*, which is not a vector for PEMV (Bonning et al., 2014). The CPs from two other luteoviruses, barley yellow dwarf virus and soybean dwarf virus, are also effective vehicles for ISVP delivery (Miller and Bonning, 2007).

A significant advantage of the GNA and CP delivery methods is that the fusion proteins can be produced cheaply by recombinant techniques, and transgenes encoding the fusion proteins can be engineered into entomopathogens and plants (see Sections 4 and 5). GNA is a highly generic delivery vehicle as it has been used to deliver insecticidal peptides to lepidopterans, hemipterans, coleopterans and dipterans. However, GNA does have deleterious effects on some parasitoids, although only at very high doses (Gatehouse et al., 2011). Thus, it will be important, especially in the context of integrated pest management programs, to establish whether attachment to GNA alters ISVP selectivity. In contrast with GNA, CP is a highly specific delivery vehicle that is currently limited to aphids and possibly other sap-sucking hemipterans. However, development of a deeper molecular understanding of the mechanism of CP recognition and uptake in the insect gut might allow this technology to be adapted to the control of other insect pests (Bonning and Chougule, 2014).

4. USE OF ENTOMOPATHOGENS FOR ISVP DELIVERY

4.1. Entomopathogens as bioinsecticides

Entomopathogens, which can be bacterial, viral or fungal, are pathogens that kill or seriously disable insects. They play a vital role in natural regulation of insect pests (Sandhu et al., 2012). A wide range of entomopathogens have been studied and used as bioinsecticides, including the soil bacterium *Bacillus thuringiensis* (*Bt*), baculoviruses and entomopathogenic fungi such as *Metarhizium* and *Beauveria* species.

Bt is a naturally occurring spore-forming bacterium from the family Bacillaceae. It is widespread in soil and lethal to a range of insects due to the production of pore-forming Cry toxins (also known as δ-endotoxins) (Gatehouse et al., 2011; Pardo-López et al., 2013; also refer to Chapters 2, 3 and 6). Cry toxins are toxic to specific species of insects, especially larval stages, and they have proved to be safe to humans and other vertebrates. Heterologously expressed Cry toxins supply the insect-resistance trait in *Bt* transgenic crops (see Section 5; Chapter 6). Most *Bt* species are only active against *Lepidoptera*, *Diptera* and *Coleoptera*, although the subspecies *Bt israelensis* is effective against the larval stage of mosquitoes (Nartey et al., 2013; Chapter 3).

Baculoviruses are pathogenic to the larval stage of a limited range of insects, mostly within the orders *Lepidoptera*, *Hymenoptera* and *Coleoptera* (Bonning and Hammock, 1996; Inceoglu et al., 2006). As for *Bt*, baculoviruses have to be ingested by the insect to cause mortality. Although wild-type baculoviruses have been used successfully for control of soybean and forest pests, they have not competed successfully with chemical insecticides. This is because baculoviruses take 5–14 days to kill infected insects, during which time the insect continues to feed and cause crop damage (Bonning and Hammock, 1996; Inceoglu et al., 2006).

There are 400–500 species of entomopathogenic fungi and some have been used, without modification, for biocontrol of pest insects for 150 years (Bailey and Boyetchko, 2010; Shah and Pell, 2003). Compared to other micro-organisms, fungi infect a broader range of insects including lepidopterans, homopterans, hymenopterans, coleopterans and dipterans (Vilcinskas et al., 1997). The EPA have registered more than 100 fungus-based insect control products since 1995 (De Faria and Wraight, 2007; Shah and Pell, 2003). Mycotrol®, for example, is a *Beauveria*-based mycoinsecticide used for control of whiteflies, thrips, mealybugs and aphids (De Faria and Wraight, 2007), while Green Muscle® contains spores of *Metarhizium acridum*, which are selectively pathogenic to locusts. Green Muscle® showed no detrimental effects on non-target organisms and it is now registered and recommended for locust control by the United Nations (Lomer et al., 2001). However, as for baculoviruses, a major disadvantage of entomopathogenic fungi compared with chemical insecticides is their slow kill time (typically >7 days).

4.2. Entomopathogens as a delivery vehicle for ISVPs

Since ISVPs are gene-encoded mini-proteins, transgenes encoding these peptides could be engineered into a variety of entomopathogens, which would mitigate two of their potential disadvantages. First, in this scenario, ISVPs would be produced systemically in the insect host after entomopathogen infection, and consequently their low level of oral activity would not be an impediment to toxin deployment. Second, the phyletic selectivity of the ISVP would become less important as the range of affected insects would be determined primarily by the host range of the selected entomopathogen. Off-target effects on predators and parasitoids could be limited by choosing entomopathogens with restricted host selectivity; for example, *M. acridum* exclusively infects locusts in the suborder Caelifera, making it valuable as a locust-specific bioinsecticide.

Transgenic baculovirus have been engineered that express insecticidal venom peptides from sea anemones, scorpions or spiders; in all cases, the toxin transgene reduced the time between virus application and cessation of feeding or death. The most dramatic improvement in insecticidal activity resulted from incorporation of a transgene encoding an ISVP (Maggio et al., 2010). The potency and speed of kill of entomopathogenic fungi can also be enhanced by engineering them to express insecticidal venom peptides. For example, the time as well as the dose required for *Metarhizium anisopliae* to kill the tobacco hornworm, *M. sexta*, and the dengue vector, *Aedes aegypti*, was reduced when the fungus was engineered to express the scorpion-venom peptide AaIT (Wang and St Leger, 2007). Entomopathogenic fungi are desirable insect control agents as they directly parasitise arthropods via penetration of the host integument, which sets them apart from bacteria and viruses, which must be ingested to gain host entry (Bailey and Boyetchko, 2010; Roberts, 1973). Engineering entomopathogenic fungi to express ISVPs removes two of the major impediments to their commercial development—the kill time is dramatically decreased and the spore dose required for effective insect control is lowered, thereby reducing application costs.

5. *IN PLANTA* EXPRESSION OF SPIDER VENOM PEPTIDES

The introduction of genetically modified (GM) crops in the mid-1990s revolutionised global crop production. As of 2012, 170 million hectares of GM crops were grown by 17.3 million farmers in 30 countries (James, 2012). More than 90% of growers are small resource-poor farmers in developing countries (James, 2012). Four GM crops currently dominate global agriculture: cotton (81% of all cotton planted globally), soybeans (81%), maize (35%) and canola (30%) (James, 2012).

The introduction of insect-resistant corn and cotton expressing insecticidal Cry toxins from *Bt* dramatically reduced insecticide use, improved crop yields and helped conserve beneficial natural enemies (Que et al., 2010; Tabashnik et al., 2013; also refer to Chapter 6). However, constitutive expression of the toxin in transgenic plants has expedited resistance development in a number of pest species; field-evolved resistance to *Bt* corn has been reported for the maize stalk borer, *Busseola fusca*; the Western corn rootworm, *Diabrotica virgifera virgifera*; and the fall armyworm, *Spodoptera frugiperda*; while resistance to *Bt* cotton has been reported for the corn earworm, *Helicoverpa zea*, and the pink bollworm, *Pectinophora gossypiella* (Gassmann et al., 2014; Tabashnik et al., 2013; this chapter). In China, 2.6% of the population of the cotton bollworm, *Helicoverpa armigera*, are now resistant to Cry1Ac cotton, which might be a prognostic indicator of future widespread resistance (Tabashnik et al., 2013). Moreover, since Cry toxins primarily target lepidopterans and coleopterans, other insects (especially sap-sucking hemipterans) that were only secondary pests prior to the introduction of *Bt* crops have now become a major pest on some crops (Bonning and Chougule, 2014).

Resistance to *Bt* crops can be delayed through the use of non-GM refuges and/or by engineering *Bt* plants to express additional insecticidal genes that act via different mechanisms, an approach known as pyramiding or trait stacking (Moar and Anilkumar, 2007; Que et al., 2010; Chapter 6). ISVP transgenes are good candidates for trait stacking with *Bt* since (i) they have completely different mechanisms of action; (ii) they are likely to be synergised by Cry toxins which lyse midgut epithelial cells (Soberón et al., 2007), a property that should facilitate translocation of ISVPs into the insect haemocoel; and (iii) whereas *Bt* toxins are largely specific for the insect orders Lepidoptera and Coleoptera, ISVPs with complementary selectivity, particularly against sap-sucking hemipterans, can be selected for trait stacking.

Attempts to engineer plants expressing ISVPs began almost two decades ago with the demonstration that transgenic tobacco expressing ω-HXTX-Ar1a, a 37-residue insect-specific calcium channel blocker from the Australian funnel-web spider, *Atrax robustus*, had enhanced resistance to *H. armigera* (Jiang et al., 1996). Transgenes encoding this ISVP (or its orthologue Hv1a) have subsequently been engineered into cotton (Omar and Chatha, 2012), tobacco (Khan et al., 2006) and poplar (Cao et al., 2010). All of these ISVP-expressing transgenic plants were shown to have significantly increased resistance to insect pests. For example, the mortality of second instar *H. armigera* fed on transgenic tobacco expressing Hv1a was 75–100% after 72 h compared to 0% for larvae fed on untransformed plants, regardless of whether the ISVP was expressed under the strong *35S* promoter (Khan et al., 2006) or weaker phloem-specific promoters (Shah et al., 2011). It has even been claimed that transgenic cotton expressing Hv1a is as effective as Bollgard II® cotton in controlling major cotton pests (Omar and Chatha, 2012).

Tobacco plants engineered to express U_7-HXTX-Mg1a (Magi-6), a 36-residue ISVP from the venom of the hexathelid spider *Macrothele gigas* were found to be significantly more resistant than wild-type plants against *S. frugiperda* (Hernandez-Campuzano et al., 2009). More recently, transgenic *Arabidopsis* expressing Hv1a fused to the CP of PEMV were shown to be resistant to a diverse range of aphid species (Bonning et al., 2014). Thus, ISVP transgenes appear to have significant potential either as a standalone insect-resistant plant trait or for trait stacking with Cry toxins.

6. ISVP MIMETICS

6.1. Synthetic mimics of venom peptides

Venom peptides present structural characteristics that are far from the norm for synthetic agrochemicals and drugs (Tice, 2001), and their size and polarity impede passive diffusion through cell membranes. Instead, their oral bioavailability to targets beyond the gut lumen relies on paracellular or active transport (see Section 2). Given the diversity and potential complexity of mechanisms that can mediate such paracellular/active transit through the physical barriers of the gut wall, it seems unlikely that a simple set of general rules could be derived that describe the oral bioavailability of venom peptides. While the oral activity of ISVPs might be improved by chemical modifications (e.g. PEGylation and acylation) or by packaging them in liposomes, microparticles and/or nanoparticles, these are complex and expensive modifications that would significantly increase production costs. An alternative approach to improving oral activity is to create peptidomimetics wherein the functionally critical chemical features of the peptide are displayed on a smaller structural scaffold with better pharmacokinetic properties.

This alternative has been most extensively pursued for the conotoxin family of venom peptides. The physicochemical properties of Ziconotide/Prialt® (ω-conotoxin MVIIA), a cone snail venom peptide used for management of intractable pain, result in insignificant oral bioavailability and consequently this drug requires intrathecal administration (Bingham et al., 2010). In order to avoid the complexity of this drug delivery method, efforts have been undertaken to design small-molecule mimetics that might show oral bioavailability and blood–brain barrier permeability (Brady et al., 2013).

Peptidomimetics that could surmount these challenges are classifiable into three types (Ripka and Rich, 1998): (Type I) molecules in which peptide bonds are replaced with more tractable bioisosteres; (Type II) molecules which use moieties that are not structurally equivalent to the native peptide to interact with the receptor; and (Type III) molecules which utilise a scaffold that can spatially display the critical amino acid residues in the same way as the native peptide. Several papers have been published on the design of Type-III mimetics of ω-conotoxins due to the medical and commercial need (Baell et al., 2001, 2004; Menzler et al., 2000). Despite early progress, the resultant conotoxin mimetics were much less potent than the native peptide (Brady et al., 2013).

Recently, we took a more aggressive approach in the design of Type-II mimetics of Hv1a (Tedford et al., 2013). Hv1a is a 37-residue ISVP that blocks insect but not mammalian voltage-gated calcium (Ca_V) channels (Fletcher et al., 1997; Tedford et al., 2004b). It is harmless when injected into newborn mice (Atkinson et al., 1998) and a closely related homologue (Ar1a) is inactive against nerve–muscle preparations from chicken and rat (Chong et al., 2007). We developed 3D models of the Hv1a pharmacophore and deployed them as search terms for an *in silico* screen of a database of millions of commercially available compounds. Hits from this screen were sorted and filtered as part of a multiple-round selection process, ultimately yielding an "Hv1a mimetic" collection that was tested for activity against *A. aegypti* and *Spodoptera exigua*. This testing identified several families of viable prototype molecules, including the VNX-000440 (VNX-440) family of triazine molecules that were further optimised by iterative synthesis and analogue testing.

6.2. Hv1a pharmacophore analysis and modelling

The 3D solution structure of Hv1a comprises a core region rich in β turns and disulfide bonds (residues 4–21) and a β hairpin (residues 22–37) that projects from the disulfide-rich core (Fletcher et al., 1997). Site-directed mutagenesis and synthetic truncates were used to elucidate the Hv1a pharmacophore (i.e. the set of key structural features contributing to insecticidal activity, presumably by directing interactions with one or more classes of insect Ca_V channels) (Tedford et al., 2001, 2004a). The primary pharmacophore residues (Pro^{10}, Asn^{27} and Arg^{35}) form a small, contiguous patch of 200 $Å^2$ on one face of the toxin (Tedford et al., 2001, 2004a). To build a viable peptide pharmacophore model, key functionalities of the critical amino acids were converted into pharmacophoric features and mapped out in three dimensions (Fig. 8.2). Analyses of the structure of Hv1a and several related peptides suggested that certain chemical features not identified previously by scanning mutagenesis, such as the Gly^8 carbonyl oxygen and the hydrophobic Val^{33} isopropyl functionality, are involved in Hv1a activity. This conclusion was taken into account while building different versions of the pharmacophore model: the Val^{33} side chain was represented in two versions of the pharmacophore model as a hydrophobic feature, while the backbone carbonyl oxygen of Gly^8 was included as a hydrogen-bond acceptor (HBA) in one of the models (Table 8.2 lists features built into each version of the pharmacophore models that were used for database searches).

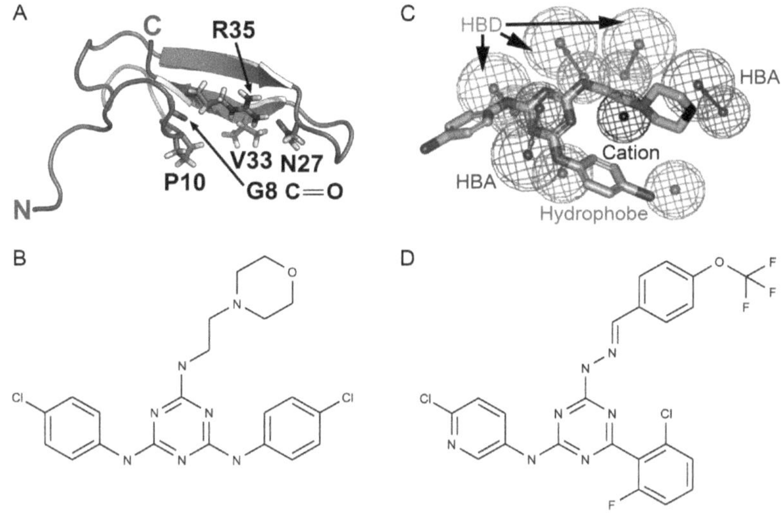

Figure 8.2 Development of small-molecule mimetics of Hv1a. (A) Richardson schematic of Hv1a (first molecule in the PDB file 1AXH) showing the key chemical features that were used for pharmacophore modelling (i.e. the side chains of Pro^{10}, Asn^{27}, Val^{33}, and Arg^{35} and the backbone carbonyl of Gly^{8}). Hydrogen, carbon, oxygen and nitrogen atoms are shown in white, orange, red and blue, respectively. The N- and C-termini are labelled. (B) Bioassay screening hit VNX-440. (C) Three-dimensional stick representation of VNX-440; atoms are coloured as in (A) except that carbons are grey and chlorine atoms are green. Superimposed on VNX-440 are representations of the eight chemical features of the residues that were included in pharmacophore models of the peptide. Cyan spheres indicate hydrophobic features matching the hydrophobic side chains of Val^{33} and Pro^{10}; magenta spheres and sticks indicate hydrogen-bond donor (HBD) features projecting from Asn^{27} and Arg^{35}; green spheres and sticks indicate hydrogen-bond acceptor (HBA) features projecting from Asn^{27} and Gly^{8} (leftmost and rightmost residues, respectively). (D) Optimised lead compound VNX-443, an analogue of VNX-440. (See the colour plate.)

Available SAR data suggested that the most functionally important residue of Hv1a is Arg^{35} (Tedford et al., 2004a). This residue contributed three features to the pharmacophore models: two hydrogen-bond donors (HBDs) and one positively ionisable group. Asn^{27} contributed two features: one HBD and one HBA. Pro^{10} and Val^{33} presumably mediate hydrophobic interactions with Ca_V channels targeted by the peptide; hence, hydrophobic features represented their side chains in the pharmacophore. Gly^{8}, which adopts an unusual backbone configuration (backbone dihedral angles: $\phi = 90°$ and $\psi = -20°$), is conformationally restricted by the cystine knot in a way that allows the Gly^{8} carbonyl oxygen to contribute to the toxin pharmacophore.

The five toxin residues—Gly^{8}, Pro^{10}, Asn^{27}, Val^{33} and Arg^{35}—whose substituents were used to derive features of the pharmacophore models—are illustrated in the context of the overall structure of Hv1a in Fig. 8.2A. The minimal set of pharmacophore features is formed by Arg^{35}, Pro^{10} and Asn^{27}, yielding a 6-feature pharmacophore (see "6 point 1", Table 8.2).

The pharmacophore model was deployed to search the Accelrys Chemicals Available for Purchasing (CAP) database. Low-energy conformers of the compounds were measured and scored for their fit to chemical features of the pharmacophore. The "8-point" model proved too stringent as a search term, so two 6-feature models were constructed and used to query the CAP database. After searches using all three versions of the pharmacophore (Table 8.2), 8773 compounds were identified as "hits" that showed a substantial structural match to the Hv1a pharmacophore model.

A set of physical properties considered relevant for this exercise were then used to filter, sort and annotate the hit compound set. This resulted in the purchase of 1370 unique compounds that were each tested for insecticidal activity against *A. aegypti* larvae. This led to the identification of several families of compounds, including one containing the screening hit VNX-440 (Fig. 8.2B). VNX-440 is a commercially available compound that was previously shown to have modest antimicrobial activity (McKay et al., 2006). The remarkable fit of VNX-440 to the pharmacophore model used for *in silico* screening, presented in Fig. 8.2C, can be described as follows: a nitrogen of the triazine core ring and the oxygen of the morpholine functionality of the compound overlap with the HBA features of the pharmacophore; two amine substituents on the triazine core of the compound overlap with the model's HBD features; the basic nitrogen of the morpholine overlaps with

Table 8.2 Chemical features of key Hv1a residues used to build 3D pharmacophore models

	Arg^{35}			Asn^{27}		Val^{33}	Pro^{10}	Gly^{8}
Model version	**HBD (Nη1)**	**HBD (Nη2)**	**PI (Nε)**	**HBA (Oδ1)**	**HBD (Nδ2)**	**HPH (Cβ, Cγ1, Cγ2)**	**HPH (Cβ, Cγ, Cδ)**	**HBA (O)**
"8-point"	Y	Y	Y	Y	Y	Y	Y	Y
"6-point 1"	Y	Y	Y	Y	Y	N	Y	N
"6-point 2"	N	Y	Y	Y	Y	Y	Y	N

HBD, hydrogen-bond donor; PI, positively ionisable group; HBA, hydrogen-bond acceptor; HPH, hydrophobe.

the cationic feature; and, finally, the remaining anilide has modest overlap with the two hydrophobic features.

6.3. Optimisation and mechanism of action of the Type II mimic VNX-440

Iterative analoguing, guided by the results of bioassays performed against *A. aegypti* and *S. exigua*, allowed us to further characterise the VNX-440 family of hits and to develop a lead molecule (VNX-443; Fig. 8.2D) with significantly improved activity compared with the initial screening hit. The structural changes between VNX-440 and the latter analogue are summarised as follows (cf. Fig. 8.2B and D): first, the amine of the anilide that roughly matched the hydrophobic features in the model was excised to provide a direct connection of a phenyl group to the triazine core. Second, the anilide that only matched a single HBD feature was replaced with heterocycles to improve water solubility. Third, the hydrophilic tail was altered to change the positioning and basicity of the amine. These alterations retained a match with HBD and cationic features of the model for this pendant group and increased insecticidal activity. In aggregate, these structural changes increased insecticidal activity as much as 2200-fold (against *S. exigua*) and resulted in a class of novel compounds (Kennedy and Steinbaugh, 2013).

To confirm that VNX-440 acted against insects via a mechanism similar to Hv1a, its effect on insect ion channels was examined using whole-cell voltage-clamp recordings from dorsal unpaired median (DUM) neurons harvested from the ventral nerve cord of the American cockroach, *Periplaneta americana* (Tedford et al., 2013). Figures 8.3A and B show examples of the effects of 100 μM VNX-440 on I_{Ba} (a measure of Ca_V channel activity) recorded from clamped DUM neurons. Control superfusion with bath solution had no effect on I_{Ba}, but superfusion with 100 μM VNX-440 reduced I_{Ba} dramatically, with maximum inhibition reached within 10 min of exposure to VNX-440.

The IC_{50} for inhibition of peak I_{Ba} by VNX-440 was estimated from the dose–response curve to be 23.1 μM (Fig. 8.3C), which is only ~20 times higher than the IC_{50} (1.08 μM) reported for inhibition of high voltage-activated Ca_V channels by Hv1a (Chong et al., 2007). Unfortunately, this activity was not replicated when VNX-443 was tested in this system. Regardless, VNX-443 and various other analogues in this series caused no adverse effects in an acute oral toxicity study in which rats were dosed at 100 mg/kg (R.M. Kennedy, unpublished results).

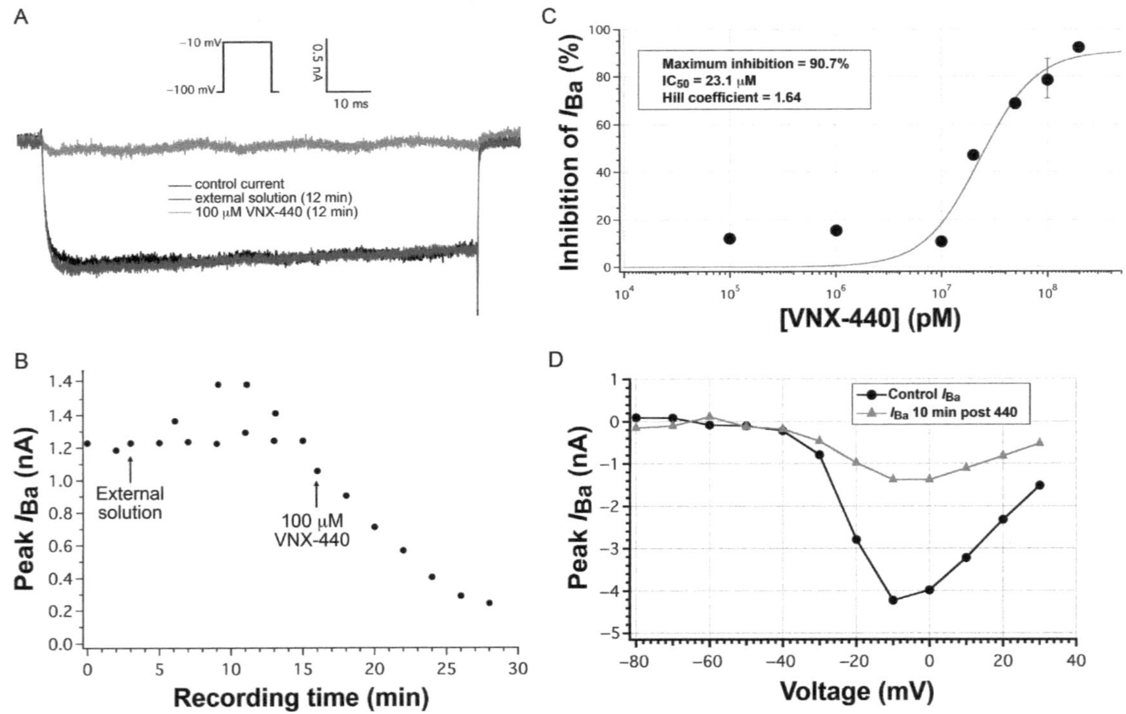

Figure 8.3 Inhibition of Ba^{2+} currents (I_{Ba}) in American cockroach DUM neurons by VNX-440. (A) I_{Ba} recorded at −10 mV: black trace shows control I_{Ba}, green trace shows I_{Ba} during superfusion with bath solution and the red trace shows I_{Ba} following with superfusion with 100 μM VNX-440. (B) Time course of peak I_{Ba} values recorded at a test potential of −10 mV after addition of 100 μM VNX-440. (C) Dose–response curve for VNX-440 inhibition of peak I_{Ba} at a test potential of −10 mV. (D) *I–V* curves for I_{Ba} obtained before, and 10 min after, application of 100 μM VNX-440. (See the colour plate.)

6.4. Conclusions from development of Type-II mimetics of Hv1a

Our *in silico* modelling and screening using the pharmacophore of Hv1a led to a small-molecule hit (VNX-440) that showed properties consistent with mimicry of the native peptide's biological activity. To our knowledge, this remains the only published example of the discovery of an insecticidal lead compound through testing of a small set of candidate compounds chosen to match the pharmacophore of a peptide. However, since the screening lead VNX-440 was optimised solely via an *in vivo* assay (with insect mortality as the read-out), it is not entirely clear if fidelity was maintained with respect to the peptide's mechanism of action, or if another property was optimised. An additional consideration is that while VNX-443 has good *per os* activity, it has only limited contact activity in adult mosquitocidal tests. The physicochemical properties enabling topical bioavailability are considerably stricter than for oral bioavailability and may not allow full coverage of the 200 Å^2 pharmacophoric hot spot on peptides such as Hv1a.

7. OUTLOOK

An unmodified ISVP developed by Vestaron Corporation was recently approved by the U.S. EPA for use against cabbage looper, *Trichoplusia ni*, marking the start of new era of peptide-based insecticides. ISVPs have many properties that make them attractive as insect control agents: they are potently insecticidal, highly selective for insects, stable to extremes of temperature and should degrade in the field into innocuous breakdown products. Although they typically have low levels of intrinsic oral activity, simple formulation can improve their oral activity sufficiently to make them competitive with chemical insecticides. The oral activity of ISVPs can also be markedly improved by fusing them to proteins that ferry them via transcytosis across the insect gut, including plant lectins and the CP of insect-vectored viruses. Transgenes encoding these fusion proteins could be incorporated into crop plants either as a standalone insect-resistance trait or they could be pyramided with *Bt* transgenes in order to reduce the likelihood of resistance development and expand the range of susceptible insects. Entomopathogens can also be engineered to express transgenes encoding ISVPs or ISVP fusion proteins with enhanced oral activity. This approach has the advantage that the selectivity of the ISVP can be tailored by taking advantage of the limited host range of many entomopathogens.

Thus, in summary, there are a wide variety of methods by which ISVPs can be deployed as natural insecticides. We also recently demonstrated for the first time that the pharmacophore of ISVPs can be used to rationally develop small-molecule mimetics with improved oral activity, thereby providing another approach by which these peptides can be exploited for the control of insect pests.

ACKNOWLEDGEMENTS

The authors would like to thank the Australian Research Council for financial support. R.M. Kennedy was funded in part by a grant from the Foundation for the National Institutes of Health through the Vector-based Transmission of Control: Discovery Research (VCTR) program of the Grand Challenges in Global Health initiative.

REFERENCES

Atkinson, R.K., Howden, M.E.H., Tyler, M.I., Vonarx, E.J., 1998. Insecticidal toxins derived from funnel web (*Atrax* or *Hadronyche*) spiders. U.S. Patent No. 5,763,568.

Audsley, N., Matthews, J., Nachman, R., Weaver, R.J., 2007. Metabolism of cydiastatin 4 and analogues by enzymes associated with the midgut and haemolymph of *Manduca sexta* larvae. Gen. Comp. Endocrinol. 153, 80–87.

Audsley, N., Matthews, J., Nachman, R.J., Weaver, R.J., 2008. Transepithelial flux of an allatostatin and analogs across the anterior midgut of *Manduca sexta* larvae in vitro. Peptides 29, 286–294.

Baell, J.B., Forsyth, S.A., Gable, R.W., Norton, R.S., Mulder, R.J., 2001. Design and synthesis of type-III mimetics of ω-conotoxin GVIA. J. Comput. Aided Mol. Des. 15, 1119–1136.

Baell, J.B., Duggan, P.J., Forsyth, S.A., Lewis, R.J., Lok, Y.P., Schroeder, C.I., 2004. Synthesis and biological evaluation of nonpeptide mimetics of ω-conotoxin GVIA. Bioorg. Med. Chem. 12, 4025–4037.

Bailey, K.L., Boyetchko, S.M., 2010. Social and economic drivers shaping the future of biological control: a Canadian perspective on the factors affecting the development and use of microbial biopesticides. Biol. Control 52, 221–229.

Bingham, J.P., Mitsunaga, E., Bergeron, Z.L., 2010. Drugs from slugs—past, present and future perspectives of ω conotoxin research. Chem. Biol. Interact. 183, 1–18.

Blackledge, T.A., Scharff, N., Coddington, J.A., Szuts, T., Wenzel, J.W., Hayashi, C.Y., Agnarsson, I., 2009. Reconstructing web evolution and spider diversification in the molecular era. Proc. Natl. Acad. Sci. U.S.A. 106, 5229–5234.

Bloom, D.E., 2011. 7 billion and counting. Science 333, 562–569.

Bonning, B.C., Chougule, N.P., 2014. Delivery of intrahemocoelic peptides for insect pest management. Trends Biotechnol. 32, 91–98.

Bonning, B.C., Hammock, B.D., 1996. Development of recombinant baculoviruses for insect control. Annu. Rev. Entomol. 41, 191–210.

Bonning, B.C., Pal, N., Liu, S., Wang, Z., Sivakumar, S., Dixon, P.M., King, G.F., Miller, W.A., 2014. Toxin delivery by the coat protein of an aphid-vectored plant virus provides plant resistance to aphids. Nat. Biotechnol. 32, 102–105.

Boyer, S., Zhang, H., Lemperiere, G., 2012. A review of control methods and resistance mechanisms in stored-product insects. Bull. Entomol. Res. 102, 213–229.

Brady, R.M., Baell, J.B., Norton, R.S., 2013. Strategies for the development of conotoxins as new therapeutic leads. Mar. Drugs 11, 2293–2313.

Cao, C.W., Liu, G.F., Wang, Z.Y., Yan, S.C., Ma, L., Yang, C.P., 2010. Response of the gypsy moth, *Lymantria dispar* to transgenic poplar, *Populus simonii* × *P. nigra*, expressing fusion protein gene of the spider insecticidal peptide and Bt-toxin C-peptide. J. Insect Sci. 10, 200.

Casartelli, M., Corti, P., Giovanna Leonardi, M., Fiandra, L., Burlini, N., Pennacchio, F., Giordana, B., 2005. Absorption of albumin by the midgut of a lepidopteran larva. J. Insect Physiol. 51, 933–940.

Chong, Y., Hayes, J.L., Sollod, B., Wen, S., Wilson, D.T., Hains, P.G., Hodgson, W.C., Broady, K.W., King, G.F., Nicholson, G.M., 2007. The ω-atracotoxins: selective blockers of insect M-LVA and HVA calcium channels. Biochem. Pharmacol. 74, 623–638.

De Faria, M.R., Wraight, S.P., 2007. Mycoinsecticides and mycoacaricides, a comprehensive list with worldwide coverage and international classification of formulation types. Biol. Control 43, 237–256.

Escoubas, P., Sollod, B., King, G.F., 2006. Venom landscapes: mining the complexity of spider venoms via a combined cDNA and mass spectrometric approach. Toxicon 47, 650–663.

Fitches, E., Woodhouse, S.D., Edwards, J.P., Gatehouse, J.A., 2001. *In vitro* and *in vivo* binding of snowdrop (*Galanthus nivalis* agglutinin; GNA) and jackbean (*Canavalia ensiformis*; Con A) lectins within tomato moth (*Lacanobia oleracea*) larvae; mechanisms of insecticidal action. J. Insect Physiol. 47, 777–787.

Fitches, E., Edwards, M., Mee, C., Grishin, E., Gatehouse, A., Edwards, J., Gatehouse, J., 2004. Fusion proteins containing insect-specific toxins as pest control agents: snowdrop lectin delivers fused insecticidal spider venom toxin to insect haemolymph following oral ingestion. J. Insect Physiol. 50, 61–71.

Fitches, E.C., Pyati, P., King, G.F., Gatehouse, J.A., 2012. Fusion to snowdrop lectin magnifies the oral activity of insecticidal ω-hexatoxin-Hv1a peptide by enabling its delivery to the central nervous system. PLoS One 7, e39389.

Fletcher, J.I., Smith, R., O'donoghue, S.I., Nilges, M., Connor, M., Howden, M.E., Christie, M.J., King, G.F., 1997. The structure of a novel insecticidal neurotoxin, ω-atracotoxin-HV1, from the venom of an Australian funnel web spider. Nat. Struct. Biol. 4, 559–566.

Gassmann, A.J., Petzold-Maxwell, J.L., Clifton, E.H., Dunbar, M.W., Hoffmann, A.M., Ingber, D.A., Keweshan, R.S., 2014. Field-evolved resistance by western corn rootworm to multiple *Bacillus thuringiensis* toxins in transgenic maize. Proc. Natl. Acad. Sci. U.S.A. 111, 5141–5146.

Gatehouse, A.M.R., Ferry, N., Edwards, M.G., Bell, H.A., 2011. Insect-resistant biotech crops and their impacts on beneficial arthropods. Philos. Trans. R. Soc. Lond. B Biol. Sci. 366, 1438–1452.

Hardy, M.C., Daly, N.L., Mobli, M., Morales, R.A., King, G.F., 2013. Isolation of an orally active insecticidal toxin from the venom of an Australian tarantula. PLoS One 8, e73136.

Hernandez-Campuzano, B., Suarez, R., Lina, L., Hernandez, V., Villegas, E., Corzo, G., Iturriaga, G., 2009. Expression of a spider venom peptide in transgenic tobacco confers insect resistance. Toxicon 53, 122–128.

Ikonomopoulou, M., King, G.F., 2013. Natural born insect killers: spider-venom peptides and their potential for managing arthropod pests. Outlooks Pest Manag. 24, 16–19.

Inceoglu, A.B., Kamita, S.G., Hammock, B.D., 2006. Genetically modified baculoviruses: a historical overview and future outlook. Adv. Virus Res. 68, 323–360.

James, C., 2012. Global Status of Commercialized Biotech/GM Crops: 2012. ISAAA Brief 44-2012. International Service for the Acquisition of Agri-Biotech Applications, Ithaca, NY.

Jiang, H., Zhu, Y.X., Che, Z.L., 1996. Insect resistance of transformed tobacco plants with the gene of the spider insecticidal peptide. J. Integr. Plant Biol. (Acta Bot. Sin.) 38, 95–99.

Kennedy, R.M., Steinbaugh, B.A., 2013. Insecticidal triazines and pyrimidines. U.S. Patent Application 8389718 B2.
Khan, S.A., Zafar, Y., Briddon, R.W., Malik, K.A., Mukhtar, Z., 2006. Spider venom toxin protects plants from insect attack. Transgenic Res. 15, 349–357.
King, G.F., 2011. Venoms as a platform for human drugs: translating toxins into therapeutics. Expert Opin. Biol. Ther. 11, 1469–1484.
King, G.F., Hardy, M.C., 2013. Spider-venom peptides: structure, pharmacology, and potential for control of insect pests. Annu. Rev. Entomol. 58, 475–496.
King, G.F., Tedford, H.W., Maggio, F., 2002. Structure and function of insecticidal neurotoxins from Australian funnel-web spiders. J. Toxicol. Toxin Rev. 21, 359–389.
Klint, J.K., Senff, S., Saez, N.J., Seshadri, R., Lau, H.Y., Bende, N.S., Undheim, E.A., Rash, L.D., Mobli, M., King, G.F., 2013. Production of recombinant disulfide-rich venom peptides for structural and functional analysis via expression in the periplasm of *E. coli*. PLoS One 8, e63865.
Kuhn-Nentwig, L., Stöcklin, R., Nentwig, W., 2011. Venom composition and strategies in spiders: is everything possible? Adv. Insect Physiol. 40, 1–86.
Lomer, C.J., Bateman, R.P., Johnson, D.L., Langewald, J., Thomas, M.B., 2001. Biological control of locusts and grasshoppers. Annu. Rev. Entomol. 46, 667–702.
Maggio, F., Sollod, B.L., Tedford, H.W., Herzig, V., King, G.F., 2010. Spider toxins and their potential for insect control. In: Gilbert, L.I., Gill, S.S. (Eds.), Insect Pharmacology: Channels, Receptors, Toxins and Enzymes. Academic Press, London, pp. 101–123.
McKay, G.A., Reddy, R., Arhin, F., Belley, A., Lehoux, D., Moeck, G., Sarmiento, I., Parr, T.R., Gros, P., Pelletier, J., Far, A.R., 2006. Triaminotriazine DNA helicase inhibitors with antibacterial activity. Bioorg. Med. Chem. Lett. 16, 1286–1290.
Menzler, S., Bikker, J.A., Suman-Chauhan, N., Horwell, D.C., 2000. Design and biological evaluation of non-peptide analogues of ω-conotoxin MVIIA. Bioorg. Med. Chem. Lett. 10, 345–347.
Michiels, K., Van Damme, E.J., Smagghe, G., 2010. Plant-insect interactions: what can we learn from plant lectins? Arch. Insect Biochem. Physiol. 73, 193–212.
Miller, W.A., Bonning, B.C., 2007. Plant resistance to insect pests mediated by viral proteins. U.S. Patent No. 7,312,080.
Moar, W.J., Anilkumar, K.J., 2007. Plant science. The power of the pyramid. Science 318, 1561–1562.
Mukherjee, A.K., Sollod, B.L., Wikel, S.K., King, G.F., 2006. Orally active acaricidal peptide toxins from spider venom. Toxicon 47, 182–187.
Nartey, R., Owusu-Dabo, E., Kruppa, T., Baffour-Awuah, S., Annan, A., Oppong, S., Becker, N., Obiri-Danso, K., 2013. Use of *Bacillus thuringiensis var israelensis* as a viable option in an Integrated Malaria Vector Control Programme in the Kumasi Metropolis, Ghana. Parasit. Vectors 6, 116.
Nyffeler, M., Sunderland, K.D., 2003. Composition, abundance and pest control potential of spider communities in agroecosystems: a comparison of European and US studies. Agric. Ecosyst. Environ. 95, 579–612.
Oerke, E.C., 2006. Crop losses to pests. J. Agric. Sci. 144, 31–43.
Omar, A., Chatha, K.A., 2012. National Institute for Biotechnology and Genetic Engineering (NIBGE): genetically modified spider cotton. Asian J. Manag. Cases 9, 33–58.
Pal, N., Yamamoto, T., King, G.F., Waine, C., Bonning, B., 2013. Aphicidal efficacy of scorpion- and spider-derived neurotoxins. Toxicon 70, 114–122.
Pallaghy, P.K., Nielsen, K.J., Craik, D.J., Norton, R.S., 1994. A common structural motif incorporating a cystine knot and a triple-stranded β-sheet in toxic and inhibitory polypeptides. Protein Sci. 3, 1833–1839.
Pardo-López, L., Soberón, M., Bravo, A., 2013. *Bacillus thuringiensis* insecticidal three-domain Cry toxins: mode of action, insect resistance and consequences for crop protection. FEMS Microbiol. Rev. 37, 3–22.

Platnick, N.I., 2013. The world spider catalog, version 13.5, American Museum of Natural History Online at http://research.amnh.org/entomology/spiders/catalog/index.html.
Que, Q., Chilton, M.D., De Fontes, C.M., He, C., Nuccio, M., Zhu, T., Wu, Y., Chen, J.S., Shi, L., 2010. Trait stacking in transgenic crops: challenges and opportunities. GM Crops 1, 220–229.
Ripka, A.S., Rich, D.H., 1998. Peptidomimetic design. Curr. Opin. Chem. Biol. 2, 441–452.
Roberts, D.W., 1973. Means for insect regulation: fungi. Ann. N. Y. Acad. Sci. 217, 76–84.
Saez, N.J., Senff, S., Jensen, J.E., Er, S.Y., Herzig, V., Rash, L.D., King, G.F., 2010. Spider-venom peptides as therapeutics. Toxins 2, 2851–2871.
Sandhu, S., Sharma, A.K., Beniwal, K., Goel, G., Batra, P., Kumar, A., Jaglan, S., Sharma, A.K., Malhotra, S., 2012. Myco-biocontrol of insect pests: factors involved, mechanism, and regulation. J. Pathog. 2012, 126819.
Shah, P.A., Pell, J.K., 2003. Entomopathogenic fungi as biological control agents. Appl. Microbiol. Biotechnol. 61, 413–423.
Shah, A.D., Ahmed, M., Mukhtar, Z., Khan, S.A., Habib, I., Malik, Z.A., Mansoor, S., Saeed, N.A., 2011. Spider toxin (Hvt) gene cloned under phloem specific RSs1 and RolC promoters provides resistance against American bollworm (*Heliothis armigera*). Biotechnol. Lett. 33, 1457–1463.
Skaer, H.B., Maddrell, S.H., Harrison, J.B., 1987. The permeability properties of septate junctions in Malpighian tubules of *Rhodnius*. J. Cell Sci. 88, 251–265.
Smith, J.J., Herzig, V., King, G.F., Alewood, P.F., 2013. The insecticidal potential of venom peptides. Cell. Mol. Life Sci. 70, 3665–3693.
Soberón, M., Fernandez, L.E., Perez, C., Gill, S.S., Bravo, A., 2007. Mode of action of mosquitocidal *Bacillus thuringiensis* toxins. Toxicon 49, 597–600.
Tabashnik, B.E., Brevault, T., Carriere, Y., 2013. Insect resistance to *Bt* crops: lessons from the first billion acres. Nat. Biotechnol. 31, 510–521.
Tedford, H.W., Fletcher, J.I., King, G.F., 2001. Functional significance of the β hairpin in the insecticidal neurotoxin ω-atracotoxin-Hv1a. J. Biol. Chem. 276, 26568–26576.
Tedford, H.W., Gilles, N., Menez, A., Doering, C.J., Zamponi, G.W., King, G.F., 2004a. Scanning mutagenesis of ω-atracotoxin-Hv1a reveals a spatially restricted epitope that confers selective activity against insect calcium channels. J. Biol. Chem. 279, 44133–44140.
Tedford, H.W., Sollod, B.L., Maggio, F., King, G.F., 2004b. Australian funnel-web spiders: master insecticide chemists. Toxicon 43, 601–618.
Tedford, H.W., Steinbaugh, B.A., Bao, L., Tait, B.D., Tempczyk-Russell, A., Smith, W., Benzon, G.L., Finkenbinder, C.A., Kennedy, R.M., 2013. *In silico* screening for compounds that match the pharmacophore of omega-hexatoxin-Hv1a leads to discovery and optimization of a novel class of insecticides. Pestic. Biochem. Physiol. 106, 124–140.
Tice, C.M., 2001. Selecting the right compounds for screening: does Lipinski's Rule of 5 for pharmaceuticals apply to agrochemicals? Pest Manag. Sci. 57, 3–16.
Vilcinskas, A., Matha, V., Gotz, P., 1997. Effects of the entomopathogenic fungus *Metarhizium anisopliae* and its secondary metabolites on morphology and cytoskeleton of plasmatocytes isolated from the greater wax moth, *Galleria mellonella*. J. Insect Physiol. 43, 1149–1159.
Vollrath, F., Selden, P., 2007. The role of behavior in the evolution of spiders, silks, and webs. Annu. Rev. Ecol. Evol. Syst. 38, 819–846.
Wang, C., St Leger, R.J., 2007. A scorpion neurotoxin increases the potency of a fungal insecticide. Nat. Biotechnol. 25, 1455–1456.
Windley, M.J., Herzig, V., Dziemborowicz, S.A., Hardy, M.C., King, G.F., Nicholson, G.M., 2012. Spider-venom peptides as bioinsecticides. Toxins 4, 191–227.

INDEX

Note: Page numbers followed by "*f*" indicate figures and "*t*" indicate tables.

R

S

T

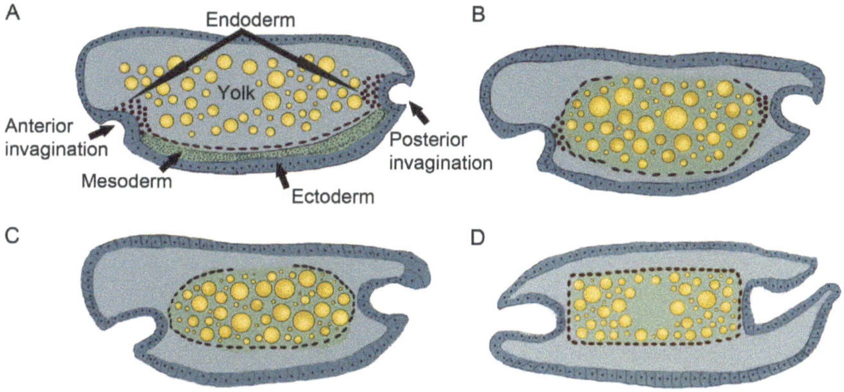

Paul J. Linser and Rhoel R. Dinglasan, Figure 1.1 Embryonic development of the main components of the alimentary canal in insects. Invaginations of the ectoderm at the anterior and posterior poles give rise to the foregut and hindgut, respectively. The midgut forms from endodermal cells adjacent to the invaginations, proliferating and migrating to enclose the central yolk. The ectodermal and endodermal tubes eventually fuse to form a contiguous alimentary canal. The sequence of developmental steps progress from "A" through "D". *Redrawn by Gabriela Marie Ferguson after several sources including Johansen and Butt (1941).*

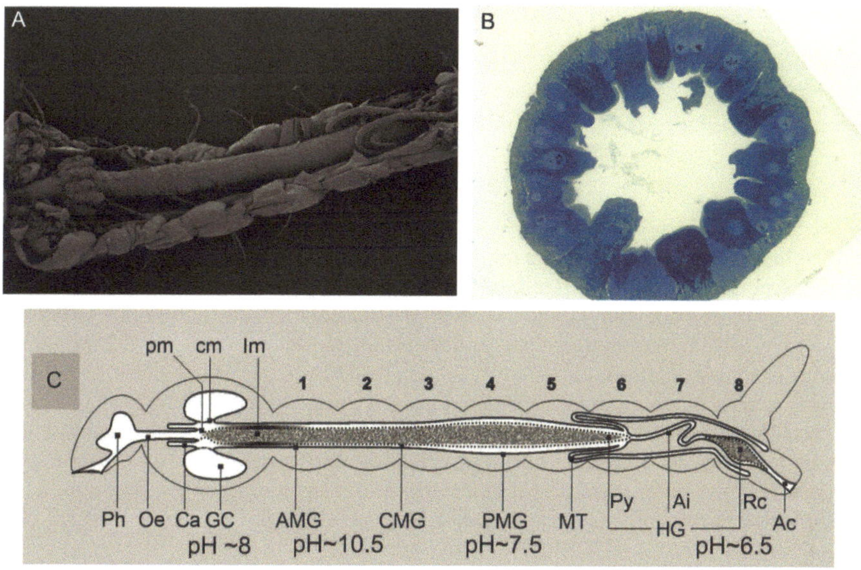

Paul J. Linser and Rhoel R. Dinglasan, Figure 1.2 Architecture of the larval mosquito alimentary canal. Panel A: a scanning electronmicrograph of a fourth instar larva of *Aedes aegypti* that has been dissected to reveal the full length of the midgut amidst the exoskeleton and integument. Panel B: a cross section of the gut epithelium in the region of the posterior midgut (PMG) showing that it is a single cell thick epithelial tube but with varying morphological characteristics of the individual cells. Panel C: a diagrammatic rendering of the larval mosquito alimentary canal with labels applied for orientation. Numbers 1–8 indicate abdominal segments. GC, gastric caecum; AMG, anterior midgut; CMG, central midgut; PMG, posterior midgut; MT, malpighian tubules; HG, hindgut; Py, pyloris; Ai, anterior intestine; Rc, rectum; Ac, anal canal; Ph, pharynx; Oe, oesophagus; Ca, cardia; pm, peritrophic membrane; cm, caecal membrane; lm, gut lumen. The approximate pH of the lumen of the gut is shown at the bottom of the cartoon. *Taken from Linser et al. (2007) with permission.*

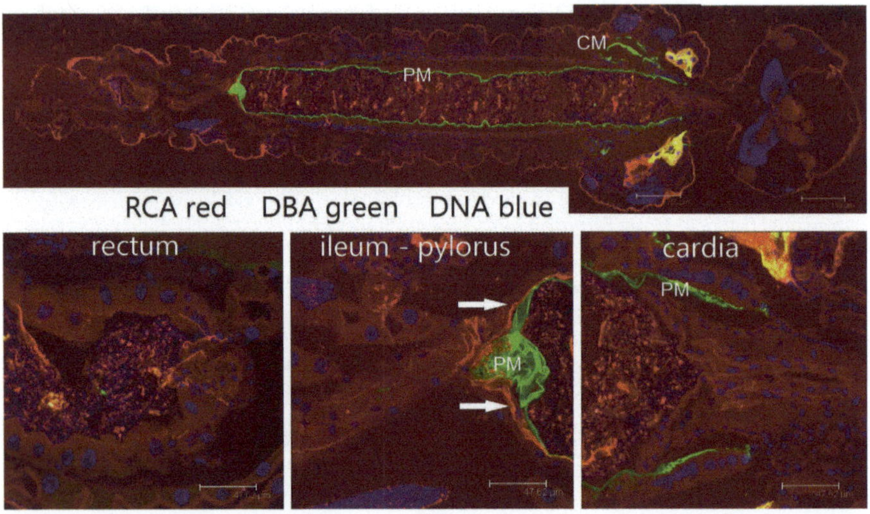

Paul J. Linser and Rhoel R. Dinglasan, Figure 1.6 Figure shows longitudinal sections of paraffin embedded *Anopheles gambiae* fourth instar larvae at low (upper montage) and high (lower three panels) magnification with the anterior (head) to the right. Labelling was with TRITC-conjugated *Ricinus communis* Agglutinin I (red), FITC-conjugated *Dolichos biflorus* Agglutinin (green) and DRAQ5 for DNA (blue). For the purpose of this discussion, note that the green DBA staining labels the peritrophic membrane (PM) and the caecal membrane (CM), and the red RCA labels cuticular structures including the exoskeleton and the lining of the foregut adjacent to the beginnings of the PM in the cardia and at the posterior end of the PM at the beginning of the hindgut at the level of the ileum-pyloris junction (arrows). Also note that the rectum is lined by red, RCA+ cuticle. *From Linser et al. (2008) with permission.*

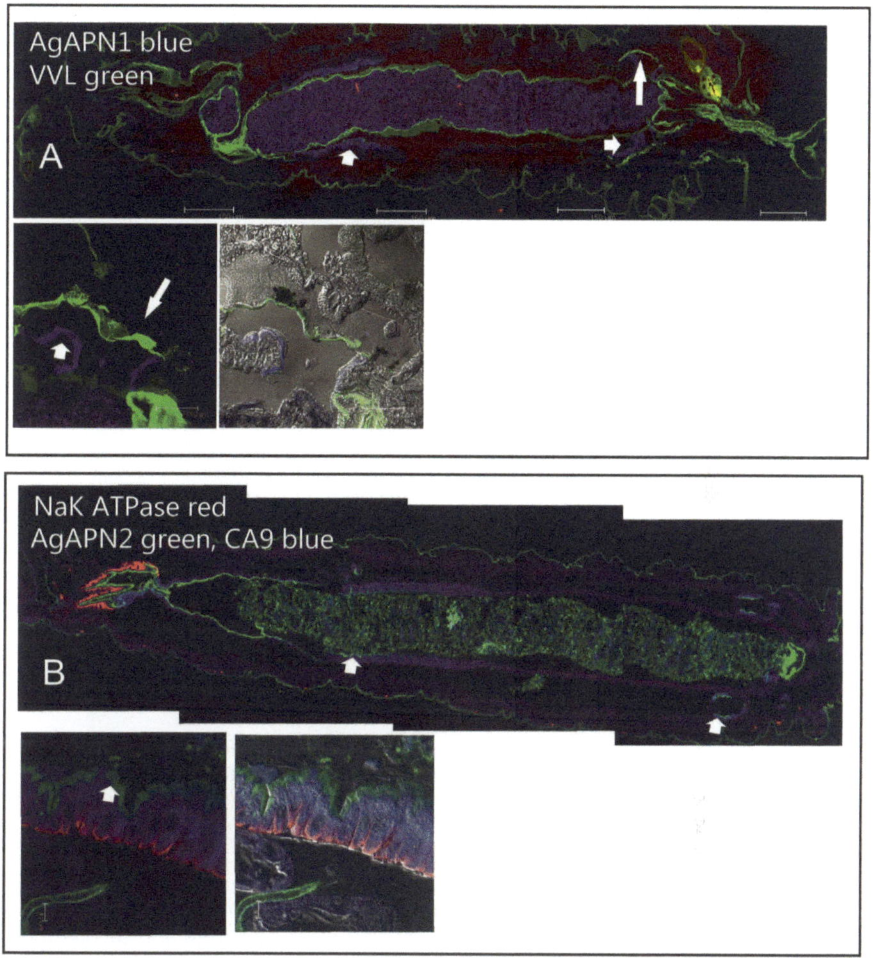

Paul J. Linser and Rhoel R. Dinglasan, Figure 1.7 Figure shows the distribution of two integral membrane amino peptidases, AgAPN1 (panel A) and AgAPN2 (panel B) in larval *Anopheles gambiae* gut sections. In (A), AgAPN1 (blue) is seen on the brush border membranes (BBM) of the posterior midgut (PMG) cells (short arrow to left) and the BBM of the neck of the gastric caeca (GC) (short arrow to right and in high mag images at bottom of panel). Note that FITC-conjugated Vicia Villosa Lectin (green) was used to highlight the cuticular structures, the peritrophic membrane (PM) and the caecal membrane (CM) (long arrows) making the limitation of AgAPN1 to the neck cells of the GC evident. In (B), AgAPN2 (green) is compared to Na^+/K^+-ATPase (red) and the cytoplasmic marker carbonic anhydrase-9 (CA9). The short arrows indicate labelling for AgAPN2 on the BBM of the PMG and on the BBM of the GC cells internal to the CM. *From Linser et al. (2008) with permission.*

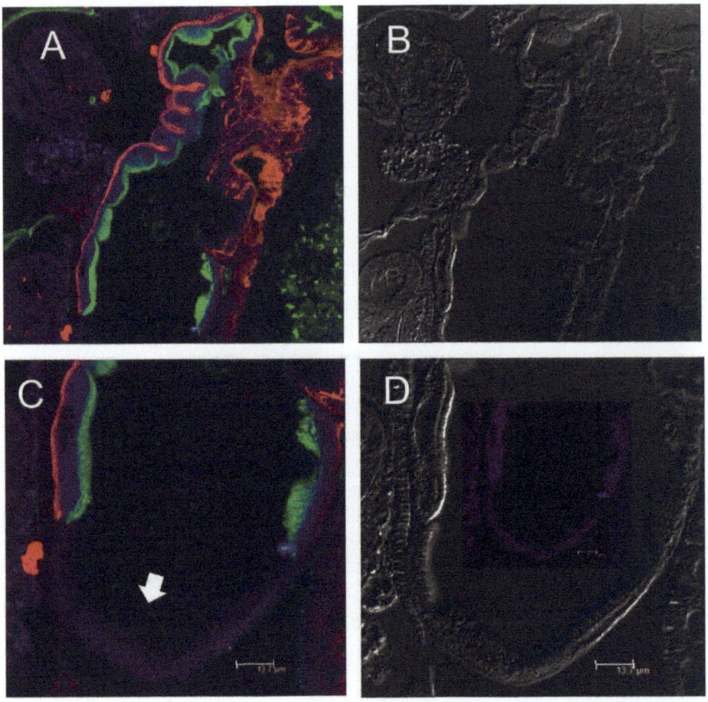

Paul J. Linser and Rhoel R. Dinglasan, Figure 1.8 Figure shows the GC of fourth instar *Anopheles gambiae* at high magnifications to highlight the distribution of AgAPN2 (green) relative to the basal membrane marker Na^{+}/K^{+}-ATPase (red) and the cytoplasmic CA9 (blue). Panels A and B show the anterior portion of a caecum at the junction with the AMG. Note that AgAPN2 is prominently localised to the BBM of the anterior GC cells. Panels C and D show the posterior portion on the same caecum. Note the complete loss of staining for both Na^{+}/K^{+}-ATPase and AgAPN2 (arrow) at the posterior extreme of the caecum, which is the CAP cells. The inset in D shows only CA9 staining at the same magnification as shown in A and B to provide reference. *From Linser et al. (2008) and Harvey et al. (2010) with permission.*

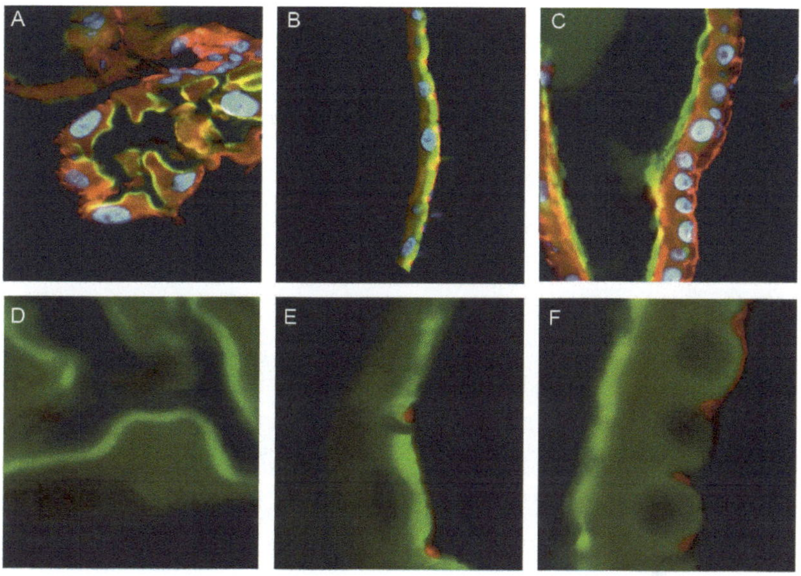

Paul J. Linser and Rhoel R. Dinglasan, Figure 1.9 Figure demonstrates the apical to basal back to apical transitions in the polarised distribution of V-ATPase in the *Aedes aegypti* larval midgut. Panels A–C show regions of the GC, AMG and PMG, respectively. Green/yellow labelling shows the localisation of V-ATPase on the apical BBM of the GC (A) and PMG (C) and on the basal membrane of the AMG (B). Red labelling identifies the basal side of the epithelium in all cases as it depicts the external layer of muscle labelled with TRITC-conjugated Phalloidin. Blue is DAPI staining of cell nuclei. Panels D–F show the same series of gut regions at higher magnification. *From Zhuang et al. (1999) with permission.*

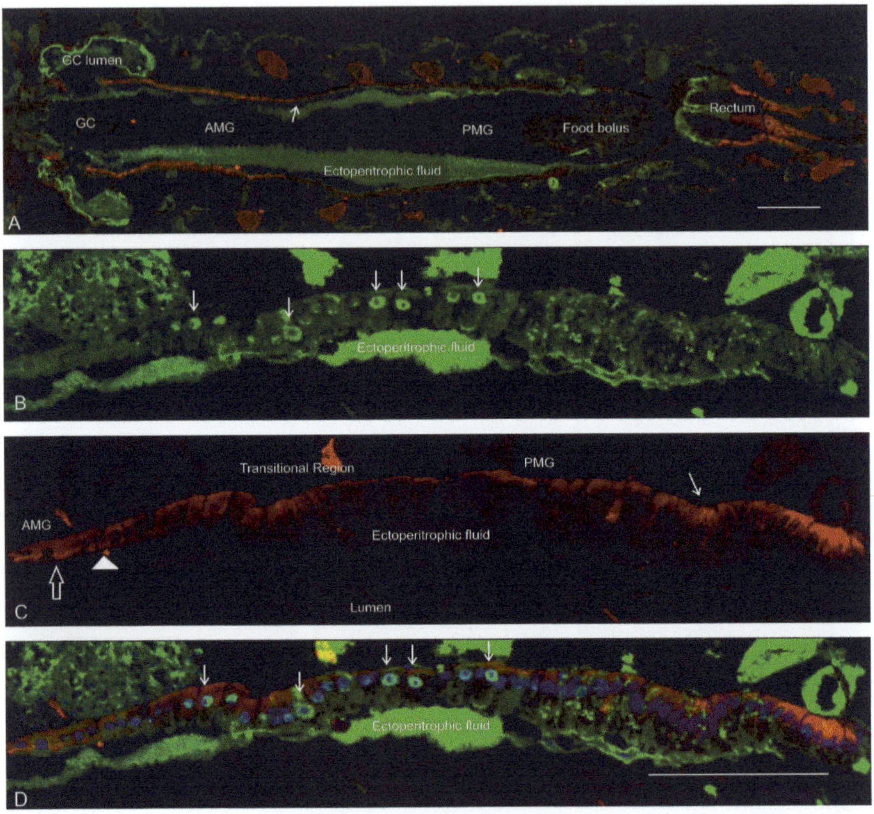

Paul J. Linser and Rhoel R. Dinglasan, Figure 1.10 Figure shows the entire length of the larval *Anopheles gambiae* alimentary canal from the cardia at the extreme to the rectum at far right in panel A. CA9 is labelled in green and Na^+/K^+-ATPase in red. The arrow in (A) indicates the transition region at which Na^+/K^+-ATPase shifts from basal in the AMG to apical in the PMG. Panels B–D show the transition region at higher magnification. CA9 is expressed by several cell types but is secreted by GC cells into the ectoperitrophic space. In the transition region, CA9 is also found in cell nuclei (arrows, B and D). Panel C isolates the Na^+/K^+-ATPase signal and thus reveals the shift in epithelial cell polarity from basal (PMG, solid arrow) to apical (AMG, hollow arrow) of this very important physiological function. Panel D overlays the red and green images with a blue (DRAQ5, DNA) signal. *From Smith et al. (2007) with permission.*

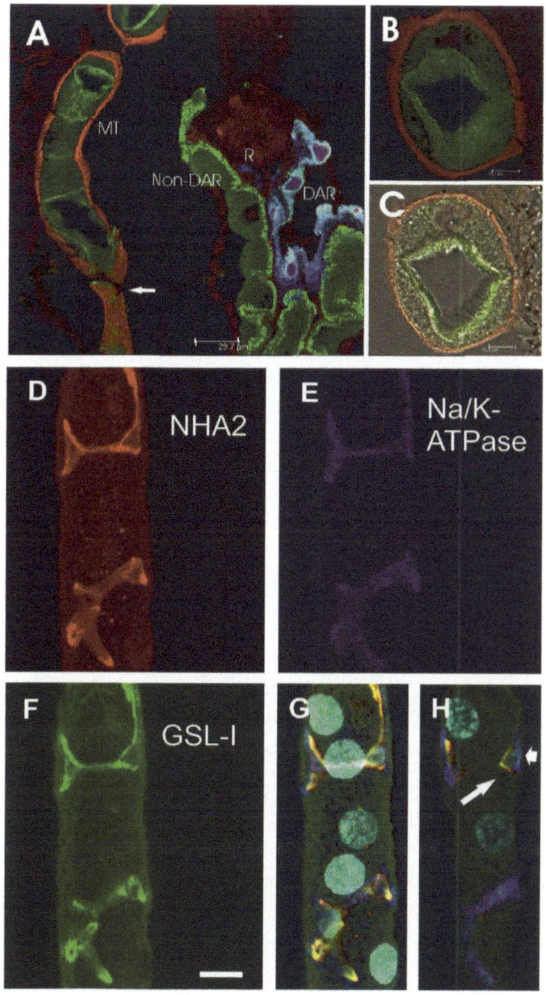

Paul J. Linser and Rhoel R. Dinglasan, Figure 1.11 Figure reveals several details of MT cell biology in the hindgut region of *Anopheles gambiae* larva sections. Panel A shows immunolabelling for the bicarbonate transporter AgNDAE1 (red), V-ATPase (green) and CA9 (blue). The MT is characterised by basal membrane staining of the principal cells for AgNDAE1 and internal and apical staining for V-ATPase. Panel A also shows a partial view of the rectum and that CA9 is characteristic of the DAR cells whereas V-ATPase is also present in and on the non-DAR cells. Panels B and C show a high magnification view of a cross section through an MT displaying the basal labelling for AgNDAE1 and the apical labelling for V-ATPase. The arrow in (A) indicates a stellate cell which shows little signal in this labelling combination. Panels D–H show high magnification views of MT whole mounts labelled for proton antiporter NHA2 (red), Na^{+}/K^{+}-ATPase (blue) and Griffonia simplicifolia I lectin (green) and in D&H, DAPI for nuclei (aqua). NHA2, Na^{+}/K^{+}-ATPase and GSL-I all label the stellate cells of the MT. In the merge images G and H, the polarity of the markers is revealed. NHA2 and GSL1 co-localise (long arrow in H) to the apical aspect of the stellate cell whereas Na^{+}/K^{+}-ATPase is basal. *Figures A–C from Linser et al. (2012) and D–H from Xiang et al. (2012) with permission.*

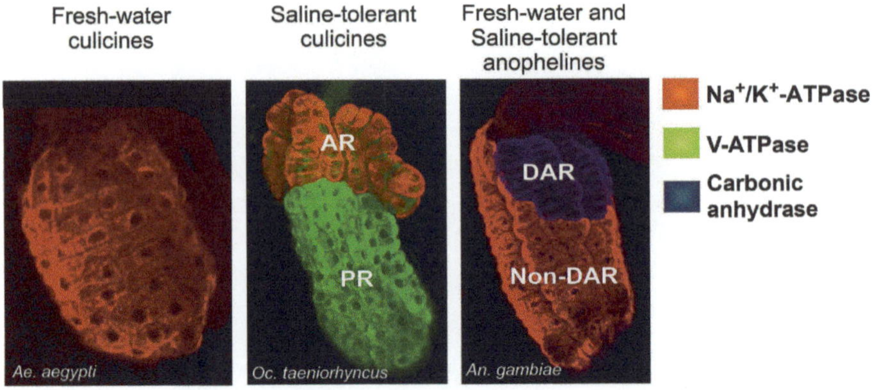

Paul J. Linser and Rhoel R. Dinglasan, Figure 1.12 Figure depicts the gross structure of the three types of larval mosquito rectum as seen in fresh-water culicenes, salt-tolerant culicenes and all anophelines. Fresh-water culicenes such as *Aedes aegypti* have a uniform rectum with one primary cell type as seen in this whole mount immunolabelled for basal NaK-ATPase (left). Salt-tolerant culicenes such as *Ochlerotatus taeniorhynchus* have a rectum with distinct anterior and posterior regions. Anophelines possess a rectum that is distinguished by an anterior patch of distinct cells termed dorsal anterior rectum (DAR) cells and a posterior majority of non-DAR cells. AR, anterior rectum; PR, posterior rectum. *Image modified from Smith et al. (2008).*

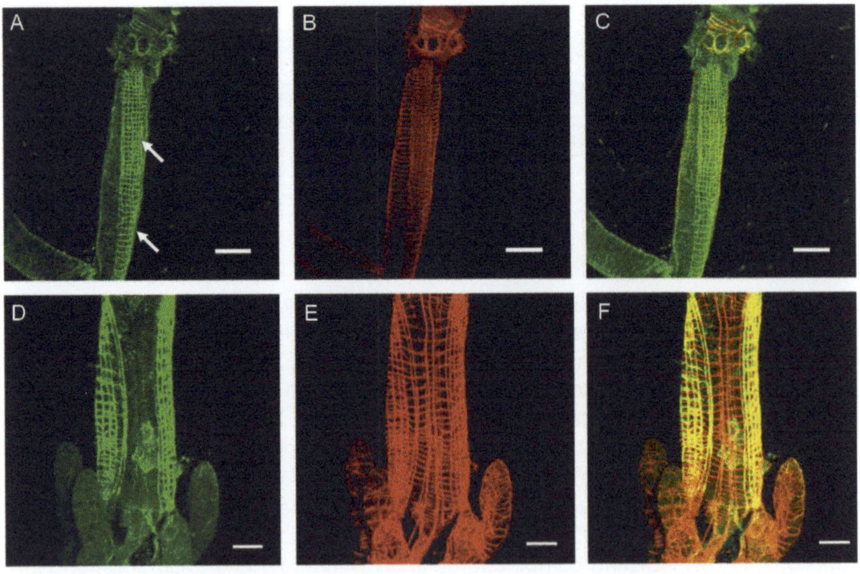

Paul J. Linser and Rhoel R. Dinglasan, Figure 1.13 Figure shows whole mount preparations of isolated midgut from fourth instar larvae of *Aedes aegypti* (A–C) and *Anopheles gambiae* (D–F) immunolabelled for the peripheral membrane protein carbonic anhydrase CA-10 (green) and Phalloidin to reveal muscle (red). The arrows in (A) and the merged images show that a specific subset of the gut musculature labels for CA-10 whereas other muscles of the gut do not. Thus the musculature of the gut exhibits more structure/function variability than simple longitudinal versus circumferential. *From Seron et al. (2004) with permission.*

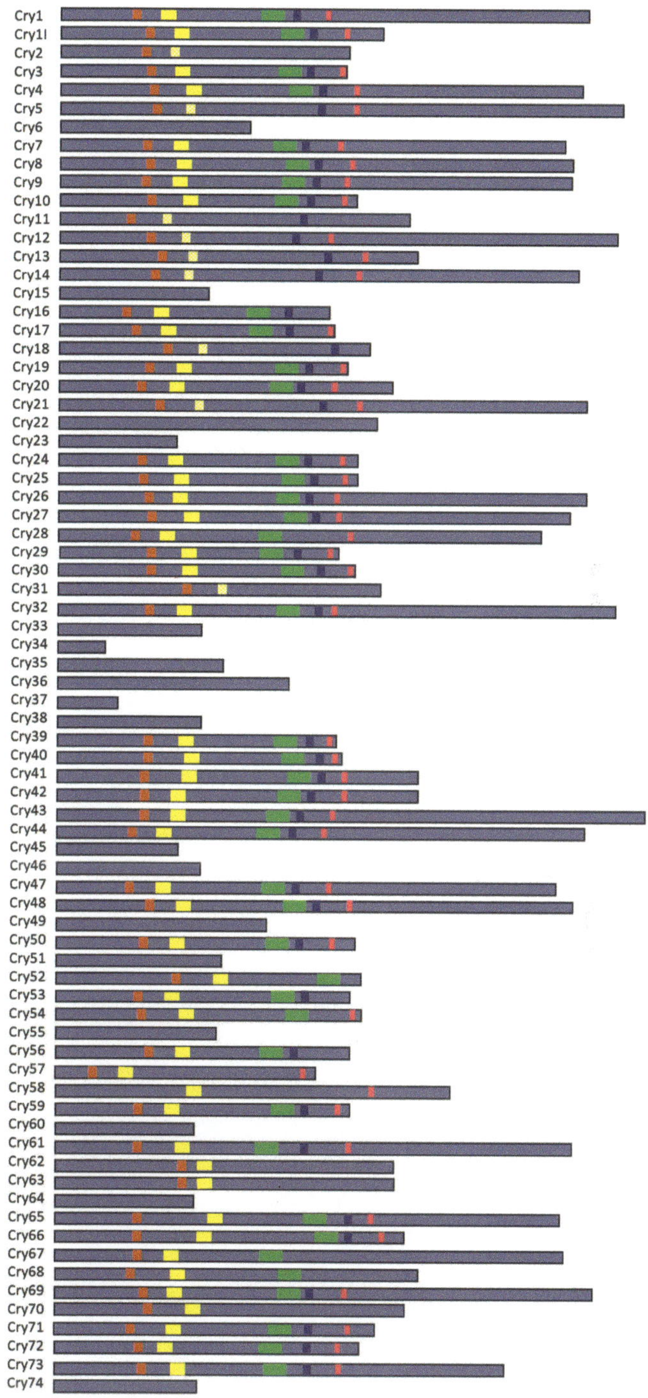

Michael J. Adang *et al.*, Figure 2.1 Graphical representation of the diversity of *Bt* Cry toxins. The length of each toxin is drawn to scale and the five conserved blocks described in Schnepf et al. (1998) are shown as coloured inserts.

Three-domain	ETX–MTX	Bin	Parasporin

Cry1Aa	Cry2Aa	Cry8Aa	Cry15Aa	Cry31Aa	Cry40Aa	Cry58Aa	Cyt1Aa
Cry1Ab	Cry2Ab	Cry8Ab		Cry31Ab	Cry40Ba		Cyt1Ab
Cry1Ac	Cry2Ac	Cry8Ac	Cry16Aa	Cry31Ac	Cry40Ca	Cry59Aa	Cyt1Ba
Cry1Ad	Cry2Ad	Cry8Ad		Cry31Ad	Cry40Da	Cry59Ba	Cyt1Ca
Cry1Ae	Cry2Ae	Cry8Ba	Cry17Aa				Cyt1Da
Cry1Af	Cry2Af	Cry8Bb		Cry32Aa	Cry41Aa	Cry60Aa	Cyt2Aa
Cry1Ag	Cry2Ag	Cry8Bc	Cry18Aa	Cry32Ab	Cry41Ab	Cry60Ba	Cyt2Ba
Cry1Ah	Cry2Ah	Cry8Ca	Cry18Ba	Cry32Ba	Cry41Ba		Cyt2Bb
Cry1Ai	Cry2Ai	Cry8Da	Cry18Ca	Cry32Ca	Cry41Ca	Cry61Aa	Cyt2Bc
Cry1Ba	Cry2Aj	Cry8Db		Cry32Cb			Cyt2Ca
Cry1Bb	Cry2Ak	Cry8Ea	Cry19Aa	Cry32Da	Cry42Aa	Cry62Aa	Cyt3Aa
Cry1Bc	Cry2Ba	Cry8Fa	Cry19Ba	Cry32Ea			
Cry1Bd		Cry8Ga	Cry19Ca	Cry32Eb	Cry43Aa	Cry63Aa	Vip1Aa
Cry1Be	Cry3Aa	Cry8Ha		Cry32Fa	Cry43Ba		Vip1Ab
Cry1Bf	Cry3Ba	Cry8Ia	Cry20Aa	Cry32Ga	Cry43Ca	Cry64Aa	Vip1Ac
Cry1Bg	Cry3Bb	Cry8Ib	Cry20Ba	Cry32Ha	Cry43Cb		Vip1Ad
Cry1Bh	Cry3Ca	Cry8Ja		Cry32Hb	Cry43Cc	Cry65Aa	Vip1Ba
Cry1Bi		Cry8Ka	Cry21Aa	Cry32Ia			Vip1Bb
Cry1Ca	Cry4Aa	Cry8Kb	Cry21Ba	Cry32Ja	Cry44Aa	Cry66Aa	Vip1Bc
Cry1Cb	Cry4Ba	Cry8La	Cry21Ca	Cry32Ka			Vip1Ca
Cry1Da	Cry4Ca	Cry8Ma	Cry21Da	Cry32La	Cry45Aa	Cry67Aa	Vip1Da
Cry1Db	Cry4Cb	Cry8Na	Cry21Ea	Cry32Ma	Cry45Ba		
Cry1Dc	Cry4Cc	Cry8Pa	Cry21Fa	Cry32Mb		Cry68Aa	Vip2Aa
Cry1Ea		Cry8Qa	Cry21Ga	Cry32Na	Cry46Aa		Vip2Ab
Cry1Eb	Cry5Aa	Cry8Ra	Cry21Ha	Cry32Oa	Cry46Ab	Cry69Aa	Vip2Ac
Cry1Fa	Cry5Ab	Cry8Sa		Cry32Pa		Cry69Ab	Vip2Ad
Cry1Fb	Cry5Ac	Cry8Ta	Cry22Aa	Cry32Qa	Cry47Aa		Vip2Ae
Cry1Ga	Cry5Ad		Cry22Ab	Cry32Ra		Cry70Aa	Vip2Af
Cry1Gb	Cry5Ba	Cry9Aa	Cry22Ba	Cry32Sa	Cry48Aa	Cry70Ba	Vip2Ag
Cry1Gc	Cry5Ca	Cry9Ba	Cry22Bb	Cry32Ta	Cry48Ab	Cry70Bb	Vip2Ba
Cry1Ha	Cry5Da	Cry9Bb		Cry32Ua			Vip2Bb
Cry1Hb	Cry5Ea	Cry9Ca	Cry23Aa	Cry32Va	Cry49Aa	Cry71Aa	
Cry1Ia		Cry9Da		Cry32Wa	Cry49Ab		Vip3Aa
Cry1Ib	Cry6Aa	Cry9Db	Cry24Aa			Cry72Aa	Vip3Ab
Cry1Ic	Cry6Ba	Cry9Dc	Cry24Ba	Cry33Aa	Cry50Aa		Vip3Ac
Cry1Id		Cry9Ea	Cry24Ca		Cry50Ba	Cry73Aa	Vip3Ad
Cry1Ie	Cry7Aa	Cry9Eb		Cry34Aa			Vip3Ae
Cry1If	Cry7Ab	Cry9Ec	Cry25Aa	Cry34Ab	Cry51Aa	Cry74Aa	Vip3Af
Cry1Ig	Cry7Ba	Cry9Ed		Cry34Ac			Vip3Ag
Cry1Ja	Cry7Bb	Cry9Ee	Cry26Aa	Cry34Ba	Cry52Aa		Vip3Ah
Cry1Jb	Cry7Ca	Cry9Fa			Cry52Ba		Vip3Ai
Cry1Jc	Cry7Cb	Cry9Ga	Cry27Aa	Cry35Aa			Vip3Ba
Cry1Jd	Cry7Da			Cry35Ab	Cry53Aa		Vip3Bb
Cry1Ka	Cry7Ea	Cry10Aa	Cry28Aa	Cry35Ac	Cry53Ab		Vip3Ca
Cry1La	Cry7Fa			Cry35Ba			
Cry1Ma	Cry7Fb	Cry11Aa	Cry29Aa		Cry54Aa		Vip4Aa
Cry1Na	Cry7Ga	Cry11Ba	Cry29Ba	Cry36Aa	Cry54Ab		
Cry1Nb	Cry7Gb	Cry11Bb			Cry54Ba		Sip1Aa

Michael J. Adang *et al.*, Figure 2.2 List of *Bt* toxins. The toxins are grouped according to their primary ranking within the nomenclature. The background colour represents likely protein tertiary structural groupings and those toxins named with red text indicate the parasporin class.

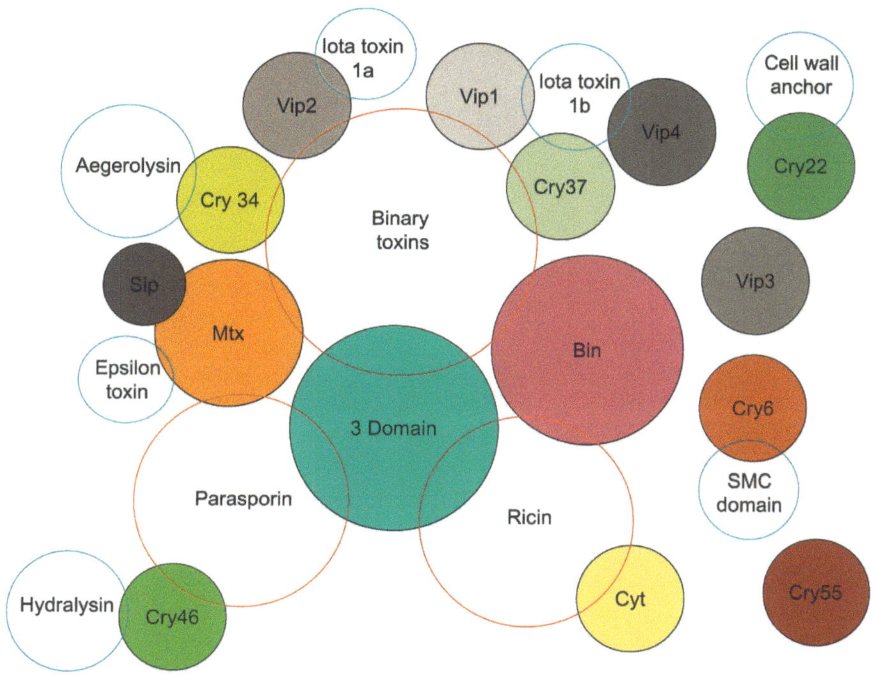

Michael J. Adang *et al.*, Figure 2.3 Venn diagram showing structural or functional relationships between toxins. The individually coloured circles represent the different structural groups identified in Fig. 2.2. Overlapping open blue circles indicate sequence similarity with non-*Bt* toxins or proteins. The open red circles indicate which of the structural groups contain members that share that particular characteristic namely binary toxin, parasporin or ricin domain containing.

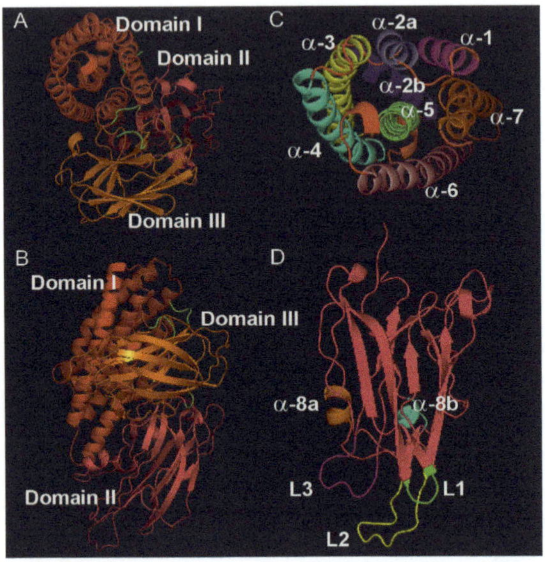

Michael J. Adang *et al.*, Figure 2.4 Views of Cry1Aa toxins from the Resource for Structural Bioinformatics Protein Data Bank (http://www.rcsb.org/pdb/) of (id: 1CIY) (Grochulski et al., 1995) modified using PyMOL Version 1.7 (http://www.pymol.org/). Domain amino acid sequence position information was obtained from NCBI data base (GenBank: AAP40639.1) and (Grochulski et al., 1995). Overall views of Cry1Aa toxin are shown (Panels A and B). Domain I is red, domain II is pink and domain III is orange. Domain I with the names of α-helices and domain II with named α-helices and loops are shown in Panels C and D, respectively.

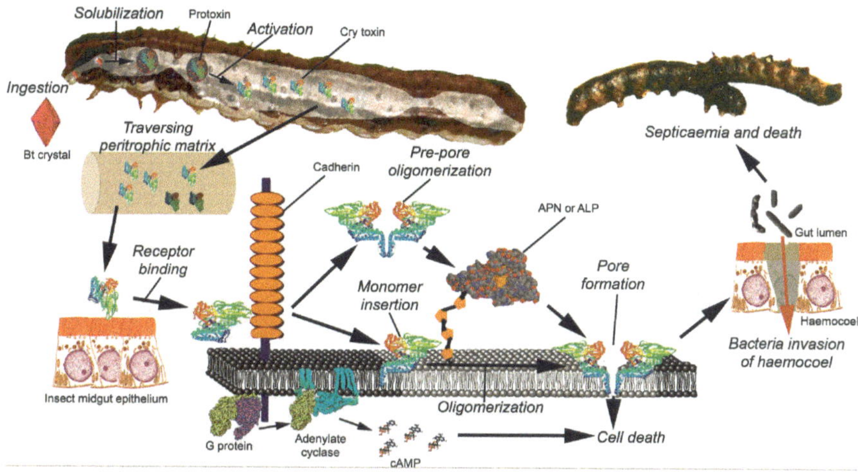

Michael J. Adang *et al.*, Figure 2.5 Representation of the current models of Cry toxin action in the insect midgut epithelium. After ingestion by susceptible larvae, toxin crystals are solubilized in the midgut fluids to yield the Cry protoxin form, which is processed to an activated Cry toxin form equivalent to the Cry toxin ingested by insects feeding on transgenic *Bt* crops. The Cry toxin core then traverses the peritrophic matrix, which is able to retain some of the toxin (darker toxin molecules in the figure). Once reaching the brush border membrane of the midgut epithelium, the Cry toxin binds with high affinity to cadherin, which results in activation of intracellular cell death pathways (represented by G protein, adenylate cyclase and cAMP in the figure), and/or according to the 'sequential binding' model, further proteolysis of the toxin monomer to result in formation of a pre-pore oligomer. This pre-pore oligomer is proposed to bind to alkaline phosphatase (ALP) or aminopeptidase (APN) to insert in the membrane forming a pore that leads to osmotic cell death. The insertion of toxin monomer and formation of a pore by oligomerization of inserted monomers is also presented as an alternative step. Disruption of the midgut epithelium barrier allows for bacterial invasion of the haemocoel, leading to septicaemia and death of the insect.

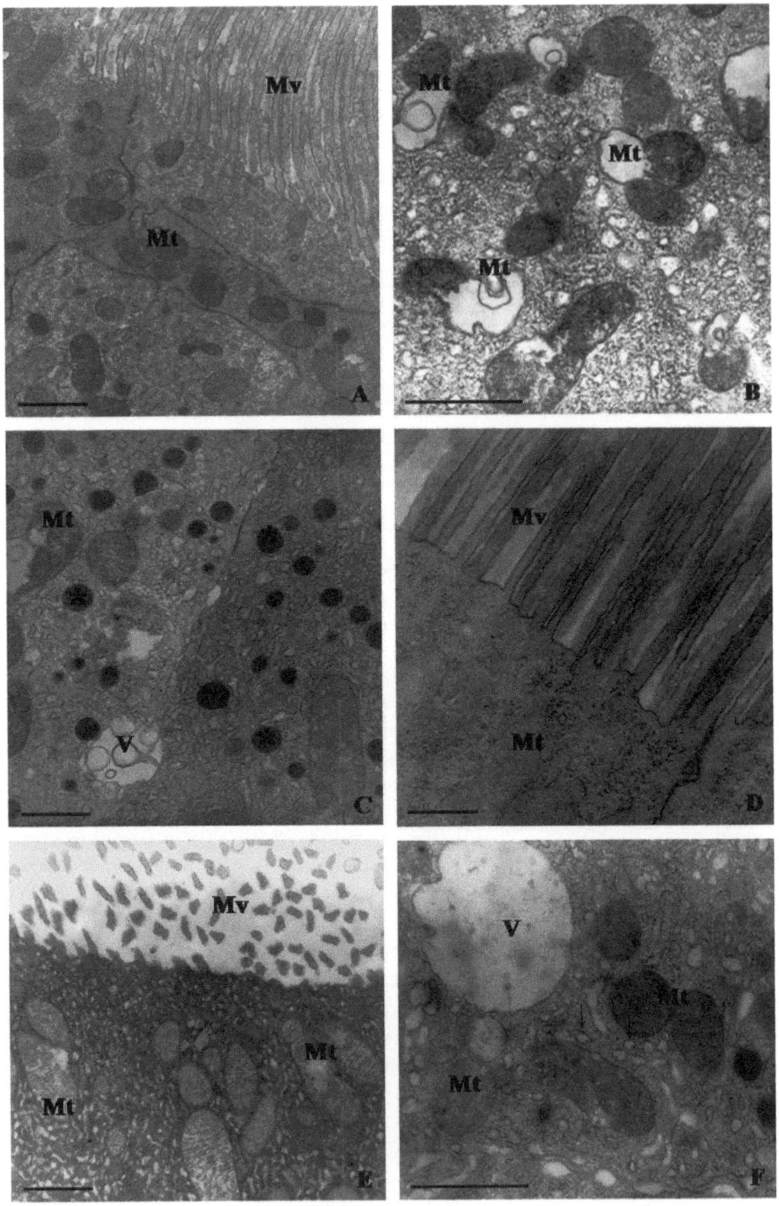

Maria Helena Neves Lobo Silva Filha *et al.*, Figure 3.2 Cytopathological effects of Cry48Aa/Cry49Aa toxin to *Culex quinquefasciatus*. Transverse ultrathin sections of the posterior midgut from fourth instar larvae treated with Cry48Aa/Cry49Aa toxin from *Lysinibacillus sphaericus* IAB59 strain. (A) Cells from a non-treated larva rich in microvilli (Mv) and mitochondria (Mt); (B) cells from 1 h-treated larva showing mitochondrial vacuolation (Mt); (C) cells from 1 h-treated larva presenting electron-dense granules (*), cytoplasmic vacuoles (V) and small vesicles from endoplasmic reticulum breakdown; mitochondria (Mt); (D) cells from 1 h-treated larva with preserved microvilli (Mv) and mitochondrial swelling (Mt) at the apical side of the cell; (E) cells from 6 h-treated larva showing evident microvilli damage (Mv) and mitochondrial vacuolation (Mt). (F) Cells from 6 h-treated larva presenting large cytoplasmic vacuoles (V) and small ribosome-coated vesicles (arrows); mitochondria (Mt). Scale bars = 1 μm. *Taken from de Melo et al. (2009).*

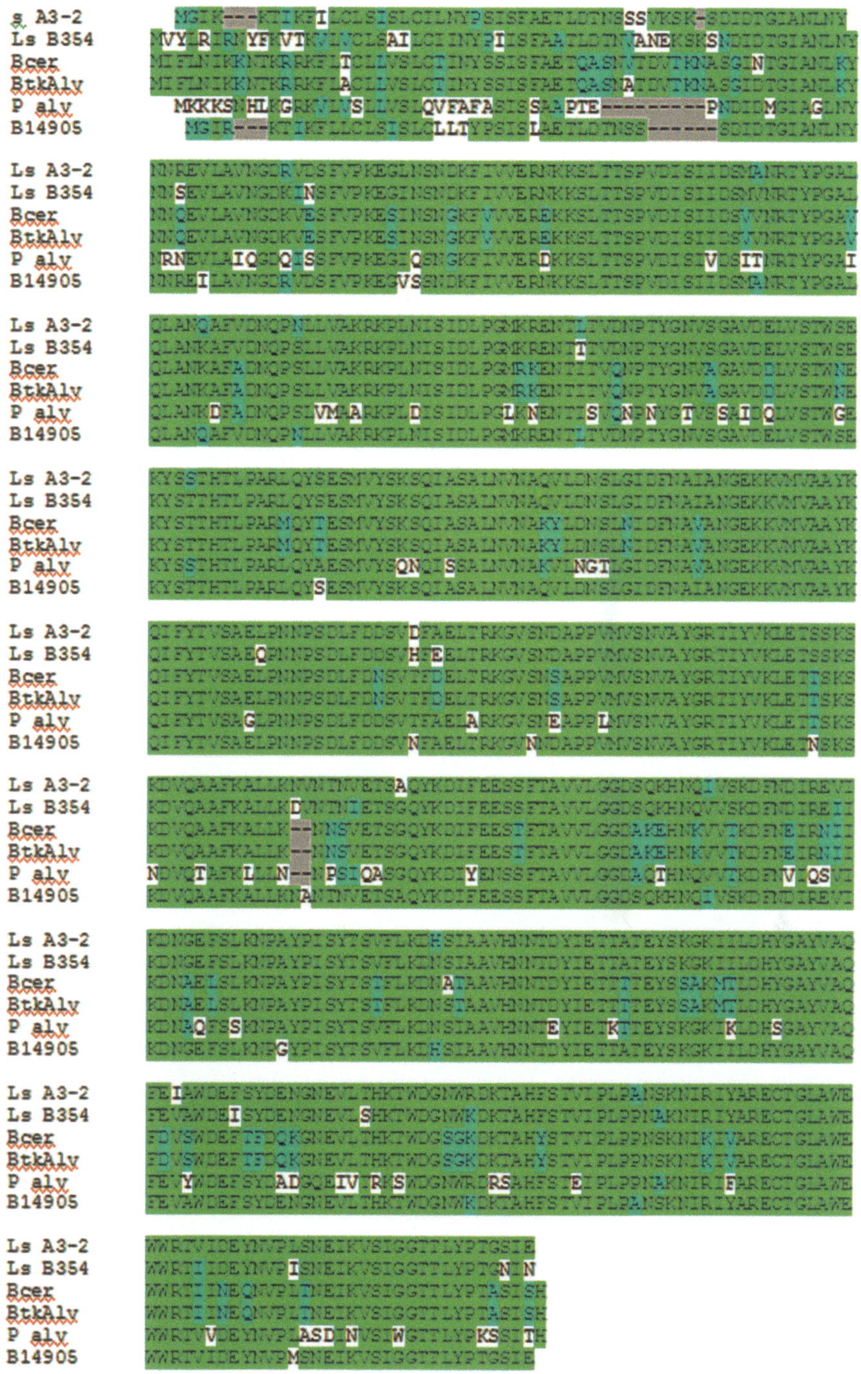

Maria Helena Neves Lobo Silva Filha *et al.*, Figure 3.3 Alignment of sphaericolysin and related proteins. Identical residues are shaded green or cyan, spaces introduced

(*Continued*)

Maria Helena Neves Lobo Silva Filha *et al.*, Figure 3.4 Modelled structure of sphaericolysin. The structure of sphaericolysin from *Lysinibacillus sphaericus* strain A3-2 was generated using Swiss model (Schwede et al., 2003) based on the structure of anthrolysin O from *Bacillus anthracis* (sequence identical to the *B. cereus* anthrolysin shown in Fig. 3.3, PDB accession number 3CQF, Bourdeau et al., 2009). Model coloured according to secondary structure: alpha helix (brown), beta sheet (yellow), other (grey). The putative cholesterol binding loop is shown in green and residue Tyr 187 in blue.

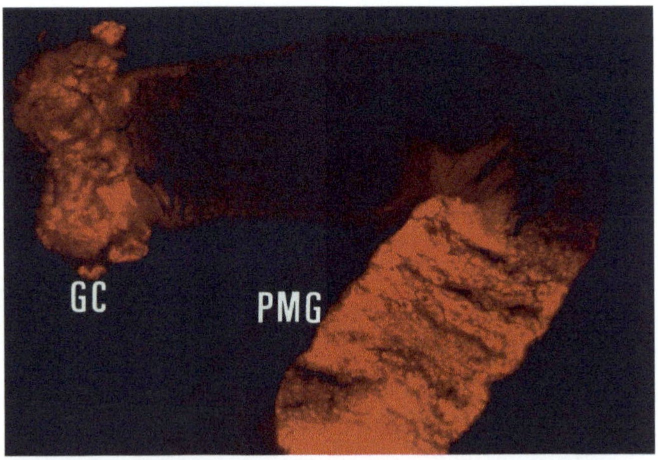

Maria Helena Neves Lobo Silva Filha *et al.*, Figure 3.5 Binding of fluorescent-labelled binary toxin regionalized in the gastric caeca (GC) and posterior midgut (PMG) of *Culex quinquefasciatus* larvae. *Figure kindly provided by Colin Berry, Cardiff University (UK).*

Figure 3.3—Cont'd to optimise the alignment are shaded grey. The sequences shown are *Ls* A3-2 (sphaericolysin from *Lysinibacillus sphaericus* strain A3-2, accession number BAF62176), *Ls* B354 (sphaericolysin from *L. sphaericus* strain B354, accession number EU043116), Bcer (anthrolysin from *Bacillus cereus* accession number ZP_03231124), BtkAlv (alveolysin from *B. thuringiensis* subsp. *kurstaki* accession number ZP_04117355), P alv (*P. alvei* alveolysin accession number ZP_10866820), B14905 (perfringolysin O precursor from *Bacillus* sp B14905 accession number ZP_01724867). Alignment generated using ClustalX (Thompson et al., 1994).

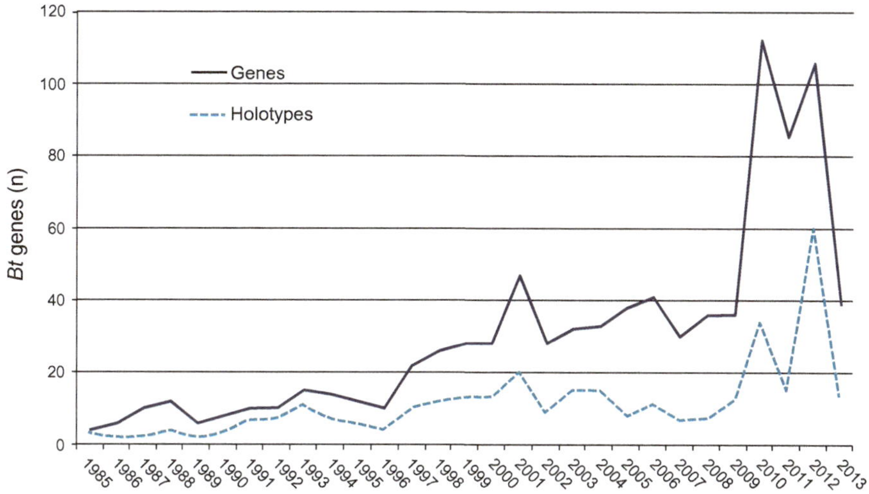

Kenneth E. Narva *et al.*, Figure 4.1 Discovery of *Bt* genes recorded on the *Bt* Toxin Nomenclature Website maintained by the *Bt* toxin nomenclature committee (Crickmore et al., 2014). The total number of new Cry, Cyt and Vip genes recognized by the committee in a given year is shown as *Genes*. The total number of new gene classes (as defined by the committee) recognized in a given year is shown as *Holotypes*.

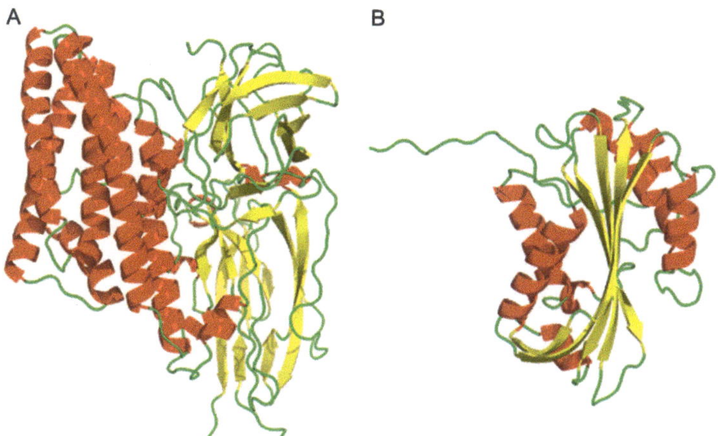

Kenneth E. Narva *et al.*, Figure 4.2 Protein crystal structures of representative *Bt* insecticidal proteins. (A) Three dimensional structure of Cry1Aa1 (PDB code: 1CIY), a three domain Cry protein. (B) Three-dimensional structure of the cytolytic crystal protein Cyt2Aa (PDB code: 1CBY).

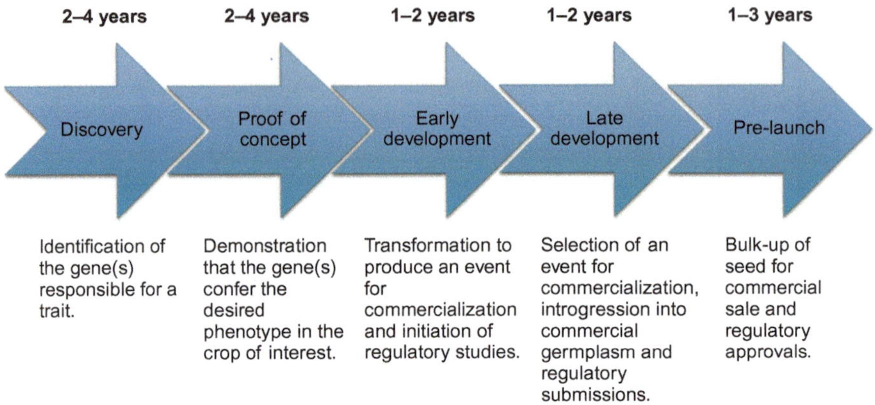

Kenneth E. Narva *et al.*, Figure 4.4 Generalized discovery and development staging system for a *Bt* crop. The websites of *Bt* crop developers are typically a good source of information on their specific discovery and development staging systems as well as the innovation that is in their pipelines.

Kenneth E. Narva *et al.*, Figure 4.5 Insect-protected *Bt* corn. (A) Contrast of corn root systems after heavy pressure from larval western corn rootworm. Top row is non-*Bt* plants showing severe root damage. Bottom row is plants from the same genetic background as the plants in the top row but containing event DAS-59122-7. (B) Contrast of ears after heavy pressure from larval western bean cutworm. Top row is ears from non-*Bt* plants showing severe damage. Bottom row is ears from the same genetic background as the plants in the top row but containing event TC1507. *Photo credit: Bader Rutter.*

Kenneth E. Narva *et al.*, Figure 4.6 Insect-protected Bt cotton. Contrast of yield after heavy lepidopteran pest pressure. Rows on the left are plants containing events 281-24-236 and 3006-210-23 (i.e. WideStrike®) showing a large number of open bolls and harvestable lint. Rows on the right are plants of the same genetic background as plants on the left but without the Bt events. *Photo credit: Eileen Crosby and Jim Steadman, Bader Rutter.*

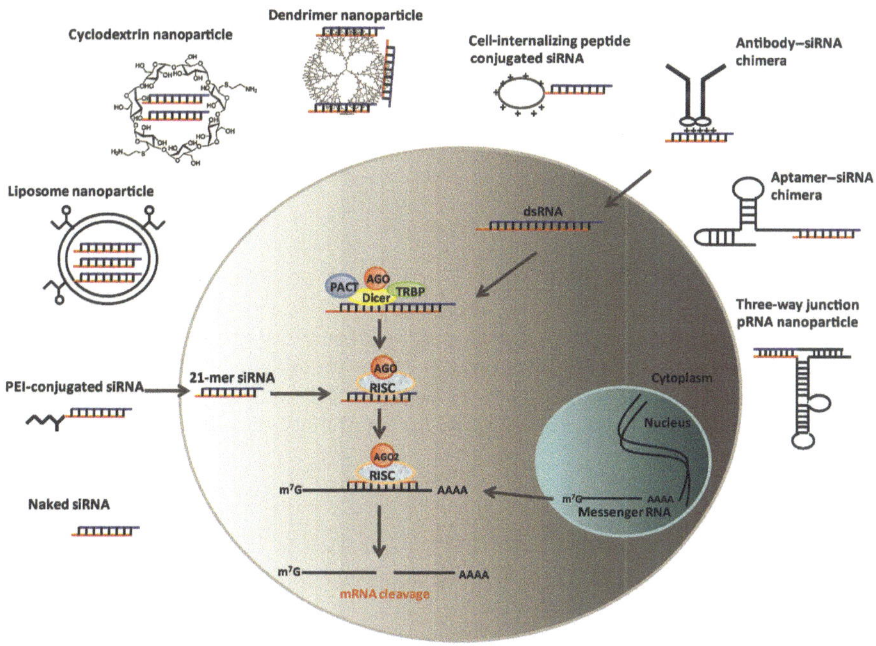

James A. Baum and James K. Roberts, Figure 5.4 The mechanism and delivery strategies for RNA interference. RNAi therapeutics (e.g. siRNA) can be internalized into cells via different delivery vehicles. *Figure reproduced with permission from Zhou et al. (2013).*

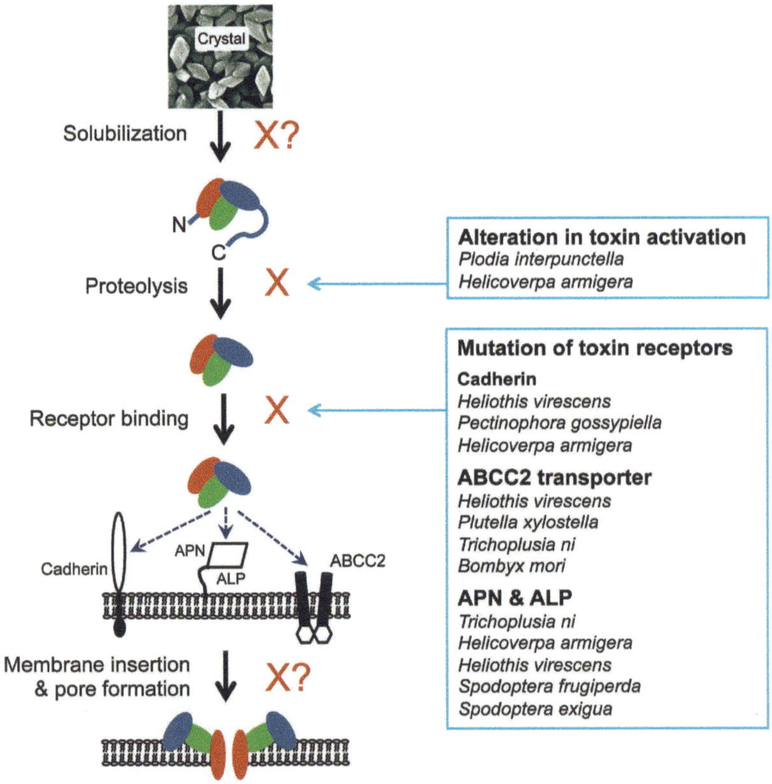

Yidong Wu, Figure 6.2 Schematic diagram of the steps of Cry toxin action and possible resistance mechanisms. Receptor mutations and/or proteolytic alterations have been identified in a number of *Bt*-resistant lepidopterans, whereas resistance mechanisms involving solubilization failure or abnormal membrane insertion/pore formation have not been reported yet.

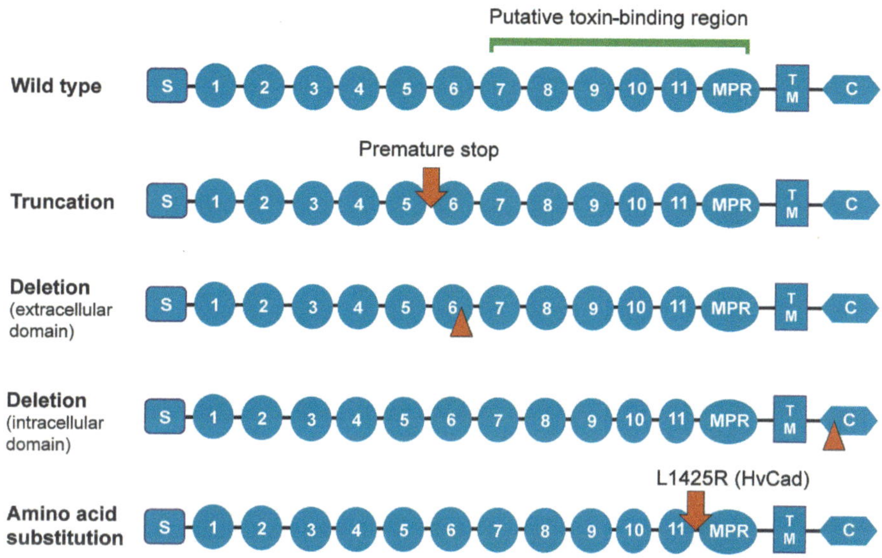

Yidong Wu, Figure 6.3 Types of cadherin mutations associated with Cry1Ac resistance in three cotton pests (*Heliothis virescens, Pectinophora gossypiella* and *Helicoverpa armigera*). (1) *Truncation*: a premature stop codon is present at any sites on the extracellular domain of cadherin. (2) *Deletion in the extracellular domain*: a stretch of amino acid residues on the extracellular domain, present within or ahead of the toxin-binding region. (3) *Deletion in the intracellular domain*: lack of 55 amino acid residues in the r_{15} allele of *HaCad*, which confers non-recessive resistance to Cry1Ac (Zhang et al., 2012b). (4) *Amino acid substitution*: the L1425R mutant of HvCad of *H. virescens* can decrease binding to Cry1Ac and has the potential to confer resistance (Xie et al., 2005). The r_{10}–r_{14} alleles of *HaCad* in *H. armigera* were suggested to confer Cry1Ac resistance by amino acid substitutions (Zhang et al., 2012a).

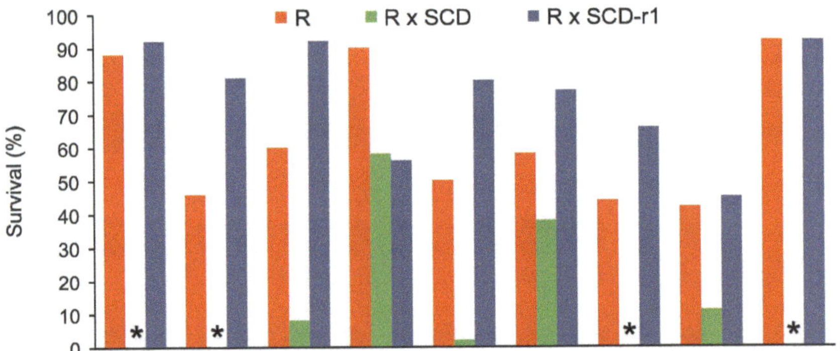

Strain	AY9	AY148	AY335	AY423	AY440	AY441	SC23	SW34	SCD-r1
R allele	*HaCad* r_1	*HaCad* r_9	*HaCad* r_{10}	Non-HaCad	*HaCad* r_{11}	*HaCad* r_{12}	*HaCad* r_{13}	*HaCad* r_{14}	*HaCad* r_1
Dominance	Rec	Rec	Rec	Dom	Rec	Dom	Rec	Rec	Rec

Yidong Wu, Figure 6.4 Diverse genetic basis of Cry1Ac resistance in field populations of *Helicoverpa armigera* from China (Zhang et al., 2012a). Survival at the diagnostic concentration of Cry1Ac of nine resistant (R) strains (red bars; eight strains isolated from the F_2 screen and the laboratory-selected strain SCD-r1 with the r_1 allele of *HaCad* at the cadherin locus), and progeny from crosses between each resistant strain and either the susceptible SCD strain (green bars) or the resistant SCD-r1 strain (blue bars). Asterisks indicate 0% survival for progeny from crosses between the SCD strains. Rec, recessive; Dom, dominant.

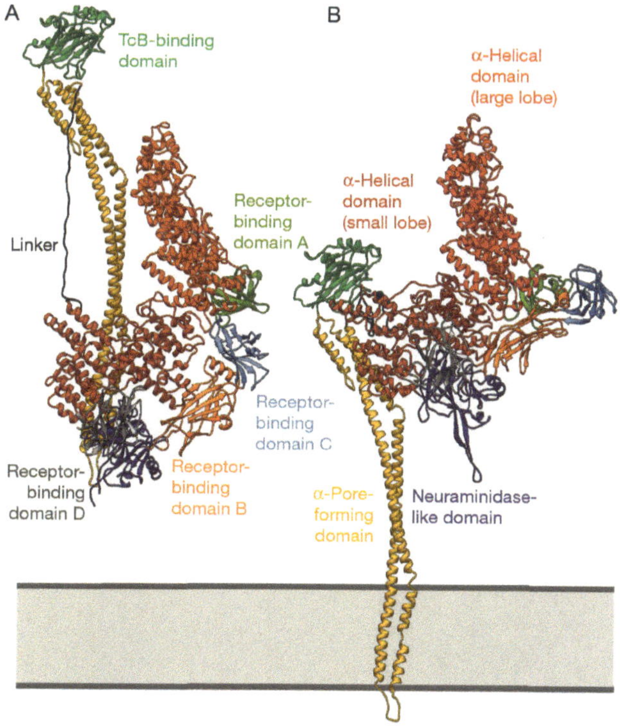

Richard H. ffrench-Constant and Andrea J. Dowling, Figure 7.1 Ribbon diagram showing the TcA protomer in (A) pre-pore state and (B) in pore-forming state. Receptor regions A–D and neuramidase-like region are labelled, along with the Tcb-binding domain, linker, pore-forming domain and α-helical domains (the small and large lobes). Note the dramatic changes that occur as the TcA protomer switches from pre-pore to pore-forming states, specifically, the collapse of the elastic linker, the extension of the pore-forming domain and the re-organization of the receptor domains A–D. *Reprinted from Meusch et al. (2014) with permission from MacMillan.*

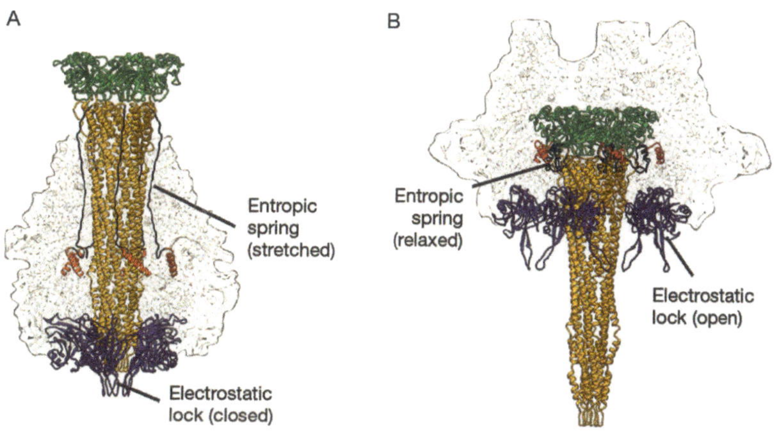

Richard H. ffrench-Constant and Andrea J. Dowling, Figure 7.2 Another view of the structure of (A) Tca pre-pore and (B) pore-forming complex. Note the entropic spring or linker (black), release of which provides the elastic energy to drive the pore-forming channel down towards the membrane. The neuramidase-like domain is shown in blue (see text for discussion). *Reprinted from Meusch et al. (2014) with permission from MacMillan.*

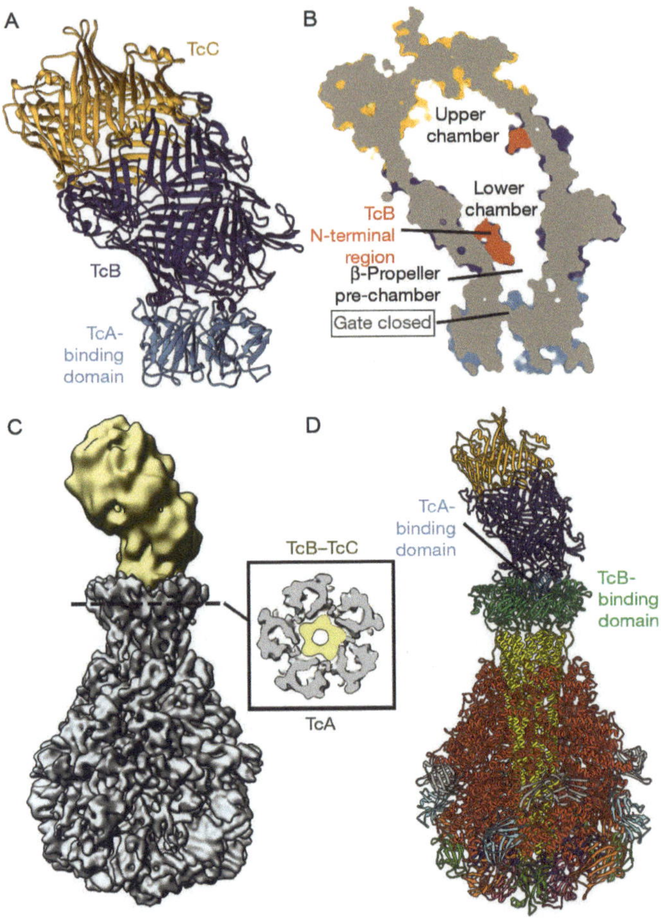

Richard H. ffrench-Constant and Andrea J. Dowling, Figure 7.3 Interaction of the B/C (TcB/TcC) subunits with the A (TcA) subunit assembly. (A) The B subunit (labelled TcB in blue) and the C subunit (TcC in yellow) interact to form a large 'cocoon-' or 'egg-shell'-shaped structure. The B subunit also comprises the A subunit-binding domain (labelled Tca-binding domain in light blue) which folds into an asymmetric six-bladed β-propeller. (B) A vertical section through the egg-shell-like structure formed by B/C to show the upper and lower chambers and the pre-chamber formed by the β-propeller. Note that the gate out of the B/C chamber is closed. Subunit domains are coloured: Tca-binding domain (light blue), TcB (dark blue), TcC (yellow) and Tcb N-terminal region (red). (C) Cryo-EM structure of the ABC holotoxin. Subunits and domains are labelled: TcA (grey) and TcB/TcC (yellow). The inset labelled TcB–TcC shows that the β-propeller gate is open here. (D) Final molecular model of the ABC holotoxin formed by fitting the crystal structure of the A subunits and the crystal structure of B/C into the overall cryo-EM structure. *Reprinted from Meusch et al. (2014) with permission from MacMillan.*

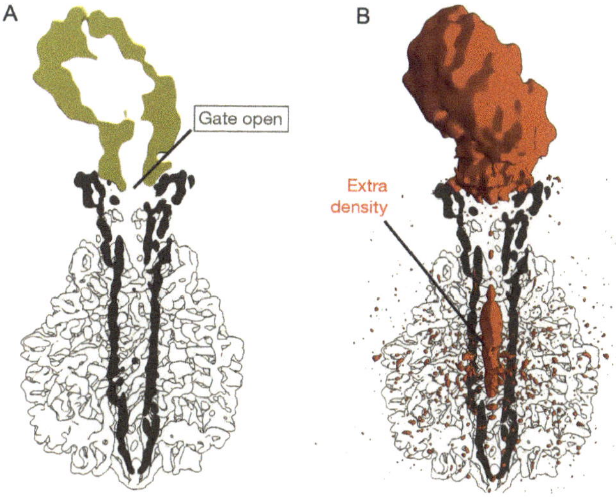

Richard H. ffrench-Constant and Andrea J. Dowling, Figure 7.4 Vertical sections through the density of the ABC holotoxin channel. (A) Section showing the continuous channel between the egg-shell-like structure of B/C in yellow and the channel of the A subunits in grey. (B) Difference map between the cryo-EM structure of the ABC holotoxin and the cryo-EM structure of the A subunits alone. The difference map is overlaid in red. Note the extra density present in the channel which may correspond to the delivered ADP-ribosyltransferase domain of the C subunit (see text above for discussion). *Reprinted from Meusch et al. (2014) with permission from MacMillan.*

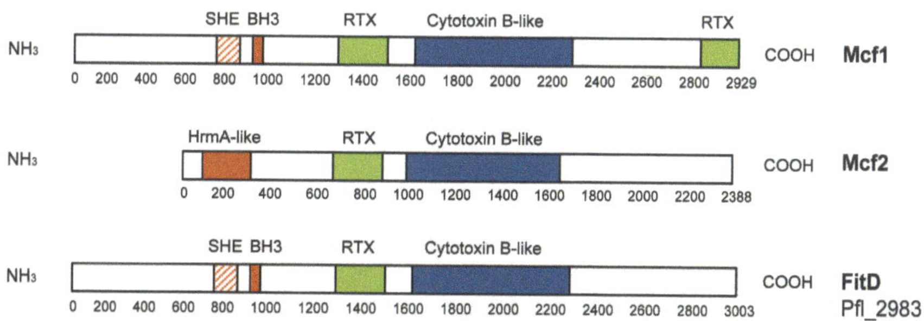

Richard H. ffrench-Constant and Andrea J. Dowling, Figure 7.5 Diagram showing the inter-relationship between the different domains in Mcf-like toxins. The *Photorhabdus* toxin Mcf1 carries similarities to the predicted MCF1/SHE proteolysis domain (domain labelled SHE), a BH3-like domain (BH3) and two regions that carry similarity to 'Repeats in toxins' (RTX) toxins. The central domain is similar to *Clostridium difficile* toxin B over a region important for toxin translocation (labelled as Cytotoxin B-like). The *Photorhabdus* toxin Mcf2 has a different N-terminal domain which carries similarity to the type III delivered plant effector HrmA (domain labelled HrmA-like). Finally, the Mcf-like FitD toxin carries the SHE, BH3 and one of the RTX-like domains but lacks similarity to RTX-like toxins at its C-terminus (see text for further discussion).

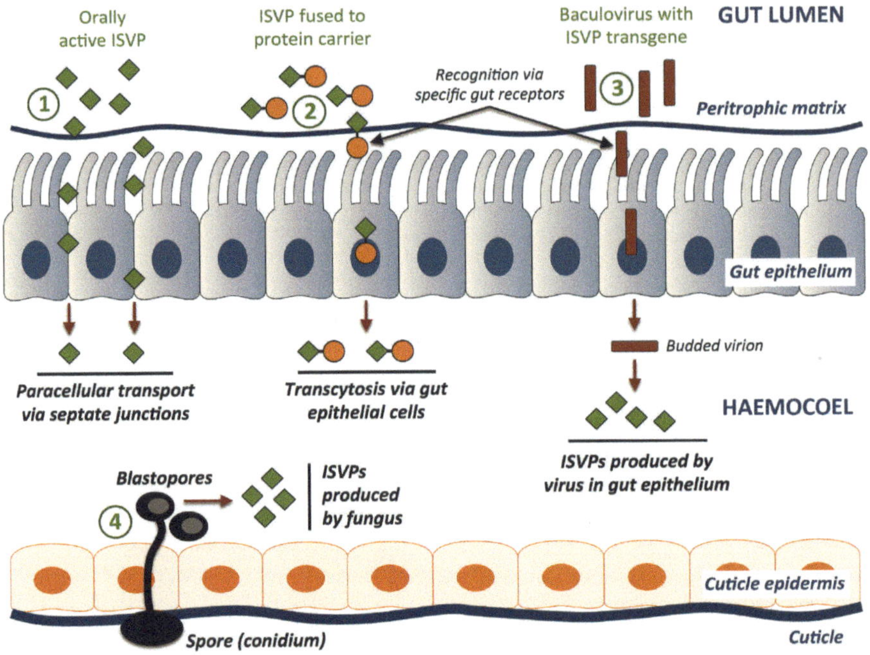

Volker Herzig *et al.*, Figure 8.1 Summary of methods for delivering ISVPs to the insect haemocoel: (1) Some ISVPs appear to be able to access the haemocoel via paracellular uptake through leaky septate junctions. (2) ISVPs can be fused to carrier proteins such as snowdrop lectin (GNA) and the coat protein of pea enation mosaic virus, which are recognised by specific receptors on gut epithelial cells. These fusion proteins can be taken up via receptor-mediated endocytosis and delivered to the haemocoel by transcytosis. (3) Baculovirus can be engineered to encode an ISVP transgene. Upon recognition by specific gut receptors, budded virions infect gut epithelial cells and other body tissues and release ISVPs. (4) Entomopathogenic fungi such as *Metarhizium* and *Beauveria* can also be engineered to encode an ISVP transgene. Fungal spores (conidia) adhere to the insect cuticle, germinate and then invade the integument. The fungus then proliferates in the haemocoel in the form of wall-less blastopores, which produce and release ISVPs. *Adapted from Bonning and Chougule (2014).*

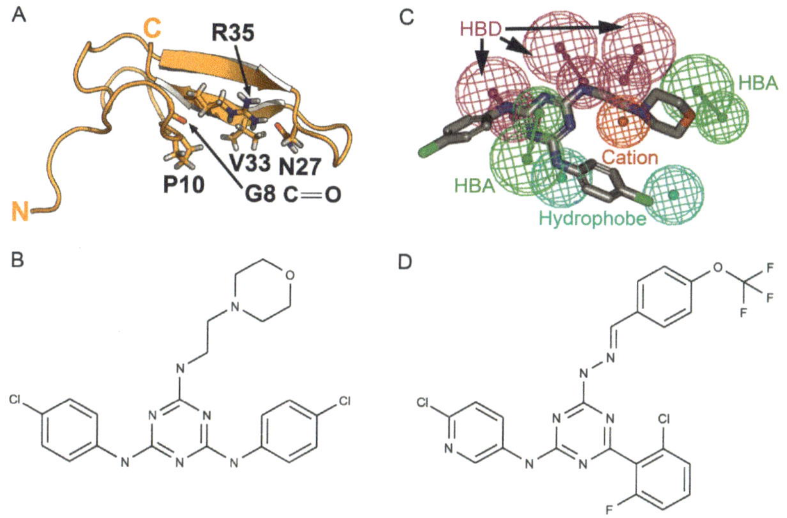

Volker Herzig *et al.*, Figure 8.2 Development of small-molecule mimetics of Hv1a. (A) Richardson schematic of Hv1a (first molecule in the PDB file 1AXH) showing the key chemical features that were used for pharmacophore modelling (i.e. the side chains of Pro^{10}, Asn^{27}, Val^{33}, and Arg^{35} and the backbone carbonyl of Gly^{8}). Hydrogen, carbon, oxygen and nitrogen atoms are shown in white, orange, red and blue, respectively. The N- and C-termini are labelled. (B) Bioassay screening hit VNX-440. (C) Three-dimensional stick representation of VNX-440; atoms are coloured as in (A) except that carbons are grey and chlorine atoms are green. Superimposed on VNX-440 are representations of the eight chemical features of the residues that were included in pharmacophore models of the peptide. Cyan spheres indicate hydrophobic features matching the hydrophobic side chains of Val^{33} and Pro^{10}; magenta spheres and sticks indicate hydrogen-bond donor (HBD) features projecting from Asn^{27} and Arg^{35}; green spheres and sticks indicate hydrogen-bond acceptor (HBA) features projecting from Asn^{27} and Gly^{8} (leftmost and rightmost residues, respectively). (D) Optimised lead compound VNX-443, an analogue of VNX-440.

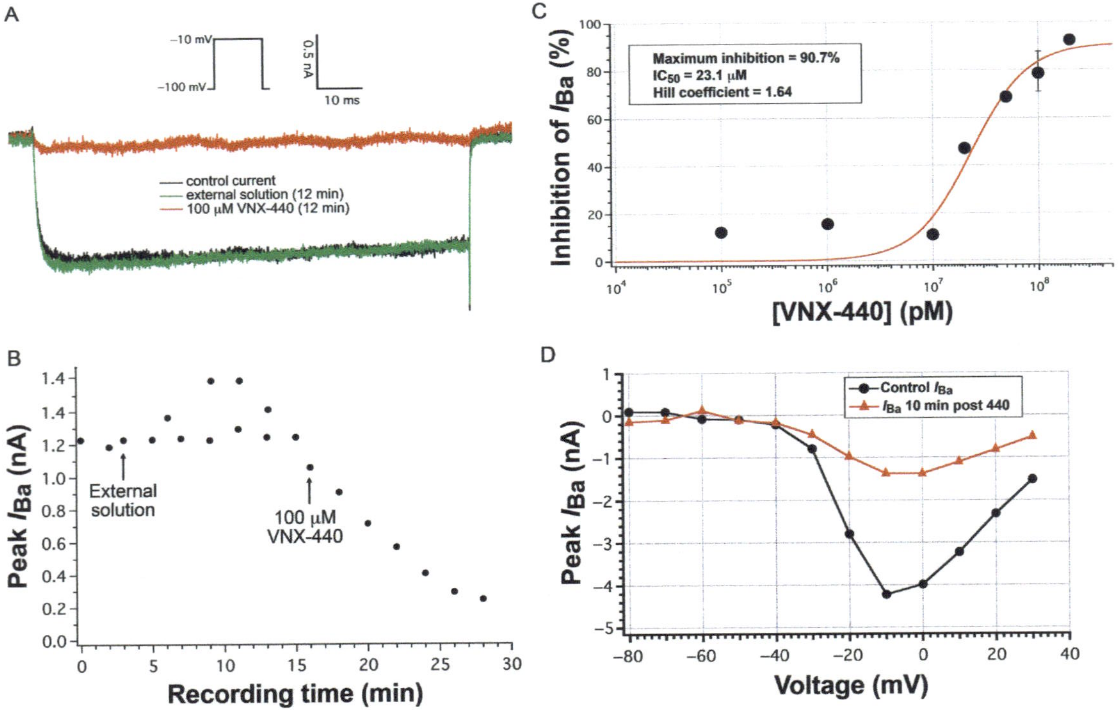

Volker Herzig *et al.*, Figure 8.3 Inhibition of Ba^{2+} currents (I_{Ba}) in American cockroach DUM neurons by VNX-440. (A) I_{Ba} recorded at −10 mV: black trace shows control I_{Ba}, green trace shows I_{Ba} during superfusion with bath solution and the red trace shows I_{Ba} following with superfusion with 100 μM VNX-440. (B) Time course of peak I_{Ba} values recorded at a test potential of −10 mV after addition of 100 μM VNX-440. (C) Dose–response curve for VNX-440 inhibition of peak I_{Ba} at a test potential of −10 mV. (D) *I–V* curves for I_{Ba} obtained before, and 10 min after, application of 100 μM VNX-440.

FSC
www.fsc.org
MIX
Papier | Fördert
gute Waldnutzung
FSC® C083411

Kolibri 360 GmbH
Ferdinand-Jühlke-Straße 7
99095 Erfurt, Deutschland
produktsicherheit@kolibri360.de